BIO-CHEMISTRY

BIO-CHEMISTRY

By

Ramesh Mathur
Meenakshi Mehta

ANMOL PUBLICATIONS PVT. LTD.
NEW DELHI - 110 002 (INDIA)

ANMOL PUBLICATIONS PVT. LTD.
4374/4B, Ansari Road, Daryaganj
New Delhi - 110 002
Ph.: 3261597, 3278000
Visit us at: www.anmolbooks.com

Bio-chemistry

First Edition, 2002

ISBN 81-261-1325-1

PRINTED IN INDIA

Published by J.L. Kumar for Anmol Publications Pvt. Ltd., New Delhi - 110 002 and Printed at Mehra Offset Press, Delhi.

CONTENTS

Phosphate Moiety of Nucleotides, The Purine Ring is attached to Ribose Phosphate during its Assembly, AMP and GMP are formed from IMP, Purine base can be Recycled by Salvage Reactions that Utilize PRPP, AMP and GMP are Feedback Inhibitors of Purine Nucleotide Biosynthesis, The Pyrimidine Ring is Synthesized from Carbamoyl Phosphate and Aspartate, Orotate acquires a Ribose Phosphate Moiety from PRPP, A single Polypeptide Chain contains the first three Enzymes of Pyrimidine Biosynthesis, Nucleoside Mono-, Di-, and Triphosphates are Interconvertible, CTP is Formed by Amination of UTP, Pyrimidine Nucleotide Biosynthesis is Regulated by Feedback Inhibition, Aspartate Transcarbamoylase Consists of Separable Catalytic and Regulatory Subunits, Deoxyribonucleotides and Synthesized by Reduction of Ribonucleoside Diphosphates, Deoxythymidylate is Formed by Methylation of Deoxyuridylate, Several Anticancer Drugs Block the Synthesis of Deoxythymidylate, ATP is a Precursor of NAD+, FAD, and Coenzyme A, Purine are Degraded to Urate in Humans, Urate is further Degraded in some Organisms, Birds and Terrestrial Reptiles Excrete Urate Instead of Urea of Conserve Water, Degradation of Pyrimidines, Excessive Production of Urate is a cause of Gout, Lesch-Nyhan Syndrome : Self-mutilation, mental Retardation, and Excessive Production of Urate.

Preface

Biochemistry is the study of the composition, the synthesis and the chemical degradation of the chemical compounds found in the living organisms. It encompasses all the chemical reactions carried out by cells. It spans an immune range from thermodynamics to cell biology, genetics and physiology. It is designed to introduce the principles of biochemistry and their application to the students of Indian Universities who are taking their first and perhaps their only course in biochemistry. Every effort has been made to incorporate the modern conception in biochemistry in a very simple process, explicit and lucid manner. Whenever possible, the subject matter has been illustrated with the aid of self explanatory diagrammatic representations.

This book is principally for students taking a first course in cell biology, by the undergraduates, graduate students, or medical students. Although we assure that most readers have had at least an introductory biology course, we have attempted to write the book so that even a stranger to this branch could follow it by starting at the beginning.

It has been the constant endeavour of the authors to furnish maximum substance, keeping in view the limitations of size of the volume. Efforts have been made to condense the matter as far as practicable. It is hoped that this book will not only meet the requirement of Indian students but will also be useful as a guideline to the teachers in their teaching.

There can be no claim to originality except in the manner of treatment and much of the information has been obtained from the books and scientific journals available in the different libraries.

The authors express their thanks to their friends and colleagues whose continue inspirations have initiated them to bring out this book.

The authors are painfully aware of the shortcomings, errors and misprints that have crept in, and shall be grateful to receive suggestion for improvement of the next edition from all the readers.

Authors

1

The Cell

The cell is a unit of biological activity delimited by a semipermeable membrane and capable of self-reproduction in a suitable non-living medium. The body of all living organisms (bacteria, blue green algae, plants and animals) except viruses has cellular organization and may contain one or many cells. The organisms with only one cell in their body are called acellular organisms (e.g., bacteria, blue green algae, some algae, protozoan, etc.). The organisms having many cells in their body are called multicellular organisms (e.g., most plants and animals). Any cellular organism may contain only one type of cell Prokaryotic cells, Eukaryotic cells.

It was *Robert Hooke* who first of all in 1665 observed under a microscope certain honey comb like structures in cork, and he applied the term 'cell' (L., *cella;* small room) for the same. Previously it was believed that there is only one component of the cell which is cell wall. But in 1831, one more peculiar structure was observed by *Robert Brown.* He gave the name 'nucleus'. Later on the 1835 the word 'Sarcode' was proposed for the jelly like material present inside the cell by *Dujardin.* In 1840 *Purkinje* replaced the word sarcode by Protoplasm which is now used universally.

Prokaryotic Cells

The prokaryotic (Gr., *pro*-primitive, *Karyon*-nucleus) cells are the most primitive cells from morphological point of view. They occur in bacteria and blue green algae. Prokaryotes are small, single cells organism, usually less than a micrometer (abbreviated μm; 1000 μm = 1 millimeter, abbreviated mm) are generally not longer than 3 μm. Their structural organization is simple than that of the eukaryotes.

They do not have a distinct nucleus, and they are lacking in many membranous structures found in the more complex cell types. The hereditary information is contained in a single circular structure that lacks many of the molecular components of the eukaryotic hereditary structure called the chromosome. In addition, the prokaryotes seem to lack any kind of special apparatus for cell division that is why they divide amitotically or by fission.

The prokaryotes are characterized by:

(i) the absence of a membrane around the nuclear material.

(ii) the absence of clearly defined membrane-limited organelles like mitochondria, chloroplast, Golgi complex and lysosomes.

(iii) the genetic material is located on a single chromosome which consists of a circular double strand of DNA.

(iv) the basic protein—histones, which are one of the most important constituents of chromosomes of eukaryotic cells, are absent in prokaryotic chromosomes.

(v) the absence of nucleolus and mitotic apparatus.

(vi) the cell wall is non-cellulosic, being formed of carbohydrates and amino acids.

(vii) the plasma membrane which lies below the cells wall is produced into the cytoplasm and acts as the mitochondrial membrane carrying respiratory enzymes, and

(viii) the cytoplasm neither exhibits streaming nor the amoeboid movement.

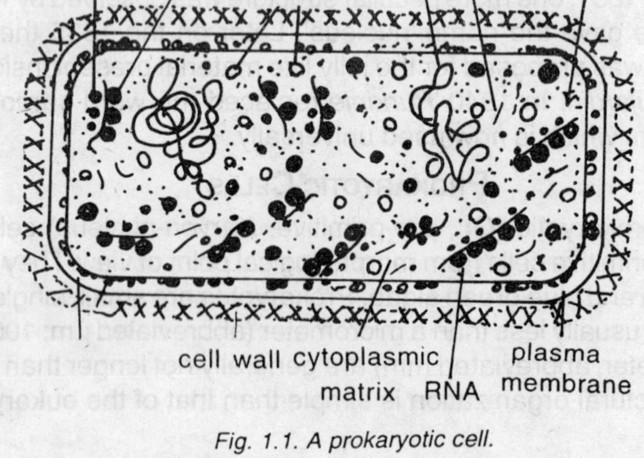

Fig. 1.1. A prokaryotic cell.

The prokaryotes include bacteria, blue green algae, spirochaete, *Mycoplasma* or pleuropneumonia like organism (PPLO).

Bacterial Cell

The bacteria are microscopic, cellular, achlorophillous, asexually reproducing prokaryotes which occur in fresh and salt water, in soil and in plants and animal. They lead either a saprophytic or parasitic mode of existence. The saprophytic species of bacteria are of great economic significance for man. Some parasitic species of bacteria are pathogenic (disease producing) to plants, animals and man.

Bacteria are very small having an average diameter of 1.25μ.

The smallest bacterium is *Dialister pneumosintes* (0.15 to 0.3μ) and largest bacterium is *Spirillum volutans* (13-15μ) in length.

One of the best studied bacteria is *Escherichia coli.* It is about 2 μm long and 1μm wide and contains several thousand different kinds of molecules. Structurally, *E. coli* is rod like and has a typical cell wall and underlying cell membrane. The membrane contains the special regions called *mesosomes,* which apparently have a role in cellular respiration and cell division. There may be one or several "nucleoid" areas, which contain a single, circular molecule of hereditary information (DNA).

The rest of the bacterial cytoplasm is full of structures called ribosomes, which differ slightly from the ribosomes of eukaryotic cells. These structures are the sites of protein synthesis. In addition, recent studies have revealed the presence of small circular DNA molecules in the cytoplasm called *plasmids.* These structures are very important in genetic engineering.

Mycoplasma

The *Mycoplasma* (previously called PPLO, pleuropneumonia like organisms) are small (0.2 to 0.3 μm) and more simply organized than bacteria. The hereditary material (DNA) is contained in a non-membrane-bound, nucleus like region, and it may exist either as strands or as a circular molecule as in the bacteria. However, there is only one tenth the amount of genetic material in Mycoplasma as there is in bacteria. Similarly, the Mycoplasma contains 1/50 to 1/100 the number of bacterial ribosomes per cell. A delimiting cell membrane exists, but no cell walls have been detected. A variety of other cytoplasmic inclusions, such as vacuoles and granules, have been detected, but their function is not known. As in other prokaryotes, there are no intracellular membranous structures.

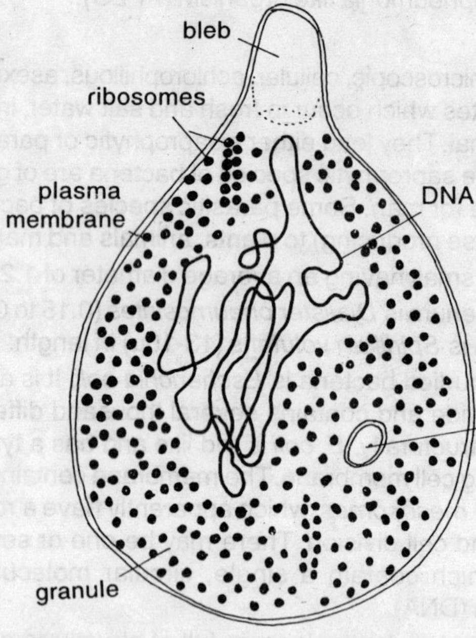

Fig. 1.2. Diagram of a Mycoplasma.

Mycoplasma has been known to cause certain respiratory diseases in animals and man. They also occur as routine contaminants in eukaryote tissue culture and currently are of much concern to researchers who use these culture procedures in their studies.

Blue-Green Algae

The blue-green algae are the third major group of prokaryotic organisms. The cellular organization is more comples than that of the other prokaryotes, primarily because of the existence of photosynthetic machinery. The blue-green algae contain chlorophyll as well as other more unusual pigments called *phycobilins.* The chlorophylls are bound to membranes in flattened rows called lamellae, which give the algal cytoplasm a rather organized and structured appearance. The phycobilins appear to be located in granular structures called *cyanosomes,* which apparently are attached to the outer lamellar membrane surface. Like the other prokaryotes, the blue-green algae do not have a membrane-bound

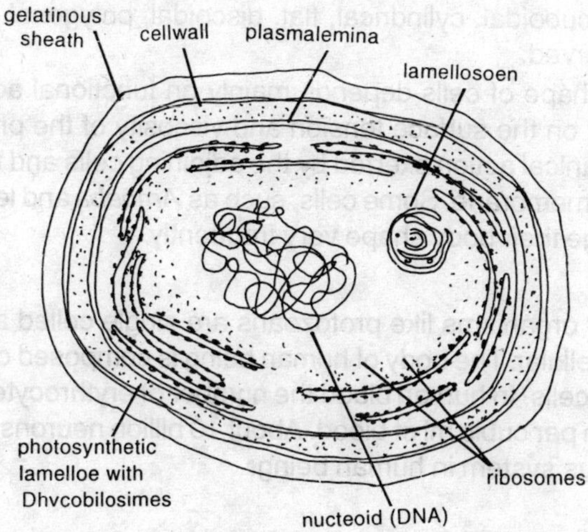

Fig. 1.3. Diagram of a blue green algae.

nucleus or other membrane-delimited organelles. The genetic material exists as free strands of DNA, and the ribosomes are free in the cytoplasm as in the bacteria and *Mycoplasma.* There appears to be other cytoplasmic granules of a protein and phosphate nature, and the entire cell is surrounded by a cell wall with an underlying plasma membrane.

Eukaryotic Cells

Eukaryotic cells may be acellular organisms, such as protozoans and acellular algae, or they may be cells that make up the tissues and organs of multicellular organisms. Though the eukaryotic cells have different shape, size, and physiology but all the cells are typically composed of plasma membrane, cytoplasm and its organelles, viz., mitochondria, endoplasmic reticulum, ribosomes, Golgi complex, etc., and a true nucleus.

Morphology of Eukaryotic Cell

Shape

Cell shapes are almost as numerous as cell types: there is not typical shape. The cells of certain unicellular forms, such as *Amoeba, Diatoms, Acetabularia* and bacteria exhibit a number of shapes. But generally the cells are rounded or spherical. Besides this the cells like oval, cuboidal, cylindrical, flat, discoidal, polygonal have also been observed.

The shape of cells depends mainly on functional adaptations and partly on the surface tension and viscosity of the protoplasm, the mechanical action exerted by the adjoining cells and the rigidity of the cell membrane. Some cells, such as *Amoeba* and leucocytes, can change their body shape very frequently.

Number

Some organisms like protozoans are single celled and others are multicellular. The body of human being is composed of about 26 trillions of cells. In human blood the number of erythrocytes is about five million per cubic ml of blood. About 10 billion neurons constitute the nervous system in human being.

Size

The size of different cells ranges within broad limits. Some plant and animal cells are visible to the naked eye, such as eggs of certain birds have a diameter of several centimeters. But the great majority of cells are visible only under microscope. The smallest living cells are found among bacteria and viruses where the size ranges 0.1μ to 1μ (1μ = 0.01 mm). Organisms like Pleuropneumonia-like organisms so called "elementary bodies" have been observed with diameters of 100 mμ. These appear to be a resting from of the bacteria which may grow into bodies of 250 mμ in diameter during the active metabolism of the organism. It would seem that 200-250 mμ in diameter is the lower limit for the size of inactive, living cells. This lower limit may be set by the minimum number and size of the component necessary for independent cellular existence.

Usually the vast majority of cells lie in the range of .5 to 20μ in diameter. The length of acellular diatoms is about 100μ. And the size of *Amoeba proteus* is 1 m.m. (1000μ) in length. The size of human R.B.C. is 7-8μ in diameter.

The long nerve is also an example of the extremes in cell sizes, the range of which is quite remarkable in the living world.

Factors Restricting Cell Size

In general, the size of the cell is correlated with its function. In no way it is associated with the size of the organism. For instance, a microscopic cell of bread mould can be several metres long, whereas cells in the body of whale or elephant may be as small as 10μ. Two major factors seem to restrict cell size.

1. The nucelocytoplasmic ratio

The nucleus controls activities of the cell by protein synthesis through RNA synthesis. In a large cell cytoplasm requires more proteins and consequently more RNA. The DNA content of a cell is constant, which can control the activities of the cell upto a specific size, beyond which proper control is not possible. For this reason a cell after attaining a specific size stops growing.

2. The surface area

As shown in fig, the small cells have more surface area per unit volume and with an increase in size, this ratio between surface area and volume gradually decreases. However, a large cell naturally requires more nutrients and oxygen for its metabolism, which diffuse into the cell through its surface. It means, the surface area of a cell affects its ability to exchange materials with its environment and puts a check on cell volume. For this reason, the cells that are metabolically more active are small in size.

Similarly, large cells face problem in the diffusion of excretory products from cell cytoplasm to the exterior.

Structure of Fixed Cells

Examination of the living cell is limited to light microscopy and is based mainly on the differences in refractive index of the different cell components. Sometimes the use of stains that act on the living organism (vital staining) facilitates observation of the living cell. However, more important in the morphologic study of the cell are methods of fixation, by which cell death results in such a way that physiologic structure and chemical composition are preserved as much as possible.

The complexity of the structural organization in a cell of the higher plants and animals is most impressive. Although there are great differences between the primitive forms of life, and the higher plant and animal cells, the similarities between primitive and advanced cells are also notable. Eukaryotic cells are characterized by a true nucleus with a nuclear membrane or envelope which divides the cell into two main compartments: nucleus and cytoplasm. The cytoplasm is turn is limited by the plasma membrane. In a plant cell the plasma membrane is covered and protected on the outside by a thicker cell wall through which there are tunnels, the *plasmodesmata,* by which the cell intercommunicates with neighbouring cells via fine cell process. In animal cells, parts of the plasma membrane are covered by a thin layer of material, which is generally described as the extraneous coat of the plasma membrane. The co-called *basement membranes* correspond to this extraneous coat.

In every stained cell usually one nucleus is present. It is central organelle of cell and is the main site of hereditary material. The nucleus is surrounded by a nuclear membrane but in bacteria and blue green algae it is without nuclear membrane. Within the nucleus a single nucleolus and network of chromatin is present. These two materials remain suspended in the nucleoplasm.

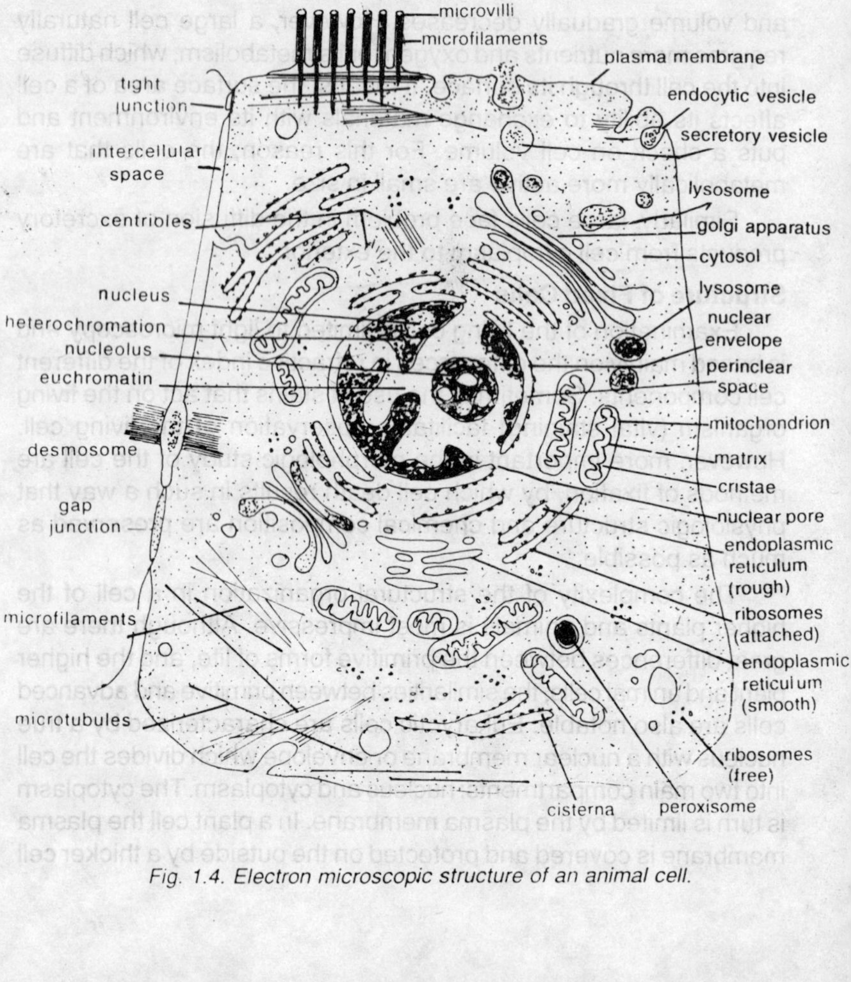

Fig. 1.4. Electron microscopic structure of an animal cell.

A small cytoplasmic organoid 'centriole' is present in the majority of animal cells and in some cells of the lower plants. The position is generally fixed for each type of cell. Generally the cell center or centriole occupies an axial position behind the nucleus. In some cells, it has the tendency to occupy the geometrical center i.e., in the leucocytes it occupies this position within the nucleus which is horseshoe shaped. The morphology of centriole changes with the functional state of the cell.

It is cytoplasm of the cell, there are several organelles out of them, only mitochondria and Golgi bodies are visible under light microscope. Mitochondria or chondriosomes are scattered in the cytoplasm of all kinds of cells. Their shape in the fixed cells depends upon the fixation and handling of the tissues before preparation they may appear as long rods, short rods or as small granules. Golgi material appears in its very prominent form in nervous and secretory cells in animals. Plastids are cytoplasmic organelles of plant cells may or may not contain pigments. In green plants chlorophyll is present in these organelles and thus termed as chloroplasts.

Electron Microscopic Structure of a Typical Cell

Under ordinary light microscope only few cell organelles like mitochondria, Golgi complex, chloroplasts and nucleus are visible. However, under electron microscope, several other cytoplasmic organelles such as endoplasmic reticulum, ribosomes, lysosomes, nuclear membrane, plasma membrane, etc. are seen.

Some important cellular components are being described here under the following heads:

Plasma Membrane

The outermost boundary of the animal cell is called the 'plasma membrane'. It is invisible under a light microscope but under electron microscope, it appears to be composed of two dense layers which are called the outer dense layer and inner dense layer. Both these layers are about 20Å in thickness, and are separated from each other by a less dense area of about 35Å in thickness. By this way the total thickness of plasma membrane is about 75Å. Before the advance of electron microscope people used to think of the plasma membrane has been stretched tightly over the cell. But now we know that a single plasma membrane can have as many as 3000 microvilli. Another peculiarity of the structure of the membrane is that, it remains connected with the endoplasmic reticulum.

Plasma membrane is lipo-proteinaceus in composition.

Functions

(i) Gives an identity to the cell.
(ii) Protects the cell organelles.
(iii) Helps in digestion through phagocytosis and pinocytosis.
(iv) Being permeable to water, water can pass through it.
(v) Help in the transport of ions.
(vi) Helps in the formation of cytoplasmic organelles.

Mitochondria

The term mitochondria was, for the first time used by *Benda* (1902) to designate the filamentous and granules of cell cytoplasm. Most of the knowledge related to mitochondrial fine structure is due to the studies of *Palade* (1952, 55, 56) and *Sjostrand* (1955). Due to these studies, now mitochondria are defined as the double walled structures of cell organelles within which the membranous structures called *cristae* are present. The outer chamber of mitochondrion is made up of two dense layers and a less dense layer having the total thickness of 140 Å to 180Å while the outer and inner dense layers vary in their measurements from 35Å to 60Å in thickness depending upon the physiological state of the cell. The less dense layer is 40-47Å in thickness.

The matrix of the mitochondria may have some fine opaque granules of various size. *David E. Green* (1964) observed in a very highly magnified electron microscope that the inside of the inner membrane and the outside of outer membrane are covered with very small particles. They measure about 90-100Å in diameter. The particles on the outer membrane differ from those on the inner membrane. The former generally appear as simple spheres packed closely together which give to the surface of the mitochondrion, a rough or pimpled appearance. The particle on the inside of the inner membrane has the more complex configuration. In some cells, but not all, this particle is seen to have a base, a stalk and a spherical head. Mitochondria represent the power generating centres of the cell. They can bring about oxidation of the intermediates of carbohydrates, fat and amino acid metabolism in the presence of O_2 releasing energy. The released energy is trapped in the form of adenosine triphosphate (ATP).

Golgi Complex

Golgi comples have probably been first observed earlier in 1867 by *La. Vallete St. George*, but any how credit goes to *Camillo Golgi*

(1898) who described it as internal reticulate apparatus' and after it, was named *Golgi apparatus.* It is variously called Golgi element, Golgi complex, Dalton's complex etc. Each Golgi complex is composed of many lamella, tubules, vesicles and vacuoles. The membranes of Golgi complex are of lipoproteins and these are supposed to be originated from the membranes of endoplasmic reticulum. The function of the Golgi complex is storage of proteins and enzymes which are synthesized by ribosomes and transported by endoplasmic reticulum to them. Further the Golgi complex has most important secretory function. It secretes many secretory granules and lysosomes. In plant cells the Golgi complex is known as dictyosome and secretes necessary materials for cell wall formation during cell division.

Lysosomes

The lysosomes (Gr. *lyso;* digestive+*some;* body) are also found scattered in the cytoplasm. By gently centrifuging lysosomes can be separated from other cell organelles. They are usually spherical but may be irregular in shape, covered with a membrane of lipoprotein. Inside structure is always variable. Lysosomes contain hydrolysing enzymes, such as acid phosphatase, acid ribonuclease, acid deoxyribonuclease and cathespins. But they do not have oxidative enzymes. This property distinguishes them from the mitochondria. Lysosomes are polymorphic in nature and are commonly known as suicide bags and are responsible for postmortem changes in cells.

Endoplasmic Reticulum

The term "Microsome" was introduced to the modern cytology by *Claude* (1943) to represent one of the submicroscopic cellular components isolated by centrifugation. These are now known to be the broken parts of the endoplasmic reticulum after the studied of *Kollman* (1953). *Porter* (1945) was the 1st man to describe their electron microscopic structure in cultured cells. He described them as lace-like reticulum. These are membrane bounded sacs in the form of double membrane or cisternae. The term cisternae was introduced by *Sjostrand* (1953) to describe long or elongated rod like parts of the endoplasmic reticulum measuring 50-200Å. *Wiess* (1953), described the vesicles of the endoplasmic reticulum which usually has a diameter of 25-500Å while *Bradfield* (1913) came across of another type of endoplasmic reticulum termed them as tubules, they measures from 50 to 100Å in diameter.

On the basis of presence or absence of ribosomes or RNP particles the endoplasmic reticulum (E.R) is distinguished in two varieties. The *rough walled endoplasmic reticulum,* and *smooth walled endoplasmic reticulum.*

Rough Walled Endoplasmic Reticulum

Rough walled endoplasmic reticulum (R.E.R.) is that variety which bears the ribosomes at the external surface of the cisternae. The distribution of the ribosomes can be circular, spiral, or rosette type. These particles may be induced to leave the surface of the

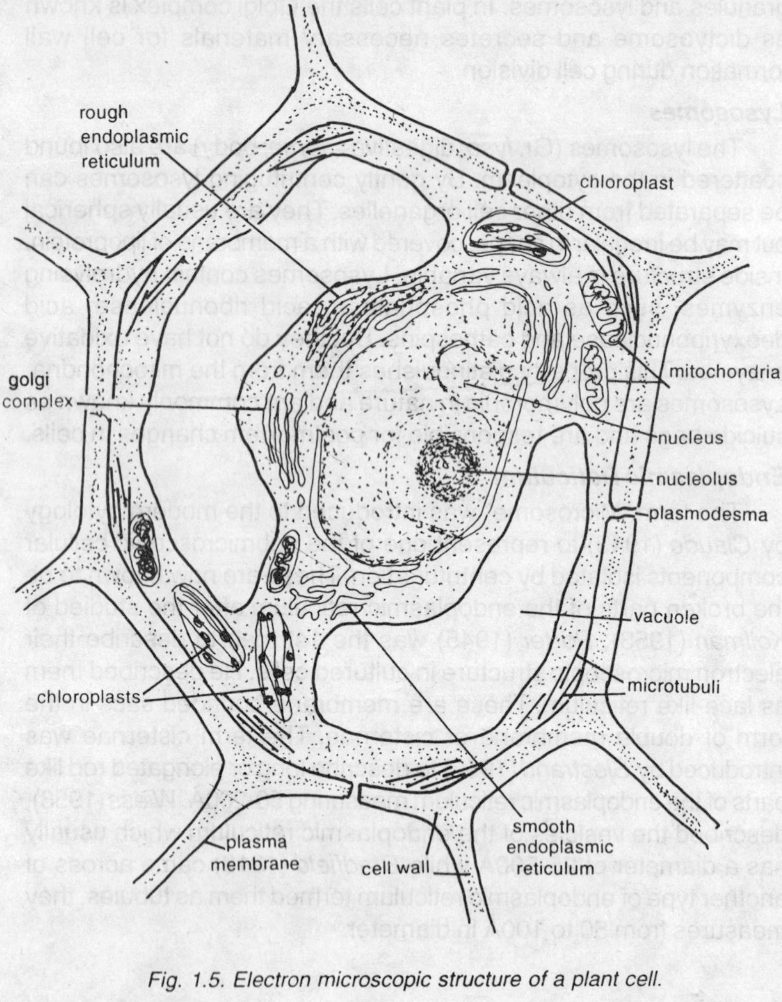

Fig. 1.5. Electron microscopic structure of a plant cell.

cisternae without damaging the structure of the cisternae or the capacity to associate again with the particles. Usually the rough walled endoplasmic reticulum is richly distributed in those cells which are engaged in the synthesis of the protein.

Smooth Walled Endoplasmic Reticulum

The other large division of the system owes its identity in parts of the absence of the particles and therefore commonly referred to as the smooth surfaced a granular form. Like the R.E.R., the S.E.R. shows a characteristic morphology which is tubular rather than cisternae. The tubules have the diameter nearly of 50-100 mµ. As studied by *Fawcett* (1960) S.E.R. is richly distributed in those cells which are engaged in the synthesis of steroids.

Continuity between the S.E.R. of a particular cell type possess a characteristic pattern of structure. Thus it is typical for rat liver cells to show groups of eight or ten slender profiles in parallel array.

Plastids

Plastids are present only in plant cells (not in animal cells). These are double-membrane structures relatively larger than the mitochondria. Some of the plastids are colourless and store starch (Leucoplasts). Other contain various pigments and impart different colours (red, brown, yellow, etc.) to different parts of the plant. These are collectively called chromoplasts. Those containing green pigment—cholorophyll, are called choloroplasts. These contain several enzymes in their matrix necessary for photosynthesis. The chloroplasts capture light energy of the sun and convert it into the chemical energy—ATP, which is used in the biosynthesis of glucose from CO_2 and H_2O.

Ribosomes

Many minute, spherical structures known as *ribosomes.* They remain attached with the membrane of endoplasmic reticulum and forms granular or rough type of E.R. The ribosomes are originated in the nucleolus and consist of mainly the ribonucleic acids (RNA) and proteins. Each ribosome is composed of two structural units, small subunit known as 40S subunit and a larger subunit known as 60S subunit. The ribosomes remain attached with the membranes of endoplasmic reticulum by the 60S subunits. The 40S subunits occur on the larger sub unit and form a cap-like structure. The ribosomes consists of 3 types of RNA's known as ribosomal RNAs or rRNA, viz., 5S, 18S and 28S rRNAs. The 28S and 5S rRNAs occur in large (60S) subunit, while 18S rRNA occurs in the smaller ribosomal subunit. The ribosomes also occur freely in the cytoplasm. They are the sites of protein synthesis.

Peroxisomes

These are also bounded by a single membrane. These contain enzymes for the breakdown of hydrogen peroxide to form water and oxygen. These are protective organelles of the cell.

Cilia and Flagella

The cells of many unicellular organisms and ciliated epithelium of multicellular organisms consists of some hair-like cytoplasmic projections outside the surface of the cell. These are known as cilia or flagella. They help in locomotion of the cell. The cilia and flagella consists of nine outer fibrils around the two large central fibrils. Each outer fibrils consists of two microtubules. The two microtubules of each peripheral pair or *doublet* remain morphologically distinct. The central fibrils are separate but remain enveloped in a common *central sheath.* Cilia and flagella are envolved in locomotion, feeding, cleansing etc.

Vacuoles

These structures are fluid-filled "bubbles" found in the cytoplasm. Vacuoles are bounded by a membrane, called *tonoplast,* which is structurally similar to, if not identical with, a cell-membrane. Inside the vacuoles, food materials or wastes are found.

Vacuoles are consipicuous in plant-cells but they are found less in animal cells. Vacuoles are absent in actively dividing cells. In a young plant-cell, there are many vacuoles but, when a cell matures, they unite to form a single, large, central vacuole.

Guilliermond (1941) was of the opinion that vacuoles are colloidal with a strong imbibing capacity. Vacuoles arise *de novo* i.e., spontaneously formed from pre-existing vacuoles.

In many protozoan, vacuoles are formed at the bottom of the gullet. In these vacuoles, food-materials, such as bacteria, are collected. Some vacuoles are formed by cell-drinking. This type of vacuoles are called *pinosomes.* Food-vacuoles formed by evagination of cell-membrane surrounding food-particles with some fluid, are called *phagosomes.* In animals cells, the vacuoles are formed by a lipoprotein membrane that helps in storage and transport of materials and in the maintenance of internal pressure of the cell.

Vacuoles are not a living part of the protoplasm but are a storing place of reserve materials and waste products.

Crystals and oil droplets

In certain plant cells the food or waste material is found deposited in the form of crystals. The oil droplets are round in both plant and animal cells, which appear as tiny glistening white spheres. These serve as a reserve supply of energy rich fuel.

Nucleus

It is the most vital part of cell which is invariably present in all the cells of higher animals and plants. Normally one nucleus is present in a cell but more than one may be present in certain cells.

Structurally it is enveloped by a porous *nuclear membrane* that separates it from the surrounding cytoplasm. It has the characteristic permeability. During cell division, it breaks up and soon after the mitosis it is reconstituted. The inside of nucleus is filled with *nuclear sap* or *karyolymph.* In nuclear sap remain embedded other nuclear constituents. The *chromatin* is the Feulgen positive material in the interphase nucleus which during division becomes the definite chromosomes. In short it represents the chromosomes substance. The sex-chromatin are the specific heterochromatic bodies, found generally near the periphery of the nucleus. In nucleus usually one sex chromatin body is found but more than one can also be observed. *Nucleolus* is relatively large, generally spherical ball-like body. It has a high concentration of RNA and thus is an important structure involved in protein synthesis. Its number in a nucleus vary in different cells and depend on either the species or the sets of chromosomes.

The nucleus co-ordinates many cell activities. DNA molecules are the source of coded instructions, which pass to RNA molecule which then act as messenger, moving out of the nucleus to one of the various structures of the cytoplasm. It also controls the cell reproduction. Here too, DNA is the source of coded instructions.

The Cytosol, Hyaloplasm or Ground Substance

The material that remains suspended under conditions sufficient to sediment even the ribosomes is referred to by biochemists as the *soluble fraction* or *cytosol.* This fraction usually contains some material lost from the organelles in the course of homogenization and centrifugation. However, it also contains other molecules, including some enzymes, that are not known to be part of the intracellular structures indentifiable in the microscope. Such molecules are usually thought to derive from the *hyaloplasm* (also called ground substance, cell matrix, or cell sap), which was originally defined by microscopists as the apparently structureless medium that occupies the space between the visible organelles. As will be evident in subsequent chapters, with the improved techniques less and less of the cell appears structureless. Most investigators continue to believe that there does exist in the cytoplasm a "soluble phase"—a solution containing inorganic ions and small molecules, materials in transit from one cell region to another and dissolved or suspended enzymes that function as individual molecules or perhaps or parts of small groups of molecules rather than as components of large, microscopically recognizable organelles. But there is a lively debate going on over proposals that many components once though to be suspended in this solution are actually bound, perhaps loosely, in structured arrangements that do not survive the disruption involved in cell fractionation.

Table 1.1. The Relative Volumes Occupied by the Major Intracellular Compartments in a Typical liver cell (Hepatocyte)

Intracellular Compartment	*Percent of Total Cell volume*	*Approximate Number per Cell*
Cytosol	54	1
Mitochondria	22	1700
Rough endoplasmic reticulum cisternae	9	1
Smooth endoplasmic reticulum cisternae plus Golgi cisternae	6	
Nucleus	6	1
Peroxisomes	1	400
Lysosomes	1	300

Table 1.2. Comparison of Prokaryotic and Eukaryotic Organisms

	Prokaryotes	*Eukaryotes*
Organisms	bacteria and cyanobacteria	protists, fungi, plants, and animals
Cell Size	generally 1 to 10 μm in linear dimension	generally 10 to 100 μm in linear dimension
Metabolism	anaerobic or aerobic	aerobic
Organelles	few or none	nucleus, mitochondria, chloroplasts, endoplasmic reticulum, etc.
DNA	circular DNA in cytoplasm	very long DNA containing many non-coding regions; organized into chromo-

		somes and bounded by nuclear envelope
RNA and protein	RNA and protein in same compartment	RNA synthesized and processed in nucleus; proteins synthesized in cytoplasm.
Cytoplasm	no cytoskeleton, cytoplasmic streaming, endocytosis, or exocytosis	cytoskeleton composed of protein filaments; cytoplasmic streaming; endocytosis and exocytosis
Cell division	by binary fission	by mitosis (or meiosis)
Cellular organization	mainly acellular	mainly multicellular, with differentiation of cells.

Table 1.3. Comparison of plant and animal cells

Plant Cell	*Animal Cell*
1. Generally larger than the animal cell.	1. Usually smaller than the plant cell.
2. Plant cells are surrounded by a rigid cellulose wall.	2. Cellulose wall is not found in animal cell.
3. Majority of the mature plant cells have a large central sap vacuole.	3. Vacuoles in animal cells are small.
4. Plastids, especially chloroplasts are found usually in cells of green plants.	4. Chloroplast is not found in animal cell except in some Protozoa.
5. In the cell division, the division of cytoplasm of plant cells take by formation of a partition called place cell plate.	5. Animal cells divide by constriction during cell division.

Table 1.4. Functions of Cellular Organelles

Cell Organelles	*Functions*
Cell or Plasma membrane	Serves as a differentially permeable membrane through which extra-cellular substances may be selectively sampled and cell products may be liberated. Gives external form. Provides mechanical support and thus protects the cell.
Cell Wall (in plants)	Thick cellulose wall around the cell membrane gives strength and rigidity of the cell.

Cytoplasm	Provides greatly expanded surface. Endoplasmic recticulum-area for various biochemical reactions which normally take place at or across membrane surfaces.
Ribosomes	Sites of protein synthesis and known as "Protein factories on the cell"
Mitochondria	Oxidation of food substances, changes potential energy into a form of energy to be used by cell. Energy production (Krebs' cycle, electron transport chain, beta oxidation of fatty acids etc.). Power house of cell.
Centrioles (usually not seen in plant cells)	Form poles for the cell division. Capable of replication usually.
Golgi body or apparatus	Produce cellular secretions known as dictyosomes in plants.
Lysosome (in animals)	Produce intracellular digestive enzymes which help in disposal of bacteria and other foreign bodies. If ruptured, may cause cell destruction.
Plastids (in plants)	Serve for storage of starch, pigments and other cellular products. Photosynthesis takes place in chloroplasts.
Vacuoles	Store excess water, waste products, soluble pigments etc.
Hyaloplasm	Contains enzymes for glycolysis and structural materials, viz., sugars, amino acids, water, vitamins, nucleotides etc.
Cilia and flagella	Serve to move the animals or to move the material.
Nucleus	Provides selective continuity.
Nuclear membrane	between nuclear and cytoplasmic materials.
Nucleoplasm	Contains materials, for building DNA and messenger molecules which serve as intermediates between nucleus and cytoplasm.
Nucleolus	May synthesize ribosomes. Disappears during cellular replication. Function uncertain.
Chromosomes	Bearers of hereditary instructions. Regulation of cellular processes (evident only during nuclear division). Control the nucleus and metabolism.

2

PROTOPLASM

Protoplasm is a highly complex mixture of some elements and compounds found in the bodies of living beings. The protoplasm is variously known as the *living matter, living substance* or *physical basis of life.* It is the basic fundamental substance exhibiting all the vital processes of the cell. The protoplasm was first observed by *Corti* in 1772. In 1835 *Dujardin* a Frenchman described it as a soft and gumy substance and named it as *scarcode* i.e. *flesh. Purkinje,* a Bohemian physiologist (1839), was the first biologist, who gave the name *protoplasm* to this living substance. *Hugo von Mohl,* a German botanist, in 1839, also suggested the name *protoplasm* for the granular and viscous substance found in plants similar to that found in the animals. The above-mentioned German botanist popularized the word protoplasm as a name given to the living matter found in plants and animals. Protoplasm is the most complex and interesting substance. It is not to be thought of a chemical compound but rather as very complex organized system. Protoplasm varies somewhat in its nature from cell to cell and from organism to organism, but basically it must be the same, as evidenced by its common manifestations of metabolism, growth, reproduction and by some other peculiarities.

PHYSICAL NATURE OF PROTOPLASM

Different workers have proposed different theories to explain the physical nature of protoplasm as given under the following heads:

1. Granular Theory

This theory was propounded by *Altmann* in 1893. According to this theory, protoplasm consists of numerous tiny granules as shown in *Amoeba. Henle, Maggi,* etc., considered these protoplasmic granules *plastidules. Altmann* reconginzed them as "elementary organisms," or *bioplasts* (or cytoplasts).

2. Alveolar Theory

The alveolar nature of protoplasm was suggested by *Butschilli* in 1892. According to him, protoplasm consists of many suspended droplets or alveoli or minute bubbles, resembling to foam of emulsion.

3. Fibrillar Theory

This theory was put forward by *Flemming.* According to him, protoplasm consists of fibres embedded in the inner mass of matrix. The fibrillae are called mitome or spongioplasm and ground substance is termed paramitome or hyaloplasm.

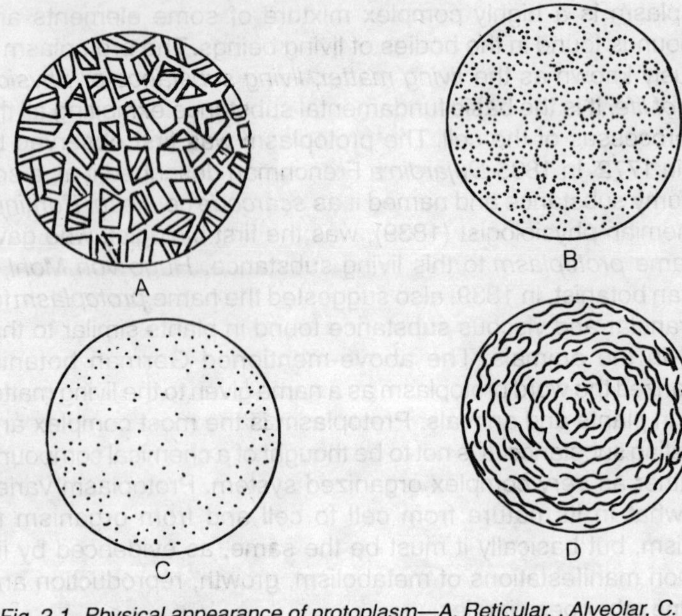

Fig. 2.1. Physical appearance of protoplasm—A. Reticular, . Alveolar, C. Granular, D. Fibrillar.

4. Reticulate Theory

This theory was proposed by *Klein, Cornoy* etc., suggesting that the protoplasm consists of a network or reticulum of fibres in the ground substance.

5. Colloidal Theory

Proposed by *Wilson* in 1925. According to this theory the

protoplasm having any of the aforesaid appearances is always a fluid-colloidal system having various chemical inclusions in *gel phase.* A gel is a semi-solid condition which presents jelly-like appearance. Here the molecules are held together by various bonds depends upon the imature and strength. By absorbing water the gel changes to more liquid-like phase. This is known as *sol* and the process is termed as *solation.* The sol can stream easily and by losing water again changes into gel-state. Both these states of colloidal matrix are interchangeable according to the various physiological, mechanical and biochemical activities of the cell. This represents phase reversal in the colloidal system. These sol-gel conditions of colloidal system are the basis for the mechanical behaviour of protolplasm.

PROPERTIES OF PROTOPLASM

1. Cohesiveness

The various particles or molecules of protoplasm are adhered with each other by forces, such as Van der Waal's bonds, that hold long chains of molecules together. These Van der Waal's bonds are weak and non-specific forces between non-polar groups of atoms. This property varies with the strength of these forces.

2. Contractility

This property is significant in various stomatal operations in plants. The contractility of protoplasm is important for the absorption and removal of water as they generally occur in protoplasm.

3. Electrical Charge

Protein molecules repel each other because their overall charge is similar. All molecules are either positively charged or negatively charged. However, if the molecules approach one another close enough so that valency forces can act, then they may be attracted to each other.

4. Precipitation

Addition of certain amount of electrolyte in a colloidal suspension causes its dispersed particles to colloide, aggregate and finally to precipitate as suspension. For example, the addition of HCl to a colloidal system of arsenic sulphide causes precipitation.

5. Viscosity

The viscosity of the ground substance of the cell varies greatly. It may be as low as that of water, or may be very high in the gelating cytoplasm of pseudopodia of *Amoeba.*

6. Streaming Movement or Cyclosis

The protoplasm exhibits various sorts of streaming movements inside the cell. These have been studied in *Amoeba, Paramecium* etc. The movement involves only the localized portions with no visible changes in the protoplasm. No complete explanation of this has been given as yet. But it is seen that its rate depends upon the rate of cell metabolism. It is due to the fact that the energy is supplied by respiration.

7. Amoeboid Movement

The amoeboid movement as exhibited by *Amoeba* and other protozoans involves the movement of entire protoplasm of the cell, where the cytoplasm moves as one mass carrying the various inclusion with it. It is due to the continuous change of gel to sol and sol to gel.

8. Brownian Movement

It is characterized by the zigzag motion of suspended colloidal particles, occurring due to the bombardment of one particle or molecule by other. This type of movement of particles was first of all observed by *Robert Brown* in 1827 in the colloidal solution and hence such movements are known as Brownian movements. The higher the temperature, more rapid the movement and thus viscosity of cell is decreased. This means that high viscosity indicates a more gel-like state of protoplasm and low viscosity, a more sol-like condition.

9. Tyndall Effect

Colloidal particles of protoplasm have the property of scattering light. When a beam of light is passed through a colloidal solution it becomes visible. This is the Tyndall effect. A colloidal solution of proteins in water shows a typical Tyndall cone.

10. Adsorption

The tendency of particles, molecules or ions to adhere to the surface of certain solids or liquids is known as adsorption and is exhibited by the particles of colloidal system. The phenomenon helps the matrix to form protein boundaries.

11. Biological Properties

Protoplasm has all the biological properties of a living organism. It is capable of nutrition, respiration, excertion, metabolism, growth and reproduction. It has the property of irritability, e.g., it responds to stimuli like heat, light and chemicals. It also has the property of conductivity, i.e., of conducting impulses produced by stimuli.

Chemical Nature of Protoplasm

For the detailed study the chemical components of the cell can be classified as inorganic (mineral salts) and organic (proteins, carbohydrates, nucleic acids, lipids and so forth) substance. Although the most prominent constitutent of protoplasm is water—the substance which gives protoplasm its characteristic structure is protein. Lipids are important in all membranes and carbohydrates serve as nutrient stores. The protoplasm of a plant or animal cell contains 75 to 85% water, 10-20% protein, 2-3% lipids, 1% carbohydrates and 1% inorganic material. The following Table 2.1. gives approximate active protoplasm.

Table 2.1. Chemical Analysis of Protoplasm

Substance	*Percent*	*Average molecules weight*	*Number of molecues in relation to protein*
Water	85	18	180
Protein	10	36000	1
DNA	0.4	10	—
RNA	0.7	4.0×10^4	—
Lipid	2	700	10
Other Organic matter	0.4	250	20
Inorganic substances	1.5	55	100

Water

Water is a substance that is justifiably called the *fluid of life.* It comprises over 90% of the chemical content of most organisms. In human body about 55% of the water (20-22 litres) is intracellular water and remains confined to the cells and the rest is found in extracellular fluids such as blood, tissue fluid and lymph. Water participates directly or indirectly in all metabolic reactions. The biologically active conformations of macromolecules and arrangement of phospholipids in the lipid bilayer of membranes are dependent on water. Water helps to keep minerals ionised in body fluids. It ionises itself to provide hydrogen ion (H^+) concentration to body fluids. It also helps in maintaining the constancy of the internal environment of an organism.

Water is a remarkable compound with unique properties that result from its molecular configuration and hydrogen bonding. Some important properties of water are as follows.

1. Polarity

The water molecule is composed of two hydrogen atoms covalently bonded to one side of an oxygen atom. Since the mean angle (105^0) between the hydrogen atoms is not rigid, water can absorb large quantities of heat and be subject to other physical stresses without breakage of the bonds. Water is a polar molecule and as with other polar molecules it has a surface charge. Water is a dipolar substance in that the hydrogen pole is positively charged and the other pole is negatively charged due to electrophilic (electron attracting) properties of oxygen. The polar nature of the water molecule causes salts to be held in solution through charge interaction (ionization; salt dissolved in water exist in the form of positive and negative charges).

2. Cohesion and adhesion

Because of the asymmetrical distribution water molecules associate with each other (cohesion) and 'wet' other substances (adhesion). The attraction of the positive hydrogen atom of one water molecule for the negative oxygen atom of another water molecule results in a *hydrogen bond.* Hydrogen bonding of water produces a dipolar molecule and favours the formation of a lattice-like structure that enables the packing of many atoms into a small area and stabilizes the molecular structure of water. Fluidity of water is maintained by very rapid formation and dissociation of hydrogen bonds between water molecules.

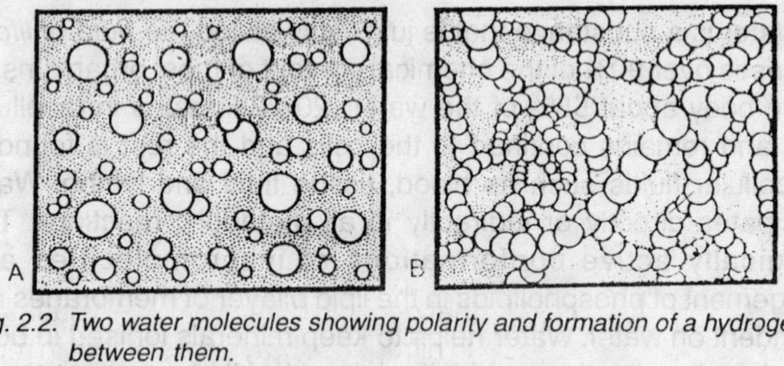

Fig. 2.2. Two water molecules showing polarity and formation of a hydrogen bond between them.

3. Latent heat and specific heat

The presence of hydrogen bonds is directly responsible for the high heat of fusion, high specific heat, and high heat of vaporization of water. The energy required to break hydrogen bonds for melting of ice, heating of water, and evaporation of water is considerably greater than the energy needed to overcome the Van der Waal's forces that are normally found in the weak association of molecules of ethane, ether and benzene. Due to high specific heat, water affords protection against sudden temperature changes in living organisms. High latent heat of evaporation of water causes elimination of excess heat through evaporation of sweat. This maintains a constant body temperature.

4. Density of freezing properties

The density of water decreases below 4^{0}C and ice therefore tends to float. It is the only substance whose solid form is less dense than its liquid form. The fact that water below 4^{0}C tends to rise helps to maintain circulation in large bodies of water. This may result in nutrient cycling and colonisation of water to greater depths.

5. Solvent action

A property of water that is very important to the living cell is its solvent action. Because it forms a solution with a vast array of compounds, water is sometimes referred to as the *universal solvent.* The solvent action of water is an effect of its ability to form hydrogen bonds due to the asymmetrical distribution of its charges. In solution with water, compounds like sugars, alcohols, and amino acids—which contain oxygen atoms, hydroxyl (–OH) groups, amino (–NH_2) groups—form hydrogen bonds with the molecules of water.

The solvent action of water is of tremendous importance for the living system. The essential elements necessary for normal growth, the compounds necessary for energy transfer and storage, and the components of structural compounds all require water as a translocation and reaction medium. Indeed, physiological processes in dilute solutions and suspensions; and the reactions are, therefore, under control of the physical and chemical laws that govern the activities of dilute solutions and suspensions.

Inorganic Compounds

A number of inorganic salts occur both in free as well as in the ionised state. The elements that occur in quantity in protoplasm are oxygen, carbon, hydrogen and nitrogen. Always present but in much smaller amounts are potassium, phosphorus, calcium, sulphur, magnesium and iron. These ten are generally referred to as *essential elements.* Chlorine and sodium are necessary components of most animal protoplasm but are apparently not essential for plant protoplasm. Copper, boron, iodine, maganese, zinc and several other elements which are present in all protoplasm but only in minute quantities, are called *trace elements.* This term must be used with care, because these elements are still important even though they are present only in very small quantities. The essential elements found in human protoplasm are listed below:

Table 2.2. Showing Percentage by Weight to the Element in Protoplasm

Elements	*Symbol*	*Percent*
Oxygen	O	65.04
Carbon	C	18.24
Hydrogen	H	10.05
Nitrogen	N	3.15
Potassium	K	1.60
Phosphorus	P	0.84
Calcium	Ca	0.25
Sulphur	S	0.20
Magnesium	Mg	0.04
Iron	Fe	0.01
Chlorine	Cl	0.27
Sodium	Na	0.26

The elements are almost always present in protoplasm in the form of chemical compounds rather than elements. Many of these compounds inorganic salts, which are usually in solution. These salts have numerous functions. They serve as source of elements to be built into other compounds, and some act as *buffer* in maintaining the proper acidity or alkalinity in the protoplasm. Because salts in solutions are electrical conductors, they also function in some way in connection with the electrical properties of protoplasm, however, this is not well understood at present.

Organic Compounds

Proteins

Protein is an indispensable constituent of the diet because it is the only source of the amino acids. Amino acids are needed to built up new tissue during he period of growth; to maintain the structure of every tissue cell including its content of protein—containing enzyme systems; to provide raw material for the manufacturing of digestive enzymes and certain hormones; and to maintain the normal concentrations of plasma proteins and haemoglobin.

The proteins are macromolecules of very high molecular weight. The polymer molecules of protein are formed by linking together of a number of amino acids. About 20 odd different kinds of amino acids are known. Each amino acid has an amino (NH_2) and a carboxyl (COOH) group. A bond is formed between the amino group of one amino acid with the carboxyl group of another amino acid. The bond is called peptide bond. With the help of such peptide bonds a variety of amino acids can form a long chain molecule called polypeptide.

The nutritional value of a protein depends upon its amino acid composition. Synthesis of amino acids from organic keto-acids is of common occurrence in bacteria and plants. Animals can synthesize only about half of the naturally occurring amino acids (20). Thus there are two categories of amino acids:

1. Essential amino acids

The amino acids which cannot synthesized by an organisms i.e. they must be obtained from the dietary proteins are called essential amino acids. These are *arginine, histidine, isoleucine, leucine, lysine, methconine, theronine, phenylalanine, tryptophan* and *valine.*

2. Non-essential amino acids

Although these amino acids are required by the animal as they are found in the protein of the tissues, but they can apparently be synthesized from alpha-keto acids, by the process of amination. These amino acids are *tyrosine, serine, alanine, asparagine, proline, hydroxyproline, asparatic acid, glycine, glautamine, glutamic acid* and *cystine.*

People who eat diets lacking only one of the essential amino acids are unable to synthesize normal body proteins and become protein deficient, despite having a normal nitrogen intake.

Most animal proteins, are *complete or first class proteins;* that is they contain all the necessary amino acids in appropriate proportions for ultilization by man as eggs, meat, kidney, fish, liver, poultry, milk etc.

Plant proteins are *incomplete* or *second class proteins,* since they lack certain essential amino acids. as. cereals, nuts, legumes etc. Thus people living on a strict vegetarian diet may become protein deficient.

Associated with protein deficiency is *Kwashiorker*, which affects millions of children in developing countries. It must commonly occurs in infants after weaning and particularly when given an inadequate and predominantly carbohydrate diet. Many of the children also have an inadequate energy intake and the disorder is known as *protein-energy malnutrition* (PEM). The affected child stops growing and loses weight. The skin and hair may become depigmented, and oedema associated with low plasma levels of albumin often develop. Untreated Kwashiorker is often fatal.

Structure of Protein

The critical determination of biological function of a protein is its conformation, which is defined as the three-dimensional arrangement of the atoms of a molecule. Four basic structural levels are assigned to proteins. These are:

1. Primary Structure

Arrangement of amino acid in a polypeptids chain is the primary structure of a protein. Amino acid sequence of the smallest protein known, that is of insulin, was studied by *F. Sanger* (1954). Insulin consists of two chains, A and B, joined together by two disulplide bonds (A chain contains 21 and B chain 30 amino acids). Normal human haemoglobin consists of four polypeptide chains (2α and 2β) to be held together by non-covalent forces. Each α-chain is composed of 141 amino acids, and each β chains 146 amino acids.

2. Secondary Structure

The polypeptides are held together to give rise a definite shape (helical pattern) of the protein molecules by hydrogen bonds. Hydrogen bonds can be broken by many physical and chemical treatments, for example, by excessive heat, pressure, *pH*, electricity, heavy metals and other agents that changes the environment of protein.

The most common type of secondary structure is α-helix. This is formed by the bending of polypeptide chain to form hydrogen bonds in the same chain. The bending of the chain is very regular and chain assumes the shape of a helix.

Pauling and *Corey* have considered an alternate type of secondary structure. In this case the polypeptide chains are lead linearly either parallel or antiparallel with respect to one another. This is called β-structure of β-helix.

3. Tertiary Structure

The arrangement and inter-relationship of the twisted chains of protein into specific loops and bends in called the tertiary structure. Such a structure enables the proteins to form specific layers, crystals or fibres. The tertiary structures is maintained by hydrogen bonds disulphide bonds, Van der Waals interaction, hydrophobic interactions and ionic bonds. The tertiary structure is important and found in globular proteins. If this structure is disrupted, the biological activity of protein would be lost.

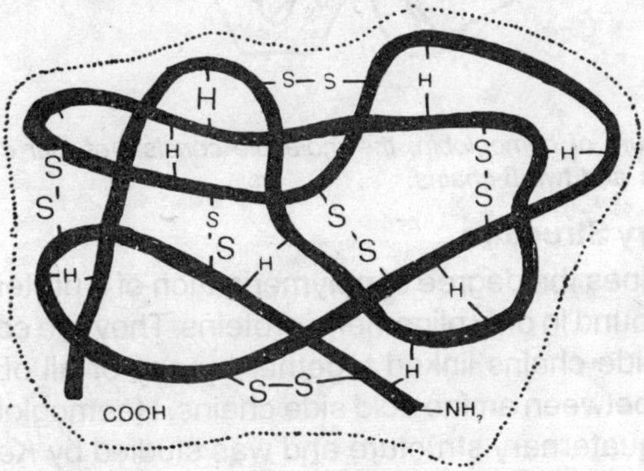

Fig. 2.3. Tertiary structure of a hypothetical protein molecule; note the extensive folding of the secondary structure which imparts a globular shape to the protein.

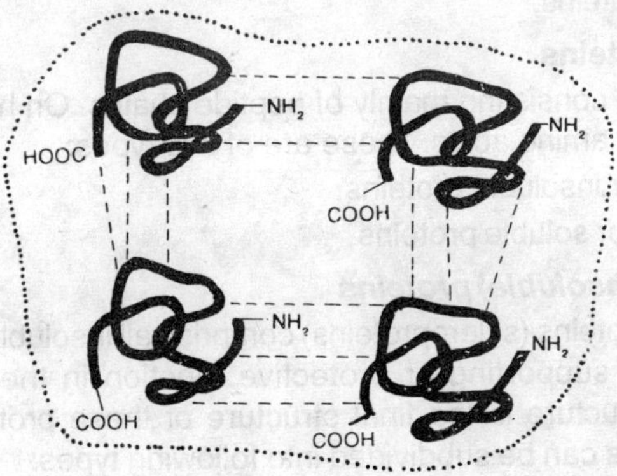

Fig. 2.4. Quaternary level of protein structure; note a diagrammatic representation showing four sub-units which are polymerised in the functional protein molecule.

Fig. 2.5. Structure of hemoglobin; the molecule consists of four chains: two α-chains and two β-chains.

4. Quaternary Structure

This defines the degree of polymerisation of a protein unit. This structure is found in only oligomeric proteins. They are composed of subunit peptide chains linked together by any or all of the forces that can act between amino acid side chains. Haemoglobin is a fine example of quaternary structure and was studied by *Kendrew* and *Perutz* (1963).

Classification of Protein

The proteins are mainly classified into two—simple and conjugated proteins.

1. Simple Proteins

These are consisting mainly of peptide chains. On hydrolyses they yield only amino acids. These are of two types:

(A) Fibrous or insoluble proteins.

(B) Globular or soluble proteins.

(A) Fibrous (insoluble) proteins

Fibrous proteins (scleroproteins) comprise all insoluble proteins which have a supporting or protective function in the animals. Secondary structure is the final structure of these proteins. The fibrous proteins can be subdivided into following types:

(i) *Collagens.* Approximately 30% of the total protein in mammalian body is collagen. They are the principle proteins of skin, tendon and bones and are resistant to peptic and tryptic digestion. They contain large amounts of hydroxyproline. They can be converted into soluble gelatin by boiling with water, dilute acids or alkalies.

(ii) *Elastins.* They are present in ligament, arteries and other elastic tissue. They cannot be converted into gelatin.

(iii) *Keratins.* They are present in animal skin, nails, hairs, horns, hooves, claws, feathers etc. they are derived from ectoderm.

(iv) *Myosin.* The major protein of muscles is also a fibrous protein.

(B) Globular or Soluble proteins

These are soluble simple proteins with more or less definite molecular weight. These are speherical or ovoid in shape. They are generally soluble in water. This group includes enzymes, oxygen carrying proteins (globulins), hormones like oxytocin, vassopressin, insulin, glucogon etc.

(i) *Albumins.* They are soluble in water, dilute acids and bases. They coagulate when heated. Common examples are albumins of egg white, blood serum and lactalbumin.

(ii) *Globulins.* They are insoluble in water but soluble in dilute salt solution. They are also coagulated when heated. Examples are serum globulin, fibrinogen, ovoglobulin in egg-yolk etc.

(iii) *Glutelins.* These are insoluble in water but soluble in dilute acids or base solution. They are commonly found in various plant seeds. Examples are wheat, cereals etc.

(iv) *Protamins.* These are basic proteins. They are very soluble, small, stable proteins and cannot be coagulated by heating. They are exceptionally rich in arginine amino acid. They are found in the ripe sperm cells of certain fishes.

(v) *Histones.* These are water-soluble basic proteins. They coagulate on heating. These are found associated with nucleic acids in nucleoproteins. They are obtainable from the thymus, spleen and nucleated erythrocytes of birds.

2. Conjugated Proteins

The conjugated proteins differ from the simple proteins in that they consist of proteins combined with some non-protein substances. The term *prosthetic group* is generally used to designate such substances (non-proteinous). They are classified according to the nature of prosthetic group as follows:

(i) *Nucleoprotins.* These are formed by the combination of nucleic acids with protein.

(ii) *Mucoproteins.* They contain carbohydrate bound to protein molecules. The protein components of *mucoids* are combined with large amount (more than 4%) of carbohydrate. Examples are ovomucoid-b from egg white; or osomucoid from blood serum.

(iii) *Glycoproteins.* The carbohydrate percentage is less than 4% in glycoproteins. The carbonhdrate in these is usually the mucopolysaccharids. Mucin of saliva, chorionic gonadotropins and some other hormones of pituitary such as follicle stimulating hormone ánd luteinizing hormones are glycoproteins.

(iv) *Lipoproteins.* They are compounds of lipid and protein. The protein part is water soluble and lipid unsoluble. Examples are cholesterol, phosphatidylglycerides, lipovitelline of egg-yolk and serum lipoporteins.

(v) *Chromoproteins.* These are the proteins which have a matelloprophyrin or some similar substance. Examples are haemoglobin which contains a basic protein globin with an iron porphyrin as Heme. Haemocyanin, haemoerythrin, erythrocuorins are other examples.

(vi) *Phosphoproteins.* In these the protein molecule is linked to phosphoric acid when treated with dilute sodium hydroxide yield inorganic phosphate. Examples are casein of milk, vitellin of egg-yolk and pepsin.

(vii) *Metalloproteins.* These proteins contain metals as an inherent part of their molecules. They include tyrosinase (Cu), carbonic anhydrase (Zn), arginase (Mn or Mg) etc.

(viii) *Flavoproteins.* The prosthetic group of flavin component remains permanently attached to the protein. Flavoproteins act as coenzymes which catalyze oxidation-reduction reactions.

Derived Proteins

These include the products obtained from proteins by the action of heat and other physical agents or by hydrolysis.

1. Denatured and Coagulated proteins

Upon treatment with certain chemical agents or heat treatment the proteins become insoluble on account of some sort of intramolecular rearrangement and it is said to have denatured and when it gets separated in the form of precipitate, it is said to have undergone coagulation.

2. Peptides

These include the various fragments of the proteins which are broken off during hydrolysis of the big protein molecules either the presence of the acids or enzymes. The larger fragments had previously been named as the proteoses and the smaller fragments as peptones. This terminology is being abandoned now.

Properties of Proteins

1. ***Colour and Taste.*** Proteins are colourless and usually tasteless. These are homogenous and crystalline.
2. ***Molecular weight.*** The proteins generally have large molecular weights ranging between 5×10^3 and 1×10^6. It might be noted that the values of molecular weights of many proteins lie close to or multiples of 35,000 and 70,000.
3. ***Colloidal nature.*** Because of their giant size the proteins exhibit many colloial propertis such as (a) their diffusion rates are extremely slow, (b) they may produce considerable light—scattering in solution, thus resulting in visible turbidity (Tyndall effect).
4. ***Denaturation.*** In refers to the changes in the properties of a protein. The process of denaturation is followed by coagulation. Denaturation may be brought by a variety of agents. The physical agents, include mechanical action, heat treatment, cooling and freezing operations, ultraviolet rays etc. The chemical agents include X-rays, acetone, alcohol (solvents), salicyclates etc.
5. ***Amphoteric nature.*** The proteins are amphoteric i.e., they act as acids and alkalies both. These migrates in an electric field and the direction of migration depends upon the net charge possessed by the molecule. The net charge is influenced by the *pH* value. Each portion has a fixed value of isoelectric point (pl) at which it will move in an electric field.
6. ***Solubility.*** It is influenced by *pH* solubility is lowest at isoelectric point and increases with increasing acidity or alkalinity.
7. ***Hydrolysis.*** The proteins are hydrolyzed by a variety of hydrolytic agents.

(a) *By acidic agents.* Proteins, upon hydrolysis which come HCl (6-12 N) at 100-110°C for 6-20 hrs yields amino acids in the form of their hydrochlorides.

(b) *By alkaline agents.* Proteins may be hydrolyzed with 2N NaOH. This process is less used as it is highly disadvantageous.

(c) *By proteolytic enzymes.* Under relatively mild conditions of temperature and acidity, certain proteolytc enzymes such as pepsin and trypsin hydrolyze the proteins. Enzyme hydrolysis is used for the tryptophan.

8. *Reactions involving—COOH group*

(a) *Reaction with alkalies* (salt formation). The carboxylic group of amino acids release a H^+ ion and forms carboxylae (Coo^-) ions. These may be neutralized by cations like Na^+ and Ca^{++} to form salts.

(b) *Reaction with alcohols (esterification).* With alcohols, corresponding esters are produced.

$$R—\underset{\displaystyle NH2}{\overset{\displaystyle H}{C}}—COOH + HOC_2H_5 \xrightarrow[\text{catalyst}]{\text{Acid}} R—\underset{\displaystyle NH_2}{\overset{\displaystyle H}{C}}—COOC_2H_5 + H_2O$$

Ethyl ester of AA.

(c) *Reaction with amines.* Forms amides.

$$R—\underset{\displaystyle NH_2}{CH}—COOH + HHN - R \longrightarrow R—\underset{\displaystyle NH_2}{CH}—CO—NH—R + H_2O$$

9. *Reaction involving NH2 groups*

(a) *Reaction with mineral acids* (Salt formation). When AA or proteins are treated with HCl the acid salts are formed.

(b) *Reaction with formaldehyde.* With formaldehyde, he hydroxy methyl derivatives are formed which are insoluble in water and resistant to the attack of microorganisms.

(c) *Reaction with Nitrous acid* (*Van Slyke reaction*). The amino acids react with HNO_2 to liberate N_2 gas and to produce α-hydroxy acid.

(d) *Reaction with fluro-dinitrobenzene* (*FDNB*) or *Sanger's reagent.* In mildly alkaline solution, 1-fluro - 2, 4-dinitrobenzene reacts with a-amino acids to produce yellow coloured derivative, DNB-amino acid.

10. *Reaction involving both COOH and NH_2 groups* (a) Reaction with triketohydrindene hydrate (*Ninhydrin reaction*). Ninhydrin in a powerful oxidizing agent causes oxidative decarboxylation

of a-amino acids producing CO_2, NH_3 and an aldehyde. The reduced ninhydrin then reacts with the liberated NH_3 forming blue-coloured Ruheman's complex.

11. ***Reactions involving—R group or side chain***

(a) *Biuret test.* Compounds containing peptide bonds produce a purple colour when treats with Biuret reagent (.2% alkaline $CuSO_4$). A substance Biuret is formed. Dipeptides do not respond to this reaction.

(b) *Xenthoproteic test.* Yellow colour develops when proteins are boiled with conc. HNO_3 due to presence of benzene ring. This reaction is due to nitration of phenyl rings. This is the test for tyrosine, tryptophan, phenylalanine.

(c) *Million's test.* When proteins are heated with $HgNO_3$ in HNO_3, a red colour develops. This reaction is specific for tyrosine. Tryptophan also responds to this reaction.

(d) *Hoplin's—Cole test* (or *Glyoxylic acid test*). Violet ring develops on addition of Conc. H_2SO_4 at the junction of protein and glyoxylic acid solution. This test is specific for tryptophan.

(e) *Folin's test.* Blue colour develops with phosphomolybdotungstic acid in alkaline soln-due to phenol group. Specific test for tyrosine.

Biological Importance of Proteins

Proteins constitute a large part of he cell-structure and are present in all the tissue. Many proteins have special physiological functions.

1. *Membrane proteins.* The integral proteins include translocases, the peripheral proteins include cytochrome—C and monamine oxidase.
2. *Enzymes.* The enzymes are biocatalysts to influence the rats of a chemical reaction. All the enzymes are proteins.
3. *Hormones.* Some of the messangers of our body, the hromones, are proteins.
4. *Blood proteins.* 6 major plasma components are albumen, α-1 globulin, α-2 globulin, β-globulin, γ-globilin and fibrinogen. All are proteins.

Carbohydrates

The class of substances, known as carbohydrates, is comprises of a large number of relatively heterogenous compounds. These are synthesized form carbon dioxide and water in chlorophyll bearing cells during the process of photosynthesis. They are especial constituents of plants (cellulose, starch etc.) but also occur and serve important functions in animals. They serve as the chief source of energy in the food of many animals. Their energy value is 4 cal per gram and this energy is provided to the various synthetic needs of a cell. In carbohydrate carbon, hydrogen and oxygen are generally found in 1:2:1 ratio.

Structure

Structurally the carbohydrate are the polyhydric alcohols of carbon possessing active aldehyde or ketonic groups, which on hydrolysis yield to such products.

CLASSIFICATION

The carbohydrates may be classified according to their complexity. They are classified into following groups:

1. Monosaccharides
2. Oligosaccharides
3. Polysaccharides.

Monosaccharides

Monosaccharides are those sugars which cannot be hydrolysed into a simple form. Their general formula is $(CH_2O)n$. The value of n ranges from 3-7. If the monosaccharide has an aldehyde group—CHO it is called an aldose e.g. glucose and if a keto group > C = O is present it is called a ketose e.g. fructose.

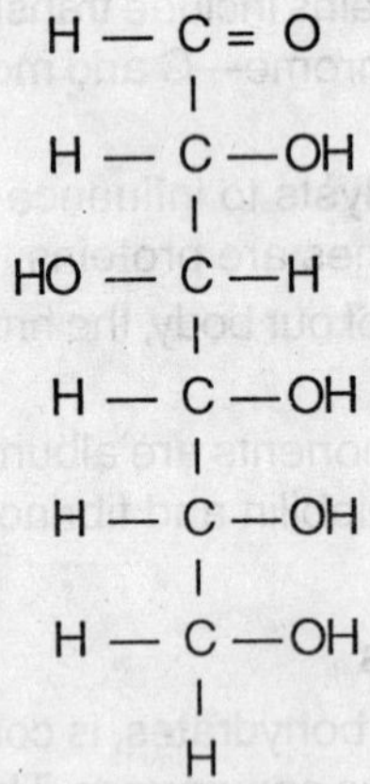

D-glucose

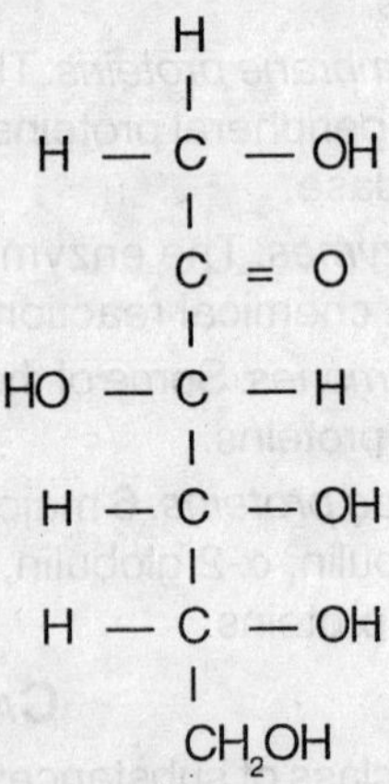

D-fructose

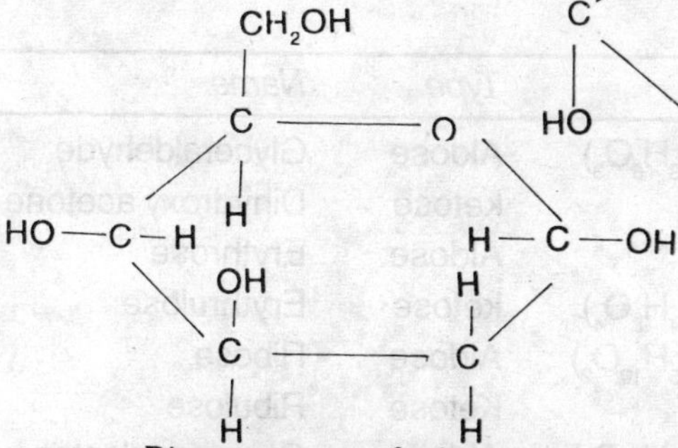

Ring sturcture of glucose

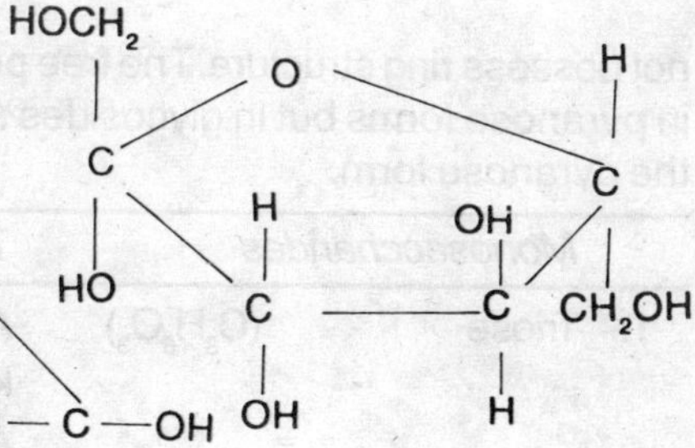

Fructose

Monosacchrides include trioses, tetroses, pentoses, hexoses, heptoses. This classification is based on the basis of carbon atoms present in their molecules. The recent trend in the classification of monosaccharides is a combination of both the system. The aldeosugars and ketosesugars are classified as below. The suffix ('-*ore*) denotes that the sugar belongs to the aldosugar grop and the suffix (-*ulose*) denotes the ketosugas.

Glyceraldehyde occurs in two forms D-glyceraldehyde and L-glyceraldelyde. In D-glyceraldelyde the hydroxyl group (OH) on right side. In L-glyceraldelyde the position of the hydroxyl group is on the left on the asymmetric carbon atom.

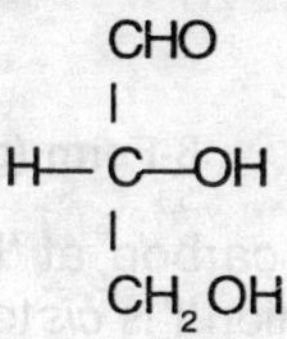

D-glyceraldehyde

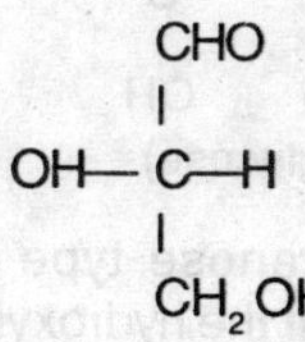

L-glyceraldehyde

It has now been well-established that in the crystalline form, the monosaccharides containing 5 or more carbon atoms exist in the *tautomeric* ring forms. Even in solutions, these sugars mostly occur in the ring forms. The free sugars mostly contain six membered ring in which five members are carbon atoms and the remaining one member is an atom of oxygen. The six membered ring is known as *pyranose ring.* In adition, there is evidence of the existence of five membered rign in the sugar in which four members are carbon atoms and the remaining one member is the atom of oxygen. The five membered rign is known as *furanose rang.* Trioses and tetroses do not possess ring structure. The free pentose sugars are largely found in pyranose forms but in glycosides and nucleic acids these exist in the pyranose form.

	Monosaccharides		*Type*	*Name*
1.	Triose	($C_3H_6O_3$)	Aldose	Glyceraldehyde
			ketose	Dihydroxy acetone
			Aldose	Erythrose
2.	Tetrose	($C_4H_8O_4$)	ketose	Erythrulose
3.	Pentose	($C_5H_{10}O_2$)	Aldose	Ribose
			Ketose	Ribulose
4.	Hexoses	($C_6H_{12}O_6$)	Aldose	Glucose, glactose
			Ketose	Fructose
			Aldose	Persenlose
5.	Heptose	($C_7H_{14}O_7$)	Ketose	Sedoheptulose

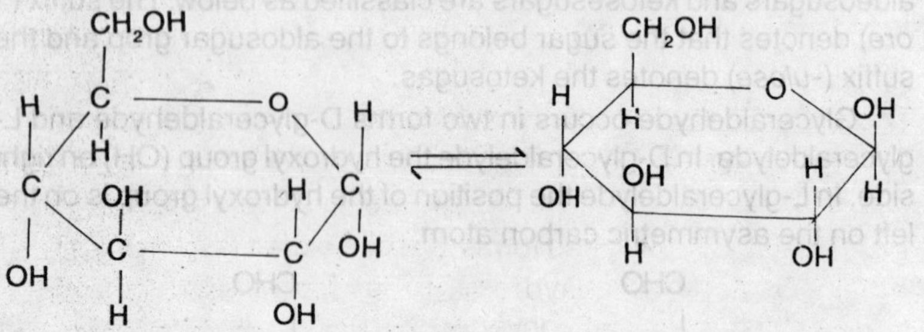

α-Form (glucose) β-Form (glucose)

In the pyranose type of hexoses, carbon at 'I' position is asymmetrical. If the hydroxyl group at position 'I' is *cis* to the hydroxyl group at position 2, it is known as α-form and if *trans* it is known as β-form. The α-form can readily pass into β-form when the sugar is brought into solution.

Properties of Monosaccharides

(a) Physical

The monosaccharides are sweet testing, colourless solids. They are soluble in water, partially in alcohols and insoluble in other.

When a polarized light is passed through a solution of these carbohydrates, the plane of light is rotated to either right or left side.

They contain asymmetric carbon atom hence exist in different isometric forms. The degree of optical rotation may change due to interconversion of isomeres. Fresh solution of glucose gives an optical rotation of +112^0 which change to + 52.7^0 on standing. The change in optical rotation is called *mutarotation.*

(b) Chemical

(i) *Oxidation.* They can easily be oxidized by oxidizing agents. Glucose yields gluconic acid after oxidation with Tollen's reagent (ammonical Ag_2O) or Fehlings solution (alkaline $CuSO_4$). The reduction of Tollen's reagent yields silver as polishing on the surface of tube whereas with the Fehling solution red ppt are obtained.

When strong oxidizing agent is used like conc HNO3, gluconic acid, is ultimately oxidizes to discarboxylic saccharic acid.

Glucuronic acid is obtained in animal body on slow oxidation of glucose. Glucuronic acid combines with hormones. It is a major component of hyaluronic acid, heparin, mucoitin sulphate and chondriotine which are found in blood, skin and cartilage.

(ii) *Reduction.* Free aldehyde and ketone groups of monosaccharides are reduced to alcoholic hydroxy groups by sodium—mercury amalgam and water. D-glucose after reduction yields a mixture of sorbitol and mannitol.

(iii) *Condensation.* Aldehyde groups of monosacharides condense with primary amines to form Schiff's base, when treated with hydroxylamine, glucose forms glucose oxime.

(iv) *Esterification.* The hydroxyl groups of alcohols in the carbohydrate may be converted to esters by treating with the appropriate acetylating agents. When D-glucose is treated with acetic anhydrine in the presence of pyridine, penta-acetyl glucose is formed.

(v) *Methylation.* Methylating agents such as Ag_2O, CH_3OH react with monosaccharide to yield glycoside.

(vi) *Fermentation.* Monosaccharides such as glucose and fructose undergo alcholic fermentation by micro-organisms such as yeasts and produce ethanol and carbon dioxide.

Oligosaccharides

The oligosaccharides are those carbohydrates which on hydrolysis give two to five simple monosaccharide molecules. The oligosaccharides are composed of two to five monosaccharide units. During union of monosaccharde units water molecule is eleminated and the units are linked through an oxygen bridge. It is a glycosidic linkage. Their general formula is $(CH_2O)_{n-1}$.

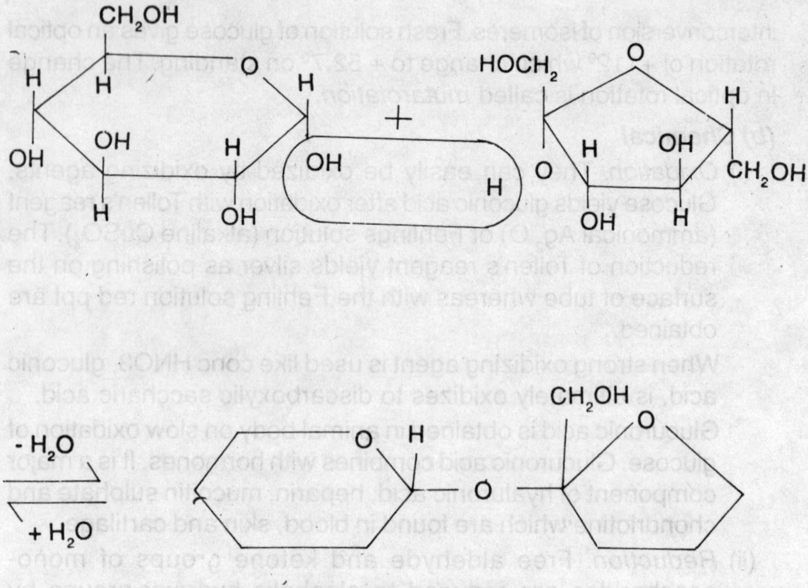

Sucrose (disaccharide)

The disaccharides have been classified into two groups namely, reducing and nonreducing. The reducing disaccharides are maltose, lactose, cellobiose, gentibiose and the non-reducing are sucrose and trehalose.

Maltose. It is composed of two units of D-glucose joined together through their 1 and 4 carbon atoms. It is obtained as a hydrolytic product by the action of amylase on strach

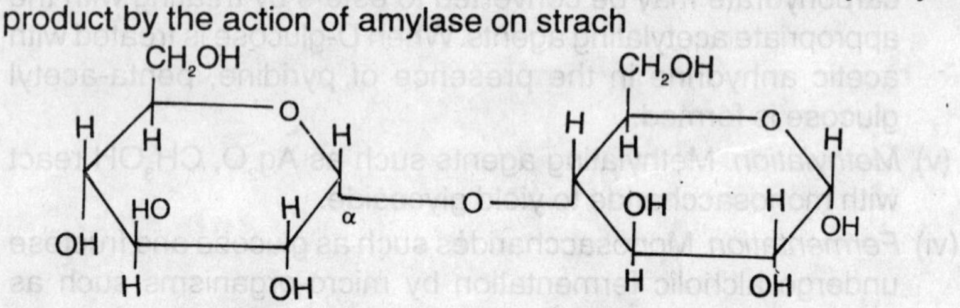

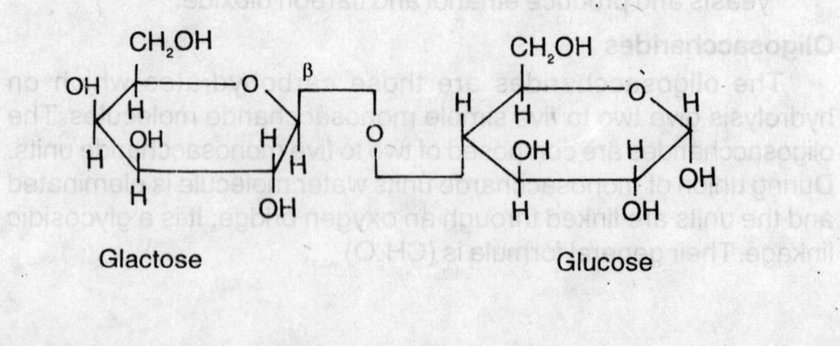

Glactose Glucose

Lactose. Lactose is a disaccharide consisting of glucose and galactose which is synthesized in the mammary glands.

Cellobiose. Incomplete hydrolysis of cellulose given cellobiose. It is composed of two molecules of glucose linked by β-1, 4 glucosidic bond.

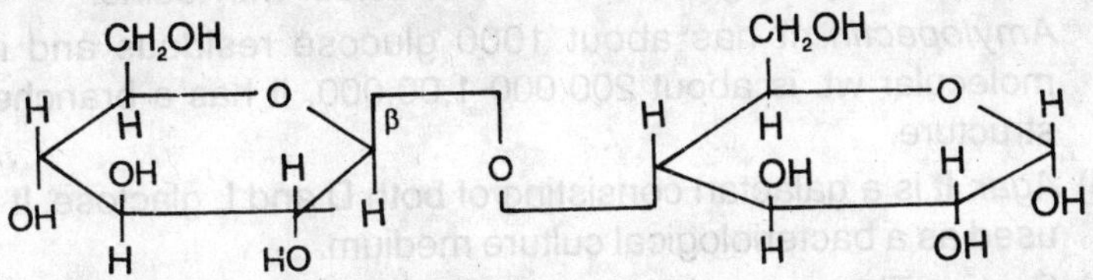

Raffinose. It is a trisaccharide consisting of fructose—glucose—galactose. On hydrolysis it yields *Melibiose* (glucose—galactose) and fructose.

Sucrose. It is formed by the union of one α-D-glucose and one β-D-fructose units with the elemination of one H_2O molecule. Hydrolysis of sucrose by dilute acid or sucrase produces a molecule of glucose and a molecule of fructose.

Polysaccharides

Polysaccharides may be regarded as carbohydrates formed by linking of a number of monosaccharides by glucosidic linkages. A carbohydrate having minimally 6 or more monosaccharide units may be regarded as polysaccharides. Their general formula is $(C_6H_{10}O_5)n$. They are tasteless, colourless amorphous powders which are insoluble in water. because of insolubility and large size, they form colloidal solutions and will not pass across natural animal membranes. They are chemically inert and do not ionize and for this reason very much suited as reserve food material such as glycogen in animals and starch in plants.

Polysaccharides consisting of only one type of monosaccharide units are called *homopolysaccharides* while those with different types of monosaccharde derivatives are *heteropolysaccharides*.

I. Homopolysacchardes

(i) *Glycogen.* It forms the carbohydrate reserve of the animal tissues and it is mainly stored in the muscles and liver. Fungi and yeast do also contain glycogen. It is a branched chain polymer having 6000—30,000 glucose units.

(ii) *Starch.* It is the most important food source of carbohydrate and is found in potatoes, rice, cereals etc. It is hydrolysed in gut yielding dextrins and maltose and eventually glucose.

Treatment of starch with hot water dissolves amylose, while amylopectine remains as such:

Amylose. Contains about 200-500 glucose units which are arranged in the form of a straight chain. The molecular wt. is about 150,000. It gives intense blue colour with iodine.

Amylopectin. It has about 1000 glucose resideus and its molecular wt. is about 200,000-1,00,000. It has a branched structure.

(iii) *Agar.* It is a galactan consisting of both D and L glactose. It is used as a bacteriological culture medium.

(iv) *Pectins.* These are abundant in fruits, particularly in the rim of citrus fruits like orange and lemons. They contain arabinose, glactose and galacturonic acid.

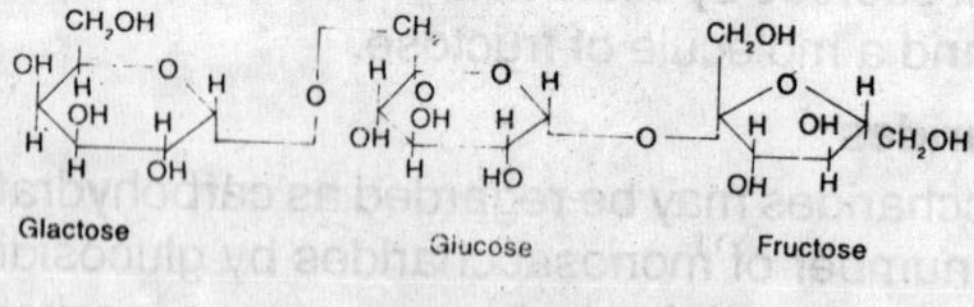

Raffinose

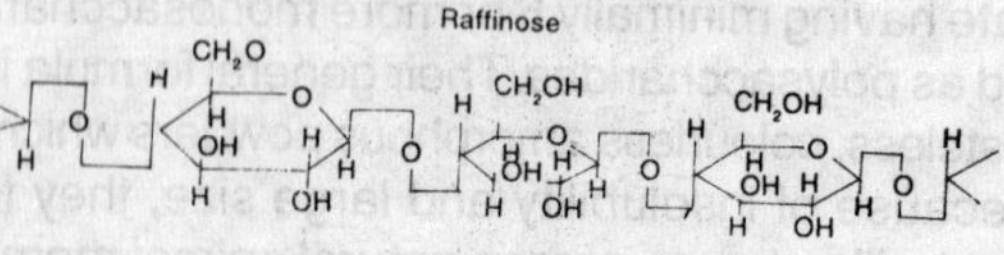

Cellulose

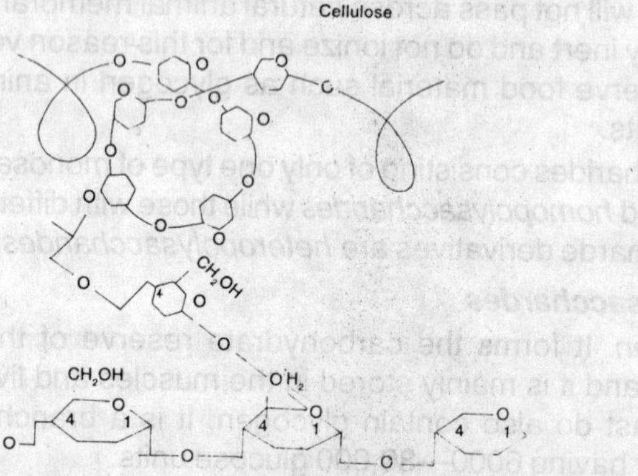

Two units of starch

Amylopectine

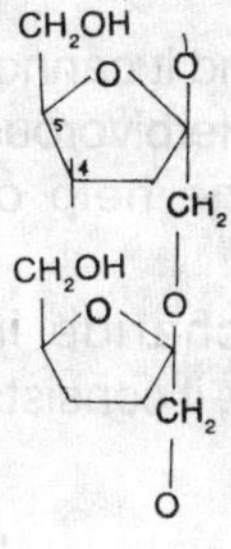

Inulin

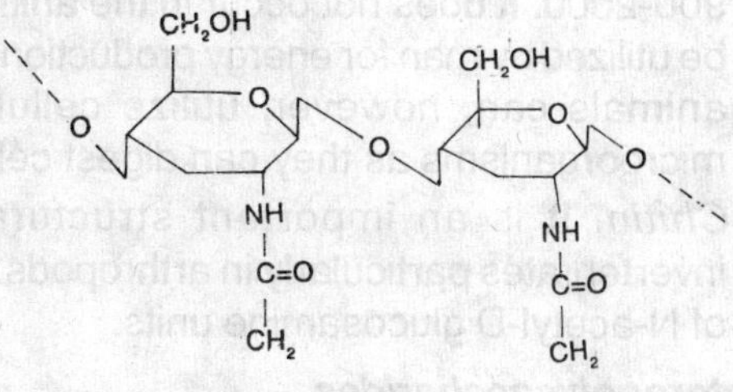

Chitiu

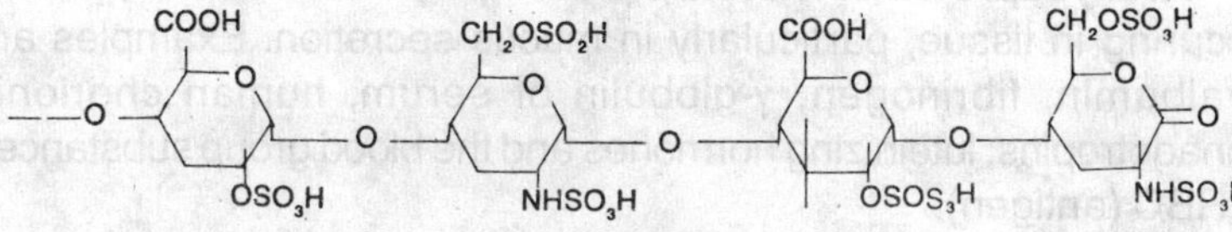

Heparin

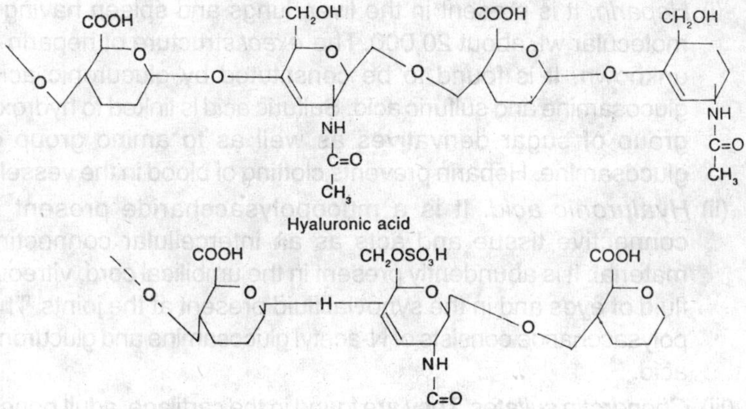

Hyaluronic acid

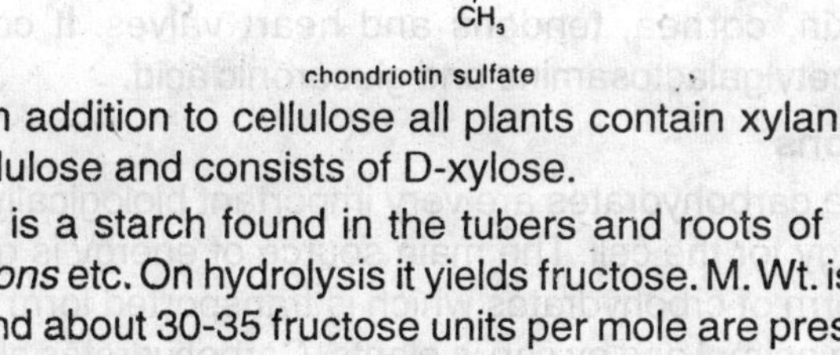

chondriotin sulfate

(v) *Xylan.* In addition to cellulose all plants contain xylan. It is a hemicellulose and consists of D-xylose.

(vi) *Inulin.* It is a starch found in the tubers and roots of *Dahlia*, *Dandelions* etc. On hydrolysis it yields fructose. M. Wt. is about 5,000 and about 30-35 fructose units per mole are present.

(vii) *Dextrins.* They are the intermediate products formed during hydrolysis of starch to the glucose.

(viii) *Cellulose.* The cell wall of plants is formed by cellulose. The cellulose consists of a lenier chain of glucose units ranging

900-2000. It does not occur in the animal tissue and it cannot be utilized by man for energy production purposes. Herbivorous animals can, however, utilize cellulose with the help of microorganisms as they can digest cellulose.

(ix) *Chitin.* It is an important structural polysaccharide in invertebrates particularly in arthropods. Structurally, it consists of N-acetyl-D glucosamine units.

II. Heteropolysaccharides

A. *Glycoproteins.* These are protein-polysaccharide compounds occuring in tissue, particularly in mucus secretion. Examples are ovalbumin, fibrinogen, γ-globulin of serum, human chorionic gonadotropins, luteinizing hormones and the blood group substances of RBC (antigen).

B. *Mucopolysaccharides.* The most important mucopolysaccharides from biological point of view are:

(i) *Heparin.* It is present in the liver, lungs and spleen having a molecular wt. about 20,000. The *exect* structure of heparin in unknown. It is found to be constituted by glucuronic acid, glucosamine and sulfuric acid. Sulfuric acid is linked to hydroxyl group of sugar derivatives as well as to amino group of glucosamine. Heparin prevents clotting of blood in the vessels.

(ii) *Hyaluronic acid.* It is a mucopolysaccharide present in connective tissue and acts as an intercellular connecting material. It is abundently present in the umbilical cord, vitreous fluid of eyes and in the synovial fluid present at the joints. This polysaccharide consists of N-acetyl glucosamine and glucuronic acid.

(iii) *Chondrotin sulfates.* They are found in the cartilage, adult bones, skin, cornea, tendons and heart valves. It consists of N-acetylgalactosamine and glucuronic acid.

Functions

The carbohydrates are very important biologically as a source of energy for the cell. The main source of energy is glucose. It the main form of crbohydrates which is transported form cell to cell by blood in animal and by sap in plants. Carbohydrates also help in the formation of cell wall in plants. Some carbohydrates may serve as the prosthetic group of certain proteins. Glycogen and starch are major storage materials in animals and plants respectively.

LIPIDS

Term lipid includes fats and fat-like substances. Strucrurally, the different substances of this group may not be similar to fats, but all are soluble in fat-solvents like alcohols, ether, chloroform, carbon tetrachloride, acetone etc. They are insoluble in water. Lipids, like carbohydrates also, contain carbon, oxygen and hydrogen. The carbon and hydrogen are present in larger quantity than oxygen. Some lipids have phosphorous and nitrogen. The lipids in the cell may serve as condensed reserve of energy as well as form the membranous structures.

CLASSIFICATION OF LIPIDS

Lipids are classified into:

I. Simple Lipids
II. Compound Lipids
III. Derived Lipids

I. Simple Lipids

These are esters of fatty acids with various alcohols.

1. Natural Fats

These are the esters of fatty acids with glycerol. The chemical term for a natural fat is *triglycerides.* The triglycerides are formed by the combination of three molecules of fatty acids joined to one molecule of glycerol. As already stated triglycerides are neurtral fats, neutral because the acid ions are neutralized during their unification with glycerol; fats because the bulk of the molecule is devoid of electro-negative elements which can unite with hydrogen to form water molecule.

$$\begin{array}{llllll} CH_2OH & & HOOC{-}R_1 & & CH_2OCOR_1 \\ | & & & & | \\ CHOH & + & HOOC{-}R_2 & \rightarrow & CHOCOR_2 \\ | & & & & | \\ CH_2OH & & HOOC{-}R_3 & & CH_2OCOR_3 \\ \text{Glycerol} & & \text{3 fatty acids} & & \text{Triglyceride} \\ \text{molecule} & & \text{molecules} & & + 3H_2O \end{array}$$

Triglycerides can exist in the solid or liquid form. It is based on the kind of fatty acid residues in its structure. The triglycerides which are blow 20°C called *oils* containing a large proportion of short-chained unsaturated fatty acids like *oleic, linoleic acid* etc. The triglycerides which are solid above 20°C are called *fats.* They contain long chained saturated fatty acids like *palmitic acid* and *stearic acid.*

2. Fatty acids

These are obtained by the hydrolysis of fats. These are monocarboxylic acids. The molecules of a fatty acid has a polar carboxyl group soluble in water and a non-polar hydrocarbon chain soluble only in fat solvents.

Table 2.3. Showing a few common fatty acids found in lipids.

Fatty acid		*Formula*
Saturated fatty acid	→	
Butyric acid	→	$CH_3(CH_2)_2$ COOH
Caproic acid	→	$CH_3(CH_2)_4$ COOH
Palmitic acid	→	CH_3 $(CH_2)_{14}$ COOH
Sterаric acid	→	CH_3 $(CH_2)_{16}$ COOH
Unsaturated fatty acids		
Palmitoleic acid		CH_3 $(CH_2)_5$ (H=CH $(CH_2)_7$ COOH
Oleic acid		CH_3 $(CH_2)_7$ CH=CH $(CH_2)_7$ COOH
Linoleic acid		CH_3 $(CH_2)_4$ CH=CH CH_2 CH=CH $(CH_2)_7$ COOH
Linolenic acid		CH_3 CH_2 CH=CH CH_2 CH=CH CH_2 CH=CH $(CH_2)_7$ COOH

The fatty acids are classifed under three groups based on their degree of saturation and unsaturation.

(i) *Saturated fatty acids.* A saturated fatty acid contains as many hydrogen atoms as its carbon chain can hold. General formula for saturated fatty acid is R—COOH, where R is CH_3 (CH_2)n. n is varying from zero in acetic acid to 86 in mycolic acid. The most abunant saturated fatty acids in nature *palmitic* (C_{18}) and *Stearic* (C_{16}) *acids.*

(ii) *Unsaturated fatty acids.* They have one or more double bonds in their carbon chain. When there is a single double bond, results in the loss of 2 hydrogen atoms, such fatty acids are called *monous saturated fatty acids.* Their general formula is CnH_{2n}—1—COOH, examples are *palmitoletic acid* and *oleic acid.* When there ae 2,3,4 or more double bonds in the carbon chain with the consequent absence of 2,6,8 or more hydrogen atoms, such fatty acids are *polyunsaturated fatty acids.* The general formula is CnH_{2n}—2 COOH or CnH_{2n}—3 COOH. Examples are *linoleic acid* and *archidonic acids.*

Linoleic, linolenic and archidonic acid are often termed as *essential fatty acids.*

3. Waxes

Wexes are another class of simple lipis containing one molecule of fatty acid and one high molecular weight alcohol. The constituent acids and alcohols have usually 24-36 carbon atoms. The waxes has high melting points. They are chemically inert as they do not have double bonds in their hydrocarbon chains and are highly insoluble in water. They serve as protective converings on leaf surface (plants). *Bees wax* is an ester of palmitic acid with myricyl alcohol ($C_{30}H_{61}OH$) and *spermaceti* from the sperm whale is an ester of palmitic acid with cetyl alcohol ($C_{16}H_{33}OH$).

II. Compound lipids

These are esters of fatty acids and alcohol with additional compounds, such as phosphoric acid, sugars, proteins etc. These are classified as follows:

(i) Phospholipids

Lipids containing phosphorous are phospholipids. They also have nitrogen containing bases and other substituents. The phospholipids are abundant in brain and nervous tissues. The different types of phospholipids are:

(a) *Phosphatidic acids.* These are compounds consisting of glycerol, two fatty acids and a phosphate group.

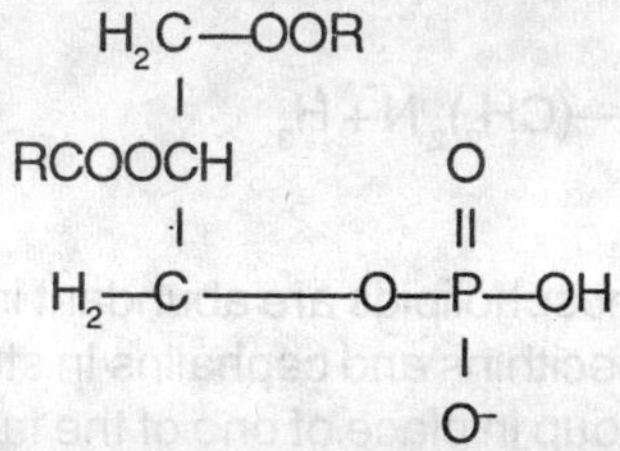

Here R_1 and R_2 represent the residues of the molecules of fatty acids. The phosphatidic acids do not found in any great quantity in tissues but these are important as in intermediate biosynthesis of triglycerides and other phospholipids.

(b) *Lecithins.* Lecithins are choline esters of phosphatidic acid. On hydrolysis they yield one molecule of glycerol, 2 molecules of fatty acids and one molecule of phosphoric acid to which a nitrogenous base choline is attached.

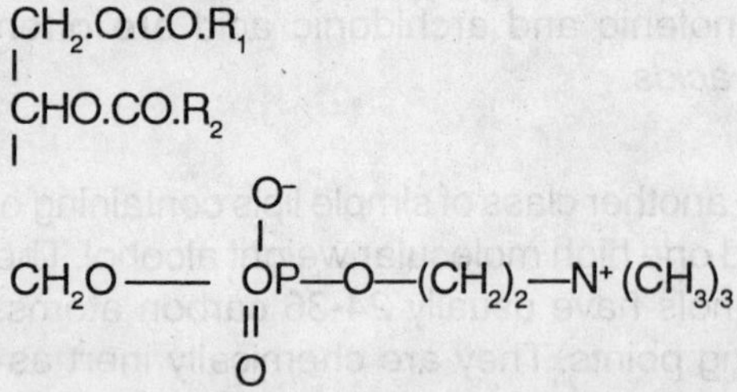

The fatty acids are palmitic, stearic, oleic, linolic and arachidonic acids. Lecithins are yellowish grey solids, soluble in ether and alcohols but insoluble in acetone. Lecithins get broken down by the enzyme lecithins to lysolecithin. This product (lysolecithin) has the ability to hemolyze the red blood corpuscles. Enzyme lecithinase occurs in the venom of snake cobra. and bee.

Lecithin is an important constituent of lipoproteins. Egg-yolk is a rich source of lecithin which also has an important role in fat metabolism in the liver.

(c) *Cephalins.* They resemble lecithins in most properties. The fundamental difference between the lecithins and cephalins is the nature of nitrogenous base. Cephalins contain ethanolamine, colamine and sometimes serine in place of choline.

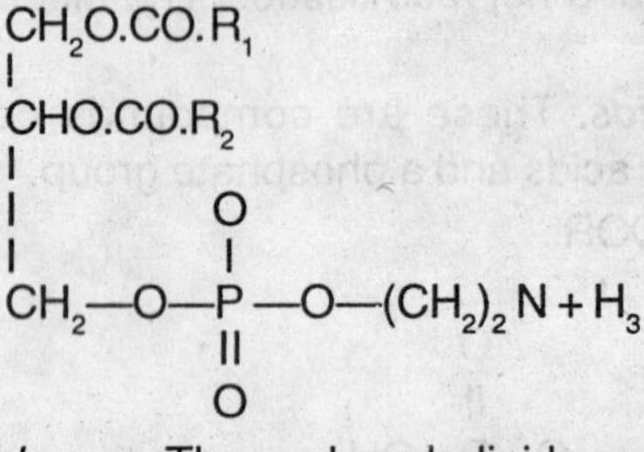

(d) *Plasmalogens.* These phospholipids are abundant in brian and muscle. They resemble lecithins and cephalins in structure but possess an aldehyde group in place of one of the fatty acids in typical phospholipid molecule.

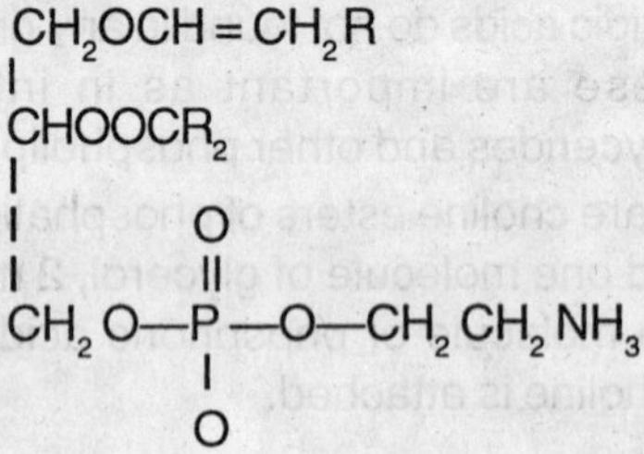

(e) *Phosphoinositides.* Phosphoinositides contain hexahydric alcohol inositol. They can be either monophosphoinositides or diphosphoinositides. An another name lipoinositol was also proposed for these substances. They occur in animal and plants but diphosphoinositides have been reported from brain tissue only.

(f) *Phosphosphingosides.* Phosphosphingosides have been found to occur in high concentration in brain, nerves, lungs and spleen tissue. These contain a nitrogenous base sphingosine or its derivative which remains attached to long-chain fatty acid by its amino group. On hydrolysis they yield fatty acids, phosphoric acid, choline and sphingosine.

(ii) Glycolipids

The glycolipids occurs in brain tissue and in the myelin sheath of nerves, lungs, kidneys, spleen, liver, retina, egg-yolk etc. The glycolipids include two major catagories, cerebrosides and gangliosides.

(a) *Cerebrosides.* They occur in the brain and myelin sheath of nerves. they are based on sphinge mine and have in addition a fatty acid and a monosaccharide sugar but no phosphoric acid or glycerol. Individual cerebrosides are differentiated by the types of fatty acids in the molecule. the important types are:

Kerasin containing the saturated lignoceric acid, $CH_3\ (CH_2)_{22}$ COOH.

Cerebron containing the cerebronic acid, $CH_3\ (CH_2)_{21}$ CHOH COOH.

Nervon containing the nervonic acid $CH_3\ (CH_2)_7\ CH = CH\ (CH_2)_{13}$ COOH.

Oxynervone containing hydroxy derivatives of nervonic acid $CH_3\ (CH_2)_7\ CH{=}CH\ (CH_2)_{12}$ CHOH COOH.

(b) *Gangliosides.* These contain N-acetylneuraminic acid (sialic acid), fatty acids, sphingosine and 3 molecules of hexose.

III. Derived Lipids

These lipids include hydrolytic products of lipids as well as other lipid like compounds like sterols, carotenoids, hydrocarbous etc.

The *sterols* are solid wax-like substances chemically, they are alcohols and occur either as such or as esters of fatty acids. They are highly soluble in fat solvents. They all contain a cyclopentanophenanthrene (sterane) nucleus (made up of three cyclohexane rings, in the phenanthrane type of arrangement and a terminal cyclopentane ring).

Cholesterol is animal origin occuring in bile, brain, spinal cord etc. It is obtained form human gall bladder stones which are deposited in the bile duct.

Ergosterol. It is similar to 7-dehydrocholesterol but differs in the side chain. It can be converted into vitamin D on exposure to the ultra violet light and, therefore, also called provitamin D.

Carotenoids. These are also included in lipids because of their solubility in fat solvent. Carontenes consists of carbon and hydrogen only where as xanthophyll contain oxygen in addition. Vitamin A is derived from carotene. *Estrogens, progesterone, testosterone* and *anderosterone* are steroidal hormoens.

Prostaglandins

Prostaglandins are derivatives of fatty acids. They were first discovered in human seminal fluid secreted by prostate gland, hence the name. It has been shown now that prostaglandins are synthesized and released by many other tissues such as kidneys, testis, placenta, lungs, liver, uterus, gastrointestinal tract, brain and heart.

The prostaglandins are 20 carbon fatty acids including a five membered ring in their molecular structure. Different prostaglandins differ with one another in the number and position of double bonds and hydroxyl group substituents.

Prostaglandins PGE, PGE_1, PGE_3, $PGF_{1\alpha}$, $PGF_{2\alpha}$ and $PGF_{3\alpha}$ are considered as *primary prostaglandins.* Most of their action appear to be binding of hormones to membrane.

Recently, new type of prostaglandin has been isolated from human seminal fluid which has been designated as PGX^2.

Table 2.4. Showing Elements that Occur in Protoplasm

S.No.	*Name of elements*	*Approximate percentage*	*Function*
Major Elements			
1.	Carbon	18	Forms backbone of all organic molecules.
2.	Hydrogen	10	Present in most organic compounds and major components of water.
3.	Oxygen	65	Cellular respiration; in organic compound and component of water.

4.	Nitrogen	03	Major components of all the amino acids, proteins and nucleic acids.
Tracer Elements			
5.	Sodium	0.2	For water balance, conduction of nerve impulse.
6.	Potssium	0.5	Conduction of nerve impulse, muscular contraction.
7.	Phosphorus	1.0	Nucleic acids formation; bones formation and in energy transfer.
8.	Magnesium	0.1	Constituent of certain enzymes (ATPase).
9.	Calcium	1.5	Blood clotting, muscle contraction, bones and teeth formation.
10.	Sulfur	0.3	Constituent of most proteins.
11.	Chloride	0.1	Negative ion of interstitial fluid.
12.	Iron	0.01	Component of haemoglobin and certain enzymes.

Amino Acids

There is a common plan of construction for the thousands kinds of proteins in living systems. The 20 kinds of naturally occurring amino acid monomers are strung together in unbranched, linear polymer chains of proteins. These are the 20 amino acids specified in the genetic code that is universal to all organisms. Some other kins of amino acids are also found in cells, but they are either degradation products or residues that have been modified form oen of the 20 commonly occurring amino acids after this latter has been inserted into the polymer chain. Hydroxyproline is a major amino acid constituent of collagen in connective tissue, but proline residues are initially included in the protein and are converted to hydroxyproline after polymerization. Hydroxyproline is not one of the encoded acid, but proline is. Many proteins contain fewer than 20 kinds of amino acids. The relative proportions and the absolute number of the amino acid repertory vary from one protein to another, as a reflection of the specific in formation in genes, which are the blue prints for protein construction.

Nucleotides and Nucleic Acids

Nucleotides are involved in at least two major cellular functions: (1) they are monomeric units from which DNA and RNA polymers are constructed, and (2) they act as agents in certain energy-transferring reactions during metabolism. A mononucleotide is made up of one nitrogen-containing organic base, one pentose sugar, and one phosphate residue derivative from phosphoric acid. When there is no phosphate group, the sugar-base combination is called a *nucleoside* (Table 2.5). For this reason, *nucleotides* (phosphate-sugar-base) are also called nucleoside phosphates. Nucleoside mono-, di-, and tri- phosphates contain one, two or three phosphate groups respectively.

Table 2.5. Showing nomenclature of nucleic acids and their constituent units.

Base	*Nucleoside*	*Nucleotide*	*Nucleic Acid*
Purines			
Adenine (A)	Adenosine	Adenylic acid	RNA
	Deoxyadenosine	Deoxyadenylic acid	DNA
Guamine (G)	Guanosine	Guanylic acid	RNA
	Deoxyguanosine	Deoxyadenylic acid	DNA
Pyrimidines			
Cytosine (C)	Cytidine	Cytidylic acid	RNA
	Deoxycytidine	Deoxyadenylic acid	DNA
Thymine (T)	Thymidine	Thymidylic acid	DNA
Uracil (V)	Uridine	Uridylic acid	RNA

The nitrogenous bases commonly found in nucleic acids and their nucleotide building blocks are derivatives of purine and pyrimidine. The commonly occuring purines *adenine* and *guanine* are found in both DNA and RNA, as in the pyrimidine compound *cytosine,* the second kind of pyrimidine in DNA is *thymine,* while its demethylated form, *uracil,* occurs in RNA.

Since each kind of nucleic acid contains one unique pyrimidine, it is convenient to study synthesis and activity of DNA or RNA using isotopically-labelled precursors containing one or the other of these bases. Usually the nucleosides uridine or thymidine, or other nucleotide forms, are added to the biological system under study.

The only difference in the pentose sugars of nucleotides is the presence of a hydroxyl group at carbon atom 2 of D-ribose in RNA, but a hydrogen at carbon-2 of 2-deoxy-D-ribose in DNA monomers and polymers. This seemingly simple difference is partly responsible for profound differences in stabilities, pairing potential, and functions of DNA and RNA.

Polynucleotides of both DNA and RNA varieties are built from mononucleotides that are linked convalently via phosphodiester

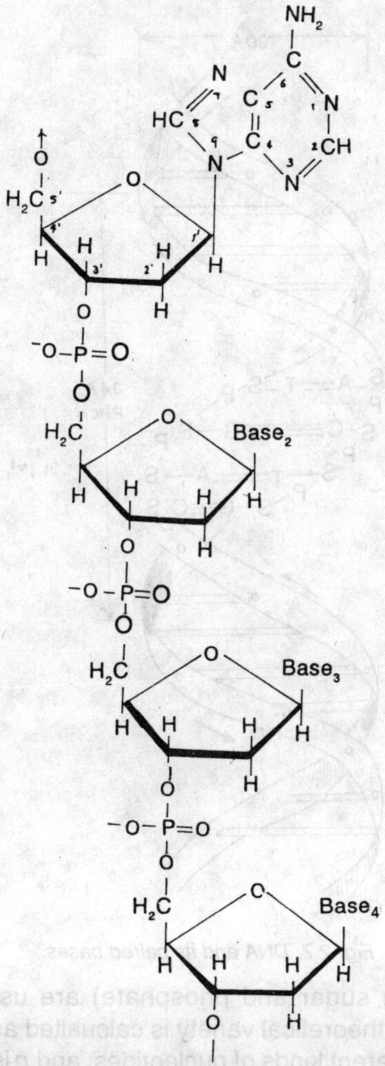

Fig. 2.6. The molecular organization of a backbone of DNA showing phosphodiester and glycosidic bonds.

bridges between the 3' position of one unit and the 5' position of hte next. Since there is no restriction on the ventrical sequence of adjacent mononucleotides in either DNA or RNA, a considerable variety of molecules is possible even though only 4 kinds of nucleotide monomers (one for each of the four kinds of bases in the

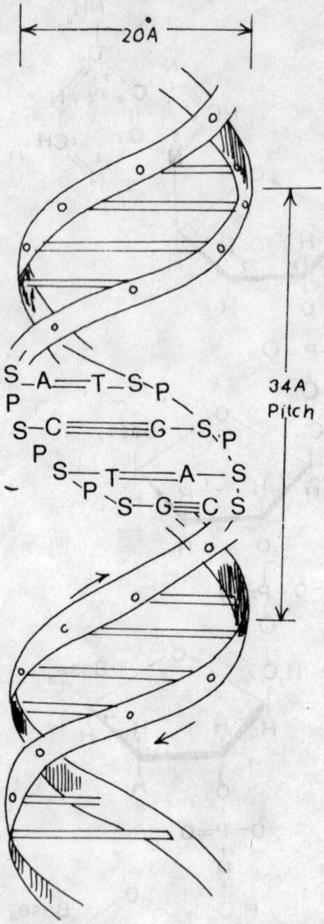

Fig. 2.7. DNA and its paired bases.

combination with sugar and phosphate) are used in polymer construction. The theoretical variety is calcualted as 4", where 4 is the number of different kinds of nucleotides, and *n* is the number of monomers in the polymer. For a molecule made of only 75 monomeric units, as in some of the smallest RNAs, there may be 4^{75} different arrangements of the constituent units. Each arrangement theoretically constitutes a moleule of different specificity. Where the average gene may include about 500 nucleotides in a DNA sequence, 4^{500} different sequences are theoretically possible and, therefore, that many different and specific genes. Despite the apparently meager number of monomer types an astronomically high number of possible genes can be constructed.

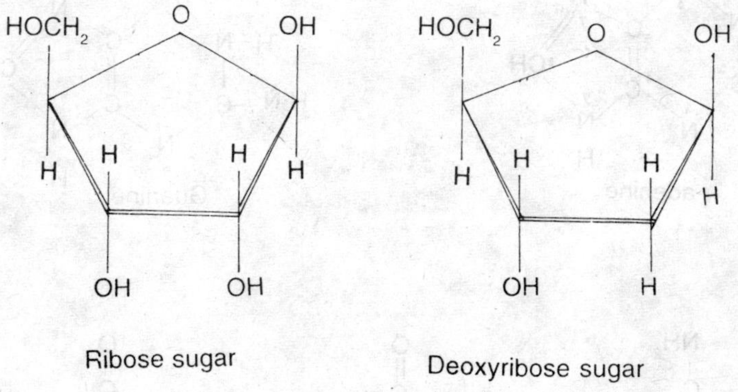

Fig. 2.8. Pentose Sugar.

Such variety can easily account for all past and present life forms.

The DNA Double Helix

DNA molecules usually have regular helical configurations because most DNA molecules consists of two *complementary polynucleotide strands.* The two strands are held together by *hydrogen bonds* between complementary pairs of purines and pyrimidines. Adenine always bind with thymine while guanine always binds with cytosine. This repeated hydrogen bonding within the double helix structure and bonding between virtually all the surface atoms in the sugar and phosphate groups with water molecules serve to stabilize the structure.

Since the purine-pyrimidine pairs are found in the center of the molecule, their flat surface can stack on top of each other and thereby limit their contact with water. In double-helical molecules, a regular structure is possible because the complementary base pairs are exactly the same size. Single polynucleotide chains could not have a regular backbone structure because pyrimidines are smaller than purines, which would cause the angle of helical rotation to vary with the sequence of bases.

DNA double helix molecules are very stable at physiological temperatures because: (1) disruption of the double helix breaks hydrogen bounds and brings hydrophobic purines and pyrimidines into contact with water, which is energetically unsatisfactory; and

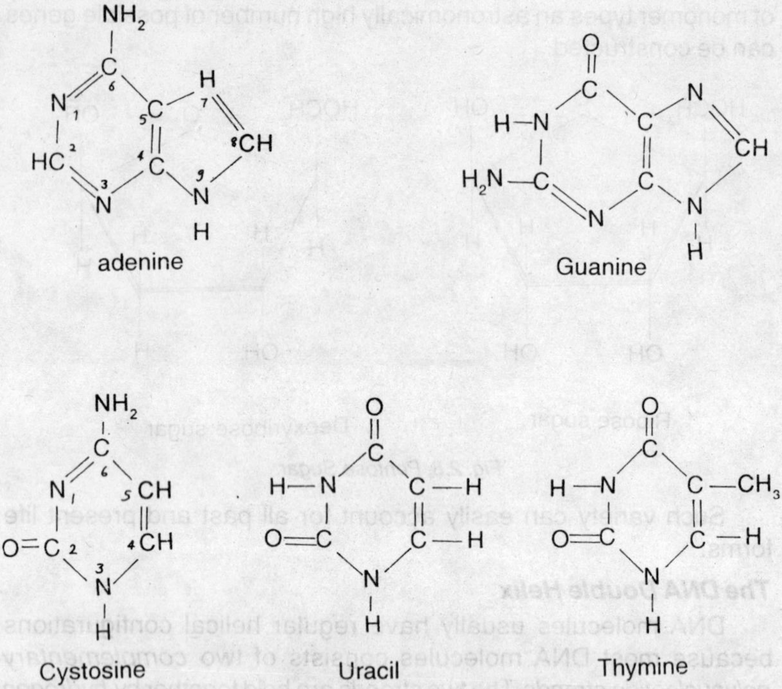

Fig. 2.9. Various heterogenous bases.

(2) there are many weak bonds within the DNA molecule, arranged so that most of them cannot break without many others breaking at the same time. Even though some hydrogen bonds may be broken by thermal motion, hydrogen bonds in the rest of the molecule remain intact and the molecule does not fall apart. In fact, when held together by more than ten nucleotide pairs the double helices are quite stable. At room temperature weak bonds is the stability of molecular shape, in proteins as well as in nucleic acids. At abnormally high temperatures there is more frequent breakage of weak bond, which become less stable as temperatures rise above physiological levels. Once a significant number of weak bonds have been broken, a protein or nucleic acid molecule usually loses its original form and changes to an inactive or denatured form.

3

CARBOHYDRATES

Carbohydrates constitute by far the greatest proportion of organic material on the face of the earth and the most abundant carbohydrate is cellulose which forms the main supporting structure of plants. They constitute important materials for the necessities of life such as food and clothing, housing and health. As reserve food material for plants, the starches are stored in grains, tubers and roots and form the staple food and main energy sources of the poorer populations of the world. The sugars are found in the nectar of flowers and in fruits and milk. Among the industrial processes based on carbohydrates may be mentioned textiles, paper, plastics, explosives, fermentation industries and alcohol. The vitamin, ascorbic acid, is a carbohydrate derivative and some of the plasma expanders belong to the class of carbohydrates. Glycosides of the digitalis-strophanthus group are used in medicine and it is recognized that the invasiveness and antigenic nature of some bacteria is due to their polysaccharide envelope. A great many of the wondrous colours of the plant kingdom are due to the glycoside pigments of the anthocyanin and flavone groups. From the theoretical standpoint, the genesis of carbohydrates in plants through the mechanism of photosynthesis, the complicated steps in the enzymatic breakdown of the simple sugar glucose to carbon dioxide and water, and the phenomena of muscular contraction and bacterial fermentation have fascinated many generations of biochemists. Inside the animal cell, the storage form of energy is present in the form of glycogen. The ubiquitous nucleic acids, deoxyribonucleic acid in the nucleus and ribonucleic acid in the cytoplasm and nucleolus, derive their name from the carbohydrate present in them. The nourishment of the cell come from the circulating glucose of the extracellular environment and the activities of the cell may be profoundly altered by the mucopolysaccharides of the surrounding extra-cellular tissue or the bloodborne glycoprotein hormones of the anterior pituitary. Mucopolysaccharides are also biological lubricants in the form of saliva, mucus and synovial fluid.

The term carbohydrate originated in the belief the naturally occurring compounds of this class could be represented by a general formula $C_x(H_2O)$. Though formaldehyde, acetic acid, and lactic acid fulfil this formula requirement, they are not carbohydrates and some of the methylpentoses and other "deoxy" sugars which belong to the carbohydrate class would be excluded by this requirement. The term carbohydrate is retained, therefore, as a matter of convenience rather than of precise definition. The term carbohydrate now encompasses a variety of substances which include polyhydroxy aldehydes and ketones, alcohols, acids, amino sugars and their derivatives as well as the numerous products formed by the condensation of these different types of compounds with each other by means of glycosidic linkages to form oligosaccharides and polysaccharides. Polyol (glycerol and ribitol) phosphate polymers like teichoic acids are also included in this class of compounds. The elucidation of their chemistry has been a saga of great effort in organic chemistry and even today presents an area of challenge and great biological significance. The structures of eight aldohexoses were carefully elucidated of which only three occur naturally while the rest were synthetic. Now over two hundred different monosaccharides are known to be present in nature. Many amino sugars, branched chain sugars and a wide variety of N-acetyl N-glycolyl neuraminic acids (sialic acids) have been isolated. Several antibiotics like streptomycin, puromycin, neomycin and amino-glycoside antibiotics like kanamycin certain unusual types of sugar residues. The glycoprotein hormones include gonadotropic hormones, thyroglobulin and thyrotropin. Anti-freeze glycoproteins in aquatic animals in cold environments have aroused attention. The characteristic bacterial cell wall peptidoglycans linked to peptides and their susceptibility to penicillin have been extensively studied. Lipid-linked sugar derivatives linked by phosphate bridges to long chain unsaturated lipids (polyprenols) offer clues to the nature of binding in glycolipid molecules. Enzyme defects and lysosomal membrane alterations present in mucopolysaccharide disorders like Hurler's syndrome have provoked attempts to induct enzymes to correct these conditions. The non-enzymatic glucosylation of hemoglobin, serum proteins, and lens proteins with an increased incidence of cateract in diabetic subjects suggests that increased formation of glycoproteins could produce changes in the basement membrane of the smaller blood vessels and give rise to complications. Purified polysaccharides like those isolated from pneumococcus type III have been shown to be antigenic. Studies using specific glycosidases and phytohaemagglutinins (lectins) with an affinity for certain sugars have revealed that blood group specificities are related to the presence of particular sugar residues on the glycoproteins associated with them. The understanding of the properties of cell membranes depends on the knowledge regarding the molecular structure of the glycoproteins and glycolipids and particularly the disposition of the sugar residues present on their surface. Sugar residues are attached in glycoproteins by glycosylamine link to the amide nitrogen of specific asparagine residues in the polypeptide chain or through the hydroxyl groups of serine, threonine or hydroxylysine. Sialylation of proteins has far-reaching effects on their biological properties. Studies in carbohydrate chemistry, now facilitated by the use of specific enzymes, chromatographic separation procedures, spectrometry of different kinds and X-ray diffraction, may illuminate the problems involved in cell regulation and immunity in general.

Carbohydrates may be broadly classified into three main groups:

(i) Monosaccharides, e.g., ribose, glucose, galactose, sedoheptulose.

(ii) Oligosaccharides with 2—10 monosaccharide units e.g., sucrose, lactose.

(iii) Polysaccharides, e.g., glycogen, starch, cellulose, insulin.

Owing to their sweet taste and crystalline character the mono-di-and trisaccharides are generally grouped together under the name of *sugars*. The pclysaccharides are tasteless, amorphous and sparingly soluble substances which on hydrolysis yield monosaccharides.

CLASSIFICATION

Carbohydrates, or *saccharides,* are polyhydroxyaldehydes or polyhydroxyketones and their derivative. The simplest of these, the *monosaccharides,* may be considered to be hydrates of carbon—$(CH_2O)_n$, for which *n* is 3 to 8. *Oligosaccharides* are carbohydrates comprised of a multiple (2 to 10) of monosaccharides joined by acetal or ketal linkages. Polymers of saccharides exceeding 10 in number are referred to as *polysaccharides;* the chains of monosaccharide units in these molecules may be linear or branched.

The carbon chain of a monosaccharide is usually unbranched. Each carbon atom, except one, carries a—OH group; the remaining carbon is a carbonyl group. If the carbonyl group is at a chain terminus, the monosaccharide is designated as an *aldose* :

$$\begin{array}{c} HC{-}O \\ | \\ HCOH \\ | \\ HCOH \\ | \end{array}$$

If the C=O is internal in the chain, the sugar is a ketone derivative and is called a *ketose* :

$$\begin{array}{c} H \\ HCOH \\ | \\ C{=}O \\ | \\ HCOH \\ | \end{array}$$

The generic names given to monosaccharides identify both the number of carbons and the nature of the carbonyl group. As examples, glyceraldehyde is an aldotriose; dihydroxyacetone is a ketotriose; ribose is an aldopentose; glucose is an aldohexose; and fructose is a ketohexose.

Configuration and Conformation

The hierachy of structure of some of the biologically important monosaccharides and their precursor relationship are described by the open-chain formulas. It will be noted that, except for dihydroxyacetone, all of these compounds are labelled with the prefix D. As is the case for the amino acids, such a label is necessary for the nomenclature of the monosaccharides because of the molecular asymmetries of the molecules.

Molecular Asymmetry and Configuration of Monosaccharides

Except for dihydroxyacetone, all of the monoaccharides described in Figures are chiral molecules. All contain one or more asymmetric carbon atoms (i.e., a carbon with four different

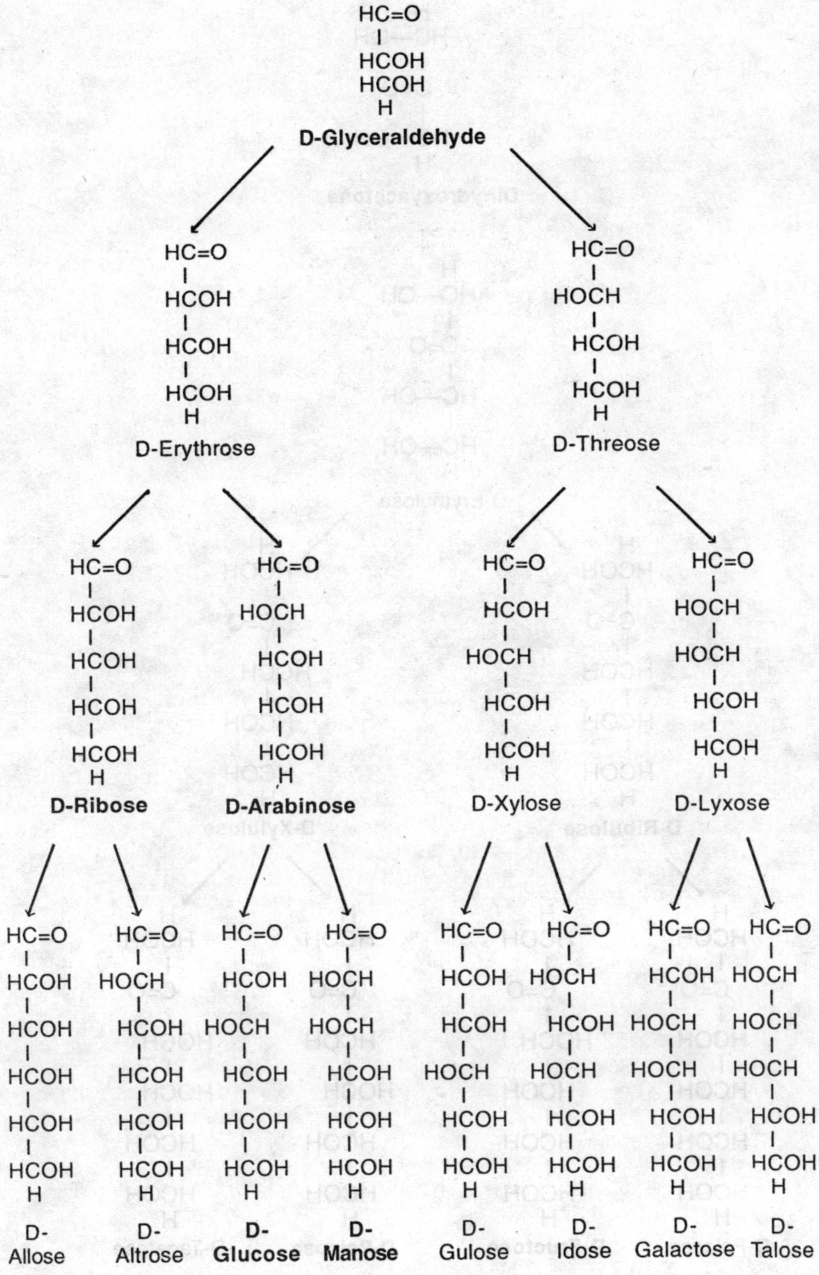

Fig. 3.1. D-Aldoses.

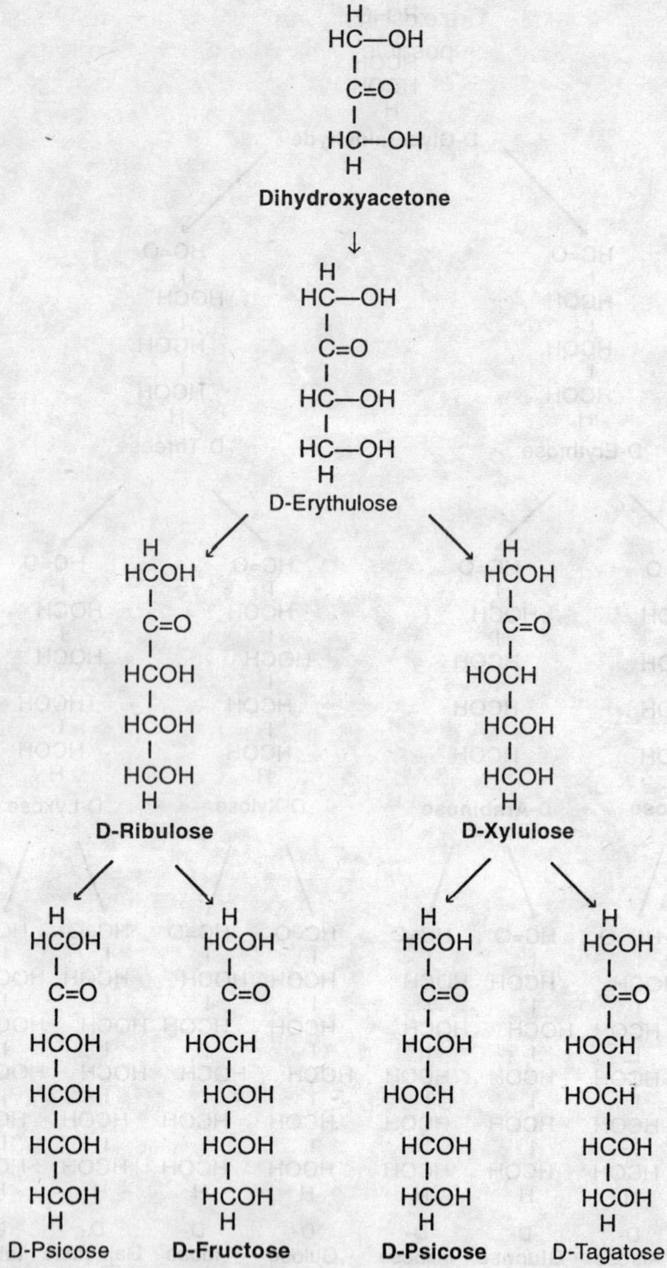

Fig. 3.2. D-Ketoses.

substituents) and can therefore exist in right and left-handed forms. Such isomers are nonsuperimposable mirror images of each other and are called *stereoisomers* or *enantiomers*. Stereoisomerism is evident by the isomers' optical activity, the ability to rotate the plane of plane-polarized light. Enantiomers are identical in all physical and chemical properties, except for their action on polarized light. Stereochemical properties influence the molecular structures and metabolic behaviour not only of carbohydrates but also of amino acids, peptides, and proteins.

Glyceraldehyde is the smallest sugar with an asymmetric carbon atom. To indicate the tetrahedral nature of the carbon atom and the spatial consequences of four different substituents on one carbon atom, consider glyceraldehyde to be tetrahedron. The molecular asymmetry of this molecule is apparent from the perspective and projection formulas shown in figure. Because of the tetrahedral character of the carbon atom, the four different substituent groups on the center carbon of glyceraldehyde can have two different spatial arrangements. These two structural configurations represent the enantiomers of glyceraldehyde since the two forms are non-superimposable. To avoid the three-dimensional projections in writing

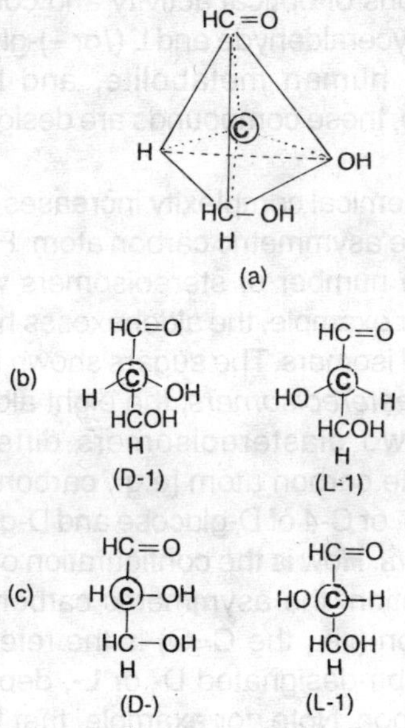

Fig. 3.3. Stereochemistry of glyceraldehyde.

formulas as stereoisomers, it is generally more convenient to use two-dimensional representations, called *fisher projections.* The formulas of the two glyceraldehydes written in this style are shown in figure. In using Fisher projections, the formulas cannot be removed from the plane of the paper when determining it two molecules are superimposable.

In describing the asymmetry of the glyceraldehyde molecule we are concerned with the *configuration* of the substituents about the center carbon. As two configurations are possible when a molecule has one asymmetric carbon. These two forms are designated as D- and L-. The D- and L-glyceraldehydes have been made the reference or parent compounds for designating the absolute configurations of all stereoisomers. By convention the projection formulas for the D- and L-glyceraldehydes are drawn as shown in figure.

All compounds whose stereochemical configurations are related to D-glyceraldehyde are labelled as D forms and all those related to L-glyceraldehyde are designated as L-compounds, irrespective of the direction of rotation of plane polarized light. Optical activities are designated as, *d*, or (+), or dextrorotatory, and *l*, or (–), or levorotatory. In the case of the glyceraldehydes, the directions of rotation are such that the designations of optical activity and configuration are similar [i.e., D (*d* or +)- glyceraldehyde and L (*l* or –)-glyceraldehyde]. L-Lactate, a common human metabolite, and L-alanine are dextrorotatory. Therefore, these compounds are designated as L (+)-lactate and L(–)-alanine.

Obviously, stereochemical complexity increases in compounds containing more than one asymmetric carbon atom. For *n* number of asymmetric centers the number of stereoisomers will be 2^n, or *n* pairs of enantiomers. For example, the aldohexoses have eight pairs of enantiomers or optical isomers. The sugars shown here which are not (+)/(–) are called *diastereoisomers*; the eight aldohexoses are diastereoisomers. If two diastereoisomers differ only in the configuration around one carbon atom [e.g., carbon 2 (C–2) of D-glucose and D-mannose, or C-4 of D-glucose and D-galactose], they are referred to as *epimers.* How is the configuration of such isomers designated? By convention, the asymmetric carbon most remote from the carbonyl carbon (i.e., the C=O) is the reference carbon. Thus, a compound will be designated D- or L-, depending on the configuration of this carbon. Note, for example, that the G-5's of all of the hexoses have the same configuration as the center carbon in

D-glyceraldehyde. Therefore, all of these sugars are of the D series. The same relationship obtains for the tetroses and pentoses. To underscore the point made earlier that configuration may not suggest the sign optical activity, it may be noted that naturally occuring glucose (D-) is dextrorotatory (+), whereas fructose (D-) is levorotatory (–).

Ring Forms of Monosaccharides

Each of the linear forms of aldohexose molecules, whose projection formulas are, contains four asymmetric carbons. The maximum number of stereoisomers for this family of monosaccharides permitted by their linear configurations is, therefore 2^4, or four pairs of enantiomers. However, in aqueous solutions each of these sugars gains an additional center of asymmetry because of an intramolecular reaction involving the carbonyl group. The formation of the fifth center of asymmetry at this carbon atom can occur by the type of reaction diagrammed. Like any other aldehyde, an aldohexose such as glucose will be in equilibrium with its hydrated form in aqueous solution. The hydrated hexose becomes the precursor of a stable, six-membered ring structure resulting from the extraction of a water molecule from C-1 and C-5. In the ring form of the aldose. C-1 becomes asymmetric because of the two possible spatial arrangements of its—OH group. Employing plane projection formulas, the isomer in which the —OH is in the same side as the ring oxygen is designated as the *α-configuration.* If the —OH group is on the side opposite to the ring oxygen the isomer is referred to as a β-*form.* The six membered ring structures of this type are related to the heterocyclic compound pyran and, hence, the sugars are labelled as pyranoses. The two structures are α-glucopyranose and β-D-glucopyranose. With the additional center of asymmetry at C-1, each D- and L- aldohexose can have an α or a β form, allowing at a total of 2^5 or 32 aldohexoses. Monosaccharides that differ only with respect to the configuration of their carbonyl carbon atoms (i.e., α or β) are known as *anomers.*

To appreciate the elementary chemistry of pyranose formation and its implications for our understanding of carbohydrate conformation, consider again the analogous and simpler reaction between an alcohol and an aldehyde or a ketone. Under conditions in which the aldehyde reacts with the equivalent of one alcohol to form a hemiacetal, the erstwhile aldehyde carbon will carry four different substituent and, therefore, becomes asymmetric. This

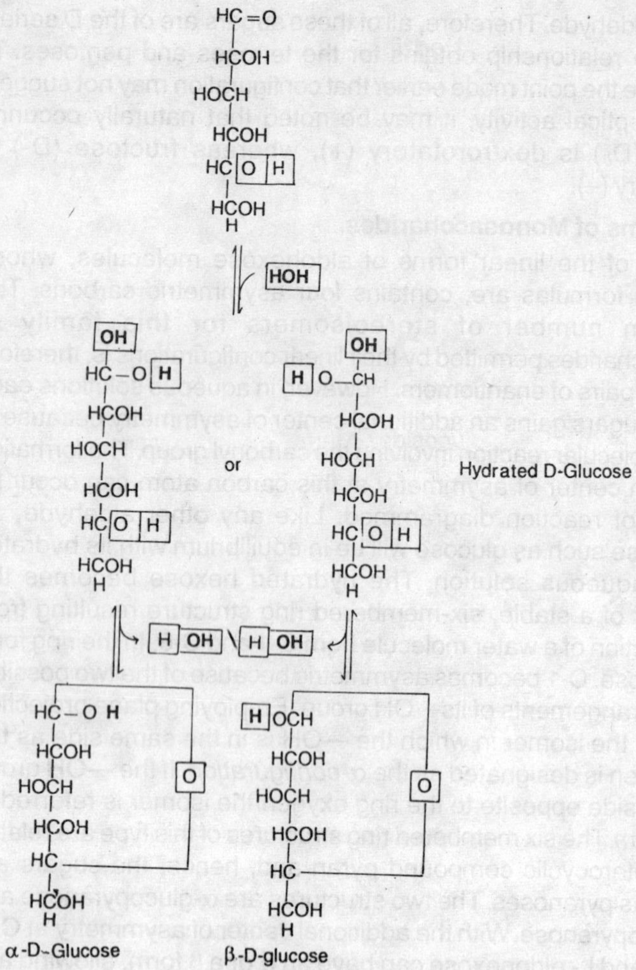

Fig. 3.4. α and β forms of D-glucose.

reaction is comparable to the intramolecular formation of the 1.5 pyranose ring of the glucose, which results in the α and β forms. If the aldehyde reacts with two molecules of alcohol of the same kind [reaction (c)], the resulting acetal is not asymmetric. If, on the other hand, the second alcohol molecule added to the hemiacètal is different reaction (c), the acetal is asymmetric. We shell see in the next section that such a reaction between two monosaccharides results in a glucosidic bond.

Glucose can be located into its α and β configurations by methylation of the —OH at C-1. For example, when glucose reacts with methanol in acid solution two methyl glucosides are produced:

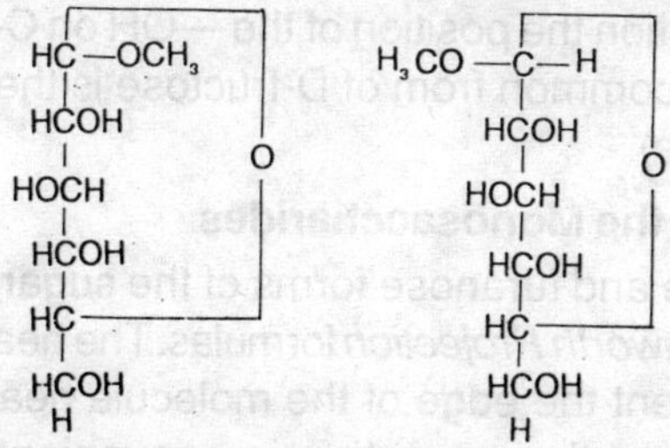

Methyl-α-D-glucoside   Methyl-β-D-glucoside

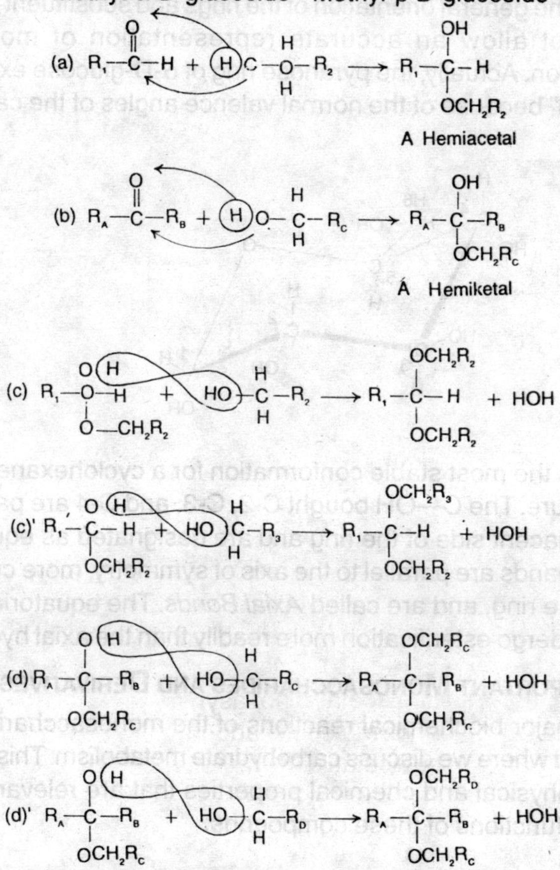

Fig. 3.5. Aldehyde hemiacetals and ketone hemiketals.

Comparable reactions involve alcohols and ketones. It is the intramolecular version of this reaction which D-fructose undergoes to gain an additional asymmetric center at C-2 cause they are related to the heterocyclic parent compound furan. The designations α and β indicate the position the position of the —OH on C-2 relative to the ring oxygen. The common from of D-fructose is the β anomer (i.e., β-D-fructofuranose).

Conformation of the Monosaccharides

The pyranose and furanose forms of the sugars are frequently represented by *Haworth Projection* formulas. The heavy lines in these structures represent the edge of the molecule nearest the reader. Although the Haworth convention is convenient shorthand for indicating the general orientation of the rings and substituent groups, it does not allow an accurate representation of molecular conformation. Actually, the pyranose ring of α-D-glucose exists in a "chair form" because of the normal valence angles of the carbons:

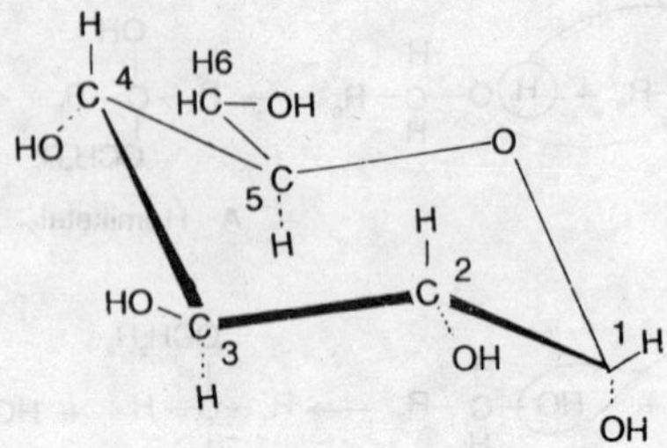

This is the most stable conformation for a cyclohexane type of ring structure. The C—OH bought C-2, C-3, and C-4 are parallel to the nonadjacent side of the ring and are designated as equatorial; the other bands are parallel to the axis of symmetry, more cut of the plane of the ring, and are called *Axial Bonds.* The equatorial —OH groups undergo esterification more readily than the axial hydroxyls.

Important Monosaccharides and Derivatives

The major biochemical reactions of the monosaccharides are considered where we discuss carbohydrate metabolism. This section stresses physical and chemical properties that are relevant to the biological functions of these compounds.

Tetroses

The structural relationships between the common tetroses, erythrose, threose, and erythulose, and the pentoses and hexoses.

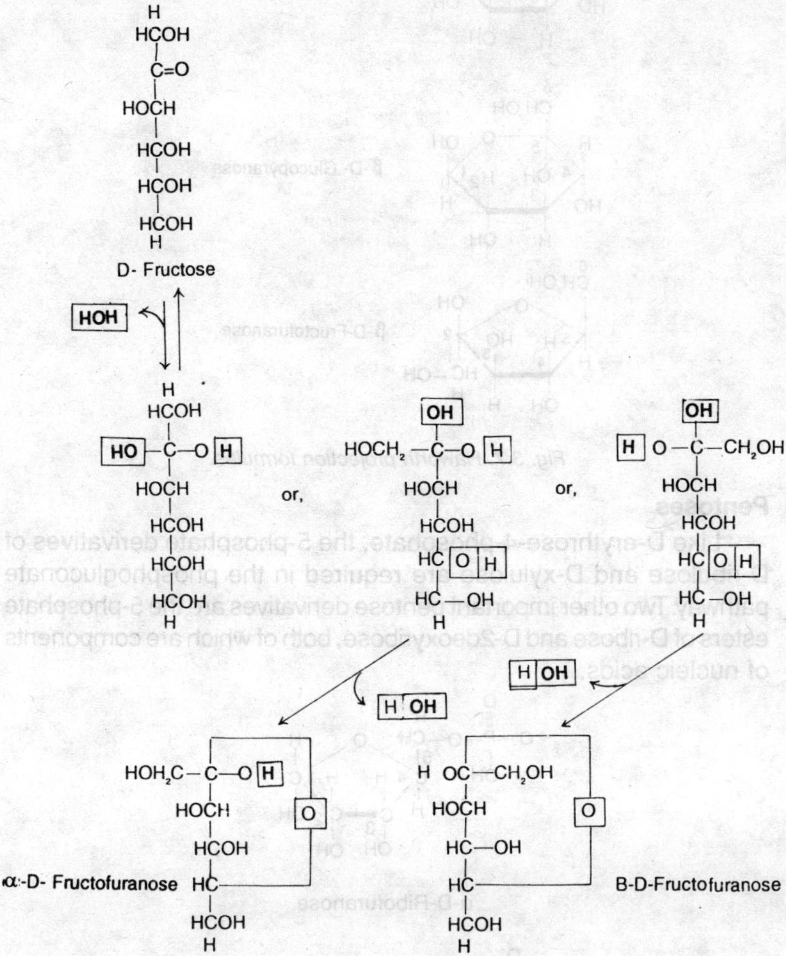

α-D-Fructofuranose α-D-Fructofuranose

Fig. 3.6. Anomeric forms of D-fructose.

D-Erythrose-4-phosphate is a component of the transaldolase reaction of the phosphogluconate pathway.

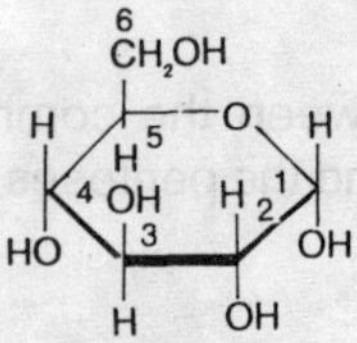

α D- Glucopyranose

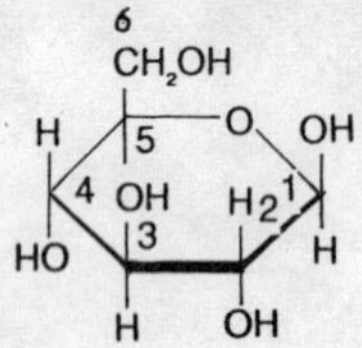

β-D- Glucopyranose

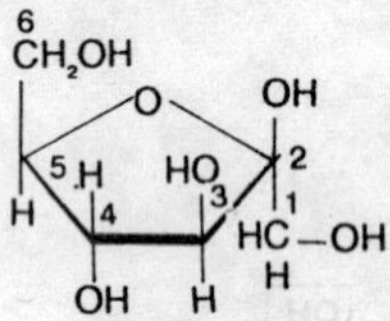

β-D-Fructofuranose

Fig. 3.7. Haworth projection formulas.

Pentoses

Like D-erythrose-4-phosphate, the 5-phosphate derivatives of D-ribulose and D-xylulose are required in the phosphogluconate pathway. Two other important pentose derivatives are the 5-phosphate esters of D-ribose and D-2deoxyribose, both of which are components of nucleic acids.

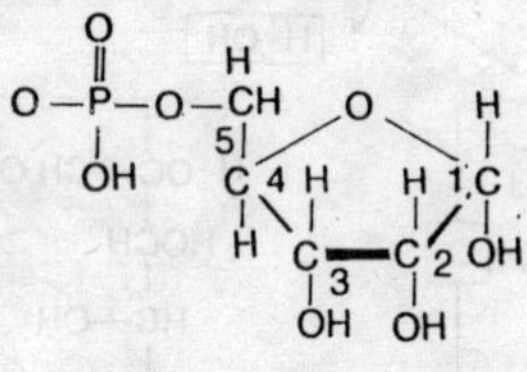

α-D-Ribofuranose

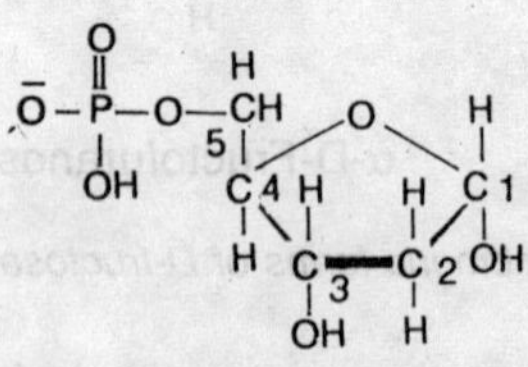

α-D-2-Deoxyribofuranose

Fig. 3.8. Structure of the phosphate esters of D-ribose and D-2-deoxyribose.

Hexoses

As in the case of the tetroses, and pentoses the phosphate esters of hexoses such as glucose, mannose, galactose, and fructose occurs as metabolic intermediates in a wide variety cells.

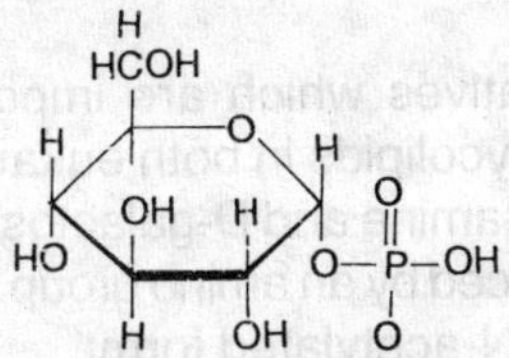

α-D-Glucose-1-phosphate

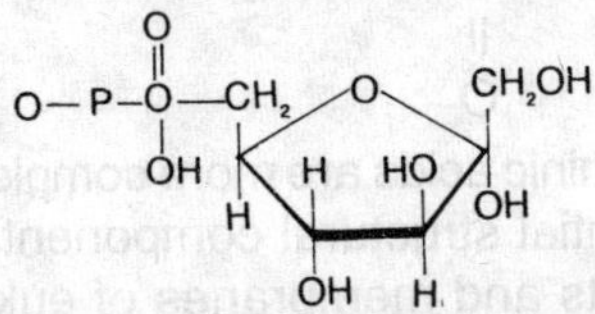

α-D-Glucose-6-phosphate

Fig. 3.9. Examples of phosphate esters of hexoses.

Hexose Alcohols

Reduction of the carbonyl group of the hexoses yields the sugar alcohols. Three important examples of this type of monosaccharide are D-mannitol, L-sorbitol, and myoinositol, whose structures.

Hexose Acids

An aldohexose such as glucose can have either or both of its terminal carbons oxidized to the carboxyl level. The structures and nomenclature of three types of hexose acids derived from D-glucose are presented. The phosphate ester of one of these acids, 6-phospho-D-gluconic acid, is an intermediate in the phosphogluconate pathway of glucose metabolism. The "onic" acids, such as D-glucuronic acid, are constituents of hyaluronic acid, chondroitin, and other mucopolysaccharides. Both gluconic acid and glucuronic acid can form intramolecular esters, or lactones, which are five- or six-membered ring structures.

Another aldonic acid important in human metabolism is L-ascorbic acid, or vitamin C. L-ascorbic acid occurs as the γ-lactone and differs from other hexose acids in that it has an enediol linkage between C-2 and C-3. Vitamin C can be reversibly dehydrogenated to dehydroascorbic acid. As will be described in more detail later, this compound is required for the establishment and maintenance of normal connective tissue structure and is also implicated in certain oxidation-reduction systems.

Amino Sugars

Two hexose derivatives which are important constituents of polysaccharides and glycolipids in both eukaryotic and prokaryotic organisms are D-glucosamine and D-galactosamine. In both sugars the —OH on C-2 is replaced by an amino group as. These compounds occur principally in the N-acetylated form:

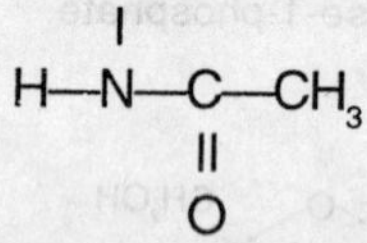

Muramic and neuraminic acids are more complex forms of amino sugars which are essential structural components of cell walls of bacteria and of cell coats and membranes of eukaryotic cells. As shown with the formulas the carbon skeleton of each of these two sugar acids may be regarded as the union of a hexose and a triose (either pyruvic or D-lactic acid); the hexose backbone of N-acetylmuramic acid is D-glucosamine and that for N-acetylneuraminic acid is D-mannosamine. The more general term "sialic acid" is frequently used in identifying N-acyl derivatives of neuraminic acid.

Properties of Monosaccharides

Physical Properties

Monosaccharides are sweet tasting colourless solids, having solubility in water but sparingly soluble in alcohol and insoluble in ether. They have asymmetric carbon atom and therefore exist in different isomeric forms. An example is glyceraldehyde which may exist in two forms which are mirror images of each other.

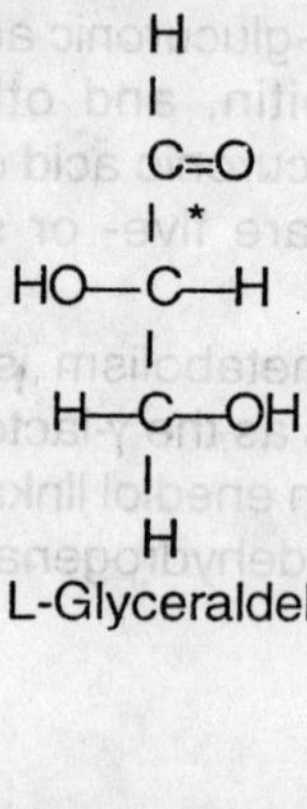

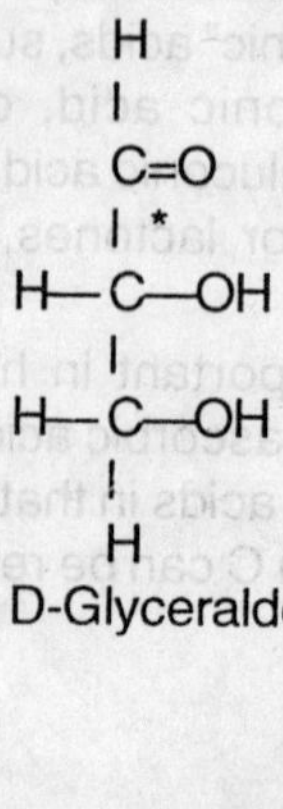

*Asymmetric carbon atom

L-Glyceraldehyde D-Glyceraldehyde

If a polarized light (light vibrating in one plane) is allowed to pass through a solution of these carbohydrates, the plane of the light gets rotated to either right or left. All monosaccharides which have a D-glyceraldehyde unit are termed as D-sugars. L-monosaccharides have a L-glyceraldehyde unit in their structure.

Carbohydrates isolated from living system are not pure compound, but a mixture of several isomers. The optical rotation may undergo a change due to interconversion of isomers. Fresh solution of glucose yields an optical rotation of +112° which gets changed to +52.7° on standing.

Similarly, fructose yields an optical rotation of –113° which gets changed to –92°. This change in optical rotation is termed as *mutarotation.*

Chemical Properties

As monosaccharides are aldehyde or ketone derivatives of the higher polyhydric alcohols, they give most, but not all, of the characteristic reactions of the carbonyl group as well as of the alcoholic group. In addition to these reactions, they also give rise to some special reactions due to the presence of a carbonyl and alcoholic group in the same molecule.

1. Reduction

The nature of the product formed by the reduction of aldoses depends upon the nature of the reagent. For example, D-glucose when reduced yields D-sorbitol and D-mannose produces D-mannitol. Sorbitol is found in the berries of many higher plants especially in the rosacea.

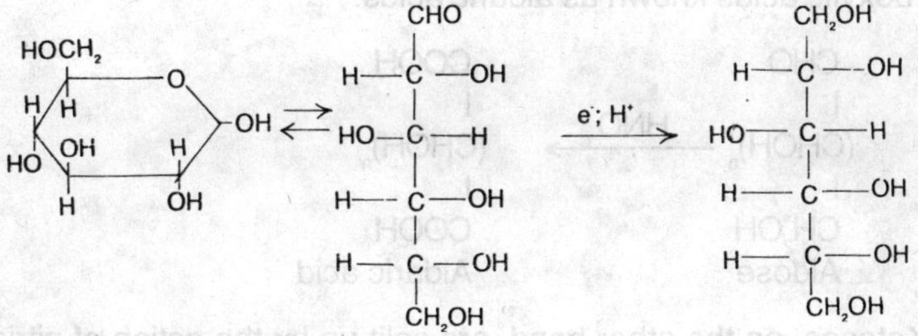

When ketoses are reduced by the reducing agents, a mixture of two alcohols is obtained in each case. For example, D-fructose always yields a mixture of two epimers, D-sorbitol and D-mannitol.

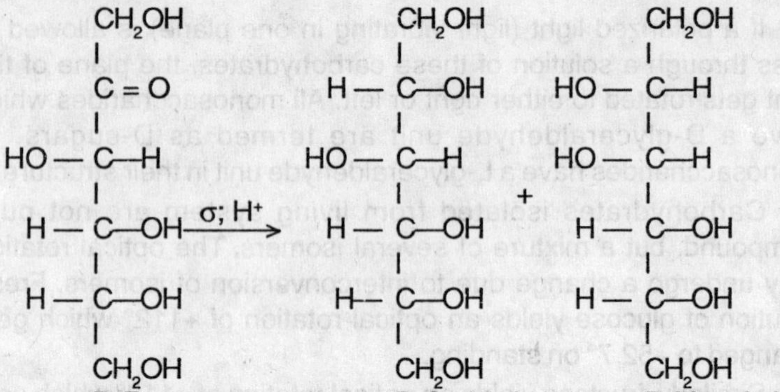

D-Fructose D-Sorbitol D-Mannitol

2. Oxidation

As with reduction, the product of oxidation depends upon the nature of the reagent. Bromine water is a general reagent which selectively oxidises the —CHO group to a —COOH group and brings about conversion of an aldose to an aldonic acid.

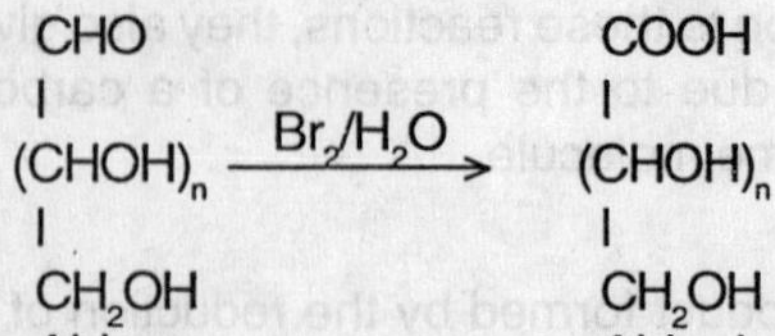

Aldose Aldonic acid

Bromine water cannot oxidise the ketoses. When oxidation of aldoses is done with strong oxidising reagents like nitric acid, both the aldehyde and primary alcoholic groups will be oxidised to yield dicarboxylic acids known as aldaric acids.

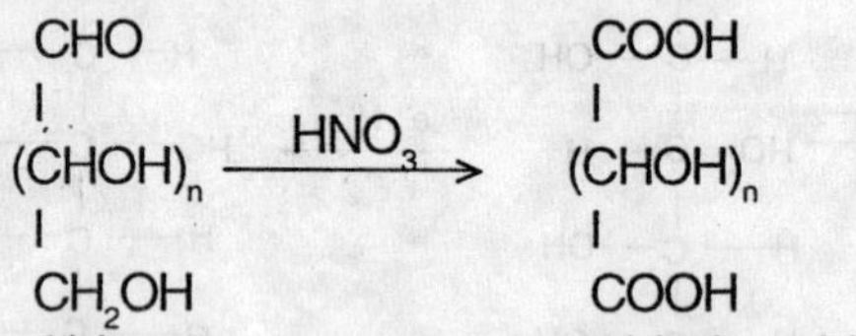

Aldose Aldaric acid

Ketoses, on the other hand, are split up by the action of nitric acid to form a mixture of acids having fewer carbon atoms. However, it has been observed that the carbon chain of sorbose is not ruptured during oxidation with nitric acid but a keto acid is the product.

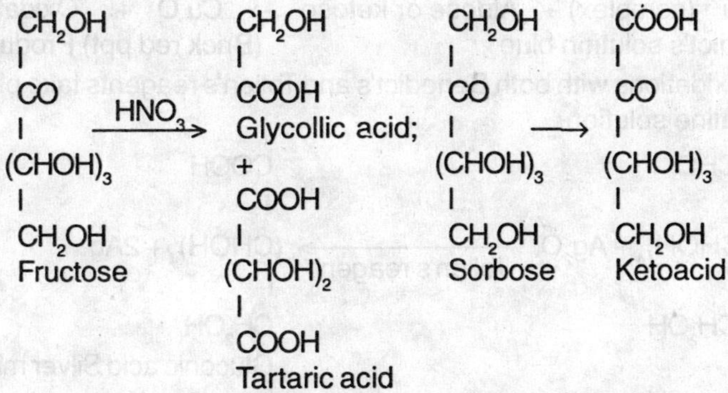

D-Glucose on periodate oxidation yields one mole of formaldehyde and 5 moles of formic acid.

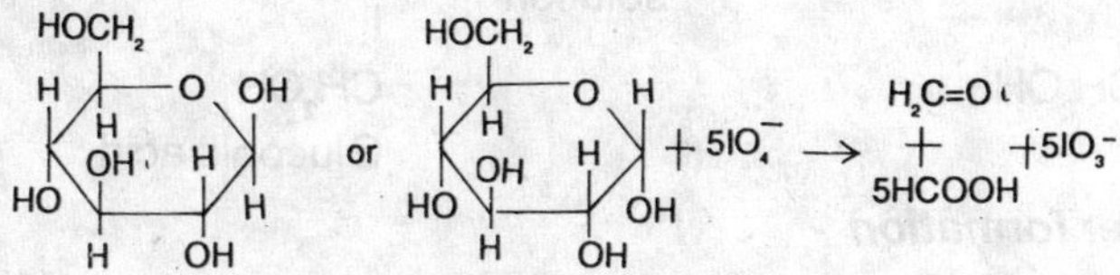

β-D-Glucopyranose α-Glucopyranose

Similarly, the glycoside methyl α-D-gluco-pyranoside will react with periodate as shown below:

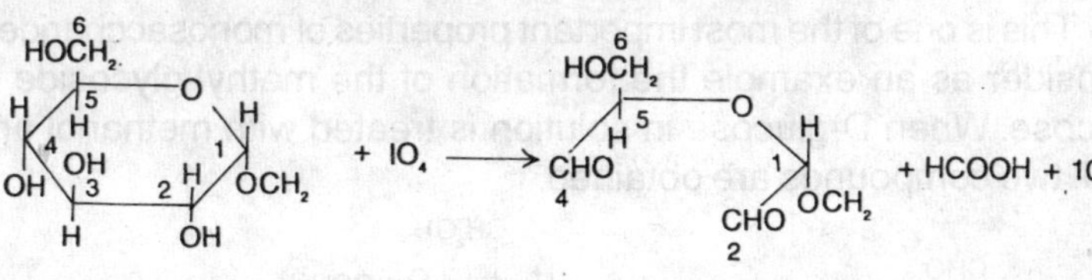

Glycoside
α-D-Gluco pyranoside

Benedict's reagent (an alkaline solution containing a cupric citrate complex ion) and Tollen's solution [$Ag(NH_3)_2OH$] oxidise and thus give positive tests with aldehydes and with α-hydroxy ketones. Both reagents, therefore, give positive tests with aldoses and ketoses. Sugars that give positive tests with Tollen's and Benedict's solutions are known as reducing sugars.

Ag^+ (complex) + aldose or ketose → Ag + Oxidation
Tollen's reagent (Silver mirror) Products

Cu^{2+} (complex) + Aldose or ketose $\rightarrow$ Cu_2O + Oxidation
Benedict's solution blue (Brick red ppt) Products

Oxidations with both Benedict's and Tollen's reagents take place in alkaline solution.

$$\begin{array}{l} CHO \\ | \\ (CHOH)_4 + Ag_2O \\ | \\ CH_2OH \end{array} \xrightarrow[\text{Tollen's reagent}]{} \begin{array}{l} COOH \\ | \\ (CHOH)_4 + 2Ag \\ | \\ CH_2OH \end{array}$$

Gluconic acid Silver mirror

$$\begin{array}{l} CHO \\ | \\ (CHOH)_4 + 2\,Cu(OH)_2 \\ | \\ CH_2OH \end{array} \xrightarrow[\text{solution}]{\text{Fehling's}} \begin{array}{l} COOH \\ | \\ (CHOH)_4 + Cu_2O + H_2O \\ | \\ CH_2OH \end{array}$$

Gluconic acid

3. Ester formation

The formation of sugar esters indicates the presence of alcohol groups. For example, the total number of acyl groups which can thus be taken up by a molecule of the sugar is a measure of the number of such alcohol groups.

4. Glycoside formation

This is one of the most important properties of monosaccharides. Consider as an example the formation of the methyl glycoside of glucose. When D-glucose in solution is treated with methanol and HCl, two compounds are obtained.

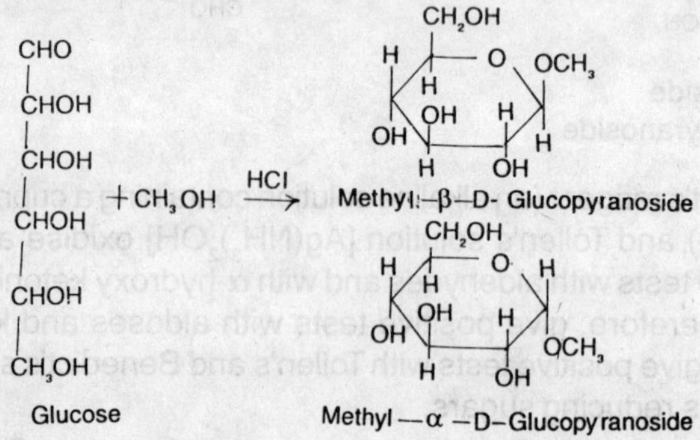

Glucosides are found in many drugs, spices and in the constituents of animal tissues. The glycosides which are important in medicines due to their action on the heart all contain steroids as the aglycone component. The examples of these include derivatives of digitalis and strophanthus.

5. Action of Acids

The monosaccharides are stable to the action of dilute (1.0 N) mineral acids but the concentrated acids decompose these compounds. However, aldohexoses and ketohexoses, when heated with strong mineral acids, are dehydrated and hydroxymethyl furfural is formed.

6. Action of Alkali

With alkali, monosaccharides react in various ways:

(a) In concentrated alkali, sugar caramelizes and produces a series of decomposition products. Yellow and brown pigments develop, salts may form, many double bonds between carbon atoms are formed, and carbon-to-carbon bonds may rupture. However, the mechanism of these changes is not clearly understood.

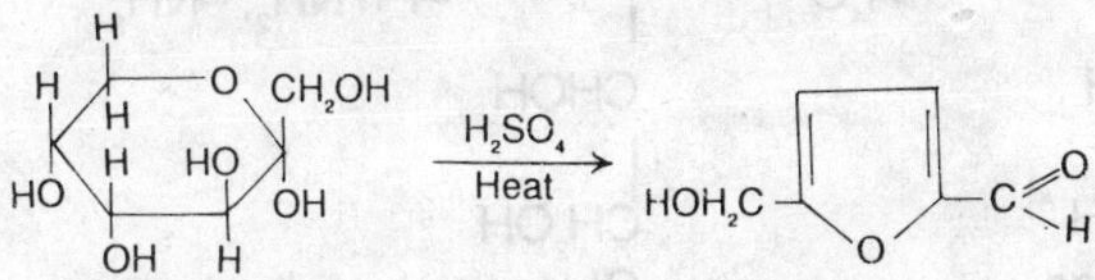

∝ – D– Fructopyranose 5 – Hydroxymethyl furfural

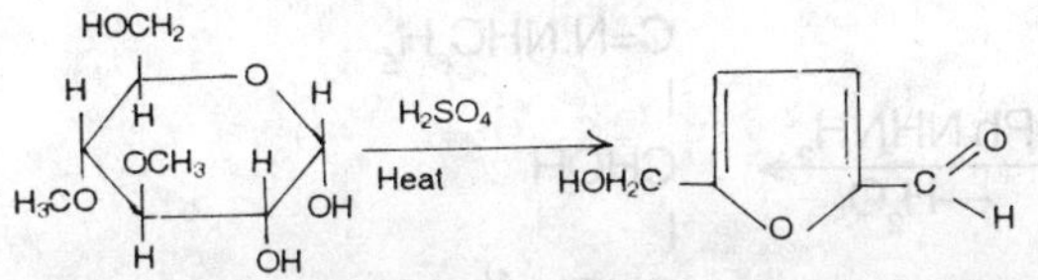

∝ – D– Glucopyranose 5 – Hydroxymethyl furfural

(b) In the presence of dilute alkali or amines, sugars undergo rearrangement to form a mixture of more than one sugar. For example, a dilute solution of D-glucose when warmed with dilute sodium hydroxide forms a mixture of D-glucose, D-mannose and D-fructose.

The mixture is almost optically inactive. Mannose and fructose also undergo the same changes to form a similar, mixture. This reaction was called the *Lobry de Bruyn-van Ekensein conversion* after its discoverer (1895).

7. Osazone formation

Aldoses as well as ketoses react with hydroxylamine and phenylhydrazine. With hydroxylamine the expected product is an oxime in aldoses and ketoses. With phenylhydrazine, however, three moles of phenylhydrazine are consumed in aldoses as well as in ketoses to form *osazones,* aniline and ammonia. The overall equation may be written as follows:

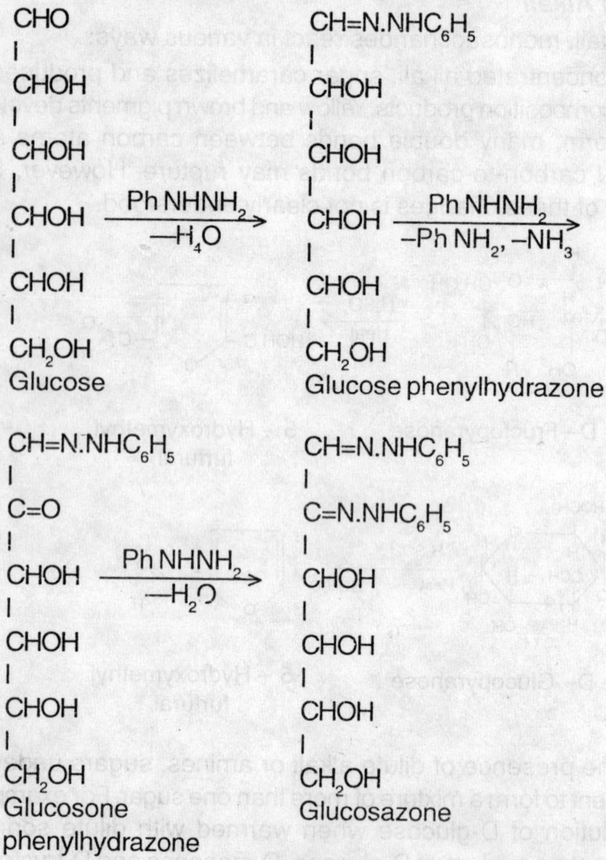

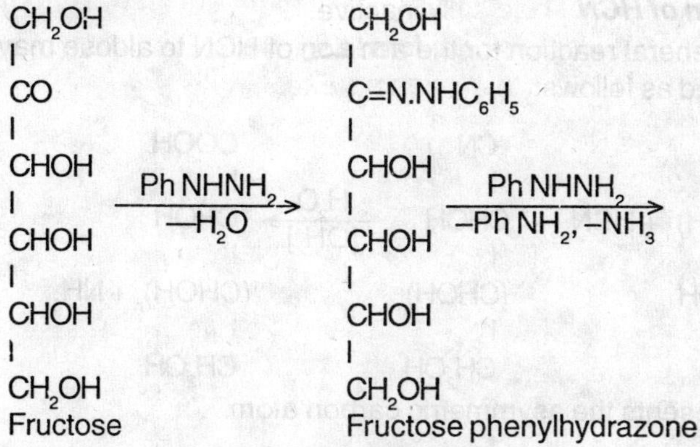

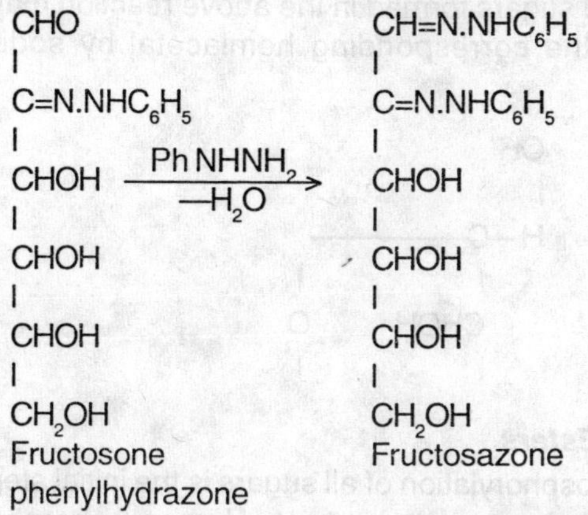

Osazones are yellow crystalline solids and are used to characterise the sugars.

When osazones are hydrolysed with hydrochloric acid, both phenylhydrazino groups are eliminated and osones are formed (osone is a dicarbonyl compound). For example, glucosazone, when hydrolysed with hydrochloric acid, yields glucosone.

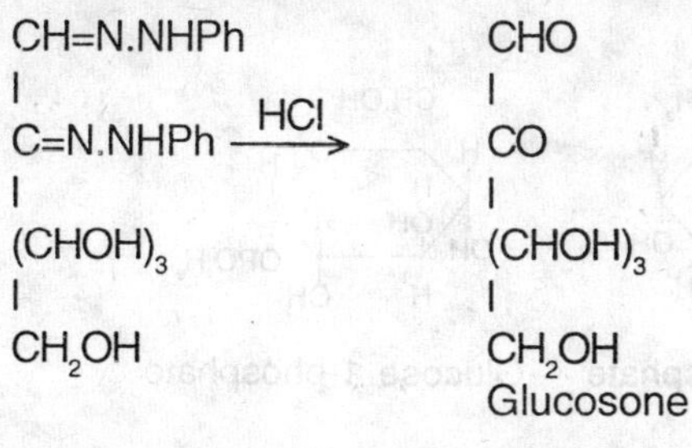

8. Addition of HCN

The general reaction for the addition of HCN to aldose may be represented as follows:

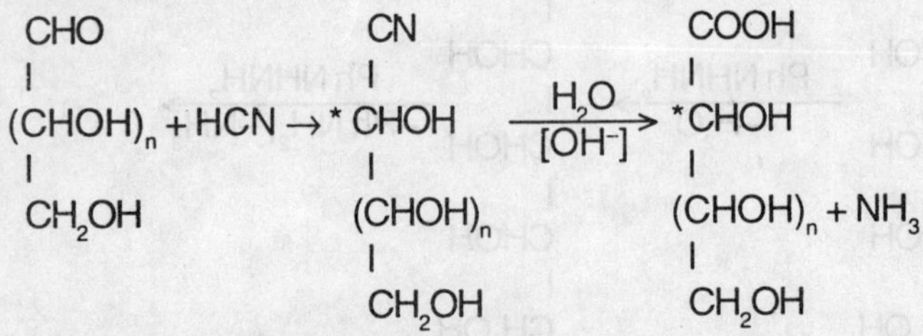

* represents the asymmetric carbon atom.

The lactone of the sugars formed in the above reaction may in turn be reduced to the corresponding hemiacetal by sodium amalgam.

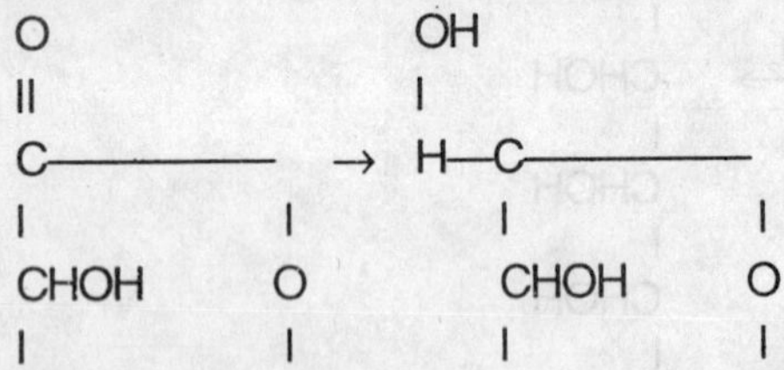

9. Phosphoric Acid Esters

In metabolism phosphorylation of all sugars is the initial step in their metabolism. Thus, glucose is converted to glucose-6 phosphate. In most of the living systems, this compound is also converted into α-glucose 1-phosphate. Glucose 6-phosphate is converted into fructose 6-phosphate which gets further phosphorylated to yield 1, 6-diphosphate. The formation of these phosphoric acid esters occupies a unique formation in biochemistry. The formulae of some phosphoric acid esters are given as follows:

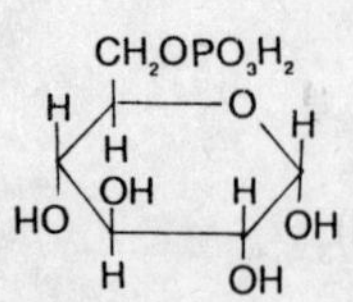

Glucose 6-phosphate

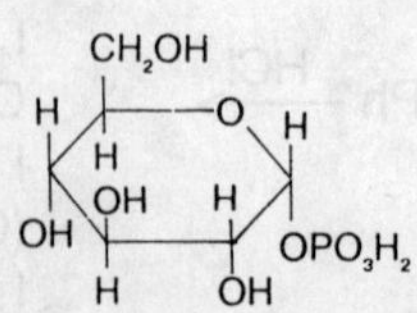

Glucose 1-phosphate

10. Reaction with Amines (N-Glucoside Formation)

When glucose is condensed with amines, Schiff's bases are not obtained but instead N-glucosides (analogous to the ordinary glucosides from alcohols) are obtained. For example, D-glucose reacts with dimethylamine in the presence of an acid to form two isomeric N-glucosides.

11. Fermentation

When a solution of glucose is fermented by an enzyme called *zymase,* the products are mainly ethyl alcohol and carbon dioxide.

$$\underset{\text{D-Glucose}}{C_6H_{12}O_6} \xrightarrow{\text{zymase}} \underset{\text{Ethyl alcohol}}{2C_2H_5OH} + 2CO_2$$

Similar to glucose, fructose also undergoes fermentation by the enzyme zymase present in yeast to yield mainly ethyl alcohol and carbon dioxide.

$$\underset{\text{D-Fructose}}{C_6H_{12}O_6} \xrightarrow{\text{zymase}} 2C_2H_5OH + 2CO_2$$

Some Derived Monosaccharides

Derived monosaccharides differ from *normal monosaccharides with respect to aldoses* and *ketoses.* They include the *glycosides, sugar phosphates, gluconic acid, glucuronic amino sugars* and *vitamin C.*

1. Glucosides

The hydroxyl group on carbon 1 in the sugar molecule can be replaced by other redicals, forming compcunds known as *glycosides.* Thus if a *methyl group* replaces the *hydroxyl group* of glucose it results in the formation of *methyl glucosides.* There are two methyl glucosides possible, α and β, corresponding to the α and β forms of glucose. Glucosides formed from *galactose* are known as *galactosides.* Among the complex glycosides occurring in nature are *digoxin,* which acts on the heart, and the antibiotic *streptomycin.*

2. Sugar phosphates

C-1 or C-6 of glucose can react with *phosphoric acid* (H_3PO_4) to yield *glucose-1-phosphate* and *glucose-6-phosphate,* respectively. These compounds play an important part in carbohydrate metabolism.

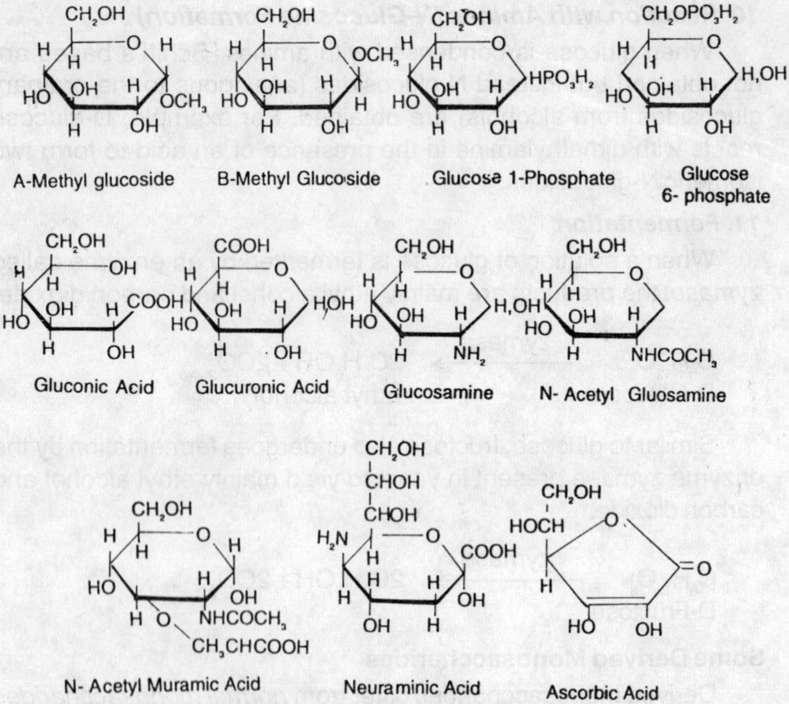

Fig. 3.10. Derived monosacchrides.

3. Gluconic acid and glucuronic acid

Oxidation of C-1 of the glucose molecule to a *carboxyl group* yields *gluconic acid.* Production of a carboxyl group at C-6 yields *glucuronic acid.* The corresponding terms for any hexoses are *hexonic acid* and *hexuronic acid.*

Glucuronic acid is an important constituent of complex polysaccharides and is also an important coupling agent. Many drugs, pesticides, and environmental pollutants and hormones are coupled with glucuronic acid and excreted in urine or bile as *glucuronides.*

4. Amino sugars or Hexosamines

Amino sugars or *hexosamines* are formed when an amino sugar is introduced into hexoses. Thus glucose with NH_2 at C-2 forms *glucosamine*. This compound is extensively found in complex polysaccharides, usually in the form of its acetyl derivative *N-acetyl glucosamine. N-acetyl galactosamine,* the amino derivative of galactose, is found in *glycoproteins* and *glycolipids. N-acetyl muramic acid* is found in the polysaccharides of the cell wall of some bacteria.

Neuraminic acid gives acetyl derivatives called *sialic acids.* These are constituents of some *glycoproteins* in bone and connective tissue and *glycolipids* of the nervous system.

5. Vitamin C (ascorbic acid)

Vitamin C (*ascorbic acid*) is a sugar acid of biological importance.

Oligosaccharides

Glycosidic linkage. A sugar molecule can combine with an identical for a different type of a sugar molecule. The linkage between two monosaccharide sugar molecules is called a *glycosidic linkage* or *glycosidic band.*

A sugar molecule has several reactive hydroxyl groups. The hydroxyl group of number one carbon atom (C-1 is known as the *glycosidic hydroxyl.* It is very reactive and readily forms a glycosidic link with a hydroxyl group of another sugar molecule. When such a linkage occurs, one hydrogen atom and one hydroxyl group are eliminated to form H_2O. This process is one of *condensation* and is in fact *dehydration synthesis.* The reverse process in which the molecule is cleaved with the incorporation of the elements of water is called *hydrolysis.*

A linkage between C-1 of one monosaccharide unit and C-4 of another is called the 1, 4, *linkage.* Such a linkage occurs in disaccharides and in the unbranched chains of polysaccharides. The 1, 4 linkage between two hydroxyl groups in the α position is called the α-1, 4, linkage (e.g., *maltose*). Similarly a linkage between two β hydroxyl groups is called the β-*1, 4 linkage* (e.g., in *lactose*). Linkage can also take place between C-1 of one monosaccharide unit and C-6 of another. Such a linkage occurs at the point of branching of a chain and is known as a *1, 6 linkage* (e.g., in *isomaltose*).

The α or β configurations can be determined by specific enzymes. The enzyme *maltase* attacks only α-glycosides, while *emulsin* hydrolyses only β-glycosides. *Starch* has α-glycosidic bonds while *cellulose* has β-glycosidic bonds. The α bonds of starch can be cleaved by enzymes which do not cleave the β bond of cellulose. This explains why starch is a valuable food while cellulose has no food value for most animals.

Maltose and *lactose* both have unlinked potential aldehyde and are therefore reducing sugars. As long as the bond is not there between two C-1 atoms, a free aldehyde or ketone group is left on the disaccharide. The disaccharide therefore shows all the reactions associated with the groups. In *sucrose* linkage is through both the potentially reducing groups of the monosaccharides. Sucrose is therefore non-reducing.

Oligosaccharides are composed of 2 to 9 monosaccharide units or their derivatives. This commonly employed definition of oligosaccharides is an arbitrary one. In general it can be said that oligosaccharides contain fewer sugar residues than polysaccharides. The oligosaccharides are named according to the number of monosaccharide units they contain. Thus *disaccharides* contain two monosaccharide units, *trisaccharides* three, *tetrasaccharides* four, *pentasaccharides* five, etc.

Disaccharides

These are formed by union of two monosaccharides. They are united by linkage between the first carbon of one monosaccharide with the second or the fourth carbon of another monosaccharide by what is known as the glycosidic linkage. The structures of the three biologically important sugars—maltose, lactose and sucrose.

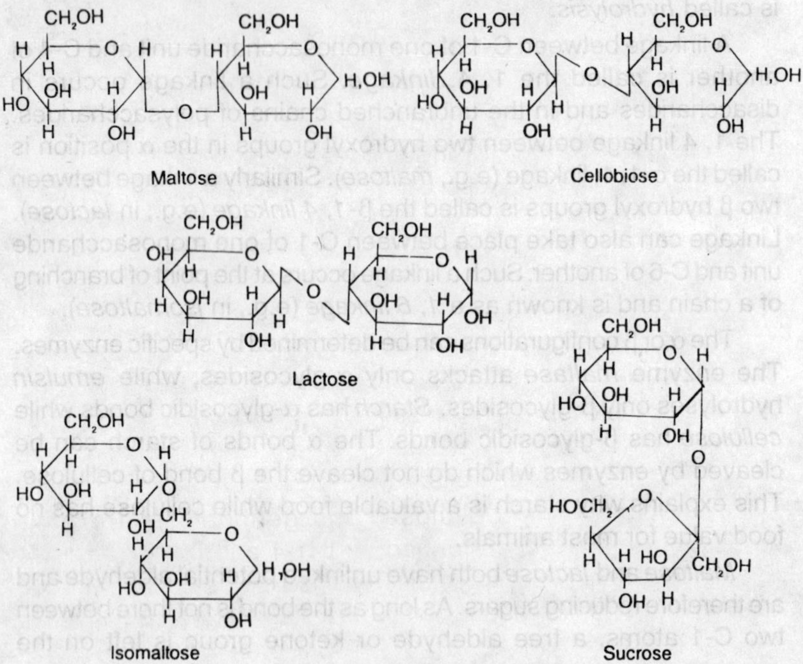

Fig. 3.11. Disaccharides.

They are colourless crystalline substances readily soluble in water and sweetish to taste.

Maltose

This is made up of two alfa D-glucose units united by the 1, 4-glycosidic linkage. Since one of the aldehyde groups remains free, it exhibits all the properties of the aldehyde group such as formation of osazones, reduction of cupric salts and others. Maltosazone crystals have a characteristic petal like appearance and a cluster of them looks like the sunflower. Maltose is formed during the digestion of starches by enzymes or by dilute acids.

Lactose

This is made up of a molecule of alfa D-glucose united by 1, 4-glycosidic linkage to a molecule of beta D-galactose. The aldehyde group of glucose remains free and exhibits the properties due to that group. Lactosazone crystals have a typical hedgehog shape and are readily identifiable. Lactose is present in milk (milk sugar) and by hydrolyzed to glucose and galactose by the enzyme 'lactase' present in intestinal juice.

Sucrose

It is made up of one molecule of alfa-D-glucose and one of beta D-fructose united by a glycosidic linkage between the aldehyde and keto groups (C_1 of glucose and C_2 of fructose). There is hence no free aldehyde or keto group in sucrose. It is a non-reducing sugar. It does not also form osazones. Solutions of sucrose exhibit a specific rotation of +62.5°. On hydrolysis with acid or with the specific intestinal enzyme 'sucrase', it is converted into an equimolar mixture of glucose and fructose whose specific activities are +52.5° and –92°. The mixture exhibits a net specific rotation of –19°. The phenomenon by which the dextrorotatory sucrose is converted to a levorotatory mixture of glucose and fructose is known as 'inversion'. Sucrose is also called the 'invert sugar' and the enzyme sucrase is also known as 'invertase'. Sucrose occurs widely in nature as cane sugar, beet sugar and in several ripe fruits.

Cellobiose (from cellulose) and gentiobiose are other disaccharides of glucose formed by β 1-4 and β 1-6 glycosidic linkages. Both are reducing sugars. Trehalose is a disaccharide formed by two glucose units combining by 1-1 linkage. It is a non-reducing sugar. It is contained in the hemolymph of some insects.

Trisaccharides

Raffinose (galactose + glucose + fructose) is found in sugar beets. Melezitose (glucose + fructose + glucose) is found in the sap of some coniferous trees.

Polysaccharides

The term polysaccharide is usually employed for polymers continuing at least 10 monosaccharide units. Polysaccharides are named after their monosaccharide monomers by changing the—*ose* ending of the monosaccharide to—*an*. Thus *cellulose*, a polysaccharide formed of *glucose* units, is called *glucan. Fructose* polymers are called *fructans* (previously *fructosans*) and *pentose* polymers *pentans* (formerly *fructosans*). Since the term *glycose* is used for any monosaccharide, the general name for any polysaccharides is *glycan.* If a polysaccharide contains more than one type of sugar units the two sugars are named in alphabetical order. Thus, a polysaccharide consisting of *D-glucose* and *D-mannose* units is termed *D-gluco-D-mannoglycan*. Some old maines which have not been changed to conform with the—*an* ending include *starch, cellulose, pectin, amylopectin, inulin, chitin* and *liquirin.*

Polysaccharides have been classified as *homopolysaccharides* (*monnoglycans*) and *heteropolysaccharides* (*heteroglycans*). *Homoglycans* contain only one type of monosaccharide while *heteroglycans* contains atleast two types. Most polysaccharides contain one or two types of glycose units. Polysaccharides containing three types of glycose units are less common. The term homoglycan is used when at least 9% of the sugars are of one type, because it is very rarely that a polysaccharide can be obtained in pure form.

Polysaccharide are formed by condensation of many molecules of monosaccharides with corresponding elimination of water molecule. The monosaccharide units are joined by *glycosidic bonds.*

A. Homopolysaccharide (homoglycans)

Some of the better known homoglycans are :

Glucans (of *glucose* monomers)—starch, glycogen, cellulose, chitin.

Galactans (of *galactose* monomers)—agar, pectin galactan from snails.

Mannans (or *mannose* monomers)—yeast mannan.

Xylans (of *xylose* monomers)—hemicellulose xylan.

Fructans (of *fructose* monomers)—inulin.

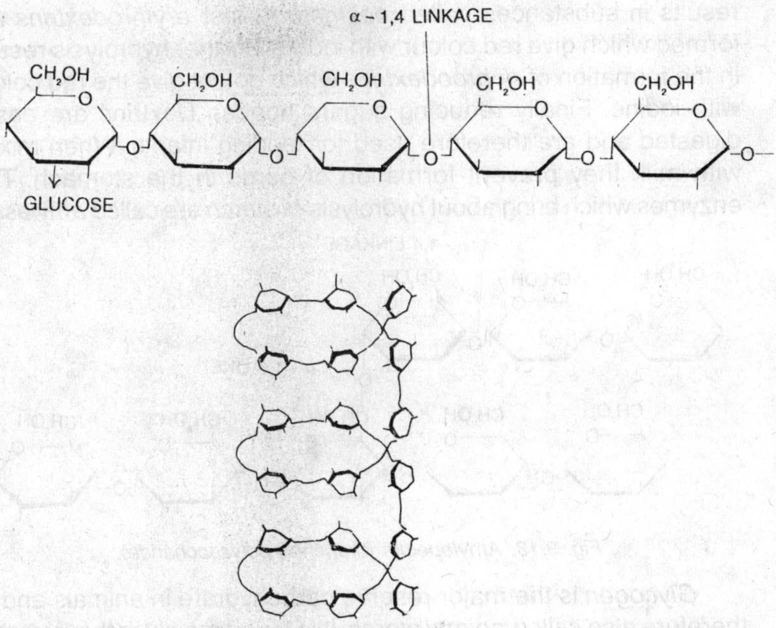

Fig. 3.12. (a) Structure of amylose showing a-1,4 linkage. (b) Helical coil of amylose on suspension in water.

Starch

Starch is the reserve substance in plant cells. It is a polymer of *D-glucopyronose* units linked by α-1-4. It consists of a mixture of *amylose* and *amylopectin* in the proportion of 1 to 4. Both are high molecular weight compounds. *Amylose is linear* while *amylopectin* is *branched.* Treating starch with hot water dissolves amylose, while amylopectin remains.

Amylose contains about 200-500 glucose units which are arranged in the form of a straight chain. Its molecular weight varies from a few thousand to about 150,000. Amylose gives intense blue colour with iodine.

Amylopectin has over 1,000 glucose residues and its molecular weight is about 200,000 to 1,000,000. It has a branched structure. Amylopectin consists of glucose units linked by α-1,4 glycosidic bonds. There are occassional α-1,6 glycosidic bonds which cause branching.

Starch on hydrolysis yields lower molecular weight polysaccharides, and finally glucose or maltose. Partial hydrolysis results in substances called *dextrins.* At first *erythrodextrins* are formed which give red colour with iodine. Further hydrolysis results in the formation of *achroodextrins* which do not give the red colour with iodine. Finally reducing sugars appear. Dextrins are easily digested and are therefore used for feeding infants. When mixed with milk they prevent formation of curds in the stomach. The enzymes which bring about hydrolysis of starch are called *amylases.*

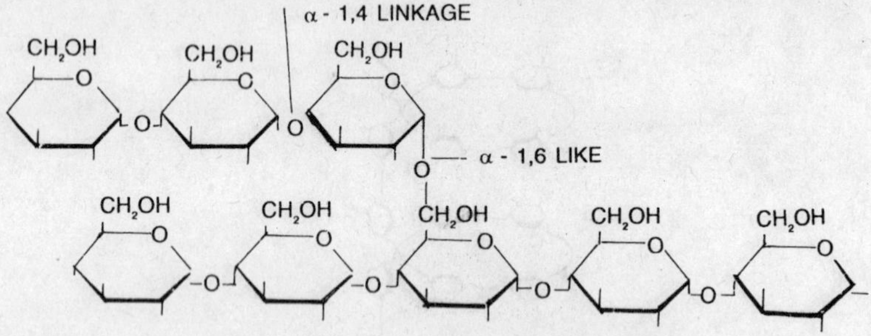

Fig. 3.13. Amylopectin (branched polysaccharide).

Glycogen is the major reserve carbohydrate in animals and is therefore also called *animal starch.* It is found mainly in the *liver* and in *muscles.* The glycogen in the *liver* supplies glucose to all tissue through the blood. *Muscle glycogen* on the other hand is available during contraction of muscle.

Chemically glycogen is analogous to starch. Both are glucans consisting of glucose units linked by α-1,4 glycosidic bonds, with branch chains formed at α-1,6 bonds. They differ mainly in their molecular weights and degree of branching. Molecular weights range from about 300,000 to 100 million. The branching chain of glycogen average about 12 glucose units as compared with the 25 or so for starch. Glycogen has branch points about every 8-10 glucose units. About 5,000 to 15,000 glucose units make up glycogen. Glycogen is nonreducing and gives a red colour with iodine.

Cellulose and Other Carbohydrates

Cellulose, which is one of the most abundant of polysaccharides, is also a glucan. On partial hydrolysis, it yields cellobiose indicating that the glucose units are connected by β-(1→4) linked glycosidic bonds in a straight polysaccharide chain. Cellulose, like cellobiose, is nutritionally useless to man as there is no enzyme in the human body which can act upon a β-glucosidic linkage. Ruminants, however,

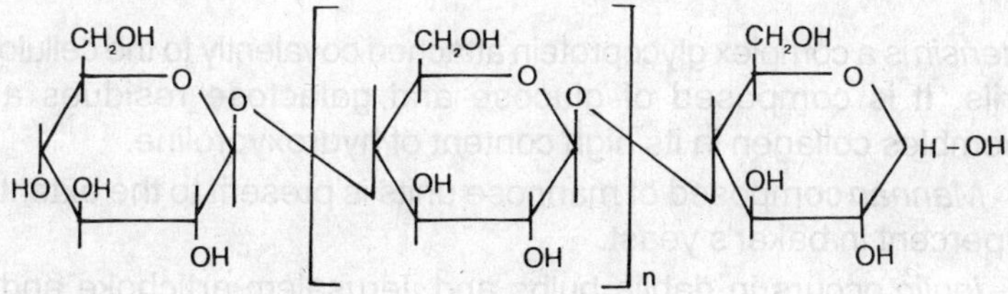

Fig. 3.14. Cellulose structure.

digest cellulose with the aid of symbiotic microorganisms which can attack β-glucosidic links and break down the glucose to organic acids.

Plant Cell wall Polysaccharides

The plant cell is distinguished from the animal cell in having a rigid cell wall composed mainly of cellulose and by the presence of a large fluid-filled vacuole in the mature cell. Most cell walls are joined together but during the processes of growth and differentiation, cells elongate as well as multiply. Meristematic cells associated with these processes have initially a thin plasma membrane which is freely permeable to water, ions and small molecules. The rate of entry of these materials determines the osmotic pressure within the cell which in turn controls the growth of the cell and the formation of the cell wall. The thick secondary cell wall which is formed later by secretion confines the enlargement and multiplication to one axis. The rigidity and strength of the cell wall are due to the extensive cross-linking of cellulose fibres with other complex polysaccharides like xyloglucan, arabinogalactan and rhamnoglucuronen. Each cellulose fibre appears to be surrounded by a layer of xyloglucan. The walls contain other carbohydrate material like hemicelluloses and pectins and non-cellulosic substances like lignin and extensin. *Lignins* are polymeric non-carbohydrates containing parahydroxy phenyl propanes as the repeating units and representing about 15-30 percent of the dry weight. They are probably derived from *shikimic acid,* which may be a precursor of the phenolic residues of lignins.

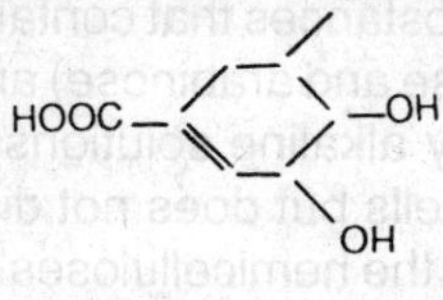

Extensin is a complex glycoprotein attached covalently to the cellulose fibrils. It is composed of glucose and galactose residues and resembles collagen in its high content of hydroxyproline.

Mannan composed of mannose units is present to the extent of 16 percent in baker's yeast.

Inulin occurs in dahlia bulbs and Jerusalem artichoke and is predominantly a fructosan which yields D-fructose on hydrolysis and a small number of D-glucose units. It is a linear polymer consisting of about 33 fructofuranose units joined together by means of β-(2→1) glycosidic linkages and having a molecular weight of around 5,200. It is an elongated molecule which is able to pass the glomerular filter of the kidney and is not secreted by the kidney tubules. It has, therefore, been used to measure the rate of glomerular filtration by injection into the blood stream as it is not metabolized and is eliminated at a constant rate by the kidney.

Agar

Natural agar from seaweeds consists mainly of the calcium salt of a sulphuric ester of a galactan. It is used as a gel-forming agent in media for culturing micro-organisms. It is not liquefied by organisms that digest gelatin. Agar is non digestible and is sometimes used to provide bulk to the faces in the treatment of constipation. Most seaweed polysaccharides are homoglycans. Alginic acid which is a linear polymer of D-mannuronic acids linked by β-(1→4) bonds has found industrial and food uses. It is used as a stabilizer of ice cream, an emulsifier in foods and salad dressings and even as an edible sausage casing.

Pectic substances are widely used in the food industry to form gels with sugar and fruit acids. They correspond to the hyaluronic acid of the ground substance of animal tissues. They are composed of three polysaccharides, a galactan, the methylester of a galacturonan and an araban. Mucilages and gums are even more complicated plant polysaccharides containing uronic acids as well as hexoses and pentoses.

Hemicelluloses

These are related substances that contain more than one sugar (glucose, galactose, xylose and arabinose) and are extractable from the cell walls of plants by alkaline solutions (17.5 percent sodium hydroxide). Cellulose swells but does not dissolve in this solution. The celluloses along with the hemicelluloses form a good part of the roughage or "bulk" in the human diet.

Chitin is the structural material of the exoskeletons of insects and the shells of crustaceans and is polymer of N-acetyl-D-glucosamine connected by β-(1→4) glycosidic bonds.

Mucopolysaccharides (Glycosaminoglycans)

The prefix "muco" was used to indicate their relationship to mucus and viscid secretions in general and this term has been loosely used to cover a large group of complex carbohydrates and carbohydrates attached to proteins or lipids (mucoproteins, mucolipids), all of which are characterized by high viscosity in solution. For convenience, however, the term mucopolysaccharide may be used to denote a complex of carbohydrate and protein whose reaction is predominantly polysaccharidic and is applied to heteroglycans which contain generally residues of both a uronic acid and a hexosamine (glucosamine, 2-amino-2-deoxy-D-glucose; and galactosamine (chondrosamine) 2-amino-2-deoxy-D-galactose). Jeanloz has suggested the name *glycosaminoglycans* to describe this group of substances. They are further subdivided into a acidic mucopolysaccharides which contain hexosamine and, usually, uronic acid with or without sulphate and neutral mucopolysaccharides which contain hexosamine and a neutral monosaccharide. The acid mucopolysaccharide include the sulphate-free hyaluronic acid and the sulphate-containing isomeric chondroitin-4- and 6-sulphates, heparin, dermatan sulphate (chondroitin sulphate B, β-heparin), heparin sulphate (heparin sulphate, hepearan sulphate), hyaluronic acid sulphate, and keratosulphate (without uronic acid).

In most connective tissues of animals, the acidic mucopolysaccharides are complexed with protein or peptide residues. They are all negatively charged and all the well-known members of the group are linear polymers which behave as anionic polymeric electrolytes. The isolation of pure entities and the avoidance of artefacts depends on the delicacy of the methods employed. One of the important components of mucoproteins, neuraminic acid, is attached by a glycosidic link that is far more labile than those linking the other monosaccharide units. This linkage can be preferentially hydrolyzed by acid or the specific enzyme neuraminidase with little alteration or damage to the remaining structure. The use of detergents like cetylpyridinium chloride and cetyl trimethylammonium bromide has enabled the intact polysaccharide-protein complex to be isolated in a number of cases, the large positively charged ions complexing with the negatively charged mucopolysaccharide. Such complexes, though insoluble in water, show a gradation of solubilites in aqueous solutions of varying ionic strength and are sometimes soluble in organic solvents. Selective precipitation of acid polysaccharides can also be achieved when acid conditions are used to suppress the ionization of carboxyl and sulphate groups. Methods for the specific splitting of mucopolysaccharides into the component carbohydrate and protein moieties are being developed with use of specific enzymes a hydrolytic agents.

Biological Function

The mucopolysaccharides are extra-cellular in location and are associated with connective tissue in different parts of the body. Three major cytological components are generally recognized in connective tissue; Cells, extracellular fibres and extracellular amorphous ground substance. Besides the fundamental similarly in the function of connective tissue to bind and support the various organs of the body, there appears to be present also a chemical similarity in the connective tissues of various parts of the body in their content of mucopolysaccharides and mucoproteins. The acid mucopolysaccharides are found in the amorphous ground substance which forms the cement that holds the cells together and comprises the medium in which the cells move and the moving parts are lubricated. By reason of its content of surface-active components, variations in the state and composition of the ground substance modify considerably the activity and permeability of cell membranes and, therefore, of cells and tissues. It has been suggested that tissue mucopolysaccharides might effects and modify various physiological and pathological processes like movement of water and electrolytes, joint lubrication, blood coagulation, fertilization, hair growth, wound healing, calcification and so on.

Among the several proteins associated with mucopolysaccharides may be mentioned the mucoproteins orosomucoid (α_1-globulin), haptoglobin (α_2-globulins), and the gamma globulins found in plasma, the glycoprotein hormones of the anterior pituitary gland which include the gonadotropins, thyrotropin, and prolactin, mucin from the submaxillary gland and proteins associated with the blood group substance as well as ovomucoid which is known to be an inhibitor of pancreatic trypsin. The blood group substances are widely distributed in the body and are present in saliva, gastric juice, other tissue extracts and also in ovarian cyst fluid. More than sixty different antigens have been identified with specific blood groups in man but haemolytic disturbances are associated primarily with the *D-antigen* of the *Rh group* and with incompatibility of ABO factors. *Blood group antigens* are glycoproteins or glycolipids. The ABO factors are associated with the transmembrane *Band 3 protein* of red blood cells. The synthesis of these antigens is determined by specific genes termed H, A, B, and so on. Gene H is concerned with the attachment of fucose to the I carbohydrate moiety consisting of galactose and N-acetyl glucosamine residues already present on the protein to establish blood group O. The transferase dependent on gene A permits attachment of N-acetyl galactosamine to form group A antigen. The transferase dependent on gene B attaches galactose which determines group B antigen. The synthesis of the antigens is completed by adding several units of fucose, the position of these fucose units determining the specificity of the blood group. The protein collagen is embedded in a mucopolysaccharide matrix. The ability of several mucoproteins to inhibit the clumping of red cells, induced by heat-treated influenza virus, appears to be associated with the sialic acid portion.

Chemistry of Mucopolysaccharides

The most abundant of the group of acid mucopolysaccharides is probably *hyaluronic acid,* the ground substance of mesenchyme. It has been isolated from the skin, synovial fluid, vitreous humour, Wharton's jelly obtained from the umbilical cord, and also from haemolytic streptococci. A hyaluronic acid-protein complex has also been isolated from human synovial fluid using zone electrophoresis. It is a high molecular weight material usually polydisperse, ranging in molecular weight from 50,000 to many millions depending on the method of isolation. The large size of the molecule and the nature of the constituent units bestow a high viscosity on solutions of the polymer constituted of the disaccharide, hyalobiuronic acid. This disaccharide is composed of N-acetyl-D-glucosamine and D-glucuronic acid in combination. Hyaluronic acid appears to bind water in the tissue spaces and in the skin. For a pure liquid and for many solutions, the viscosity is generally independent of the velocity gradient and this is referred to as "Newtonian viscosity". Asymmetric molecules, however, tend to take up an average position with their longer dimensions more or less oriented with the direction of flow. In

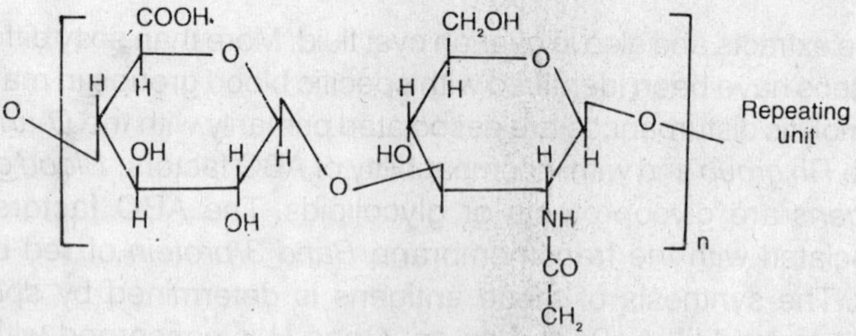

this position they make less contribution to the viscosity than when they are randomly oriented. The viscosity of such solutions tends to fall as the velocity gradient increases and this is termed "*non-Newtonian viscosity*". Such viscosity changes make synovial fluid an ideal lubricant in joints which have to bear weight as well as facilitate movement. These viscosity changes have been attributed to the presence of hyaluronic acid.

Hyaluronidase is the enzyme that catalyzes the depolymerization of hyaluronic acid and by reducing its viscosity facilitates the diffusion of material into tissues. The enzyme is present in spermatic fluid and is usually prepared from bull or ram tests. It is found also in pneumococci and streptococci, snake and bee venoms, skin autolysates and extract of leech heads. The testicular preparation cleaves the hyaluronic acid to the tetrasaccharide stage and affects only the glucosaminidic linkages between C_1 of the glucosamine moiety and C_4 of the glucuronic acid, the uronidic linkages not being attacked. The leech enzyme hydrolyzes the glucuronidic bond. The bacterial hyaluronidase cleave the tetrasaccharide obtained by mammalian hyaluronidase activity into two disaccharides. Many micro-organisms produce the enzyme and the presence of the enzyme may facilitate spread of infection and movement of bacilli into the surrounding tissues. Purified enzyme preparations are used in clinical medicine to facilitate diffusion of fluids and drugs when administering fluids by the subcutaneous route in states of dehydration or in producing local anaesthesia. It is claimed that it helps the removal of adhesions encounted during surgery.

Chondroitin-4- and 6-sulphates (Chondroitin sulphates A and C) occur in cartilage and adult bone and as minor constituents in other tissues.

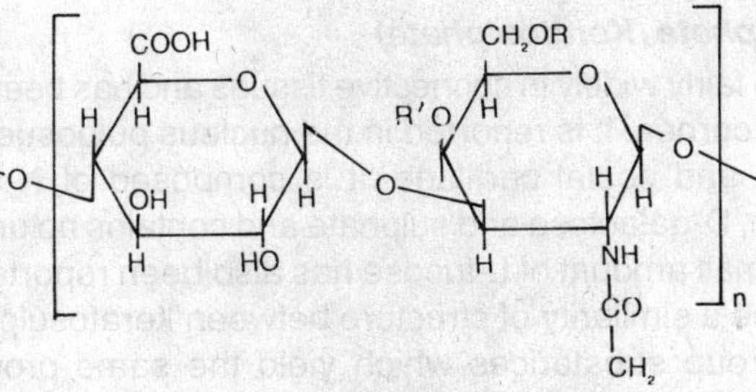

Chondroitin – 6-sulphate R = $-SO_2OH$; R' = -H

Chondroitin-4- sulphate R = -H ; R^1 = $-SO_2OH$

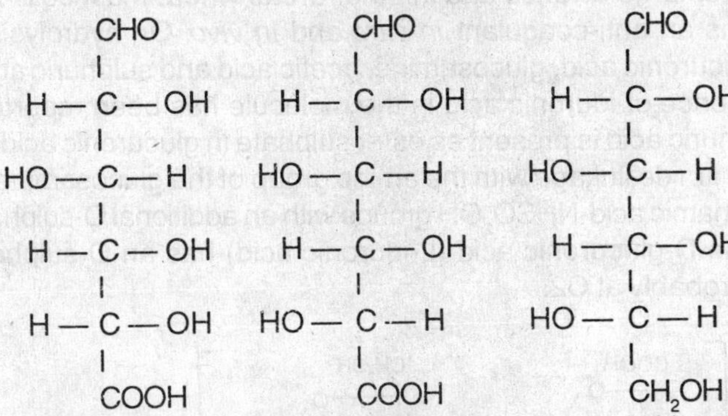

D-glucuronic acid L-iduronic acid L-idose

The linkage is assumed to be the same in both mucopolysaccharides in view of the fact that the rate of breakdown by bacterial or animal beta-hexosaminidases is similar. These compounds are associated with collagen and may account for the physical characteristics of the cartilaginous tissue in which they are present.

Dermatan sulphate (β-heparin, chondroitin sulphate B). It has been isolated from tendons, aorta, heart valves, sclera, lung parenchyma and is present in large amounts in pig skin. It is distinguished from the chondroitin-4- and 6-sulphates in not being degraded by testicular hyaluronidase. Acid hydrolysis of the compound yields D-galactosamine, acetic acid, sulphuric acid and a uronic acid which is believed to be L-iduronic acid which is an epimer of D-glucuronic acid at the fifth carbon from the aldehydic group.

Keratan sulphate (Keratoslphate)

It occurs fairly widely in connective tissues and has been isolated from bovine cornea. It is reported in the nucleus pulposus, the wall of the aorta and costal cartilage. It is composed of N-acetyl-D-glucosamine, D-galactose and sulphate and contains no uronic acid residue. A small amount of L-fucose has also been reported. There appears to be a similarity of structure between keratosulphate and the blood group substances which yield the same products on hydrolysis except for the polypeptide protein.

Heparin

Originally obtained from the liver, it is present also in the lung, the walls of large arteries and in other areas where mast cells are found. It is an anti-coagulant *in vitro* and *in vivo.* On hydrolysis it yields glucuronic acid, glucosamine, acetic acid and sulphuric acid. The presence of iduronic acid in the molecule has been reported. The sulphuric acid is present as ester sulphate in glucuronic acid as well as in amide linkage with the amino group of the glucosamine to form sulphamic acid-$NHSO_2OH$ groups with an additional O-sulphate at C3. The D-glucuronic acid (L-iduronic acid) has an O-sulphate linkage probably at C2.

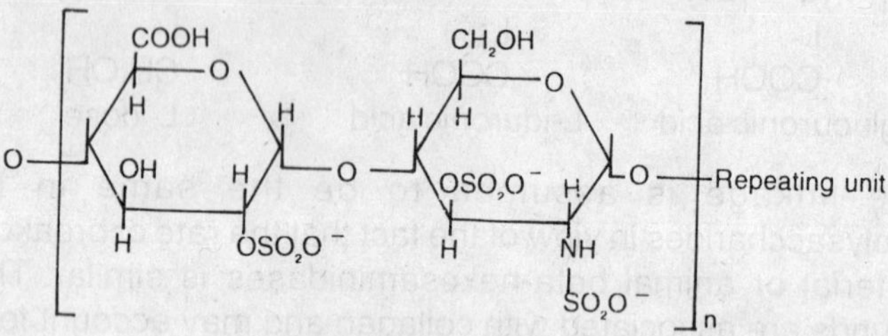

HEPARIN

Heparitin sulphate has been isolated from normal human and cattle aorta, from urine, liver and spleen of patients with gargoylism and also from amyloid liver. Heparitin sulphate differs from heparin

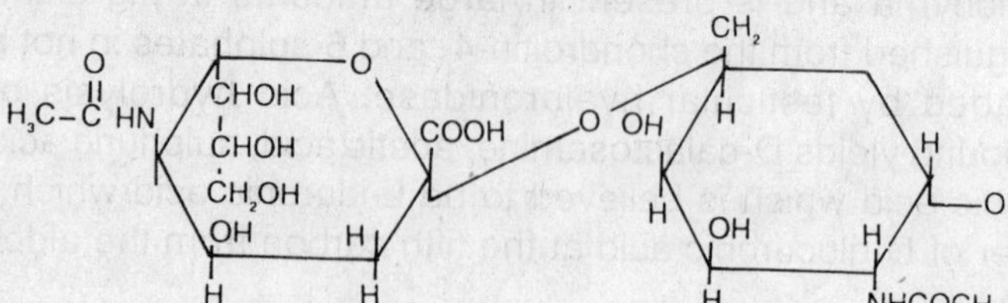

in that a variable fraction of the amino groups bears an N-acetyl residue instead of a sulphate.

Salivary mucin (*mucus*) contains numerous molecules of the disaccharide unit 6-(N-acetyl-α-neuraminyl) N-acetyl-D-galactosamine as the prosthetic group in association with protein. Other components of mucin (mucus) are L-fucose, D-galactose and N-acetyl glucosamine. This material is present throughout the gastrointestinal tract. It has a high water content (around 96 percent) and is associated with a large glycoprotein (about 2 million daltons). The carbohydrate chains are arranged on the glycoprotein like bristles on a test-tube brush and are able to enmesh large quantities of water and foreign matter. Prostaglandins stimulate mucus secretion.

Bacterial Cell Wall Polysaccharides

The chemical composition of the bacterial cell wall has been worked out in great detail since it is concerned with he virulence of bacteria, the presence of specific antigens on the cell surface and the mode of action of some antibiotics, like pencillin, which prevent the synthesis of the cell wall. The Danish physician Gram used gentian violet followed by iodine hoping to stain the nuclei of kidney cells violet and the tubules brown. After this treatment, the section was rapidly decolourised by alcohol. Certain bacteria present in the section, however, remained blue-black and this chance observation led to the development of a stain which could distinguished bacteria into Gram-positive bacteria and cocci (like the tetanus bacillus, staphylococci and streptococci) and Gram-negative bacteria and cocci like *E.coli* and the gonococcus. Ehrlich's great contributions to chemotherapy were based on the postulate that bacteria could be attacked by a form of differential staining or toxicity which would not affect the host.

Bacteria are surrounded by a tough envelope which preserves their integrity and shape. When Gram-positive bacteria are subjected to the action of lysozyme, and enzyme present in nasal secretions, tears, saliva and egg white, they swell and rupture of the cell membranes takes place. Such rupture is prevented by the presence of isotonic sucrose in the medium, though breakdown of the cell wall occurs, and the bacterial cell covered by a delicate thin membrane (protoplast) is left behind. Gram-positive bacteria and cocci are composed of complex polymers with interwoven strands of carbohydrate and amino acid residues, some of them being D-amino acids. The carbohydrates include as major components N-acetyl derivatives of glucosamine and muramic acid, ribose or glycerol phosphates united by phospho-diester linkages, and other minor constituents like glucose, galactose, mannose, fucose or rhamnose. *Teichoic acids* are polymers of glycerol or ribitol phosphates combined with D-alanine and N-acetyl D-glucosamine units. They make up 20-40 percent of the dry weight of the cell wall. The *peptidoglycans* in Staphylococcus aureus include as major components the N-acetyl derivatives of glucosamine and muramic acid with horizontal strands of pentaglycine and vertical struts of tetrapeptides composed of L-alanine, D-glutamic acid, L-lysine and D-alanine residues. Both teichoic acids and peptidoglycans are antigenic determinants. The presence of D-amino acids in the peptidoglycan confers protection from attack by peptidases.

Peptide synthesis of this type does not involve the ribosomal system and the production of D-amino acids requires the presence of pyridoxai phosphate dependent racemases. The carbohydrate residues are united through the intervention of uridine nucleotides but as UDP is too polar to pass through the membrane, a carrier phospholipid functions in the transport system. It is believed that penicillin blocks the synthesis of cell walls by interfering with the action of the glycopeptide transpeptidase concerned with the cross-linking reaction between the carbohydrate and amino acid residues. Some organisms can acquire resistance to the action of penicillin by producing an enzyme, penicillinase, which splits penicillin at the amide link of the betalactam ring to form an inactive compound, penicilloic acid. Such bacterial resistance is conferred by a plasmid with extra chromosomal DNA which can be transmitted to other bacteria to produce resistant strains. Cultures of both sulphonamide and amethopterin-inhibited bacteria accumulate in the medium 5-amino 4-imidazole carboxamide and its ribotide which are intermediates in the formation of nucleotides.

Other Heteropolysaccharides

Glycoproteins

These consist of *carbohydrate* linked to a *peptide* portion by *covalent* bonds only. *Mucopolysaccharides* may associate with peptides by *ionic* as well as by *convalent* bonds. Moreover, in contrast to mucopolysaccharides, glycoproteins contain no uronic acids only a few sulphate esters. The glycoproteins include some *plasma proteins, blood group substances* and *mucoproteins*.

Plasma Proteins include *lipoproteins* and *glycoproteins.* Among the important plasma proteins are *albumin, thyroxine binding globulin, haptoglobin, fibrinogen* and the *immunoglobins.*

Blood group substances are located in the erythrocytes and are responsible for agglutination. The human types A, B and O yield amino acids and the sugars *D-galactose, glucosamine, galactosamine* and *L-fucose* on hydrolysis.

Mucoproteins are characteristic of bone, connective tissue and salivary glands. *Galactosamine* and *glucosamine* are the sugars most commonly found in mucoproteins. They also contain *L-fucose* and *sialic acid.*

Murein

It is a *polysaccharide* linked to short *peptide chains.* It is found in the cell walls of many bacteria. It consists of alternating units of *N-acetylglucosamine* (NAG) and *N-acetylmuramic acid* (NAM) linked by its carboxyl group to the peptide chain consisting of four amino acids (*L-alanine, D-glutamic acid, L-lysine,* and *D-alanine*). Five *glycine* residues link each peptide chain to its neighbour. The enzyme *lysozyme* hydrolyses the chain into NAG-NAM disaccharides, resulting in the rupture of the cell wall.

Function of Carbohydrates

The most important function of carbohydrates is to provide *energy* to the body. They are also *structural components* of tissues. Other functions include *regulation of fat metalism, protein sparing functions,* and function in the *digestive tract.*

1. Storage Substance of Potential Energy

Carbohydrates are storage substances of potential energy in animals. About 60% of the total energy requirement of man is provided by the breakdown of carbohydrates. One gram of carbohydrate on oxidation yields on an average four calories. Glucose supplies the immediate energy needed by tissues. Glucose is the sole form of energy for the brain and other nervous tissue. Lack of glucose, or of oxygen for its metabolism, leads to rapid damage to the brain. Carbohydrate is stored in the body in the form of *glycogen.* About 100 gm of glycogen is stored in the liver and by its breakdown maintains the glucose level in the blood. Muscles contain about 200 to 240 gm glycogen. This glycogen is, however, utilized only by the muscles and is not available for regulating the blood sugar-level.

2. Structural Component

Carbohydrates are important components of some structural materials of living organisms. *Monosaccharides* are important constituents of nucleic acids, coenzymes, flavoroproteins and blood group substances, *Vitamin C* (ascorbic acid) is related to sugars *Immunopolysaccharides* play a part in the resistance of infections. *Hyaluronic acid* is the viscous substance in the matrix of connective tissue. *Heparin* prevents the clotting of blood. *Glucuronic acid,* which occurs in the liver, acts as a detoxifying agent by combining with toxic substances and bacterial by-products. *Chondroitin sulphotes* are found in cornea, cartilage tendons, skin, heart valves and saliva. *Glycosides* are component of steroid hormones. *Galactolipins* are constituents of nervous tissue. The cell wall of plants and the polysaccharides in the capsule of bacteria are carbohydrates.

3. Regulation of Fat Metabolism

Some carbohydrates are essential for normal oxidation of fats. When carbohydrates are restricted in the diet there is more rapid metabolization of fats. This results in the accumulation of incompletely oxidized intermediate products leading to *ketosis.* This is common in uncontrolled diabetes mellitus.

4. Protein-sparing Function

Carbohydrate is preferentially metabolized in the body as a source of energy as long as it is present in the required quantity. This spares protein for building of tissue. When there is deficiency of calories in the diet, however, fat and then protein are utilized for supplying energy.

5. Role is Gastrointestinal Function

Indigestible substances like cellulose, hemicellulose and pectins provide the bulk or roughage of food, and thus help the peristaltic movements of the digestive tract. Lactose promotes the growth of desirable bacteria in the small intestine. These bacteria synthesize certain B-complex vitamins. Lactose also increases calcium absorption.

The demand for high-energy phosphate is not transient, and fuels must always be available to supply the processes generating high-energy phosphate. In order for an animal to interrupt its food intake—to eat meals rather than small snacks and to sleep for long intervals—it must store fuels at times of excess and be able to recover them at times of deprivation. These stores include a polymer of glucose : glycogen. (Plants are starch in a comparable way).

Glycogen appears in muscle cells as discrete particles with mass up to 2×10^7 daltons and more. Liver contains both individual particles and covalently linked aggregates as large as 10^9 daltons. The many hydroxyl groups make glycogen quite polar, and the particles contain 1.1 grams of bound water per grams of polysaccharide. Another 1 to 3 grams of water may interact with the particles. The particles also contain the enzymes responsible for the synthesis and later breakdown of glycogen.

The amount of glycogen varies widely, not only among different tissues but also within a tissue, depending upon the supply of glucose and the metabolic demand for energy. Most of the body's glycogen is in the liver and muscles. We shall mainly be concerned with these two tissues, but glycogen is also important to the economy of other tissues. In general, the content of glycogen in the liver varies greatly with the diet, while that in the muscles varies greatly with exercise.

Much of our knowledge about the glycogen concentration of human tissues comes from analysis by Swedish investigators of biopsy specimens form normal volunteers. (We are all indebted to university students and members of the fire department in Stockholm for contributing pieces of themselves of satisfy our curiosity). The data have some limitations, because only readily accessibly regions of the organs can be explored in living subjects, but the values are consistent with other quantitative information on human fuel metabolism. Results under various conditions from several laboratories are summarized.

A typical value for the liver glycogen content in a well-fed human in near 400 millimoles of glucosyl residues (65 grams dry weight) per kilogram of tissue; there is less after fasting and more upon eating a high-carbohydrate meal.

The skeletal muscles typically have 85 millimoles of glucosyl residues (14 grams) per kilogram of tissue, which does not change much with overnight fasting or with a high carbohydrate diet. However, the content falls to 1 millimole per kilogram, or lower, after an hour or two of heavy work. After depletion in this way, high-carbohydrate diets on subsequent days can cause the level to rise as high as 300 millimoles per kilogram.

Even though the liver usually has a higher concentration of glycogen than do the skeletal muscles, the muscles contain the bulk of the total glycogen store, owing to their large mass. A man weighing 70 kilograms will have about 28 kilograms of skeletal muscle, but only about 1.6 kilograms of liver. Given typical contents per kilogram of tissue, 400 millimoles in the liver and 85 millimoles in the muscles, the total store is 0.6 moles in the liver and 2.4 moles in the muscles. The total in the body, considering all organs, will therefore range somewhat over 3 moles in a well-fed individual and near 3 moles after an overnight fast (the postabsorptive state).

MECHANISM OF GLYCOGEN STORAGE

Glycogen is made by adding one glucosyl residue at a time to an existing glycogen molecule, creating amylose chains that are then rearranged to make branches. The overall process is conveniently thought of in three stages : (1) the conversion of glucose 6-phosphate to *uridine diphosphate glucose* (*UDP-glucose*); (2) the transfer of glucosyl units from UDP-glucose onto residues in glycogen chains, so as to make growing amylose chains composed of α-1, 4-linkages; and (3) the creation of branches by shifting portions of the chains onto the C6 hydroxyl groups of adjacent chains.

The Formation of UDP-Glucose

Since it is the C1 of glucose that will later be attached to glycogen, the sequence begins by transferring the phosphate group of glucose 6-phosphate to form glucose 1-phosphate. This freely reversible reaction is catalyzed by *phosphoglucomutase,* which utilizes the corresponding bisphosphate compound, glucose 1, 6-bisphosphate, is small concentrations as an intermediate.

Glucose 1-phosphate then reacts with UTP to form UDP-glucose and inorganic pyrophosphate. As is the case with other reversible reactions involving pyrophosphate, hydrolysis by inorganic pyrophosphatase keeps the pyrophosphate concentration low, and pulls the reaction in the direction of UDP-glucose formation. The UTP utilized in this reaction is generated by a *nucleoside diphosphokinase* reaction :

$$\text{ATP + UDP} \leftrightarrow \text{ADP + UTP}$$

so the ultimate source of the high-energy phosphate is ATP.

Formation of Amylose Chains

UDP-glucose donates glucose residues used for extending the terminal branches of glycogen in a reaction catalyzed by *glycogen synthase.* The reaction is specific for the hydroxyl group on C4 of the glycogen residues, so the new glucosyl residue simply extends the 1, 4-chain. Since the nature of the chain is not changed by extension, the *glycogen synthase* reaction can be repeated indefinitely. A molecule of UDP is released for each glucosyl residue added.

Formation of Branches

If nothing else happened, the result of the glycogen synthase reaction would be long amylose chains composed only of 1, 4-bonds. However, cells storing glycogen also have a branching enzyme, a *glycosyl-4 : 6-transferase,* that catalyzes the transfer of a segment of amylose chain onto the C6 hydroxyl of a neighbouring chain. This enzyme moves a block of seven 1, 4-residues from a chain at least 11 residues in length and transfers it into another segment of amylose chain at a point four residues removed from the nearest branch. (Transfer of seven residues is favoured, but specificity is not absolute, and this number is emphasized for clarity). The new branch is therefore usually seven residues in length, and the remaining stub from which it was removed is at least four residues in length, but is more commonly six to nine residues long. 1, 6-Glucosides have about 4,800 joules per mole less standard free energy content than do 1, 4-glucosides, so the equilibrium of reaction favours branching.

Further Growth

The new branch and the remaining stub from which it was obtained can grow in length by addition of more 1, 4-glucosyl residues from UDP-glucose through the action of glycogen synthase. When enough have been added to extend them to at least 11 residues beyond the branch, a further transfer by the branching enzyme may create still another branch, which also can grow in length. What limits the size of the molecule and the number of particles is not known. There is little information to justify speculation, although large glycogen particles appear to be made of subunits of approximately the theoretical limit for tightly packed spheres. What is certain is that skeletal muscle does not accumulate glycogen indefinitely, even with a high concentration of blood glucose. The content of glycogen in the liver can rise markedly under heavy dietary loading, with the formation of large aggregates, but even in this tissue glucose is absorbed slowly enough with more usual conditions so that an increased fraction of the excess supply is used for making fat after the glycogen content rises above 50 to 60 grams per kilogram of tissue. The mashed livers of force-fed Strasbourg geese, from which the pate de fore gras favoured by epicures is made, owe their special properties to high contents of both glycogen and fat.

Priming Glycogen Synthesis

Glycogen storage is mainly an extension of already existing polysaccharide chains, but there must be some way of making the core molecules that are to be enlarged. It is possible that reactions yet to be discovered put priming carbohydrate residues on a protein core. There is some evidence that this is not necessary, that the synthetic reactions used to enlarge glycogen will also make new starter molecules, although slowly.

UTILIZATION OF GLYCOGEN AS FUEL

Phosphorolytic Cleavage

When glycogen is mobilized, all of the glucose residues appear as glucose 1-phosphate, except for one molecule of free glucose released from each 1, 6 branch. The primary reaction is a simple phosphorolysis of 1, 4 bonds at the end of each branch. This reaction, which is catalyzed by *phosphorylase,* shortens an outer branch by one residue, which appears as glucose 1-phosphate. The reaction is freely reversible, with the equilibrium ratio of P_i to glucose 1-phosphate concentrations near 3.5. However, the concentration ratio in animal tissues is always greater than that, so the reaction always goes in the directions of formation of glucose 1-phosphate.

When a glucose residue is removed by phosphorolysis, the shortened 1, 4-branch on the glycogen molecules can be attacked again to remove another residue, followed by further attack on the remainder. However, phosphorylase by itself will not catalyze the removal of all of he terminal residues. In anthropomorphic terms, it won't go near the branches. Expressed more mechanistically, the specificity of the enzyme is such that the 1, 4-bond attacked must be at least four residues removed from a 1, 6-branch. Therefore, phosphorylase alone will not make the outer branches any shorter than four glucosyl residues. The function of this specificity is not clear. Glycogen completely sheared by phosphorylase to the four-residue stubs is called the phosphorylase limit dextrin, or ϕ-dextrin. (Dextrin is a term borrowed from starch chemistry, describing partially hydrolyzed glucose homopolymers, and it is widely used for describing intermediates obtained during chemical investigation of structure, but this does not imply the occurrence of these discrete types of molecules in cells).

Removal of Branches

Further degradation of glycogen requires a second enzyme, *amylo-1, 6-glucosidase.* This enzyme catalyzes two successive reactions. In the first reaction, it acts as a *glucosyl transferase* akin to the branching enzyme functioning during glycogen synthesis, but

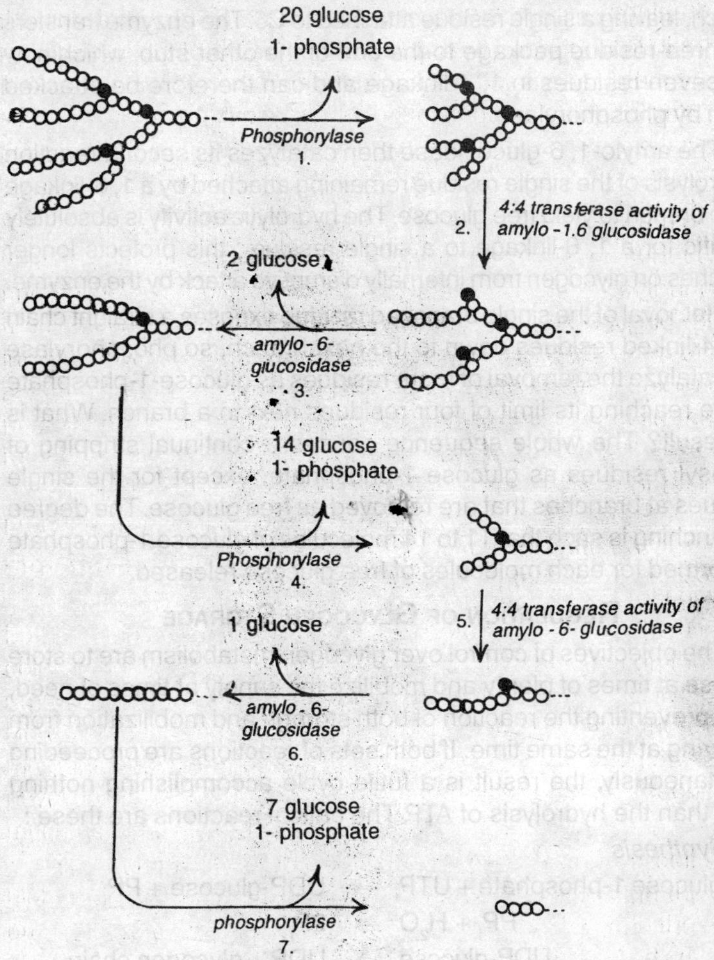

Fig. 3.15. Summary of a sample sequence during glycogen breakdown. The diagram shows the fate of four outer chains and the next two tiers of branches from which they arise. Each chain is shown with nine residues beyond the outermost branch. The residues linked by 1, 6 bonds and released as free glucose are indicated by solid circles.

differing in that it transfers three glucosyl residues from a shortened branch onto another chain terminus so as to lengthen it. Suppose that adjacent chains are shortened as much as possible at a branch by phosphorylase. Each will then be four glucosyl residues in length. The glucosyl 4: 4-transferase activity in amylo-1, 6-glucosidase catalyzes the removal of three of the residues of the stub of the branch, leaving a single residue attached to C6. The enzyme transfers the three-residue package to the end of the other stub, which now has seven residues in 1, 4-linkage and can therefore be attacked again by phosphorylase.

The amylo-1, 6-glucosidase then catalyzes its second reaction : hydrolysis of the single residue remaining attached by a 1, 6-linkage at the branch to yield free glucose. The hydrolytic activity is absolutely specific for a 1, 6-linkage to a single residue : this protects longer branches on glycogen from internally disruptive attack by the enzyme.

Removal of the single branched residue exposes a straight chain of 1, 4-linked residues down to the next branch, so phosphorylase can catalyze the removal of more residues as glucose-1-phosphate before reaching its limit of four residues next to a branch. What is the result? The whole sequence causes a continual stripping of glucosyl residues as glucose 1-phosphate, except for the single residues at branches that are removed as free glucose. The degree of branching is such that 11 to 14 molecules of glucose 1-phosphate are formed for each molecules of free glucose released.

Regulation of Glycogen Storage

The objectives of control over glycogen metabolism are to store glucose at times of plenty and mobilize the supply at times of need, while preventing the reaction of both storage and mobilization from occurring at the same time. If both sets of reactions are proceeding simultaneously, the result is a futile cycle accomplishing nothing more than the hydrolysis of ATP. The critical reactions are these :

Synthesis

1. glucose 1-phosphate + UTP → UDP-glucose + PP_i
2. $PP_i + H_2O$ → $2P_i$
3. UDP-glucose → UDP + glycogen chain
4. UDP + ATP → UTP + ATP

Mobilization

5. glycogen chain + P_i → glucose 1-phosphate

SUM ATP + H_2O → ADP + P_i

The strategy of control is to make glycogen phosphorylase and glycogen synthase respond to similar signals, but in opposite directions. When the glycogen phosphorylase becomes active, the synthase becomes inactive, and *vice versa.* In addition, UDP-glucose is made a competitive inhibitor of its own formation so that it will not accumulate when the synthase is inactive.

The nature of the signals regulating glycogen metabolism and the responses to them differ among tissues. In the *muscles,* the objectives is to mobilize glycogen as a fuel for contraction and to store it at rest if the glucose supply is adequate. Consequently, the signals are the *intracellular concentration of* Ca^{2+} *and AMP,* which rise when contraction begins, and the *blood concentrations of adrenaline and insulin.* Adrenaline signals a state of alert, perhaps requiring fuel expenditure, while insulin signals that glucose is available for storage.

The liver regulates glycogen supply so as to buffer the blood glucose concentration. After glucose is absorbed by the small bowel, it first goes to the liver *via* the portal vein and part is stored as glycogen. The liver can also store glucose residues synthesized by gluconeogenesis. When the blood glucose concentration falls, the liver breaks down glycogen and releases glucose into the blood to maintain the concentration. Important signals for glycogen metabolism in the liver include the *glucose concentration* within hepatic cells and the blood concentrations of two pancreatic hormones, *insulin* to indicate an adequate glucose supply and *glucagon* to indicate a need for more glucose. In addition, the liver responds to stimulation of its *sympathetic nerve supply,* and to other hormones to be considered later.

Many of these signals act to regulate both glycogen phosphorylase and glycogen synthase. These enzymes are regulated in two ways : by combination with allosteric effectors and by covalent modification through phosphorylation and dephosphorylation.

Control by Phosphorylation

Glycogen synthase and glycogen phosphorylase each exists in active and inactive conformations. Allosteric effectors and phosphorylation regulate these enzymes by stabilizing one conformation. Phosphate groups are added by transfer from ATP in reaction catalyzed by kinases, and they are removed by hydrolysis in reactions catalyzed by phosphatases.

The kinases and phosphatases affecting the enzymes of glycogen metabolism also act on enzymes responsible for other metabolic processes that are discussed later. In general, *phosphorylation activates catabolic enzymes and inactivates anabolic* (*synthetic*) *enzymes.* This is true for the enzymes of glycogen metabolism. Phosphorylation of glycogen phosphorylase locks it in the active conformation. That is the P-form (phosphorylated form) is glycogen phosphorylase a ("a" *for intrinsically active*), but phosphorylation locks glycogen synthase in the inactive conformation, designated glycogen synthase *b*:

glycogen synthase *a* is *not* phosphorylated
glycogen synthase *b* is phosphorylated
glycogen phosphorylase *a* is phosphorylated
glycogen phosphorylase *b* is *not* phosphorylated.

The mechanism for regulating the interconversion of active and inactive conformations of glycogen phosphorylase and glycogen synthase by the combination of allosteric effectors and phosphorylation-dephosphorylation reactions are considerably more complex than anything we have encountered before. Let us first dissect the responses in mammalian skeletal muscles.

Regulation of Phosphorylase in Muscle

Phosphorylase is an enzyme with multiple binding sites on each of its two identical subunits. The enzyme clings to a glycogen particle like a barnacle on a rock, wrapping itself around a terminal amylose chain. This holding site has no catalytic function. Another crevice on the same concave face fits the terminal amylose chain to be cleaved with inorganic phosphate. This catalytic site has a deeply buried molecule of *pyridoxal phosphate.* Pyridoxal phosphate is usually employed in reactions involving combination of its aldehyde group with amino groups on substrates, but here it is its phosphate group that is activated, probably to form a transient pyrophosphate-like intermediate, and no other example of this function is known.

The opposite face of glycogen phosphorylase, the convex face, is the control face. It has sites for binding AMP and for phosphorylation.

AMP Activates Glycogen Phosphorylase

Only the active conformation of the enzymes tightly binds/AMP. Phosphorylase a, the P-form, is already in the active conformation, so combination with AMP does not affect its activity. Most of glycogen

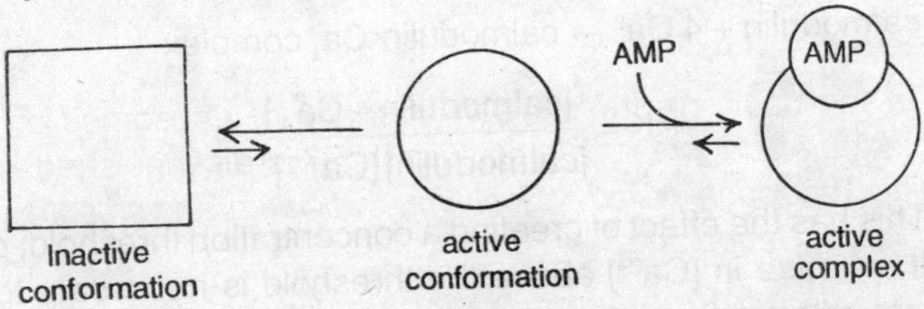

phosphorylase *b*, the dephospho form, is in the inactive conformation in the absence of AMP, but a rice in the concentration of AMP pulls the equilibrium toward the active conformation, even though the enzyme is not phosphorylated. Since a rising AMP concentration is a sensitive indicator of a fall in high-energy phosphate concentration, which will accelerate the Embden-Meyerhof pathway, its effect on glycogen phosphorylase is an important means of matching fuel supply and demand.

Calcium ions Indirectly Activate Glycogen Phosphorylase

They affect the activity of another enzyme, phosphorylase kinase, which in turn catalyzes the phosphorylation of glycogen phosphorylase:

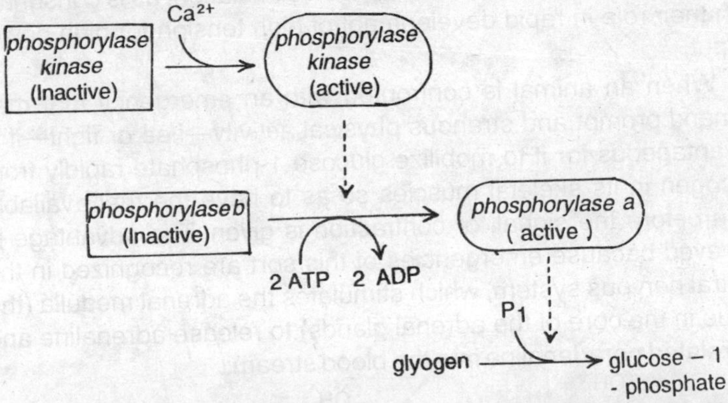

Phosphorylase kinase has the subunit composition $\alpha_4\beta_4\gamma_4\delta_4$. It is unusual in that the δ subunit is *calmodulin*, a protein that also exists separately as an important regulatory protein in many kinds of cells.

Calmodulin Binds for Calcium ions per Monomer, and its active conformation only appear when all the binding sites are filled

The activation of calmodulin therefore depends upon the fourth power of the calcium ion concentration:

calmodulin + 4 $Ca^2 \rightarrow$ calmodulin-Ca_4 complex

$$K = \frac{[\text{calmodulin} - Ca_4]}{[\text{calmodulin}][Ca^{2+}]^4}$$

This has the effect of creating a concentration threshold; only a small increase in $[Ca^{2+}]$ above the threshold is necessary to fully activate calmodulin.

When the calmodulin subunits within phosphorylase kinase are activated, the entire molecule shifts to an active conformation. The kinase then catalyzes the conversion of glycogen phosphorylase *b* to glycogen phosphorylase, a fixing it in its active conformation, and this is the way in which the release of calcium ions in a stimulated muscle fiber causes an accelerated breakdown of glycogen.

White and red skeletal muscle fibers have different isozymes of phosphorylase kinase. Both contain calmodulin subunits, but the white muscle isozyme also binds additional active calmodulin from the sarcoplasm to further augment its catalytic activity. This makes the white fibers sensitive to lower $[Ca^{2+}]$ than are red fibers; they mobilize glycogen more promptly upon stimulation, as is consonant with their role in rapid development of high tension for high power output.

When an animal is confronted with an emergency that may demand prompt and strenous physical activity—flee or fight—it is advantageous for it to mobilize glucose 1-phosphate rapidly from glycogen in its skeletal muscles so as to have the fuel available even before the signal for contraction is given. The advantage is achieved because emergencies of this sort are recognized in the central nervous system, which stimulates the adrenal medulla (the tissue in the core of the adrenal glands) to release adrenaline and the related noradenaline into the blood stream:

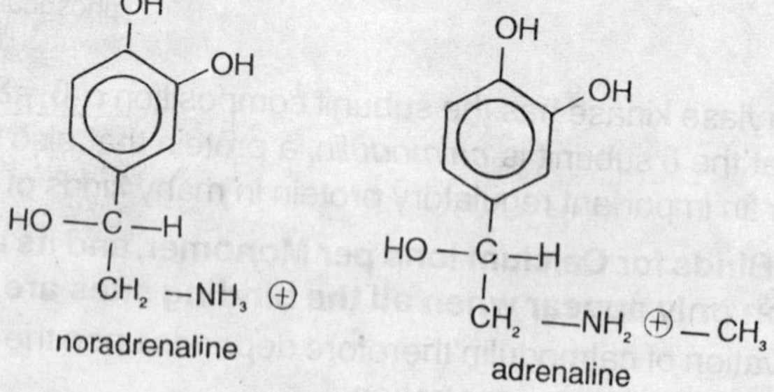

noradrenaline adrenaline

The human adrenal medulla releases more adrenaline than noradrenaline, but the reverse is true in some mammals. Both compounds are frequently lumped under the general name of *catecholamines.* They are frequently called *epinephrine* and *norepinephrine* in American medical circles because Adrenalin is a trademark. However, the tissue receptors for catecholamines are called *adrenergic receptors* by all.

Adenylate Cyclase and Cyclic AMP

Adrenaline stimulates the breakdown of glycogen in skeletal muscle by tripping a series of events that causes phosphorylase kinase to be locked in an active conformation by phosphorylation. These events begin with activation of an enzyme, *adenylate cyclase,* in the plasma membrane, of the muscle fibers. This enzyme occurs as a complex spanning the plasma membrane, with the adrenaline receptors protein on the outside and the catalytic site on the inside. (Activation involves binding of GTP by a regulator protein in the complex). Adenylate cyclase catalyzes in intramolecular condensation of ATP to produce *adenosine 3', 5'-monophosphate,* more familiarly known as *cyclic AMP.*

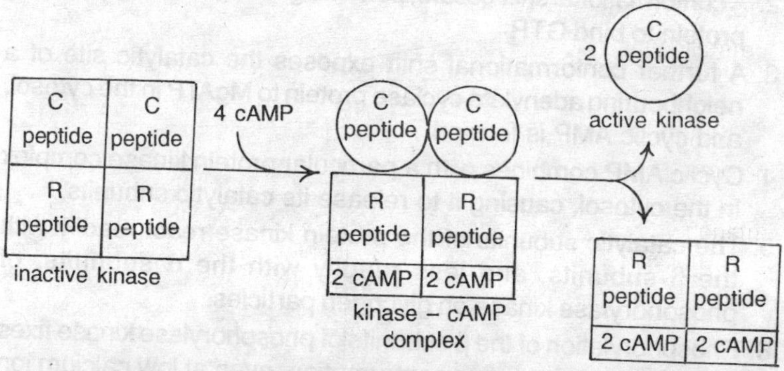

Fig. 3.16. The activation of the cAMP-dependent protein kinase.

The compound is sometimes designated as *3', 5'-AMP* and sometimes as *cAMP.* Cyclic AMP acts as a "*second messenger*", conveying the signal received in the form of adrenaline on the plasma membrane to the responsive enzymes within the cell.

Activation of cAMP-Dependent Protein Kinase

The major effect of cyclic AMP is eukaryotic cells is the activation of a particular protein kinase. The activation occurs in this way : The protein kinase consists of two pairs of subunits : R_2C_2, in which R is a regulatory subunit and C is a catalytic subunit. The combined tetramer is inactive. Each of the regulatory subunits will combine with two molecules of cAMP, but when it does, the association with the catalytic subunits is weakened, and they are related as fully active monomeres.

Effect of Protein Kinase on Phosphorylase Kinase

The cAMP-dependent protein kinase catalyzes the transfer of phosphate from ATP to eight serine side chains in phosphorylase kinase. Each of the four β subunits has one site that is rapidly phosphorylated, stabilizing the active conformation of the enzyme in the presence of Ca^{2+}. Each α subunit has one site that is less rapidly phosphorylated; the function of this phosphorylation is discussed later.

The phosphorylation of phosphorylase kinase is the final link in a multienzyme cascade by which an elevation in blood adrenaline concentration causes conversion of glycogen to glucose 1-phosphate (and lesser amounts of glucose).

1. Adrenaline is bound by a cell-surface receptor protein.
2. A conformational shift occurs, permitting a neighbouring regulator protein to bind GTP.
3. A further conformational shift exposes the catalytic site of a neighbouring adenylate cyclase protein to MgATP in the cytosol, and cyclic AMP is formed.
4. Cyclic AMP combines with a particular protein kinase complex in the cytosol, causing it to release its catalytic subunits.
5. The catalytic subunits of the protein kinase react readily with the β subunits, and less readily with the α subunits, of phosphorylase kinase on glycogen particles.
6. Phosphorylation of the β subunits of phosphorylase kinase fixes the enzyme in its active conformation, even at low calcium ion concentrations, so that it catalyzes transfer of phosphate groups from ATP to glycogen phosphorylase.
7. Glycogen phosphorylase begins the breakdown of glycogen to glucose 1-phosphate.

Since there is a magnification by successive enzyme action, a minute amount of adrenaline has a very large effect, so large that 2 micromoles (364 μg) is a potent dose for humans. The normal blood concentration of adrenaline is in the nanomolar range. It has been estimated that a rise in the concentration of cyclic AMP to 1 micromole per kg of muscle causes the formation of 25,000 times much glucose 1-phosphate per minute.

An increased calcium ion concentration itself can activate phosphorylase kinase in muscles at rest or during initial contractions. The governing factors is pH. Phosphorylase kinase *b,* the dephosphorylated form, is not activated by calcium ions unless the pH is higher than 6.8. This value is exceeded in resting fibers, in which the pH is near 7.0, and there is a transient rise of pH 7.2 or more when contraction begins, owing to the net hydrolysis of phosphocreatine. As contractions continue, the pH of the muscle falls as formation of CO_2 and lactate accelerate, and phosphorylase kinase must be phosphorylated to the a form if it is to be active. Phosphorylation does not diminish its requirement for calcium ions, but permits it to be active in a more acidic environment.

Reversal of Phosphorylase Activation

When the demand for fuel has passed, the breakdown of glycogen is halted by restoring glycogen phosphorylase to its inactive conformation. A muscle recovering from work will continue accelerated synthesis of ATP until the phosphate potential is restored to the resting level. As this occurs, the AMP concentration declines below the level at which AMP can fix glycogen phosphorylase *b* in its active conformation. However, glycogen phosphorylase a remains active in the absence of AMP, and several additional events occur in a resting muscle to convert glycogen phosphorylase a back to glycogen phosphorylase *b* and prevent its reactivation in the absence of further excitation of the muscle. These events include :

1. Activation of protein phosphatases to hydrolyze the phosphate groups from glycogen phosphorylase a and phosphorylase kinase *a;*
2. Inactivation of the cAMP-dependent protein kinase through hydrolysis of cyclic AMP by cyclic nucleotide phosphodiesterases;
3. Cessation of further formation of cyclic AMP, after the adrenaline concentration falls, through prompt hydrolysis of GTP by the regulatory protein in the adenylate cyclase complex.
4. Rapid destruction of circulating adrenaline to lower its concentration in the absence of continued stimulation of the adrenal medulla.

The activation of protein phosphatases also involves the enzyme cascade. Protein phosphatase-1 is an enzyme capable of hydrolyzing phosphate groups on many proteins. It occurs in many tissues, including the muscles, where it will convert glycogen phosphorylase a to glycogen phosphorylase *b,* and also will remove phosphate groups from the β subunits of phosphorylase kinase, provided that the α subunits are also phosphorylated. This phosphatase is regulated by the phosphorylated form of another protein known as protein inhibitor-1.

Protein inhibitor-1 is phosphorylated by the same cAMP-dependent protein kinase that converts phosphorylase kinase *b* to phosphorylase kinase *a.* Using these facts, we can expand the events in the cascade regulation of glycogen breakdown :

1. The cAMP-dependent protein kinase rapidly phosphorylates not only the β-subunits of phosphorylase kinase, making it active, but also protein inhibitor-1, which combines with protein phosphatase-1 to inactive it.
2. The protein kinase slowly phosphorylates the α-subunits of phosphorylase kinase, making it susceptible to attack by active protein phosphatase. During the interval between phosphorylation of the β- and α-subunits, phosphorylase kinase will remain fully active, ensuring some minimum response to the adrenaline signal.
3. Protein phosphatase-1 can catalyze the removal of the phosphate group from its own inhibitor, and it begins to do so, causing the inhibitor to dissociate.
4. As the active protein phosphatase is released, it removes the phosphate groups from the β-subunits of phosphorylase kinase, Marking it inactive.
5. The active protein phosphatase-1 will not attack glycogen phosphorylase a that has AMP bound to it. If the AMP concentration is low, then the protein phosphatase will convert glycogen phosphorylase a to glycogen phosphorylase *b,* which is inactive at low AMP concentrations. A high AMP concentration thus not only activates glycogen phosphorylase *b,* it also maintain phosphorylase a in its inherently active phosphorylated state, once it is formed.
6. A protein phosphatase-2 is also present to catalyze the removal of the phosphate groups from the α-subunits of phosphorylase

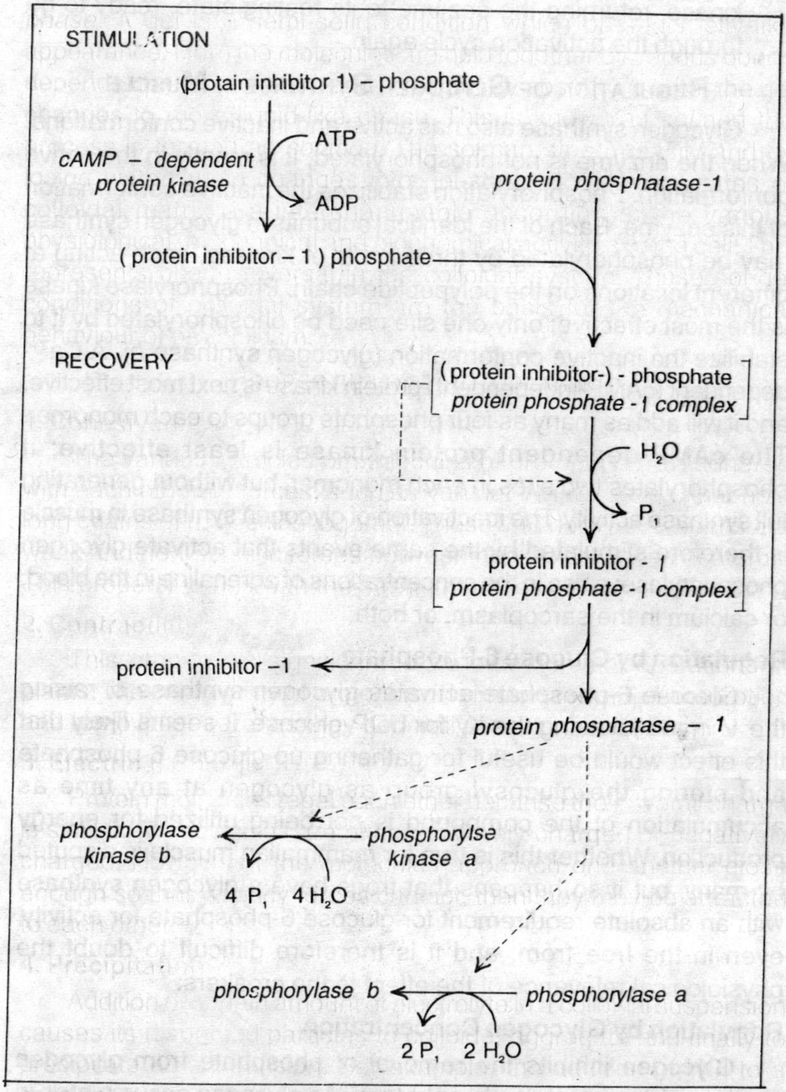

Fig. 3.17. Regulation of protein phosphatase-1, the principal enzyme inactivating phosphorylase a. (top) Stimulation of the cAMP-dependent protein kinase causes phosphorylation of a protein inhibitor, which then combines with protein phosphatase-1 and prevents its action on other proteins. (Bottom) The phosphatase can catalyze hydrolysis of the phosphate on the inhibitor bound to it, causing the protein inhibitor to dissociate. The freed phosphatase can then act on phosphorylase kinase a and phosphorylase a to inactivate them.

kinase, returning the enzyme to its resting state, ready to go through the activation cycle again.

Regulation of Glycogen Synthase in Muscle

Glycogen synthase also has active and inactive conformations. When the enzyme is not phosphorylated, it is mostly in the active conformation. Phosphorylation stabilizes the inactive conformation of this enzyme. Each of the identical subunits in glycogen synthase may be phosphorylated by three kinds of kinases, each acting at different locations on the polypeptide chain. Phosphorylase kinase is the most effective; only one site need be phosphorylated by it to stabilize the inactive conformation (glycogen synthase *b*). A Ca^{2+}-dependent (cAMP-independent) protein kinase is next most effective, and it will add as many as four phosphate groups to each monomer. The cAMP-dependent protein kinase is least effective; it phosphorylates two sites in each monomer, but without generating full synthase activity. The inactivation of glycogen synthase in muscle is therefore stimulated by the same events that activate glycogen phosphorylase; a rise in the concentrations of adrenaline in the blood, or calcium in the sarcoplasm, or both.

Regulation by Glucose 6-Phosphate

Glucose 6-phosphate activates glycogen synthase *b,* raising the V_{max} and lowering the K_M for UDP-glucose. It seems likely that this effect would be useful for gathering up glucose 6-phosphate and storing the glucosyl group as glycogen at any time as accumulation of the compound is not being utilized for energy production. Whether this is true for mammalian muscle is disputed by many, but it so happens that frogs have a glycogen synthase with an absolute requirement for glucose 6-phosphate for activity, even in the free from, and it is therefore difficult to doubt the physiological relevance of the effect to the croakers.

Regulation by Glycogen Concentration

Glycogen inhibits the removal of phosphate from glycogen synthase *b* by a protein phosphatase. As glycogen accumulates in the muscle, it therefore shuts off its own production by decreasing the conversion of glycogen synthase *b* to its more active free form. This effect is the most likely explanation of why animals on high carbohydrate diets don't store massive quantities of glycogen in their muscles.

Phosphorylation of Myosin

Phosphorylation and dephosphorylation are also used to control the contractile apparatus, at least in smooth muscles. The light chains of myosin are partially phosphorylated in all muscles by a specific protein kinase requiring the calcium-calmodulin complex for activity. The physiological function of the phosphorylation in striated muscles is unknown, but it is absolutely necessary for contraction of smooth muscles, in which it provides a locking device to sustain the contracted state without additional expenditure of ATP. There is a corresponding protein phosphatase and inhibitor protein to control the removal of the phosphate groups.

Glycogen Metabolism in the Liver

The turnover of glycogen in the liver is regulated through the same kinds of mechanism seen in striated muscles. However, several of the enzymes involved are isozymes of those occurring in muscles and are regulated by metabolites in different ways. One of the important modulations of enzyme activity in the liver is a combination of glucose with phosphorylase a, which inactivates the enzyme. This means that an elevated concentration of glucose inhibits the breakdown of liver glycogen to form more glucose. In addition, the glucose-glycogen phosphorylase a complex is a better substrate for protein phosphatase-1 than is glycogen phosphorylase a alone, so the presence of glucose accelerates the conversion of glycogen phosphorylase a to the less active *b* form.

Glucagon and Insulin

Secretion by Pancreas. The pancreas of higher animals is both an exocrine organ that elaborates enzymes to digest food and an endocrine organ that secretes hormones into the blood. The pancreas contains organized nests of cells known as the *islets of Langerhans,* which play a major role in regulating metabolism by secreting insulin, and glucagon. We have already encountered insulin, which consists of two polypeptide chains linked by disulfide bonds, as an example of modification of a polypeptide after its synthesis the cells forming insulin are known as *beta cells.* Glucagon is also a polypeptide but has only some 29 residues in the known examples, and it is secreted by different cells in the islets, the *alpha cells.* (More is said about pancreatic endocrine function).

These hormones have opposing effects on glucose metabolism. The alpha cells are stimulated to release glucagon when the blood glucose concentration falls, and the beta cells release insulin when the glucose concentration rises.

The regulation of glycogen metabolism in the liver illustrates a general principle; *Adenylate cyclase is activated by different hormones in different tissues. Adenylate cyclase in the liver is complexed with receptors for glucagon, whereas it is not in muscles. Therefore, glucagon secretion accelerates the breakdown of glycogen in liver* through activation of the cAMP-dependent protein kinase, but it has no direct effect on glycogen metabolism in the muscles. When glycogen is being broken down in the liver, most of the resultant glucose 6-phosphate is hydrolyzed to glucose by the action ofglucose-6-phosphatase. The glucose moves out of the liver into the blood and is carried to other tissues.

On the other hand, *insulin promotes the storage of glucose as glycogen in both the liver and the muscles.* It is now over 70 years since a graduate student at the University of Chicago, E.L. Scott, demonstrated that the pancreas secreted a factor lowering the blood glucose concentration, and 60 years since an insulin preparation was first used to save a diabetic patient, but we still do not know the details of how insulin works. We know that it causes more glucose transporter proteins to be moved to the plasma membrane of cells in at least some sensitive tissues, but this does not explain its internal actions on cellular metabolism. Relief may be in sight. There is evidence that combination of insulin with its receptors cause the release of a small protein from the plasma membrane into the cytoplasm, where it acts as a second messenger. This messenger is not derived from insulin directly; it may be a fragment of a larger protein in the plasma membrane. It is believed to have three effects : an inhibition of the cAMP-dependent protein kinase, an activation of a protein phosphatase-2 that attacks glycogen synthase *b* (not the one attacking glycogen phosphorylase *a*), and an activation of the protein phosphatase component of the mitochondrial pyruvate dehydrogenase complex. Therefore, *insulin causes glycogen synthase and pyruvate dehydrogenase to become more active and suppresses activation of glycogen phosphorylase.* The result is increased synthesis of glycogen in the liver and muscles, and an increased conversion of pyruvate to acetyl coenzyme A in these organs. These effects are in addition to the action of insulin in accelerating glucose transport into muscles.

In general, insulin secretion rises when the secretion of glucagon falls, and *vice versa*. Suppose both are being secreted. There is some protection against simultaneous activation of the synthase and glycogen phosphorylase because glycogen phosphorylase a, but not the glucose-glycogen phosphorylase a complex, is a potent inhibitor of the action of protein phosphatase-2 on glycogen synthase *b.* Glucagon seemingly can override the effects of insulin on glycogen turnover in the liver, provided that the glucose concentration is low.

Summary of Glucose Homeostasis

Let us pull all this together by recapitulating the events that occur when the blood glucose concentration rises and falls.

Excess Glucose

We eat and the blood glucose concentration rises. There are three important direct results :

1. The pancreatic islets put out less glucagon and more insulin. Less glucagon means less cyclic AMP formed in the liver, and less active protein kinase. With less active protein kinase, there is less formation of active glycogen phosphorylase and less inactivation of glycogen synthase. More insulin also results in more active glycogen synthase in both the liver and the muscles and more transport of glucose into the muscles.
2. Glucose combines with glycogen phosphorylase in the liver. This causes increased loss of active glycogen phosphorylase.
3. Increased blood glucose concentration results in the net formation of glucose 6-phosphate in the liver, since the liver glucokinase is not saturated with substrate and therefore can respond to changes in blood glucose concentration.

All of these things taken together mean an elevation in blood glucose concentration is itself a sufficient signal to cause further storage of glucose as glycogen in both the liver and muscles. The increased glycogen synthase activity in the muscles causes increased removal of glucose 6-phosphate : glucose 6-P$\leftrightarrow$glucose 1-P$\rightarrow$UDP-glucose$\rightarrow$glycogen. Since glucose 6-phosphate is an inhibitor of its own formation by the hexokinase reaction, its increased removal to form glycogen also promotes increased uptake of glucose in the muscle.

The deposition of glycogen in the muscles will proceed only until a characteristic level is reached even if elevated concentrations of glucose persist. However, the liver will continue to store more glycogen, although at an increasingly slower rate. We shall see in the following chapters that more and more glucose is converted to fat storage as the glycogen reserves begin to be filled.

Deprivation of Glucose

The blood concentration falls as more time elapses from the preceding meal. Here we have the opposite effects :

1. The pancreatic islets put out more glucagon and less insulin. More glucagon means more cyclic AMP formed in the liver and a more active protein kinase. The protein kinase causes the conversion of glycogen synthase to its inactive phosphorylated form, the conversion of glycogen phosphorylase to its active phosphorylated form, and the phosphorylation of protein inhibitor-1, which inactivates protein phosphatase-1.
2. The fall in the amount of glucose slows the hydrolysis of glycogen phosphorylase a because phosphorylase a without glucose is a poorer substrate for protein phosphatase-1 than is its glucose complex.
3. The low concentration of glucose results in a low rate of the liver glucokinase reaction so that liver adds glucose to the blood rather than removing it.

What are the results? The changes in enzyme activity in the liver stop the formation of glycogen and accelerate its degradation. The resultant rise in the concentration of glucose 6-phosphate will accelerate the hydrolysis of this compound to form free glucose, which will diffuse into the blood to prevent a further drop in concentration. The maintenance of the blood glucose concentration will enable the brain to continue using this fuel at its normal rate, and the supply for this purpose is conserved by the simultaneous slowing of glucose uptake and of glycogen formation in the muscles, owing to the declining insulin levels.

The Efficiency of Glycogen Storage

What is the price the organism pays for storage of excess glucose as glycogen? We can show it is only 3 percent of the potential for generating ATP. Let us put the question in this way : How much ATP will be generated per glucosyl group that disappears during combustion if the glucose has been stored as glycogen, compared to the amount generated when glucose is used directly without storage?

It requires two high-energy phosphates to store one glucose residue as glycogen according to the balance :

$$\text{glucose} + \text{ATP} \rightarrow \text{glucose 6-phosphate} + \text{ADP}$$
$$\text{glucose 6-phosphate} \rightarrow \text{glucose 1-phosphate}$$
$$\text{glucose 1-phosphate} + \text{UTP} \rightarrow \text{UDP-glucose} + \text{PP}_i$$
$$\text{UDP-glucose} + \text{glycogen} \rightarrow \text{glucosyl-glycogen} + \text{UDP}$$
$$\text{PP}_i + \text{H}_2\text{O} \rightarrow 2\text{P}_i$$
$$\text{UDP} + \text{ATP} \rightarrow \text{UTP} + \text{ADP}$$
$$\text{SUM : glucose} + 2\text{ATP} + \text{glycogen} + \text{H}_2\text{O} \rightarrow \text{glucosyl-glycogen} + 2(\text{ADP} + \text{P}_i)$$

Where is the high-energy phosphate to be obtained? It may be from combustion of any fuel, but we can approximate the net cost by assuming that glucose is the only fuel available, and that part of its supply must be burned immediately in order to store the remainder, and the complete combustion of one glucose molecule will generate 37.5 ATP in white fast-twitch fibers. From this, we can estimate that 5.3 percent of the glucose will be burned to store the remaining 94.7 percent as glycogen :

$$.947 \times 2\text{ATP consumed} \cong .053 \times \text{ATP produced}$$

When the stored glycogen is later used, approximately 95 percent of it will be converted to glucose 1-phosphate; the remaining 5 percent will be released as free glucose from hydrolysis of the 1, 6 branches.

Suppose that there were 100 millimole of glucose originally available. 94.7 millimoles were stored as glycogen, while 5.3 millimoles were burned to provide the energy for storage. When the 94.7 stored millimoles are recovered, 5 percent, or 4.7 millimoles, is released as free glucose, and 90 millimoles are obtained as glucose 1-phosphate. Oxidation of glucose 1-phosphate generates 36.5 ATP; therefore 90 × 36.5 = 3,285 millimoles of ATP will be formed.

Out of the original 100 millimoles of glucose, 4.7 millimoles reappeared as glucose during the breakdown of glycogen; therefore, only 95.3 have disappeared to generate 3,285 millimoles of ATP, for a yield of 34.5 ATP per glucose molecule disappearing. This compares with 35.5 that could be obtained by direct oxidation of glucose. The cost of glycogen storage is therefore only 3 percent of the potential ATP—a small price for an immediately available supply of rapidly combustible fuel. (This analysis neglects the cost of carrying the extra weight of the stored fuel.)

Efficiency of the Cori Cycle

A similar analysis can be made for the Cori cycle, in which part of the lactate made from glycogen in white muscle fibers is converted to glucose in the liver and returned to the muscles for re-storage as glycogen. This calculation is again a simplification of complex balances in which it is assumed that combustion of lactate by the liver provides the high-energy phosphate for gluconogenesis. It goes like this :

Skeletal muscles : 100 millimoles of glucose residues in glycogen yield 200 millimoles of lactate plus 300 millimoles of ATP.

Liver : 200 millimoles of lactate arrive in the blood from the muscles.

1. 31-3 millimoles are oxidized to CO_2 to generate 548 millimoles of ATP (17.5 millimoles per lactate, utilizing the malate-asparate electron shuttle).
2. The remaining 168.7 millimoles of lactate are converted to 84.4 millimoles of glucose, utilizing the 548 millimoles of ATP generated in step 1.

Skeletal muscles : 84.4 millimoles of glucose arrive in the blood from the liver.

1. 94.7 percent of the glucose or 79.9 millimoles can be stored as glycogen by burning the other 5.3 percent to provide the energy.
2. This replaces all but 20.1 millimoles of the original glucose residues used in the muscle; therefore the muscle gained the 300 millimoles of ATP used initially for contraction at the expense of 20.1 millimoles of glucose, for a yield of 14.9 ATP per glucose residue consumed.

This value is 40.9 percent of the 36.5 ATP generated by total oxidation in the muscle of a glucose residue in glycogen.

Genetic Defects in Glycogen Metabolism

A number of humans have been discovered who are deficient in one of the enzyme activities necessary for glycogen metabolism. The combined incidence of these defect is about 1 in 40,000 births. One case might arise per year in Los Angeles. However, they add to the catalogue of genetic defects that does demand the attention of every pediatrician. In addition, their effects have promoted our understanding of normal carbohydrate metabolism.

A defect may occur in the formation of any of the proteins necessary for either the breakdown or the synthesis of glycogen, so

Table 3.1. Glycogen storage diseases

Deficient Enzyme	Name	Type	Site of Deficiency	Incidence
glucose 6-phosphatase	von Gierke's disease	I	liver, kidneys	1 : 200,000
lysosomal α-glucosidase	Pompe's disease	II	all tissues	1 : 200,000
amylo-1. 6-glucosidase	limit dextrinosis	III	all tissues	1 : 200,000
branching enzyme	amylopectinosis	IV	all tissues	very low
glycogen phosphorylase	McArdle's disease	V	skeletal muscle	low
glycogen phosphorylase	Hers' disease	VI	liver	?
phosphofructokinase		VII	skeletal muscle, erythrocytes	very low
adenyl cyclase (?)		VIII	brain, liver	very low
phosphorylase-kinase		IX	liver, other tissues (not muscle)	1 : 100,000
cAMP-sensitive protein kinase		X	liver, muscles	very low

Note : Type numbers VI, VIII, and IX have been assigned to different deficiencies by some authors.

that the result could be either a high or a low content of glycogen with a normal structure, or the presence of glycogen with abnormalities in the degree of branching. A classification of the resultant clinical conditions in given in Table. However, it ought to be remembered that genetic errors can affect the formation of a protein in various ways; the rate of synthesis of normal protein may be changed, or mutations of coding may affect amino acid composition at any point in the molecule. The result is that there may be many types of deficiency of a given enzyme, some more severe in their consequences that others, so the rigid classification of the table does not adequately convey the actual spectrum of diseases.

A genetic defect may affect all tissues with an active glycogen metabolism. However, in some cases, a defective gene will disturb glycogen metabolism in only a few tissues because normal genes for different isozymes are still being expressed in other organs. Thus, defects in glycogen phosphorylase may be peculiar to the liver or to the skeletal muscles, because different isozymes are made in these organs.

Two decades ago it was believed that glycogen is synthesized by glycogen phosphorylase, because its reaction is readily reversed in the test tube. A young man was discovered who was weak and who developed severe pain in his muscles after modest exercise, and he was shown to have high concentrations of glycogen and an absence of glycogen phosphorylase in his muscles. This proved that he was not synthesizing glycogen by the phosphorylase reaction and provided strong supporting evidence for the then newly discovered UDP-glucose pathway. (The condition is known as *McArdle's disease,* after its discoverer. A psychiatrist engagingly confesses that he thought McArdle's patient was displaying classic hysteria owing to an unhappy childhood. We are not told how much the phosphorylase deficiency may have contributed to the unhappy childhood).

McArdle's disease tells us something else. Impairment of glycogen utilization in skeletal muscles is not fatal. The impairment is very real; there isn't any unsuspected route for utilizing the polysaccharide, and this is easily demonstrated by shutting off the blood supply to an arm with a tourniquet. Clenching of the first causes a sharp rise in lactate concentration in a normal individual, ending in painful tentany, but there is no rise in a patient with McArdle's disease. However, the importance of carbohydrate metabolism for full efficiency is shown by the weakness of the patient and is supported by the uncommon occurrence of the genetic lesin in the population—this is a mutation that is eliminated rapidly.

This view of glycogen metabolism as a convenience for full efficiency of muscles rather than as an absolute necessity is reinforced by the discovery of individuals with a deficiency of phosphofructo kinase in skeletal muscles, who have the same clinical picture found in MCArdle's disease. They not only accumulate glycogen because of an inability to use it, they can't even use glycose taken up from the blood stream directly and they still survive.

Accumulation of normal glycogen in the liver causes massive enlargements of the organ, so much so that it may occupy a large fraction of the abdomen in affected children. This in itself causes surprisingly little difficulty; infants without glycogen phosphorylase in the liver grow to be adults. However, if he failure is due to an absence of glucose-6-phosphatase, which prevents the use of stored glycogen to maintain the blood glucose concentration between meals, the consequences without treatment are grave. The blood glucose concentration falls, while the lactate concentration rises. The brain is usually damaged, perhaps owing in part to the absence of glucose as a fuel and in part to the lactic acidosis. The treatment is repeated feeding of carbohydrate at two to three hour intervals night and day. If the infant can be brought through the first four years, his chances for gradual adjustment to a less heroic feeding schedule are good.

Defects in phosphorylase kinase are interesting because two patterns of inheritance are known. One is an autosomal recessive and the other is an X-linked recessive. If we had only this information, we could infer that the kinase contains at least two polypeptide chains coded by genes in different chromosomes, and we have already seen that the kinase contains some three different chains. In addition, another kind of phosphorylase kinase deficiency has been discovered that affects both the liver and the muscles, indicating that these isozymes have some common subunit.

One of the most damaging storage diseased results from accumulation of glycogen by lysosomes. The enzyme affected is an α-glucosidase that attacks either 1, 4 or 1,6 linkages and is not involved in the major routes of glycogen metabolism. It probably is a device for removing glycogen trapped during the normal scavenging function of lysosomes, which become filled to be bursting point when the enzyme is missing. All tissues are affected, but damage to the heart is usually the immediate cause of death, which occurs in infancy.

4

CELLULAR RESPIRATION

Respiration is one of the most fundamental phenomenon of life and indeed life itself is inseparable from respiration in some forms. Respiration involves the oxidative breakdown of certain organic substrates present in the cells. Respiration, therefore, is not the concern of special organs or tissues, it is an integral part of the life of every cell. Cellular respiration is the aerobic oxidation of carbohydrates in the cell with the help of atmospheric oxygen, in the course of which energy is liberated and carbon dioxide is given out. In glycolysis and fermentation, glucose is oxidized in the absence of oxygen, i.e. *anaerobically*. Organisms that live in the presence of oxygen are aerobes, those that live in its absence are anaerobes.

The first series of reactions in oxidative break-down of glucose is called the *Embden-Meyerhof (E-M) pathway*. This was described by two German biochemists *Embden* and *Mayerhof*. The term *glycolysis* is sometimes used as a synonym for the E-M pathway, but biochemists usually apply the term glycolysis to the E-M pathway plus a form of fermentation, which together lead from glycolysis to lactic acid. The E-M pathway is a several-step process that occurs in the cytoplasmic matrix of cell. It can proceed in either the presence or absence of oxygen but makes available only a small portion of the potential energy that could be expected from glucose.

I. The E-M Pathway

From six-carbon glucose molecule the E-M pathway starts and the formation of 2-three-carbon molecules of pyruvic acid is the final step of this pathway. Following steps take place in the entire pathway.

1. *Phosphorylation.* In the first step glucose reacts with adenosine triphosphate (ATP) to form glucose-6-phosphate in the presence of *hexokinase* and Mg^{++} ions.

$$\text{Glucose + ATP} \xrightarrow[\text{Mg}^{++}]{\text{hexokinase}} \text{Glucose-6-phosphate + ADP.}$$

This reaction is exergonic in nature so that it is irreversible.

2. *Isomerization.* Glucose-6-PO_4, in the presence of the enzyme *phospho-hexose-isomerase,* is isomerized into fructose-6-PO_4.

$$\text{Glucose-6-phosphate} \xrightarrow{\text{Phosphohexose -isomerase}} \text{Fructose-6-phosphate.}$$

3. *Phosphorylation.* Fructose-6-PO_4 reacts with one molecule of ATP in the presence of Mg^{++} ions and *phosphofructokinase* to form fructose-1, 6-diphosphate. This reaction is also irreversible due to its strongly exergonic nature.

$$\text{Fructose-6-phosphate + ATP} \xrightarrow[\text{Mg}^{++}]{\text{Phospho fructokinase}} \text{Fructose-1-6-disphosphate.}$$

4. *Cleavage.* Fructose-1, 6-diphosphosphate is cleaved under the influence of an enzyme *aldolase* with the formation of dihydroxy acetone phosphate or phosphodihydroxy acetone (DHAP) and 3-phosphoglyceraldehyde or glyceraldehyde-3-phosphate (PGAL). Both the products are triose phosphates. The two trioses can be reversibly changed into one another by enzyme *Triosphosphate isomerase.*

$$\text{Fructose-1-6-diphosphate} \xrightarrow{\text{Aldolase}} \left[\begin{array}{l}\text{Dihydroxyacetone phosphate (DHAP)} \\ \text{Triose phosphate isomerase.} \\ \text{3-phosphoglyceraldehyde (PGAL)}\end{array}\right.$$

5. *Phosphorylation and oxidative dehydrogenation.* This is the first reaction in the energy-yielding phase of the E-M pathway. 3-phosphoglyceraldehyde (PGAL) reacts with phosphoric acid (H_3PO_4) to form 1-3-diphosphoglyceraldehyde in the presence of enzyme *glyceraldehyde-3-phosphate dehydrogenase.*

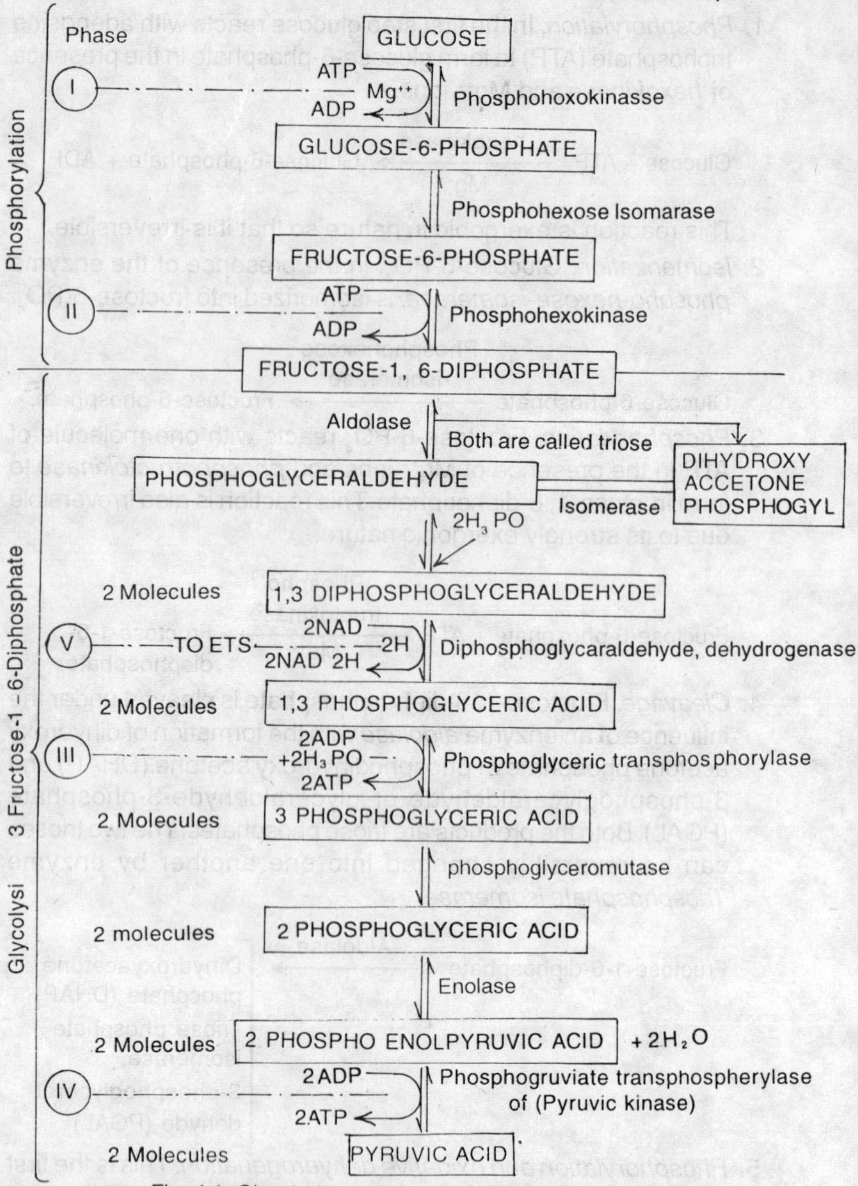

Fig. 4.1. Glycolysis of Embeden Meyerhof Pathway.

3-phosphoglyceraldehyde + $H_3PO_4 \rightarrow$ 1, 3-diphosphoglyceraldehyde.

(2 molecule) (2 mole.) (2 mole.)

Coenzyme nicotinamide adenine dinucleotide (NAD) reacts with 1-3-diphosphoglyceraldehyde removing 2 electrons and 2 protons (the equivalent of 2 H atoms) forming 1, 3-diphosphoglyceric acid.

$$\underset{(2\text{ mole})}{\text{1, 3-diphosphoglyceraldehyde} + 2NAD^{+}} \xrightarrow[]{\text{Glyceraldehyde-3-phosphate dehydrogenase}} \underset{(2\text{ mole})}{\text{1, 3-diphosphoglyceric acid} + 2\ NADH + H}$$

6. *ATP generation.* The high-energy phosphate group in 1-3-diphosphoglyceric acid is transferred to ADP resulting in the formation of 3-phosphoglyceric acid (3 PGA) and ATP. This reaction takes place in the presence of enzyme *phosphoglycerate kinase* and Mg^{++} ions.

$$\underset{(2\text{ mole.})}{\text{1, 3-diphosphoglyceric acid} + 2\ ADP} \xrightarrow[Mg^{++}]{\text{phosphoglycerate kinase}} \underset{(2\text{ mole})}{\text{3-phosphoglyceric acid} + 2ATP}$$

7. *Isomerization.* 3-phosphoglyceric acid (3-PGA) is isomerized into 2-phosphoglyceric acid (2-PGA) with the help of enzyme *phosphoglyceromutase.* This enzyme catalyse the transfer of a phosphate group from one carbon to another carbon of the same compound.

$$\underset{(2\text{ mole.})}{\text{3-phosphoglyceric acid}} \xrightarrow[]{\text{Phosphoglycero-mutase}} \underset{(2\text{ mole.})}{\text{2-phosphoglyceric acid}}$$

8. *Dehydration of 2-PGA.* The 2-phosphoglyceric acid molecule is dehydrated by the action of enzyme *enolase* to yield phosphoenol pyruvic acid (PEPA). Presence of Mg^{++} ions is also necessary.

$$\underset{(2\text{ mole.})}{\text{2-phosphoglyceric acid}} \xrightarrow[Mg^{++}]{\text{Enolase}} \underset{(2\text{ mole.})}{\text{2-phosphoenol pyruvic acid} + 2H_2O}$$

9. *ATP generation.* In the final step of E-M pathway the high energy phosphate group in 2-phosphoenol-pyruvic acid (PEPA) is transferred to ADP in the presence of *pyruvate kinase* and Mg^{++} ions with the formation of ATP and pyurvic acid.

$$\text{2-phosphoenolypyruvic acid + 2ADP (2 mole.)} \xrightarrow[\text{Mg}^{++}]{\text{Phruvative kinase}} \text{Pyruvic acid + 2ATP (2 mole.)}$$

Energy Yield of the E-M Reactions

How much energy or ATP has the cells gained after all these reactions? That depends upon whether the reactions have taken place under anaerobic or aerobic conditions. In general, the steps of energy production in E-M pathway are (1) during conversion of 1,3-diphosphoglyceric acid into 3-phosphoglyceric acid and (2) during conversion of 2-phosphoenol pyruvic acid into pyruvic acid. If ADP is present in these steps, energy of energyrich compounds is captured with the formation of ATP. One molecule of ATP at each of the above two steps in synthesized per molecule of 1, 3-diphosphoglyceric acid or 2-phosphoenolpyruvic acid utilized for forward reaction. Thus, in all, 4 molecules of ATP are formed at these two steps per molecules of glucose used under anaerobic conditions.

Besides these two steps, these exists one more site of energy production which comes into play under aerobic conditions. This step takes place during conversion of 3-phosphoglyceraldehyde into 1, 3-diphosphoglyceric acid when one molecule of reduced NAD, is oxidized with the formation of 1, 3-diphosphoglyceric acid.

In order to evaluate the net gain of energy the energy expenditure has also got to be taken into account and must be substracted from total energy production. Under anaerobic conditions 2ATP molecules are used up during reactions and there is a net gain of only 2ATP molecules.

II. Oxidative Decarboxylation

Pyruvic acid, when enters the mitochondria does not itself enter Kreb's cycle reactions. Entry into the Kreb's cycle requires preparation of a 2-carbon fragment called *acetyl group*. This takes place in several steps. In the first step there is a decarboxylation (a carbon dioxide molecule is released), that shortens the carbon chain by one carbon. A molecule of NAD is also reduced. The remaining 2-carbon fragment i.e., acetyl group is attached to a carrier molecule, *coenzyme A*. This complex, made up of two carbon fragment and coenzyme A is called *acetyl CoA*. The acetyl CoA is in an active form ready to be fed into Kreb's cycle.

$$\underset{\text{(Pyruvic acid)}}{CH_3\ COCOOH} + \underset{\text{(CoA)}}{CoASH} + NAD \xrightarrow[\substack{\text{Thymine pyrophosphate} \\ \text{Lipoic acid amide}}]{\substack{\text{Pyruvic dehydrogenase} \\ \text{multienzyme complex}}} \underset{\text{(acetyl-S-CoA)}}{CH_3COS\ CoA} + NADH + H^+ + CO_2$$

III. Krebs cycle or Tricarboxylic Acid Cycle (TCA)

The krebs cycle completes the oxidation of carbohydrates and supplies electrons to be passed through electron transport system. Enzymes catalyzing Kreb's cycle reactions are located in the mitochondrial matrix.

The cyclical nature of the reactions by which further oxidations beyond pyruvic acid occur was established after a series of experimental analyses by *Albert Szent-Gyorgyi, Hans Krebs* and others. *Krebs* was awarded Nobel Prize in 1953 for this work.

The Kreb's cycle is a *cycle* because it is series of reaction events, each catalyzed by a specific enzyme, that ends up back at its original point. The reactions involve a series of organic acids with different number of carbon atoms.

(i) *Formation of citric acid.* Acetyl Co-A, formed from pyruvic acid condenses with oxaloacetic acid forming citric acid. This reaction is catalyzed by enzyme *citrate synthatase* or condensing enzyme.

$$CH_3CO\text{-}S\text{-}CoA + Oxalicacid + H_2O \xrightarrow[\text{CoASH}]{\substack{\text{Citrate} \\ \text{synthetase}}} \text{Citric acid} + CoASH$$

(ii) *Dehydration.* Citric acid is converted into *cis*-aconitic acid by the enzyme *aconitase* with release of one molecule of water.

$$\text{Citric acid} \xrightarrow{\text{Aconitase}} \textit{Cis}\text{-aconitic acid} + H_2O$$

(iii) *Hydration I.* By hydration of *Cis*-aconitic acid, isocitric acid is formed. This reaction is catalyzed by enzyme *aconitase.*

$$\textit{Cis}\text{-aconitic acid} + H_2O \xrightarrow{\text{Aconitase}} \text{Isocitric acid.}$$

(iv) *Dehydrogenation I.* Isocitric acid is dehydrogenated to form oxalo succinic acid in the presence of *isocitric dehydrogenase* and one pair of hydrogen atoms removed is accepted by NAD+ or

NADP+ which forms NADH + H+ or NADPH + H+ (Mitochondria have been found to contain two types of isocitric dehydrogenases, one is NAD specific and other i NADP specific).

$$\text{Isocitric acid + NAD+ or NADP} \xrightarrow{\text{Isocitric dehydrogenase}} \text{Oxalosuccinic acid + NADH + H}^+ \text{ or NADPH + H}^+$$

(v) *Decarboxylation I.* In the presence of enzyme *carboxylase*, the oxalosuccinic acid undergoes decarboxylation to form a-Ketoglutaric acid. One CO_2 is removed. Thus a-Ketoglutaric acid is 5-carbon molecule.

$$\text{Oxalosuccinic acid} \xrightarrow{\text{Carboxylase}} \alpha\text{-Ketoglutaric acid} + CO_2$$

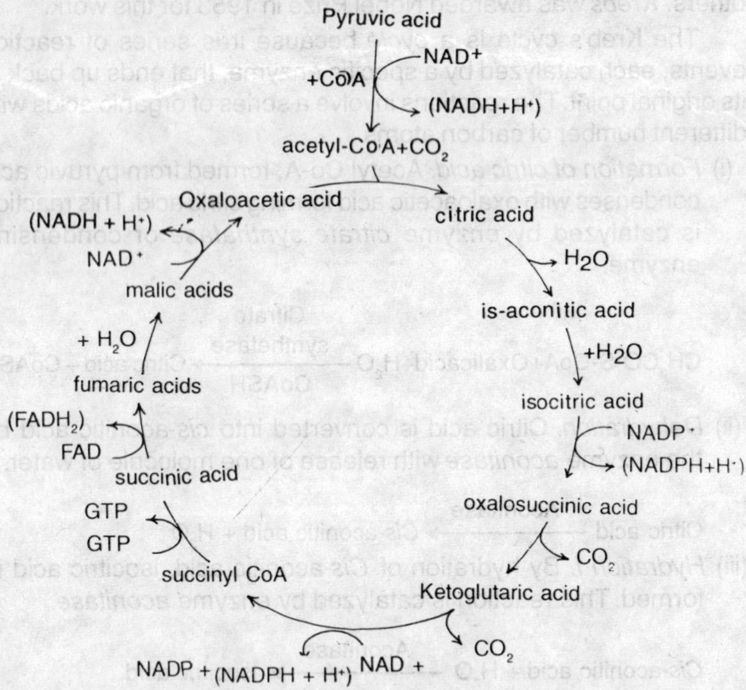

Fig. 4.2. The Krebs cycle.

(vi) *Dehydration II and decarboxylation II.* a-Ketoglutaric acid undergoes simultaneous decarboxylation and dehydration and joins with coenzyme A to form succinyl Co-A, a 4-carbon atom derivative of coenzyme A. This reaction is catalyzed by enzyme a-*Ketoglutaric dehydrogenase* in the presence of thiamine pyrophosphate (TPP) and lopoic acid. The hydrogen atoms released are accepted by NAD^+ to form NADH + H^+

$$\alpha\text{-Ketoglutaric acid+Co-ASH+NAD}^+ \xrightarrow[\text{dehydrogenase}]{\alpha\text{-Ketoglutaric}} \text{Succinyl CoA} + CO_2 + \text{NADH} + H^+$$

(vii) *Phosphorylation.* Succinyl CoA on hydrolysis yields to the formation of succinic acid and CoA. Succinyl CoA possesses a high-energy thiol group. In the presence of guanosine diphosphate (GDP) and inorganic phosphate the energy of thiol group is transferred to GDP with the formation of guanosine triphosphate (GTP). This reaction is catalyzed by enzyme *succinic thiokinase.*

$$\text{Succinyl- CoA + GDP} + H_3PO_4 \xrightarrow{\text{Succinic thiokinase}} \text{Succinic acid + GTP + CoASH.}$$

(vii) *Dehydrogenation III.* Succinic acid undergoes dehydrogenation under the catalytic action of *succinic dehydrogenase* to form fumaric acid. This dehydrogenase contains flavin adenine dinucleotide (FAD^+) as its prosthetic group which accepts the pair of hydrogen atoms removed from succinic acid.

$$\text{Succinic acid} + FAD^+ \xrightarrow{\text{Succinic dehydrogenase}} \text{Fumaric acid + FADH} + H^+$$

(ix) *Hydration II.* Fumaric acid is hydrated to form malic acid. This reaction is catalyzed by the enzyme *fumarase.*

$$\text{Fumaric acid} + H_2O \xrightarrow{\text{Fumarase}} \text{Malic acid.}$$

(x) *Dehydrogenation IV.* This is the final reaction. The Kreb's cycle. The malic acid is transferred to oxaloacetic acid in the presence of *malate dehydrogenase* NAD^+ accepts hydrogen atoms to form NADH + H^+.

$$\text{Malic acid} + NAD^+ \xrightarrow{\text{malate dehydrogenase}} \text{Oxaloacetic acid + NADH} + H^+$$

It should be noted that two molecules of pyruvic acid are formed per molecule of glucose during E-M pathway. Therefore, two molecules of oxaloacetic acid per molecule of glucose are formed at the end of Kreb's cycle. In the figure only one pyruvic acid molecule fate has been traced.

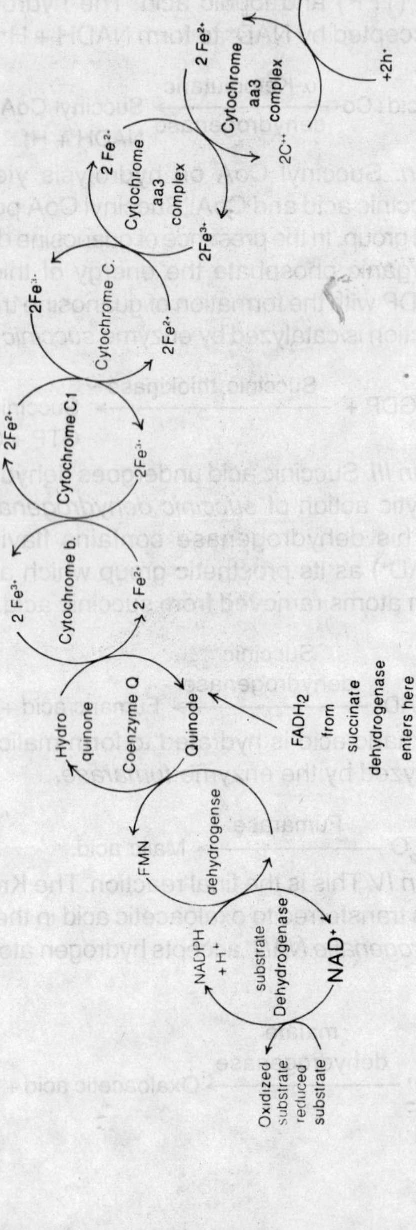

Fig. 4.3. Electron Transport system enzymes.

During transfer of hydrogen atoms from one enzyme carries in electron transport system, large amount of energy is released (as shown in Table). The enzymes which participate in the process are called respiratory chain enzymes or electron transport system enzymes. All such enzymes are found in the inner membrane of mitochondria. These are arranged in the following sequence—succinic dehydrogenase which remains associated with coenzyme NAD, flavoprotein (FAD), non-haern ion protein or iron-sulphur protein, flavoprotein (FAD or FP), cytochrome b, ubiquinone or Co Q, cytochrome c_1, cytochrome c, cytochrome a cytochrome a_3 and three coupling sites where phosphorlylation coupled with oxidation leads to the production of ATP (*Hall et. al.* 1974).

The respiratory chain system manufactures the ATP molecules during oxidation of reduced coenzymes in the following steps:

1. *Dehyderogenase* enzymes accept hydrogen from different substrates of Krebs' cycle. The hydrogen atom becomes ionized into protons ($2H^+$) and electrons (2e).

   Protons or hydrogen ions reduce the oxidized NAD

   $NAD^+ + 2H^+ \longrightarrow NADH + H^+$

   The NADH molecule passes through M-side of inner mitochondrial membrane and enters the respiratory chain system.
2. The NADH is oxidized into NAD^+ by transferring its proton (H^+) to flavoprotein FAD. The FAD acts as a hydrogen carrier. From the flavoproteins each proton is discharged into perimitochondrial space through the C-side of inner mitochondrial membrane and electrons are passed onto non-haem iron protein.
3. In normal course, reduced FADH is reoxidized by ubiquinone (Co Q) which is the next member of ETS. The ubiquinone is reduced to dihydroubiquinone which is oxidized by ferric (Fe^{++}) cytochrome—C—(Fe^{+++}) which is reduced to ferrocytochrome. At this level of oxidation, hydrogen atoms undergo ionization to form $2H^+$ and $2e^-$. The protons cannot be further carried by ETS, however, the electrons can be carried easily by the cytochromes. One molecule of each cytochrome is capable of carrying only one electron at a time. Thus, for each molecule of hydrogen which releases two electrons, two molecules of each cytochrome are required. At the level of cytochrome-b, two protons per molecule of hydrogen are released in the mitochondrial system. Ferrocytochrome b is oxidized in the presence of ferric cytochrome c_1. From ferrocytochrome c_1 these electrons are

taken by cytochrome a and than to a_3 (Both cyto-a and a_3 together called cytochrome oxidase.

4. These electrons from cytochrome oxidase react with one atom of oxygen making it reactive.
5. This activated oxygen react with $2H^+$ (protons) forming molecule of water.

$$O + 2e \longrightarrow O^- \xrightarrow{2H^+} H_2O$$

With the release of 2 electrons and 2 H atoms from every molecule of NADH + H^+ and their reaction with oxygen, 3ATP molecules are synthesized. The position of ATP molecules. Energy released at these positions reach and stored in ATP molecules. The process is known as oxidative phosphorylation because it occurs in respiration in the presence of oxygen hence it is called oxidative phosphorylation.

Now we can summarize the energy retum from the complete oxidation of a glucose molecule under aerobic conditions. Under aerobic conditions, the NADH produced in the E-M pathway can transfer electons to the ETS inside mitochondria, these by salvaging more usable energy than is possible under anaerobic conditions.

However, reduced NAD cannot enter a mitochondrion directly, so it cannot pass its electrons directly to an ETS in the inner mitochondrial membrane. Instead, the electrons are passed to a shuttal carrier that takes them into the mitochondrion and donates them to an electron transport system.

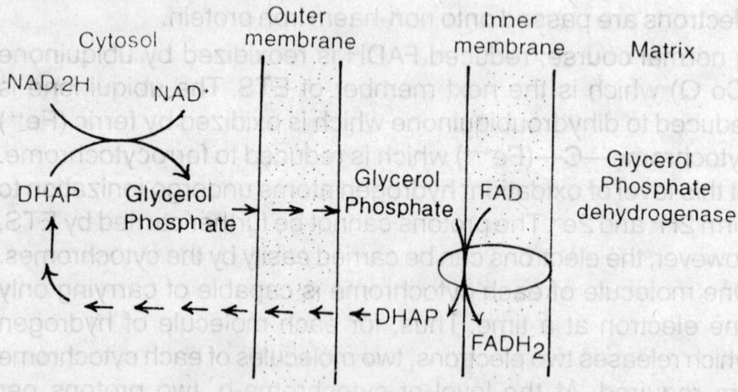

Fig. 4.4. Glycerol-phosphate shuttle.

These are several such electron shuttles for example malate-asperate shuttle, glycerol phosphate shuttle etc., but the common one is *glycerol-phosphate shuttle,* which picks up electrons from reduced NAD, carries them into mitochondrion and donates them not to NAD, but to ubiquinone. In this case the electrons do not start at the 'top' of the ETS, so the possibility for one phosphorylation is bypassed. If mitochondria use this particular mechanism, a reduced NAD in the cytoplasm ends up being worth only 2ATP molecules. In this case complete oxidation of glucose will result in a gain of 36 ATP molecules.

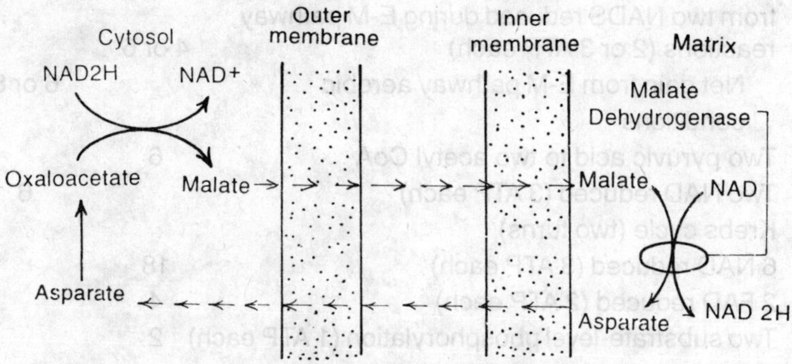

Fig. 4.5. Malate-aspartate shuttle.

In *malate-asperate shuttle,* the electrons are transferred from NADH in cytoplasm to malate which traverses the inner mitochondrial membrane and reoxidized to form NADH + H^+ thus resulting in the formation of oxaloacetate. Oxaloacetate does not readily cross the inner mitochondrial membrane 20 a transmission reaction is needed to form asperate which does traverse this barrier. 3ATP molecules are generated for each pair of electrons. In this case complete oxidation of glucose will result in a gain of 38 ATP molecules.

Oxidative Phosphorylation Mechanism

Oxidative phosphorylation was discovered in 1939 but the mechanism of coupling between ETS and ATP synthesis is not yet known. However, three principal hypothesis have been proposed for the coupling of oxidation and phosphorylation. These are:

1. Chemical Coupling hypothesis.
2. Cheiniosomotic Coupling hypothesis.
3. Conformational hypothesis.

Table 4.1. Showing ATP Formation During Oxidation of Glucose.

Embden-Meyerhof Pathway	*ATP calculations*	*Sum*
ATP expanded to phosphorylate sugar molecule	– 2	
Substrate-level phosphorylation	4	
ATP formed as a result of electron transport from two NADS reduced during E-M pathway reactions (2 or 3 ATP each)	4 or 6	
Net gain from E-M pathway aerobic conditions		6 or 8
Two pyruvic acid to two acetyl CoA	6	
Two NAD reduced (3 ATP each)		6
Krebs cycle (two turns)		
6 NAD reduced (3 ATP each)	18	
2 FAD reduced (2 ATP each)	4	
Two substrate-level phosphorylation (1 ATP each)	2	
Total from Kreb's cycle (2 turns)		24
Total ATP formation per molecule of glucose		36 or 38

Table 13.2. Free-Energy Change at Different Steps in the Electron Transport Chain.

Steps of Transport	*Difference in Redox-potential (ΔE^0), (Volt)*	*Free energy change (Kcals)*
NAD $\xrightarrow[2e^-]{2H^+}$ FAD	+0.22	–10.1
FAD′ $\xrightarrow[2e^-]{}$ cyt.b	+0.15	–6.9
2Cyt.b $\xrightarrow[2e^-]{}$ 2 cyt.c	+0.19	–8.77
2 Cyt.c $\xrightarrow[2e^-]{}$ 2 cyt.a	+0.05	–2.30
2 Cyt.a $\longrightarrow$ Oxygen	+0.52	–24.00
Total	1.13 Volts	52.07 Kcals.

1. Chemical Coupling Hypothesis

According to this hypothesis the ET is coupled to ATP synthesis by a sequence of consecutive reactions in which a high-energy covalent intermediate is formed by electron transport and subsequently is cleaved and donates its energy to ADP to form ATP. For example, when an oxidoreduction reaction occurs between A^{red} and b^{oxi}, the factor I is incorporated into the formation of an energy rich $A^{oxi} \sim I$ complex (here ~ represents a linkage having an energy-rich nature.)

$$A^{red} + I + B^{oxi} \rightleftharpoons A^{oxi} \sim I + B^{red}$$

In the following reaction an enzyme (E) replaces A^{oxi} from the above complex forming an energy rich E ~ I complex. Later on inorganic phosphate (P_i) reacts with E ~ I complex, forming phosphoenzyme complex (E ~ P). The phosphoenzyme complex reacts with ADP to form ATP liberating enzyme free.

$$A^{oxi} \sim I + E \rightleftharpoons A^{oxi} + E \sim I$$

$$E \sim I + P_i \rightleftharpoons E \sim P + I$$

$$E \sim P + ADP \rightleftharpoons ATP + E$$

An extensive search has been made for the nature of E ~ I and E ~ P, but non have been found in mitochondria. The chemical coupling hypothesis is only of historical importance today.

2. Chemiosmotic Coupling Hypothesis

This hypothesis was proposed by *Peter Mitchell* in 1961. He proposed that electron transport and ATP synthesis are coupled by a proton gradient, rather, than by a covalent high-energy intermediate or an activated protein. This theory suggests that a transmembrane electrochemical gradient is critically important to the phosphorylation process.

Chemiosmotic coupling hypothesis suggests that during electron transport, a gradient of H^+ ions is produced across the inner mitochondrial membrane. The mobile electron carrier that gains H^+ ion is reduced one side of the membrane and loses proton as it is oxidized near the other side. These reactions lead to the transport of a pair of H^+ across the inner mitochondrial membrane from the matrix to the space between the mitochondrial membrane. Thus there is a H^+ ion gradient across the membrane. There is a strong tendency for these H^+ ions to move back toward the matrix because of the excess of +ve charge is the space between the two membrane.

Mitchell further suggests that there are special channels through which the H^+ ions can move in response to the forces acting on them (electrical as well as concentration differences). In same way, the movement of H^+ ions back through the membrane is coupled in ATP synthetase complex to the phosphorylation of ADP to form ATP.

Oxidation-reduction loop

Mitchell has proposed that reducing equivalents are transferred as H atoms by some of the electron carriers (ubiquinone) and as electrons by other (Fe-S center and cytochromes). According to him hydrogen carrying and electron carrying proteins alternate in the respiratory chain to form three functional loops called *oxidation-reduction loops* (O/γ loops).

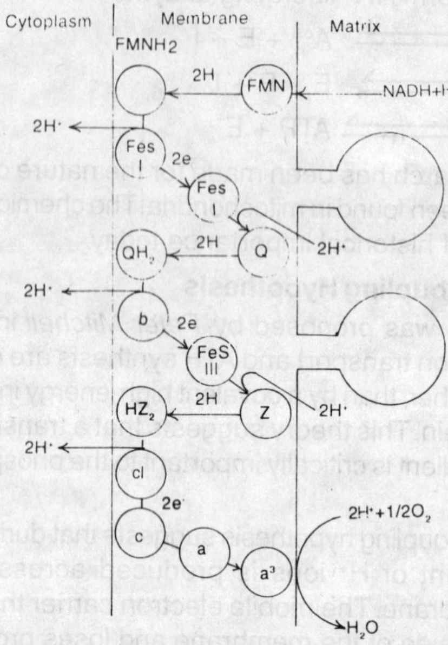

Fig. 4.6. Showing Michell's chemiosmolic coupling hypothesis.

A single loop consists of a hydrogen carrier and an electron carrier. In each loop two H^+ are carried outward through the inner membrane and delivers two H^+ to the C side, the corresponding pair of electron is then carried back from outer to the inner surface of the membrane. Thus, each pair of reducing equivalents passing through such loop carries $2H^+$ from the matrix to outside. Each loop is thought to provide the osmotic energy to form one molecule of ATP.

Mitchell's hypothesis is supported by a wealth of evidences.

3. Conformation Coupling Hypothesis

According to this proposed by *Boyer* (1965), the electron transport and phosphorylating systems are in molecular contact and that information is transmitted by short-arange interactions which may, in part, be electrostatic in nature. It is postulated that upon acceptance of the two electrons, each electron transfer protein undergoes a conformational change which is then transmitted to the ATPase by short range interactions. However, the mode of conformational changes that take place in the inner mitochondrial membrane is not yet clearly understood.

According to some recent views (*Ernster,* 1977) the conformational and chemiosmotic mechanisms of electron transport linkage to phosphorylation do not represent mutually exclusive alternative but may indeed be two aspects of the same mechanism.

5

Bioenergetics and Cellular Homeostasis

Bioenergetics is the field of biochemistry deals with the transformation and use of energy by living cells. Cells constantly manufacture new substances, they perform the mechanical work of motion, they transport substances and they generate heat. All these activities are controlled by the enzymes. Enzymes speed up the chemical reactions to certain extent which is enough, but not too much, of the products of the reaction. The ability of enzymes to regulate chemical reactions provide the cells with a homeostatic system i.e., one that maintains proper functioning of cell inspite of changing environment and need.

Concept of Energy

The term *energy* refers to the capacity for acting or to perform work. It is one of the fundamental components of any system. Physicists and chemists recognize many kinds of energy such as *thermal energy,* in which a hot system can be made to do work as in a steam-engine; *gravitational energy,* in which work is obtained from an object as it falls from high place to a low one; *electric energy, light energy* and so on. These different forms of energy are interconvertable and during conversion of one form of energy to another, there is always some loss. Moreover, when a work is done, there is also a loss of useful energy. The kind of energy in which we are most interested here, however, is called *Gibbs free energy* or *free energy* in short.

Thermodynamic Principles

Many quantitative observations made by physicists and chemists on the interconversion of different forms of energy have led to two fundamental laws of thermodynamics. These laws helps in understanding the direction of a reaction, the accomplishment of work and the source of energy.

I. The first law of thermodynamics or Principla of Conservation of Energy

According to this law, "in any physical or chemical change, the total amount of energy in the universe remains constant." Thus according to this law the energy cannot be created nor destroyed, whenever energy is used to do work or is transformed from one kind to another the total amount of energy is unchanged.

This law can be mathematically represented as:

$$\Delta E = E_B - E_A = Q - W \tag{1}$$

where ΔE is change in internal energy, ΔE_A is the energy of a system at the start of work, ΔE_B is energy of a system at the end of work, Q is heat absorbed by the system and W is work done by the system.

II. The second Law

According to this law, "All physical and chemical changes tend to proceed in such a direction that useful energy undergoes irreversible degradation into a randomized form called *entropy.* They come to a stop at an equilibrium point, at which the entropy formed is the maximum possible under the existing conditions."

This simple, somewhat abstract defination requires some explanation of free energy, heat energy and entropy.

The *free energy* is the kind of energy that can do work at constant temperature and pressure.

The *heat energy* can do work only through a change of temperature.

Entropy is the energy in a state of randomness or disorder. It is unavailable useless energy.

A quantitative rigorous defination of entropy requires mathematical statements of randomness.

There is another aspect of second law which is necessary to understand its operation particularly in biological system. Firstly, it is necessary to identify the reaction system, the collection of matter that is undergoing a particular physical or chemical process (this may be a call, an animal or two reacting components). Secondly, the surrounding with which the reacting systems can exchange energy must be defined. The system plus the surroundings together constitute the universe. Universe literally includes the entire earth rather even the outer space. Some physical or chemical process can be made to take place in isolated systems, which are unable to exchange energy with their surroundings. But in the biological world, reacting system do exchange energy, and often they exchange matter with their surroundings.

Changes in free energy, heat energy and entropy in chemical reactions at constant temperature and pressure, the conditions existing in biological systems, are related to each other quantitatively by following equation

$$\Delta G = \Delta H - T\Delta S \quad (2)$$

here ΔG is the change in free energy of a reacting system, ΔH is the change in its heat content or *enthalpy* of this system, T is the absolute temperature at which the process is taking place and ΔS is change in the entropy of this system or universe.

When any chemical reaction proceeds to its equilibrium point, the entropy of universe always increases and ΔS, therefore, always has a positive sign in the biological world. The minus sign indicates release of useful energy. The Δ is equation by definition means final state minus initial state. For example, $\Delta H = H_2 - H_1$, in the transition form a heat content of H_1 to a heat content of H_2. Hence, a system that releases heat will have ΔH i.e. $\Delta O < O$. Such reactions are called *exothermic*. The exothermic reactions are characterized by a warming of their environment at heat is liberated from reacting system as heat leaves the reacting system. On the other hand, reactions that cool their surrounding do so because they are absorbing heat energy. Such reactions are called *endothermic* ($\Delta H > O$). If the ΔG is zero, the reaction system is at equilibrium and no net change takes place.

With regards to the free energy change, ΔG of a reacting system, the points need to be emphasized are:

(i) The ΔG of a reaction system depends only on the free energy of the products minus that of the reactants. Obviously, the mechanism of a reaction has no effect on ΔG. For example, the value of ΔG is the same for the oxidation of glucose to carbon dioxide and water whether it takes place by enzymatic-catalytic reactions or combination.

(ii) The value of ΔH does not provide any information about the rate of a reaction. A–ΔG indicates that a reaction can occurs spontaneously. As already pointed out, that the rate of reaction rather depends on the free energy of activation ($\Delta G \neq$), which is unrelated to ΔG.

We must emphasize that the formation of entropy or disorder, is not a totally useless activity. The increase in entropy of the universe during biological process provides their driving force and gives direction to all biological activities. Living systems constantly produce entropy in their surrounding as a necessary cost of maintaining their own internal order.

For example

$$C_6H_{12}O_6 + 6O_2 \rightarrow 6CO_2 + 6H_2O$$

if above reaction taking place at temperature 25°C or 298 Kelvin and at a pressure 760 mm Hg (1.0 atm.) which are standard conditions in thermodynamic calculations, the following energy changes take place per molecule of glucose oxidized.

$\Delta G = -\ 686{,}000$ cal/mol (Free energy of the system)

$\Delta H = -\ 673{,}000$ cal/mol (the reacting molecules have released heat).

$$\Delta S = \frac{\Delta H - \Delta G}{T} \qquad (3)$$

$$= \frac{-673000 - (-686000)}{298}$$

$= +\,44$ cal/deg (The entropy of the universe has increased).

Relationship between standard free energy change and equilibrium constant.

Each chemical reaction has a characteristic standard free energy change ΔG^0. The ΔG^0 is not the same as ΔG given in equation. When a reacting system has achieved its state of lowest free energy, the system is said to be in *equilibrium.* At that pont free energy change for the reaction has declined to zero ($\Delta G = O$). In other words, if the reaction is started with all participants already at their equilibrium concentration, the reaction goes no-where and cannot be made to do work. To apply this concept to chemical reactions, consider the isomerization.

$$A \rightleftharpoons B$$

$$\alpha\text{-D glucose} \rightleftharpoons \beta\text{-D glucose.}$$

Let us give the name state A to a situation in which the chemical bond is intact and state B is the situation where the same bond is broken. It we examine this system when it has reached equilibrium at any absolute temperature (T) the ratio of broke to intact bond will be given by the equation.

$$\Delta G^0 = -2.303\ RT \log \frac{(B)_{eg}}{(A)_{eg}}$$

This expression is a form of *Boltzmann equation.* Here R is universal gas constant (8.314) Joules/mol/degree kelvin), 2.303 allows us to express the equation in term of common (base 10) logarithms instead of natural (base-e) logarithms and, eg is remind us the equation is valied only at equilibrium. ΔG^0 is standard state free energy change which indicates where the equilibrium position of reaction will be.

The ratio $(B)_{eg}/(A)_{eg}$ is an equilibrium constant designated as K_{eg}. The above reaction can be now written as

$$\Delta G^0 = -2.303\ RT \log K_{eg}$$

In replacing 2.303 RT with 5936, the temperature was assumed to be that of the human body (37°C or 310 K). ΔG^0 in above equation is expressed in the term of Joules per molecule.

It is clear that by knowing ΔG^0 should permit one to K_{eg} or *vice-versa*. The importance of ΔG^0 is two-fold, however, for it also permits one to calculate the amount of useful work (Gibbs free energy change) that can be obtained from a system as it moves toward the equilibrium.

$$\Delta G + \Delta G^0 + 2.303\ RT \log \frac{(B)}{(A)}$$

ΔG^0 has Characteristic Values for Different Chemical Reactions

$$\text{Glucose-1-phosphate} \xrightarrow{\text{Phosphogluco-mutase}} \text{Glucose-6-phosphate.}$$

The chemical analysis shows that if the reaction starts with, say 0.0200 M glucose-1-phosphate and allow to go in forward direction or if we start with 0.0200 M glucose-6-phosphate to go in the reverse direction, the final equilibrium mixture in either case will contain 0.00100 M glucose-1-phosphate and 0.019M glucose-6-phosphate at 25°C and *pH* 7.0. From these datas we can than calculate the equilibrium constant to three significant figures

$$K_{eg} = \frac{\text{Glucose-6-phosphate} = 0.0190}{\text{glucose-1-phosphate } 0.00100} = 19.0$$

from this value of K_{eg} we can calculate the standard free energy change

$$\Delta G^0 = -2.303 \text{ RT} \log K_{eg}$$
$$= -2.30\ (1.987)\ (298) \log 19.0$$
$$= -1360\ (1.28)$$
$$= -1740 \text{ cal/mol.}$$

ΔG of adenosine triphosphate (ATP) hydrolysis can be calculated as

$$\Delta G = \Delta G^0 + 2.303 \text{ RT} \log \frac{(\text{ADP})\ (\text{P}_1)}{(\text{ATP})}$$

$$\text{DG} = -7300 + 1360 \log \frac{(2.50 \times 10^{-4})\ (1.65 \times 10^{-3})}{2.25 \times 10^{-3}}$$
$$= -7300 + 1360 \log 1.83 \times 10^{-4}$$
$$= -7300 + 1360\ (-3.74)$$
$$= -7300 - 5100$$
$$= -12400 \text{ cal/mole}$$

or 12.4 kcal/mole.

In the above calculation, seen in human RBC, the concentration of ATP, ADP and *Pi* are 0.25, 0.25, and 1.65 mM respectively. The $\Delta G^0 = -7300$.

Thus the actual free energy (ΔG) is – 12.4 kcal/mol.

Difference Between ΔG and ΔG^0

The actual free energy change (ΔG) of a chemical reaction is a function of the condition of *pH*, concentration and temperature under which the reaction is taking place. Moreover, it is always negative in an outgoing chemical reaction. It becomes less negative as the reaction proceeds and becomes zero at the point of equilibrium. This indicates that no more work can be done by the reaction. Thus, in every spontaneous chemical or physical, process, the free energy (ΔG) of the reaction system is always decreases i.e., it is negative.

On the other hand, the value of standard free energy change (ΔG^0), for a chemical reaction is characteristic and unchanging and may be positive, negative or zero depending on the equilibrium constant of the reaction.

Energy Rich Compounds

In order to maintain biological activities, the cells must get free energy from their environment. The autotrophs get supply of free energy by coupling their metabolism to some exergonic process in their surrounding. For example, green plants utilize the energy of sun, on the other hand, animals get free energy by coupling their metabolism to the breakdown of complex organic molecules in their environment.

A large number of compounds are found in biological system. On hydrolysis they yield free energy changes of the order of –2 to – 5 kcal per molecule under close to physiological conditions. These compounds ae catagorized as *low energy bond compounds.* These compounds are very stable. Glucose-6-phosphate, glucose-1-phosphate, 3-phosphoglyceric acid, α-glycerophosphate, ethylacetate are some examples of this category. There is an another category of compounds, which on hydrolysis yield -6 to 13 kcal per molecule under physiological conditions. This category of compounds is called *high energy bond compounds.* The number of such compounds is relatively small. A dinosine triphosphate, adinosine diphosphate, phosphenol pyruvic acid, acetyl Co-A creatine phosphate, arginine phosphate etc., are the examples of this category.

In the energy rich compounds, the high energy is trapped in a particular bind. This bond is known as *energy rich bond.* Such bonds are represented by a sigmoid (~) in order to differentiate from the ordinary bonds, such bonds are represented by *covalent bonds* (–).

Classification

Energy rich compounds may be broadly classified into two categories-namely phosphate group and thioester group.

1. *Phosphate group.* This group includes:

(i) *Phosphate anhydrites.* In this catagory adinosine triphosphate, adinosine diphosphate and 1-3 diphosphoglyceric acid are included.

(ii) *Enolic phoshpate.* It include phosphoenol pyruvic acid.

(iii) *Guanidinium phosphate.* It includes creatine phosphate and orginine phosphate.

2. *Thioester group.* This group has acetyl Co-A, 5-adensyl-methionine etc.

1. Phosphate Group Compounds

(i) Adinosine triphosphate

Fiske and *Subbarow* (1929) first of all discovered ATP in the muscle extract. At the same time *Lohamann* (1929) also discovered it independently. *Lipmann* (1944) presented a detailed hypothesis of its function as an energy rich compound.

At neutral *pH,* the two terminal phosphate groups are in an ionized form and form a complex with the divalent magnesium ions (Mg^{++}). In most of the enzymatic reactions, where ATP participate as phosphate donor, its active form is magnesium—ATP Complex.

On hydrolysis of each of the terminal phosphate group—8.8 kcal/mole of free energy (ΔG^0) is obtained. In this reaction ATP is converted into ADP with the liberation of inorganic phosphate. On further hydrolysis of second phosphate group, — 8.8 kcal/mole energy is liberated. The ADP is converted into AMP with inorganic phosphate liberation. The hydrolysis of the first phosphate which is linked by an ester bond to ribose sugar, releases only-3 kcal/mole energy. To distinguish the terminal two pyrophosphate bonds from the first ester bond, the terminal bonds are represented by (~) high energy bond and the ester bond is represented by (~) straight bond.

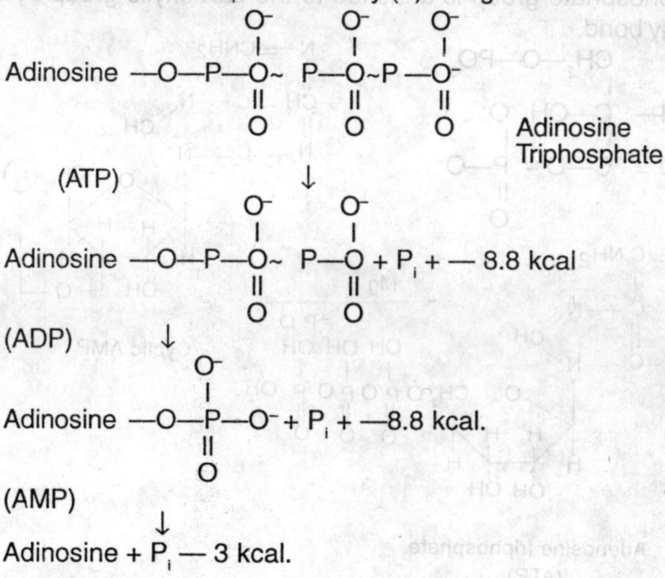

In the cell ATP and ADP both occur. ATP is utilized, but soon regenerated through a coupled reaction where ADP accepts high energy phosphate to form ATP. In reality, an ATP/ADP cycle connects these process which generate ~ P to these processes which utilize ΔP.

(ii) Phosphocreatine

It is the phosphagen found in the muscles of vertebrates and some invertebrates. It is the storage form of energy in muscles.

Although ATP is also present in muscles, but it is present in small amount and is used up in muscle contraction to form ADP. It is replanished by transfer of energy rich phosphate from creatine phosphate to ADP.

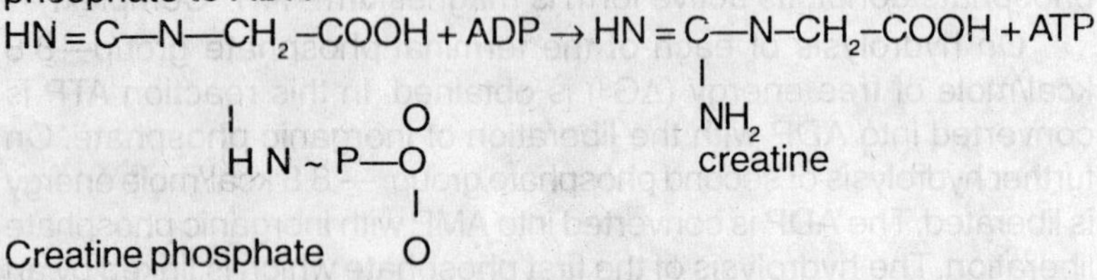

In the invertebrates the energy rich phosphagen is arginine phosphate.

(iii) 1, 3-diphosphoglyceric acid

It is formed during Embden—Mayerhoff pathway or glycolysis. The phosphate group is attached to the carboxylic group by high energy bond.

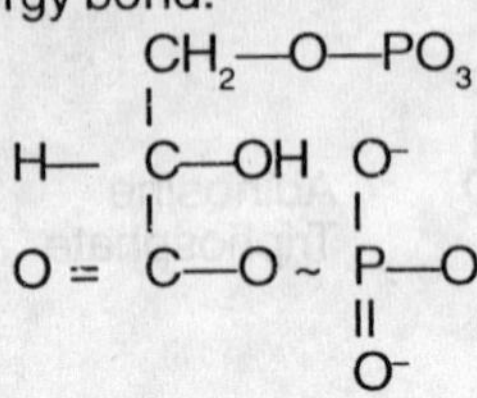

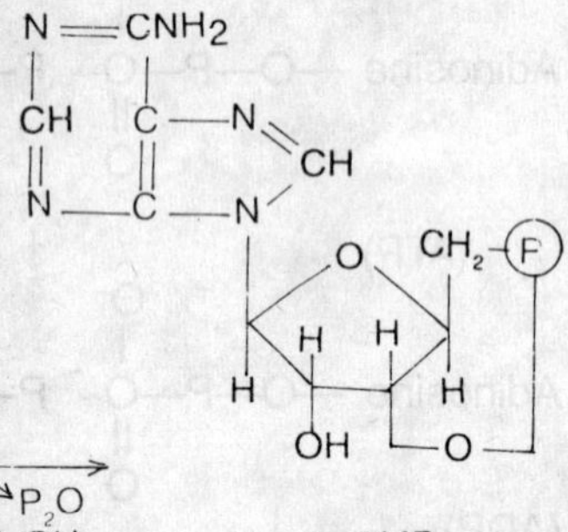

Cyclic AMP

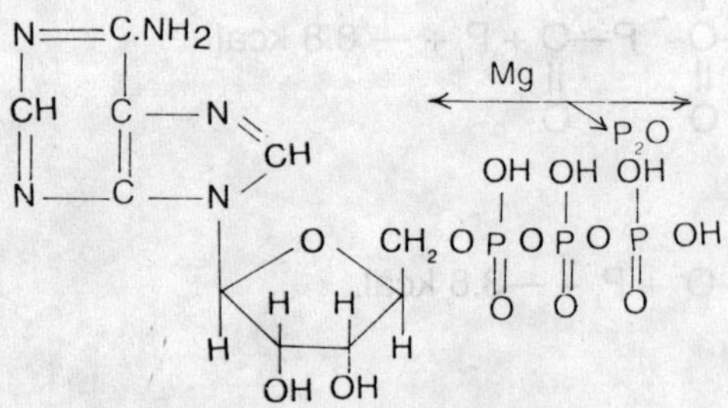

Adenosine triphosphate
(ATP)

Fig. 5.1. Structure of cAMP.

(iv) Phosphoenol pyruvic acid

This is also formed during glycolysis.

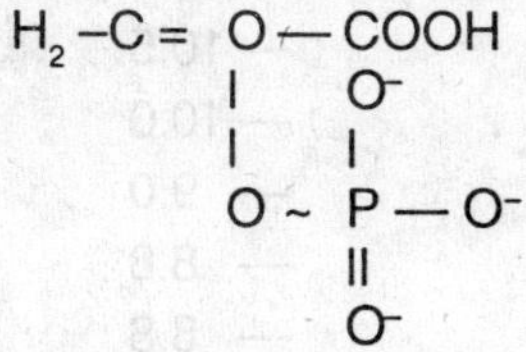

Besides above high energy rich compounds, two more such compounds are present in every cell of organisms. These are cyclic nucleotides namely — adinosine 3', 5'-cyclic monophosphate (cAMP) and other is guanosine 3', 5'-cyclic monophosphate (cGMP). The first one is formed from adinosine triphosphate and other from guanosine triphosphate. The reactions are catalized by adenylate cyclase and guanylate cyclase.

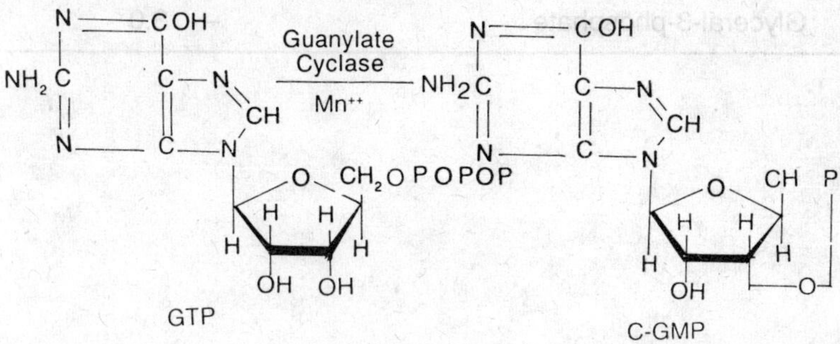

Fig. 5.2. Structure of cGMP.

On hydrolysis cyclic nucleotides yield—11.9 kcal/mole free energy. The significance of this compounds lies not in their high energy character, but in the regulatory roles in metabolism.

2. Thioester Group Compound

S-adenosyl methionine yields —10 kcal/mole free energy on hydrolysis.

Table 15.1. Standard Free Energy (ΔG^0) of Hydrolysis Some Energy Rich Compounds at pH 7.

Compounds	*ΔG^0 kcal/mole.*
Acetyl adenylate	— 13.3
Phospho enol pyruvic acid	— 13.0

cAMP	—11.9
Acetyl phosphate	—10.5
Acetoacetyl CoA	—10.5
S-adenosyl methionine	—10.0
Phosphocreatine	— 9.0
ATP (Terminal bond A)	— 8.8
ATP (Inner bond B)	— 8.8
Acetyl Co A	— 7.7
Palmityl Carnitine	— 7.7
Uridine diphosphate glucose	— 7.6
Phosphodiesters	— 6.0
Glucose -1-phosphate	— 5.0
Alanyl glycine	— 4.0
Glutamine	— 3.4
Glyceral-3-phosphate	— 3.0

6

Classification of Lipids

Lipids that yield salts of fatty acids (or "soaps") upon hydrolysis are called *Saponifiable.* The alcohol components of such compounds may be glycerol, glycerol derivatives, or long-chain alcohols. If lipids do not contain fatty acids in ester or amide linkage, they are usually categorized as *nonsaponifiable.* This group includes the terpenes, steroids, and the prostaglandins.

Fatty Acids and the Prostaglandins

Fatty acid is the name given to any acid comprised of a hydrocarbon chain with a terminal carboxyl group. Free fatty acids are found in only negligible amounts in mammalian cells and tissues. They occur principally in ester form. The most abundant have an even number of carbons, ranging from 14 to 22. The structures of the 18 fatty acids found in significant quantities in mammalian lipids are listed in table.

The acid strengths of the saturated fatty acids are not influenced significantly by the length of the hydrocarbon chain. The pK' values of the carboxyl groups are approximately 4.8, comparable with that of acetic acid. On the other hand, melting point and solubility in water are markedly affected by the number of carbons. For example, the solubility of butyric acid is 5.6 percent, whereas that of caproic acid is only about 0.4 percent. The melting points are also influenced by chain length, as seen in table 6.1.

The presence of a carbon-carbon double bond makes a fatty acid distinctly different from its fully saturated counterpart. The melting points of the unsaturated acids are lower and also their

Table 6.1. Fatty Acids of Mammalian Tissues

Formula	Systematic Name	Common Name	Symbol	Melting Point (°C)
Saturated fatty acids				
$CH_3(CH_2)_2COOH$	*n*-Butanoic	Butyric	4:0	
$CH_3(CH_2)_4COOH$	*n*-Hexanoic	Caproic	6:0	
$CH_3(CH_2)_8COOH$	*n*-Decanoic	Capric	10:0	
$CH_3(CH_2)_{10}COOH$	*n*-Dodecanoic	Lauric	12:0	44.2
$CH_3(CH_2)_{12}COOH$	*n*-Tetradecanoic	Myristic	14:0	53.9
$CH_3(CH_2)_{14}COOH$	*n*-Hexadecanoic	Palmite	16:0	63.1
$CH_3(CH_2)_{16}COOH$	*n*-Octadecanoic	Stearic	18:0	69.6
$CH_3(CH_2)_{18}COOH$	*n*-Eicosanoic	Arachidic	20:0	76.5
$CH_3(CH_2)_{22}COOH$	*n*-Tetracosanoic	Lignoceric	24:0	86.0
Unsaturated fatty acids				
$CH_3(CH_2)_5CH{=}CH(CH_2)_7COOH$	Hexadec-9-enoic	Palmitoleic	$16:1^{\Delta 9}$	–0.5
$CH_3(CH_2)_7CH{=}CH(CH_2)_7COOH$	Octadec-9-enoic	Oleic	$18:1^{\Delta 9}$	13.4
$CH_3(CH_2)_4CH{=}CHCH_2CH{=}CH(CH_2)_7COOH$	Octadeca-9, 12-dienoic	Linoleic	$18:2^{\Delta 9,12}$	–5
$CH_3CH_2CH{=}CHCH_2CH{=}CHCH_2CH{=}CH(CH_2)_7COOH$	Octadeca-9, 12, 15-trienoic	Linolenic	$18:3^{\Delta 9,12,15}$	–11

Table 6.1. (Contd.)

$CH_3(CH_2)_4(CH{=}CHCH_2)_3{=}CH(CH_2)_3COOH$	Eicosa-5,8,11, 14-tetraenoic	Arachidonic	$20{:}4^{\Delta 5,8,11,14}$	–49.5
$CH_3(CH_2)_9CH{=}CH(CH_2)_{12}COOH$	Docos-13-enoic	Erucic	$24{:}1^{\Delta 13}$	
$CH_3(CH_2)_7CH{=}CH(CH_3)_{14}COOH$	Tetracos-15-enoic	Nervonic	$24{:}1^{\Delta 15}$	
Hydroxy fatty acids				
$CH_3(CH_2)_{21}{-}CH(OH){-}COOH$	αOH-tetracosanic	Cerebronic	24:α-OH	
$CH_3(CH_2)_7CH{=}CH(CH_2)_{13}{-}CH(OH){-}COOH$	α-OH-tetracos-15-enoic	α-OH-nervonic	$24{:}\alpha\text{-}OH^{1\Delta 15}$	

solubility in a solvent such as ethanol is greater. This difference in physical properties of saturated and unsaturated lipids is due in part of the conformational peculiarities of a hydrocarbon containing carbon-carbon double bonds. Because a C=C bond cannot be rotated, the substituents on these carbon atoms may be either proximal (cis) or distal (trans) with respect to their spatial orientation. The chain of a fatty acid of the *cis* configuration can have a link produced by the C=C, resulting in a conformation distinctly different from the transform. The predominant form of unsaturated fatty acids in nature is the *cis* type.

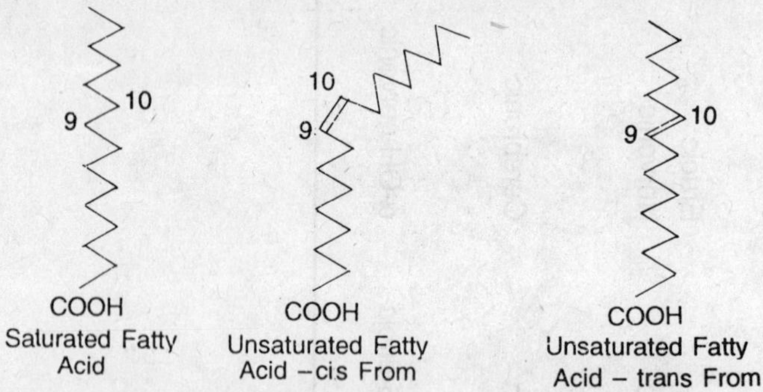

Fig. 6.1. Cis and trans forms of unsaturated fatty acids.

As in the case of all aliphatic compounds containing —C=C— groups, the unsaturated fatty acids can undergo addition reactions (e.g., with H_2 or I_2). The double bond is also vulnerable to attack be ozone to yield aldehydic products;

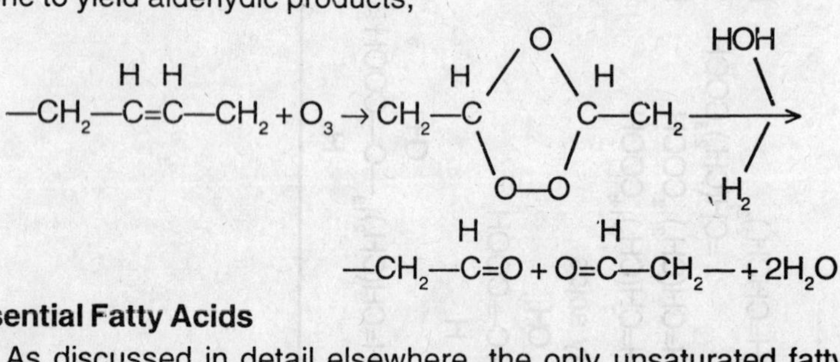

Essential Fatty Acids

As discussed in detail elsewhere, the only unsaturated fatty acids that can be made de novo in humans are the monoenic acids, palmitoleic acid and oleic acid. Linoleic, linolenic, and arachidonic acids must be supplied in the diet and are, therefore, designated as *essential fatty acids.* Although linoleic can be converted to linolenic and arachidonic acids to some extent, all three acids are most generally designated as essential.

Prostaglandins

As described in detail in the prostaglandins are a class of fatty acids which are derivatives of prostanoic acid:

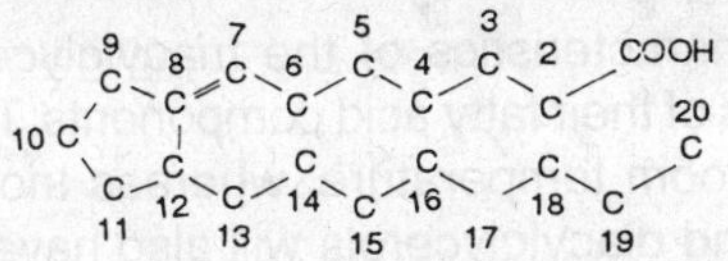

The origin of the prostaglandin family is the essential fatty acid, linoleic acid :

Linoleic acid→linolenic acid→

$$\text{arachidonic acid} \xrightarrow[-H,\ +O_2]{\text{Cyclization}} \text{prostaglandins}$$

Fatty Acid Esters of Glycerol

Glycerol, a trihydroxyalcohol, is a prominent component of animal lipids because it can form esters with inorganic acids (e.g., phosphate) as well as with organic acids such as the fatty acids. The tri-fatty acyl glycerols, the triglycerides, constitute the most abundant group of lipids in human tissue. The di- and monoglycerides are present in much smaller amounts.

Mono-, Di-, and Triacylglycerols

Three types acylglycerols or glycerides. The acylglycerols can exist in many isomeric forms, depending on the number of esterified acyl groups, the identity of the acyl moieties, and their position(s) on the glycerol. For example, a triglyceride containing palmitic (P) and oleic (O) acids can have six forms:

PPO, OOP, POP, OPO, OPP and POO

Carbon 2 of the glycerol becomes asymmetric when the acyl groups on C-1 and C-3 are different. The stereoisomers of this group will be :

C—P	C—O	C—O	C—P
C—P	C—P	C—O	C—O
C—O	C—P	C—P	C—O.

The triacylglycerols occur as complex mixtures in animal tissues and include fats in which there is only one kind of fatty acid esterified with the glycerol as well as those which contain two or three different fatty acyl groups. In naturally occuring mixed triacylglycerols, a saturated fatty acid is usually esterified at C-1 and an unsaturated acid at C-2. The third position may contain either.

The physical characteristics of the triacylglycerols generally reflect the properties of their fatty acid components. Thus, tripalmityl glycerol is solid at room temperature, whereas trioleyl glycerol is liquid. The mono- and diacylglycerols will also have lower melting points and, because of the polarity of their free hydroxyl groups, will have significant hydrophilic character.

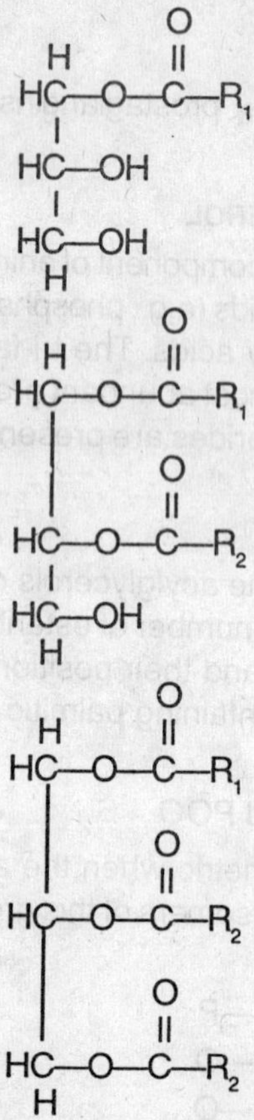

1-or α-Monoacylglycerol

1, 2- or α, β-Diacylglycerol

Triacylglycerol

Fig. 6.2. Mono-, di-, and triglycerides.

Alkyl Ether Acylglycerols

Human neoplasms have been found to contain glycerol derivatives in which the 1-, or α-, hydroxyl group is joined in ether linkage to a long alkyl or alkenyl group. Two examples of such glycerol ethers, are those derived from the alcohols corresponding to steric and oleic acids. The remaining two hydroxyl groups of the glycerol may be esterified with fatty acids.

Another class of alkyl ether acylglycerols, the plasmalogens are found in muscle, neural, and erythrocyte membranes. A representative structure of this type of lipid. In addition to phosphate, the polar end of this amphipathic lipid usually contains the ethanolamine moiety. The other distinguishing feature of the plasmalogens is that one of the two long hydrocarbon tails is contributed by an aliphatic chain in *cis* α, β-unsaturated ether linkage at position 1 of the glycerol.

Glycosylacylglycerols

Neural tissue has been found to contain a type of 1, 2-diacylglycerol in which the third hydroxyl group is in glycosidic linkage, usually with galactose as depicted.

Phosphoglycerides

As pointed out in the section on the acylglycerols the presence of just one free hydroxy group in these compounds makes them appreciably polar. The phosphoglycerides constitute a class of diacylglycerols whose polarity is even more pronounced because they contain a negatively charged phosphate group. The parent compound of this family of lipids is L-phosphatidic acid, whose structure. In most cases the naturally occurring phosphatidic acids contain a saturated fatty acid in position 1 and an unsaturated acyl group on C-2.

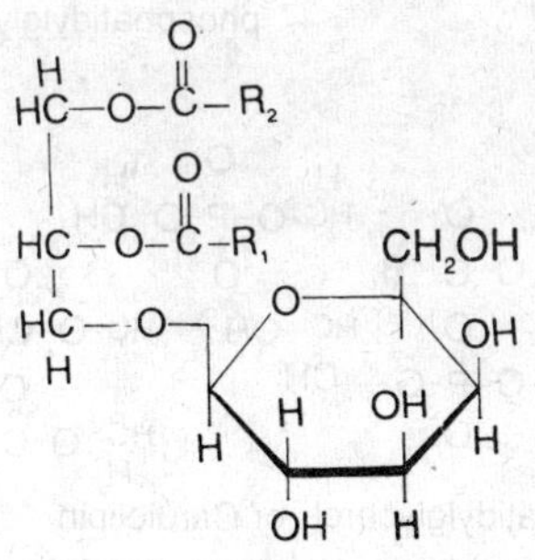

Fig. 6.3. A galactosyldiacylglycerol.

The phosphatidic acids occur in only negligible quantities in animal tissues. The most abundant forms of these lipids are those in which the phosphate on C-3 is in a second ester linkage with another alcohol. Such diesters are designated as phosphoglycerides. Particularly prominent among the phosphoglycerides of animal tissues are those containing ethanolamine, choline, serine, inositol, glycerol, and glycerol derivatives. The structures of these compounds. Because of the presence of polar and lipophilic groups within the same molecule, a phosphatidyl glyceride will have amphipathic character. It is reasonable that these compounds will be abundant in

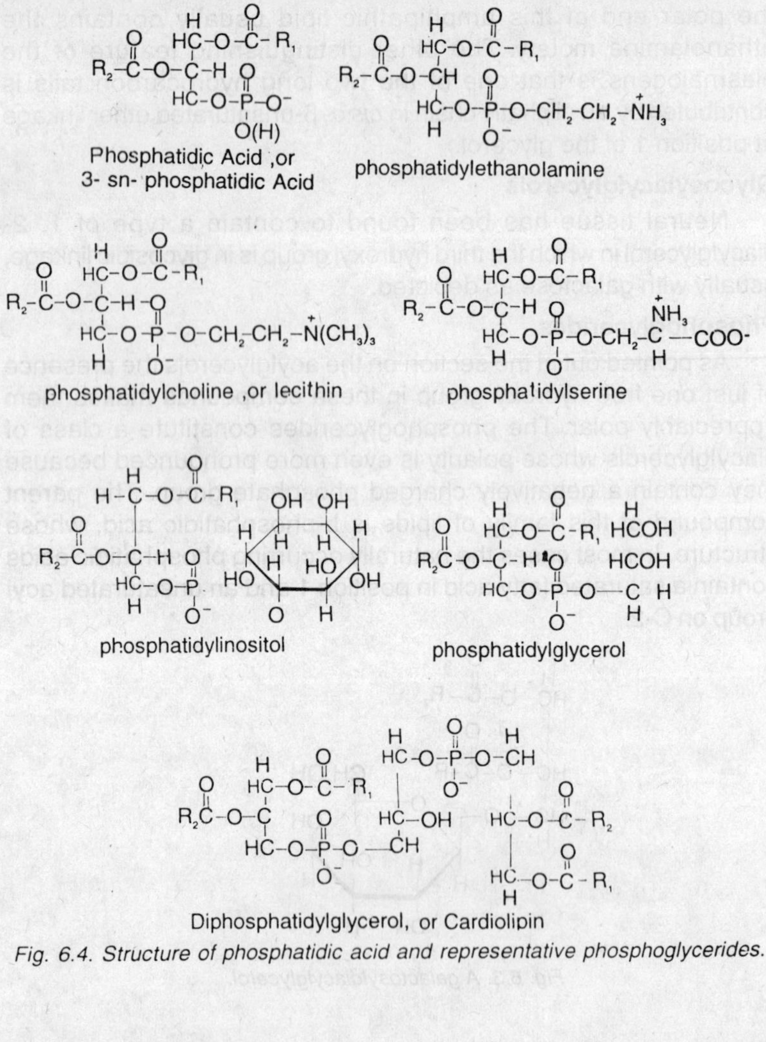

Fig. 6.4. Structure of phosphatidic acid and representative phosphoglycerides.

various types of cell membranes. Their detergent properties are also important, as illustrated by the surfactants required for normal lung function. Additional details of structure and function of the phosphoglycerides are presented in the section on lipid synthesis.

Partial hydrolysis products of the phosphoglycerides in which an acyl group is removed from position 1 or 2 of the glycerol are not found in high concentrations in animal tissues. These compounds, known as lysophosphoglycerides, injure cells by damaging membranes. The lysophosphoglycerides are the products of specific phospholipases which selectively remove acyl groups in phosphoglycerides.

Fatty Acid Esters Not Containing Glycerol

In this survey of lipid structure which began with a consideration of the fatty acids and triacylglycerols, it may be noted that we have progressed from those molecules that serve as forms of metabolic fuel to those that have more specialized functions and that are essential for normal structure in a cell or tissue. In the more complex lipids described below, the glycerol backbone is replaced by a much longer aliphatic chain. The biosynthesis of these compounds is described elsewhere.

Sphingolipids

The backbone of the sphingolipids is a long-chain, aliphatic amino alcohol. The two major forms of such a base found in humans are sphingosine and dihydrosphingosine, whose structures. The C_{18} sphingosines are the most abundant in animal tissues. The amino group of the sphingosine can be joined in amide linkage to the carboxyl of a C_{18}—C_{26} saturated or unsaturated fatty acid. The resulting lipid, called a *ceramide,* is the parent structure of the sphingolipids. The *sphingomyelins* are derivatives of ceramide in which various polar groups have been linked to the hydroxyl at its first carbon.

An example of sphingomyelin containing a charged group is one in which either phosphorylethanolamine or phosphorylcholine is esterified at the ceramide hydroxyl. Like the phosphoglycerides, these compounds are amphipathic.

Sphingomyelins containing a hexose in β-glycosidic linkage with the ceramide hydroxyl group are called *cerebrosides* or *neutral glycosphingolipids.* These complex lipids are most abundant in the myelin sheath of nerves and the predominant hexose attached to the ceramide is *d*-gelactose. The structure of a typical galactocerebroside is presented in the figure given below. The cerebroside found in nonneural tissues contain *d*-glucose and are called *glucocerebrosides.* A prominant fatty acid found in the glycosphingolipids is cerebronic acid. In additiòn to cerebrosides with only a single galactose or glucose linked to the ceramide hydroxyl, there are glycosphingolipids containing di-, tri-, and tetrasaccharides. These lipids are essential as cell surface components and are particularly important in determining the specificity of red cell membranes. An example of such a tetrasaccharide derivative is cytolipin K, or

$$\text{N-Acetylgalactosaminyl-}(1 \xrightarrow{\beta} 3)\text{-galactosyl-}(1 \xrightarrow{\alpha} 4)\text{ galactosyl-}$$

$$(1 \xrightarrow{\beta} 4)\text{-glucosyl-}1 \xrightarrow{\beta} \text{ceramide}$$

In addition to the uncharged glycosphingolipids, neural tissue also contains a variety of glycosphingolipids carrying a net negative charge. These compounds, called the gangliosides are glycosphingolipids with oligosaccharide heads containing sialic acid, or N-acetyl neuraminic acid (NANA) :

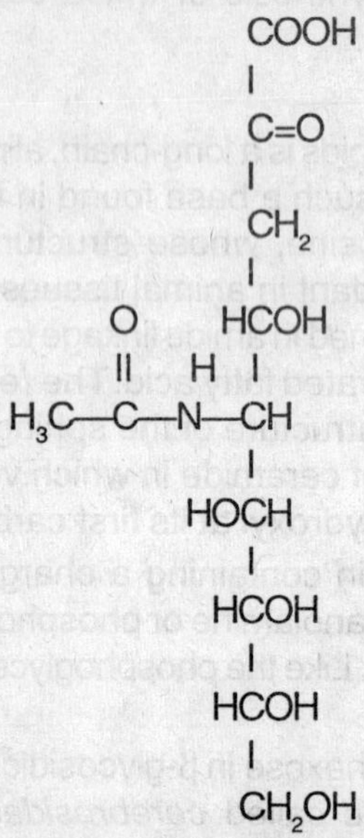

Gangliosides can be distinguished by differences in relative amounts and structural relations of *d*-glucose, *d*-galactose, NANA, and N-acetyl-*d*-galactosamine.

Waxes

Waxes are esters formed between long-chain aliphatic alcohols and fatty acids. An example is a cetyl palmitate :

$$CH_3(CH_2)_{14}—C—O—CH_2(CH_2)_{14}CH_3$$

For example, ear wax, or cerumen, contains such esters.

COMPLEX LIPIDS CONTAINING AMINO ACIDS, PEPTIDES, AND OLIGOSACCHARIDES

We have been that lipids have the capacity of becoming integrated into complex biological structures. Ultimately, it becomes irrelevant whether the building blocks of such structures are labelled as lipoproteins or proteolipids, as lipopolysaccharides or glycolipids. As illustrations of such hybrid molecules, we shall examine briefly the structure of a lipoamino acid, the lipopolysaccharides, and the lipoproteins.

Lipoamino Acids

The lipoamino acids and the lipopolysaccharides to be discussed in the next section are particularly abundant in cell membranes and walls of bacteria and are necessary for both protection of the cell and its antigenic specificity. A typical lipoamino acid derivative is that containing lysine. In this structure, the lysine carboxyl is esterified to the position 3' of the glycerol head group of the phosphoglyceride.

Lipopolysaccharides

The rigid basal wall of all bacteria is comprised mostly of a glycopeptide, or peptidoglycan. In gram-negative bacteria this basal layer is covered by a lipopolysaccharide covalently to the polysaccharide and can be removed by mild acid hydrolysis.

The lipopolysaccharides serve as the characteristic somatic O antigens, one of the three kinds of specific surface antigens of coliform bacteria. In their overall structure, these complex lipopolysaccharides consist of a core with "side-chain whiskers." Irrespective of differences in their O-antigen specificity, the core regions of the lipopolysaccharides from different strains of a particular bacterium appear to contain a common set of four sugars; heptose, glucose, glucosamine, and galactose. The lipid portion is located in the core of the polysaccharide and in Enterobacteriaceae has been determined to have the general structure.

The difference in O-antigen specificity of the bacterial strains can be ascribed to variations in the sugars located in the terminal

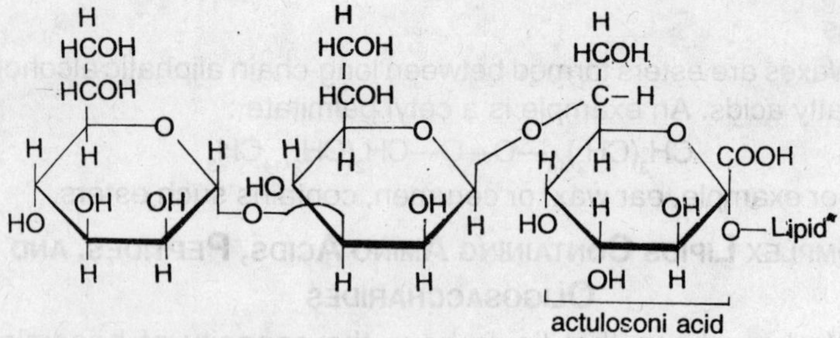

* Contains glucosamine, phosphate, acetyl groups, and β-hydroxymyristic acid

$$CH_3(CH_2)_{10}\overset{OH}{\underset{H}{C}}—CH_2—COOH$$

Fig. 6.5. Portion of "core" region of a lipopolysaccharide.

portions of the side-chain whiskers of the complex. For example, in Enterobacteriaceae the outermost regions of the side chains consist of repeating units such as the following:

$$\left(\begin{array}{c} \text{abequose} \\ | \\ \text{— mannose—rhamnose—galactose} \end{array}\right)$$

Lipoproteins

Irrespective of their origin, whether it be the diet or endogenous sources in the body tissues, lipids are transported in the plasma in association with proteins. Four major classes of lipoproteins of human plasma can be distinguished by ultracentrifugal and electrophoretic analyses as summarized in the table.

The interaction between the lipids and their proteins (apolipoproteins) is noncovalent. The solubility of the resultant lipoproteins suggests that the polar polypeptide chains of the proteins are arranged on the exterior of the complexes so as to provide hydrophilic surfaces, whereas the apolar portions of the lipids are directed toward the interiors of the lipoproteins. As shown the Table the four classes of lipoproteins differ in their proportions of protein and lipid. For the very low density lipoproteins and chylomicrons, it can be calculated that the amount of protein is not sufficient to envelop their surfaces. To account for the water solubility of these lipoproteins, it must be assumed that their hydrophilic, polar heads project from the surface of the particles.

There are five major groups of apolipoproteins (designated as A, B, C, D and E), which are distinguished by the lengths of their

Table 6.2. Composition and physical properties of plasma lipoproteins.

Characteristic	Chylomicrons	Very low density lipoproteins	Low-density lipoproteins	High density lipoproteins
Density ($g\ ml^{-1}$)	0.92-0.96	0.95-1.006	1.019-1.063	1.63-1.21
Range or average of molecular weights	5×10^8	$5\text{-}100 \times 10^6$	2.3×10^6	
Diameter (nm)	75-1000	30-75	17-26	7-10
Composition (% of dry weight)				
Protein	1-2	9-10	25	45-55
Triacylglycerols	80-95	55-65	10	3
Phospholipids	3-8	15-20	22-24	21-30
Cholesterol				
Free	1-3	10	8	3-5
Esters	2.4	5	35-37	15
Type of apoprotein				
Major	B, C	B, C, E	B	A, D
Minor	A	A		B, C, E

peptide chains and amino acid compositions. As shown in table, the four classes of lipoproteins may share several apolipoprotein components. It should also be pointed out that the four groups of lipoproteins are not homogeneous, one reason being that the A and C families of apolipoproteins consists of two three different polypeptide chains, respectively.

The changes in plasma concentrations of the four classes of lipoproteins which reflect the body's need for mobilization and transport of lipid are described in another section.

Nonsaponifiable Lipids

The nonsaponifiable lipids do not contain fatty acids and neither do they yield any other products under the usual conditions of alkaline hydrolysis. The structurally related families of compounds, the terpenes and the steroids, are major nonsaponifiable lipids in animal tissues. The carbon skeletons of both classes are derived from the five-carbon unit, isoprene, or c-methyl-1, 3-butadiene.

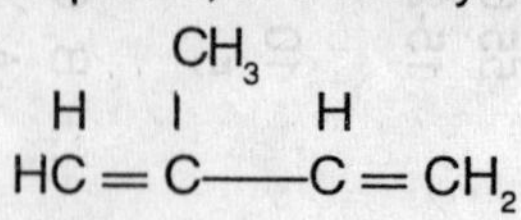

Terpenes

Terpene structures may be linear, cyclic, or a combination of both. Presents some representative compounds which illustrate the hierarchy of complexity of terpene structure. Note that isoprene units may be linked head to tail,

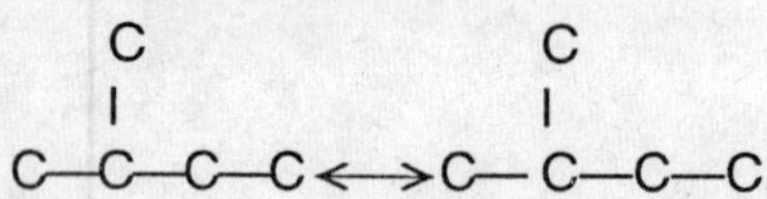

or tail to tail,

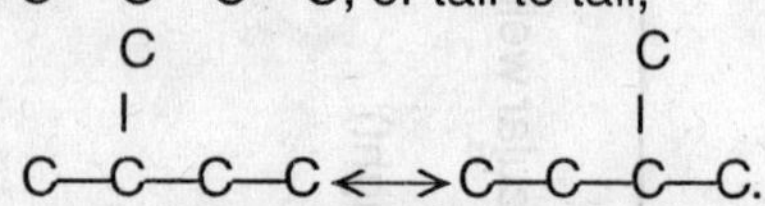

Geraniol and limonene are typical of the monoterpenes found in the essential oils of plants and trees (e.g., geranium and lemon). Phytol, containing four isoprene units and referred to as a diterpene, is associated with chlorophyll, the photosynthetic pigment. Compounds containing six terpene units are designated as triterpenes. Squalene, the precursor of cholesterol is an example of this class of terpenes. It will be noted that in squalene the head-to-tail arrangement of the isoprene units is reversed at the center of the molecule, making the two halves mirror images. The higher

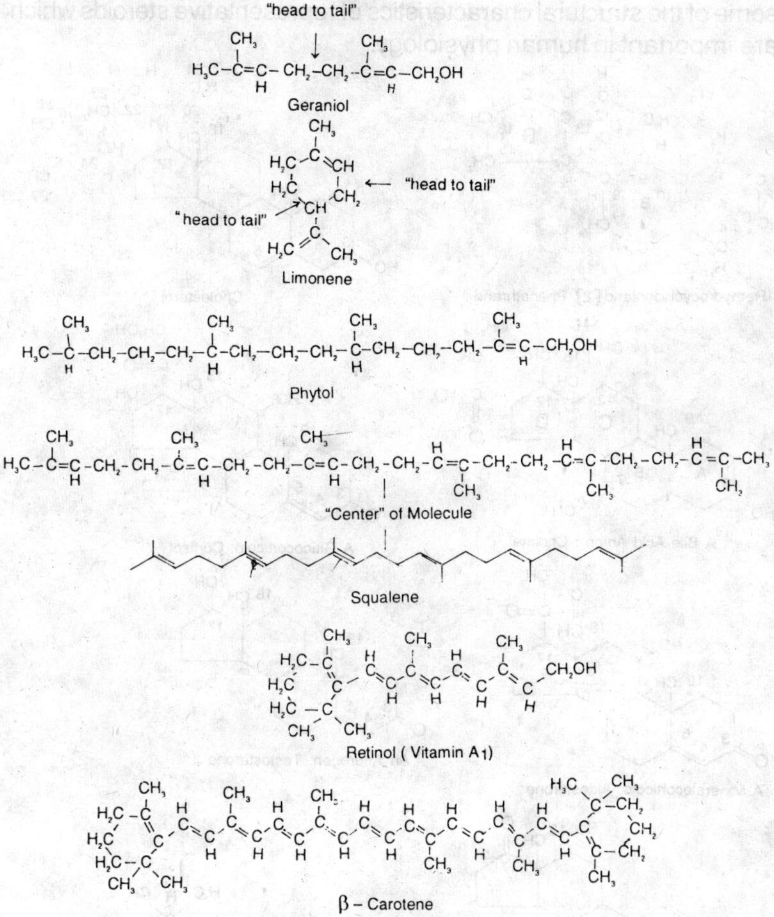

Fig. 6.6. Terpene structures.

terpenes include the tetraterpenes, those containing eight isoprene units. Representative of these compounds are the carotenoids : β-carotene, is the precursor of vitamin A. Two other fat-soluble vitamins, E and K, are also terpene derivatives. Another example of a terpenoid derivative essential for human metabolism is ubiquinone, or coenzyme Q. The quinone ring of this electron carrier in the mitochondria is attached to a long isoprenoid side chain.

Steroids

The details of biosynthesis and function of this large group of nonsaponifiable lipids are presented. In this section we highlight some of the structural characteristics of representative steroids which are important in human physiology.

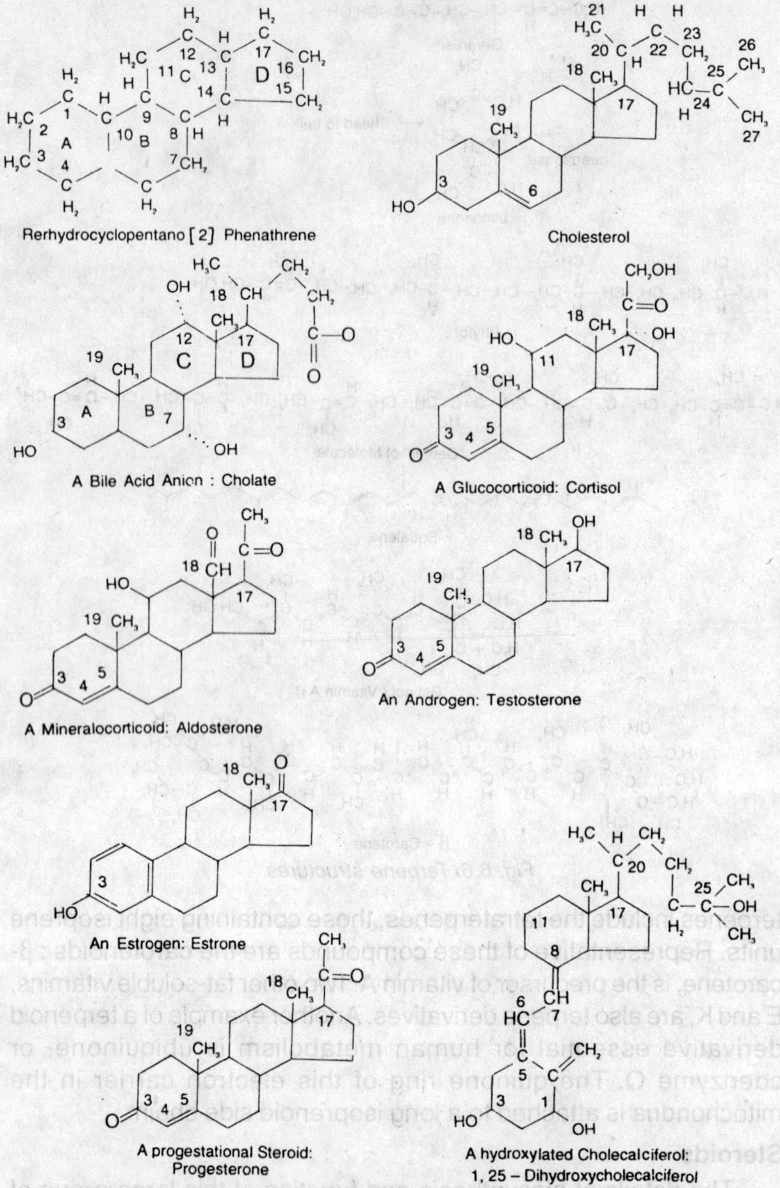

Fig. 6.7. Structure of representative steroids.

All steroids are derivatives of a fused, tetracyclic ring system, perhydrocyclopentan [α] phenanthrene. There are six centers of asymmetry in this molecule and substituted derivatives contain more. The ring junctions are usually all trans. Steroids most often contain substituent on C-3, C-11, C-17, C-18, and C-19.

In the biosynthesis of the steroids, the first intermediate resulting from the cyclization of squalene is lanosterol. Lanosterol is the precursor of cholesterol, which is the most abundant steroid is human tissues. Lanosterol and cholesterol are commonly referred to as sterols because of the alcoholic hydroxyl at position. The methyl groups at C-18 and C-19 and the —OH at C-3 are all on the same side of the fused ring system and can be regarded as projecting forward toward the viewer. This configuration of the three substituents is referred to as the β-orientation. In some sterols the 3–OH group is oriented α. In structural formulas of the type shown in figure dotted lines usually indicate α-oriented substituents and solid lines are used for β-oriented groups.

As discussed in detail elsewhere, the steroids can be grouped according to their physiologic activities. Figure shows the structures of representatives of seven classes of steroids : bile acids, glucocorticoids, mineral corticoids, androgens, estrogens, progesterone, and hydroxylated cholecalciferols.

7

FATTY ACID SYNTHESIS

The biosynthesis of long-chain fatty acids in animal tissues, plants, and microorganisms is not a direct reversal of fatty acid oxidation, although acetyl CoA is the starting material. Instead, acetyl CoA is carboxylated to form malonyl CoA in the presence of the enzyme *acetyl CoA carboxylase* in a reaction dependent on ATP and the vitamin *biotin.* The role of biotin in CO_2 fixation involves the formation of a complex between the two compounds in which CO_2 is attached to the ureido group of biotin, and the entire complex is attached to the enzyme :

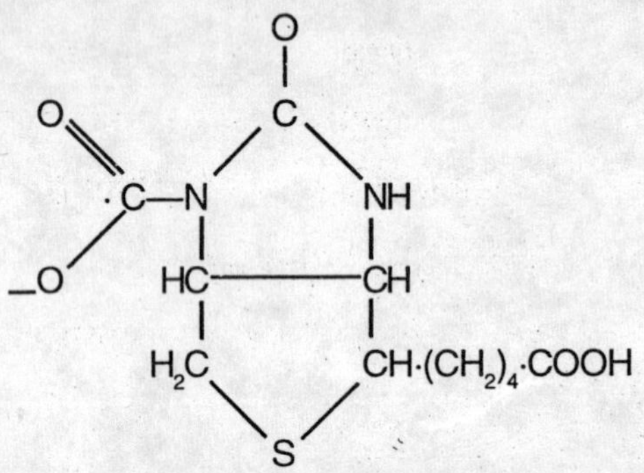

Biotin-CO_2 compound

Most of the enzymes responsible for fatty acid synthesis are contained in the soluble portion of the cell cytoplasm and can be demonstrated in tissue extracts prepared from liver or brain. Fatty acid synthesis requires NADPH, and the system, in the presence of malonyl CoA, will incorporate ^{14}C labeled β-keto-, β-hydroxyl-, and α,β-unsaturated CoA derivatives into long-chain fatty acids (mostly palmitic with some stearic and myristic acids).

Acyl Carrier Protein

The synthesis of fatty acids in many different species is accomplished by a group of enzymes which are associated together in the form of a multienzyme aggregate found in the cytoplasm. The complex form yeast has a particle molecular weight of 2.3×10^6, while that from pigeon liver is 4.5×10^5. The complex prepared from *E.coli* can be dissociated into the individual enzymes and the latter separated. However, the complex from yeast or mammalian cells is tightly bound.

One of the proteins in this complex is not an enzyme but has the specific task of binding the acyl intermediates during fatty acid synthesis—it is called the *acyl carrier protein* (ACP). It has a molecular weight of 100,000 and contains a prosthetic group at its binding site whose structure resembles that of coenzyme A. This group is 4'-phosphopantetheine linked to the protein by a serine residue, leaving its —SH group free to function as the acyl binding site.

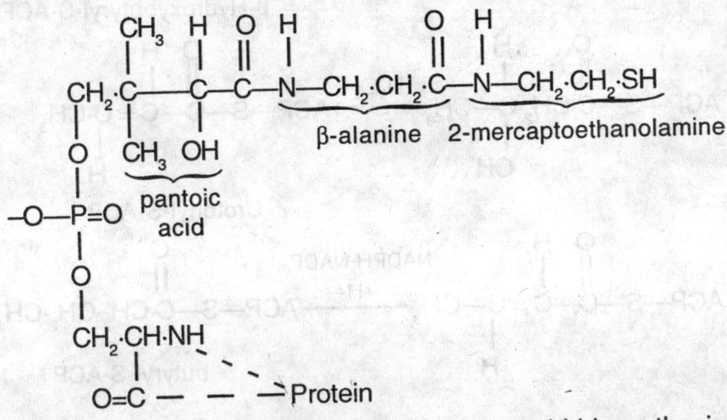

The sequence of reactions during fatty acid biosynthesis are described in the following scheme Asterisks are used to help follow the fate of specific carbon atoms:

$$1.\ \underset{\text{Acetyl CoA}}{CH_3\cdot\overset{O}{\overset{\|}{C}}\cdot S\cdot CoA} \xrightarrow[ATP \rightarrow ADP + P]{+biotin + CO_2} \underset{\text{Malonyl CoA}}{HOOC\cdot CH_2\cdot\overset{O}{\overset{\|}{C}}\cdot S\cdot CoA}$$

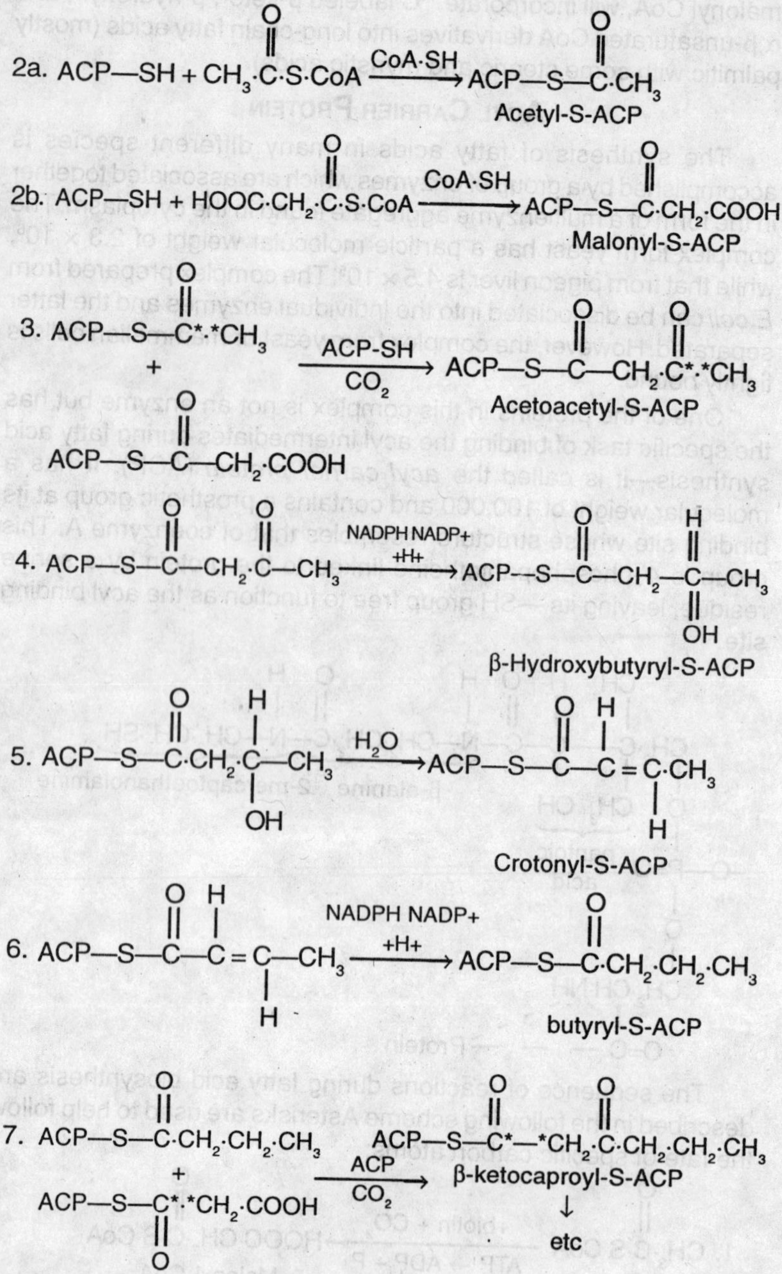
2a. ACP—SH + CH3·C·S·CoA
CoA·SH
ACP—S—C·CH3
Acetyl-S-ACP
2b. ACP—SH + HOOC·CH2·C·S·CoA
CoA·SH
ACP—S—C·CH2·COOH
Malonyl-S-ACP
3. ACP—S—C*·*CH3
+
ACP—S—C—CH2·COOH
ACP-SH
CO2
ACP—S—C—CH2·C*·*CH3
Acetoacetyl-S-ACP
4. ACP—S—C—CH2·C—CH3
NADPH NADP+
+H+
ACP—S—C—CH2—C—CH3
OH
β-Hydroxybutyryl-S-ACP
5. ACP—S—C·CH2·C—CH3
OH
H2O
ACP—S—C—C = C·CH3
Crotonyl-S-ACP
6. ACP—S—C—C = C—CH3
NADPH NADP+
+H+
ACP—S—C·CH2·CH2·CH3
butyryl-S-ACP
7. ACP—S—C·CH2·CH2·CH3
+
ACP—S—C*·*CH2·COOH
ACP
CO2
ACP—S—C*—*CH2·C·CH2·CH2·CH3
β-ketocaproyl-S-ACP
etc

Table lists the names of the enzymes in fatty acid biosynthesis. 'It should be noted that whereas fatty acid oxidation involves he action of the enzymes on CoA derivatives as substrates, during fatty acid biosynthesis the substrates are ACP-S derivatives. Also, during fatty acid oxidation it is the CoA derivative of the L(+) β-hydroxyacyl compound that is the substrate, whereas during fatty acid biosynthesis it is the D(–) β-hydroxyacyl-S-ACP derivative that is the substrate.

Table 7.1. Enzymes in Fatty Acid Biosynthesis

Reaction	Systematic name	No.	Trivial name
1	Acetyl CoA:CO_2 ligase (ADP)	6.4.1.2	Acetyl CoA carboxylase
2a	Acetyl CoA:(ACP-) acetyl-transferase	2.3.1.9	Acetyl CoA acetyl-transferase
2b	(ACP-) Acetyl CoA:malonate CoA transferase	2.8.3.3	Malonate CoA transferase
3, 7	(ACP-) Acyl CoA:(ACP-)acetyl CoA C-acyltransferase	2.3.1.16	3-Ketoacyl CoA thiolase
4	(ACP-)D-3-Hydroxyacyl CoA: NADPH oxidoreductase	1.1.1.36	Acetoacetyl CoA reductase
5	(ACP-)D-3-Hydroxyacyl CoA hydro lyase	4.2.1.17	Enoyl CoA hydratase
6	(ACP-) Butyryl CoA:NADPH oxidoreductase	1.3.99.2	Butyryl CoA dehydrogenase

It can be seen that the C atom s of acetyl CoA will be converted to C atoms 15 and 16 of the resulting palmitic acid, whereas acetoacetyl CoA comprises C atoms 13 through 16 of the resulting fatty acid. The overall reaction may be summarized as follows:

$$\text{Acetyl CoA} + 7\ \text{malonyl CoA} + 14\text{NADPH} + 14\text{H}^+$$
$$= \text{palmitate} + 14\text{NADP}^+ + 7\text{CO}_2 + 8\text{CoA} + 6\text{H}_2\text{O}$$

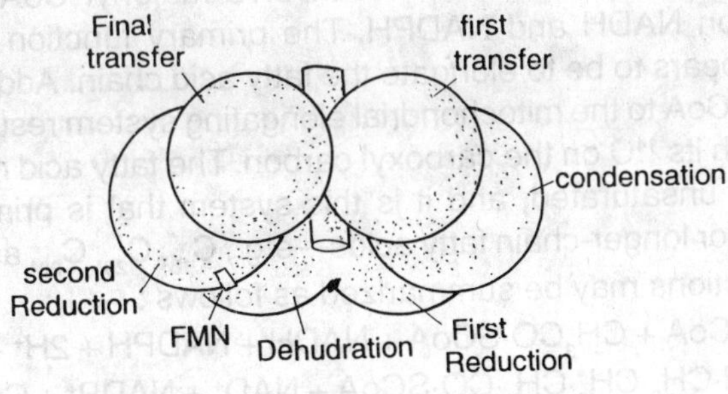

Fig. 7.1. A hypothetical structure for the "fatty acid synthetase" complex.

Acetyl Coenzyme A Carboxylase

This enzyme catalyzes the first reaction in fatty acid synthesis, the conversion of acetyl CoA to malonyl CoA by the addition of CO_2 in the presence of biotin. It is also important because it is a regulatory step; citrate acts as an allosteric activator for the animal enzyme (but not in plant or microbial systems).

In the absence of citrate the enzyme from bovine adipose tissues has a molecular weight of 550,000 and an activity of 0.16μ moles of acetyl CoA carboxylated per minute per mg of enzyme protein. In the presence of 10 mM citrate, the enzyme undergoes aggregation to form a filamentous structure with a sedimentation constant of 68S, corresponding to a molecular weight of several million, and its activity increases eight-fold. The filament lengths vary from 500 to 5000Å. The unaggregated protein consists of four nonidentical subunits of 100,000 molecular weight each. Activation by citrate involves primarily an effect on the maximum velocity, V_{max} and not on K_m. The high degree of structural organization of the animal carboxylases, which is absent in their counterparts in plants, yeast, and *E.coli*, suggests a possible structural role in addition to their known catalytic and regulatory functions. It could serve as an organizing matrix for a supramolecular (multienzyme) complex with other enzymes which take part in lipid biosynthesis. The filaments have also been observed in thin sections of developing and mature adipose tissue cells surrounding triglyceride droplets.

Mitochondrial and Microsomal Fatty Acid Synthesis

In addition to the soluble nonmitochondrial system just described, a fatty acid synthesizing mechanism is also associated with mitochondria. The enzymes catalyze the incorporation of ^{14}C-acetyl CoA into long-chain fatty acids by adding to existing fatty acid CoA derivatives. This system does not involve malonyl CoA but is dependent on NADH and NADPH. The primary function of this pathway appears to be to elongate the fatty acid chain. Addition of acetyl-1-^{14}C CoA to the mitochondrial elongating system results in a fatty acid with its ^{14}C on the carboxyl carbon. The fatty acid may be saturated or unsaturated, and it is this system that is principally responsible for longer-chain fatty acids—e.g., C_{18}, C_{20}, C_{22}, and C_{24}.

The reactions may be summarized as follows :

$$R{\cdot}CH_2{\cdot}CO{\cdot}SCoA + CH_3CO{\cdot}SCoA + NADH + NADPH + 2H^+ \rightarrow R{\cdot}CH_2{\cdot}CH_2{\cdot}CH_2{\cdot}CO{\cdot}SCoA + NAD^+ + NADP^+ + CoA{\cdot}SH$$

There is also a fatty acid synthesizing system in microsomes, and this system is also involved in chain elongation only. However, unlike the mitochondrial system, the microsomal system makes use of a reaction with malonyl CoA:

$$R{\cdot}CH_2{\cdot}CO{\cdot}SCoA + HOOC{\cdot}CH_2{\cdot}CO{\cdot}SCoA + 2NADPH + 2H^+ \rightarrow R{\cdot}CH_2{\cdot}CH_2{\cdot}CH_2{\cdot}CO{\cdot}SCoA + 2NAD^+ + CoA{\cdot}SH + CO_2$$

The NADPH required for fatty acid synthesis arises via the hexose monophosphate pathway and also by the action of the malic enzyme that catalyzes the conversion of malate to pyruvate and carbon dioxide associated with the formation of NADPH.

ROLE OF CARNITINE IN FATTY ACID METABOLISM

The compound β-hydroxy-y-butyrobetaine (carnitine)

$$(CH_3)_3{\cdot}N{\cdot}CH_2{\cdot}CH(OH){\cdot}CH_2{\cdot}COOH$$

has been known for many years as a constituent of meat extractives, although its function had been unknown. Its activity as a vitamin was first noted as necessary constituent of the diet of certain insects—e.g., the yellow mealworm *Tenebrio molitor*. It is widely distributed in microorganisms, plants, invertebrates, and vertebrates. Carnitine occurs in the free state and also bound to lipid.

The biosynthetic pathway of carnitine is suggested by the fact that its concentration in tissue is markedly influenced by dietary choline and methionine, and that the ^{14}C of methionine-methyl ^{14}C is incorporated into carnitine, and that γ-butyrobetaine is hydroxylated *in vitro* to form carnitine. The administration of ethionine (a biological antagonist of methionine) leads to a decrease in carnitine content in tissues. On a carnitine free diet, the rat excretes approximately 1 mg per day, suggesting that this is the minimal quantity synthesized per day.

Acetylcarnitine and the enzyme *carnitine acetyltransferase* have been identified in rat tissues. The acetyl group arises from acetyl CoA in the following reaction :

$$\begin{array}{l}\text{Carnitine} \longrightarrow \text{Acetylcarnitine}\\ \text{Acetyl CoA} \longrightarrow \text{CoA}\end{array}$$

It has been suggested that carnitine serves as a carrier of activated fatty acyl groups across the mitochondrial membrane, which is impermeable to fatty acyl CoA. This would explain how long-chain fatty acyl CoA compounds synthesized in the microsomal fraction of the cell are transported into the mitochondria, where they are oxidized. It has also been suggested that carnitine plays a role

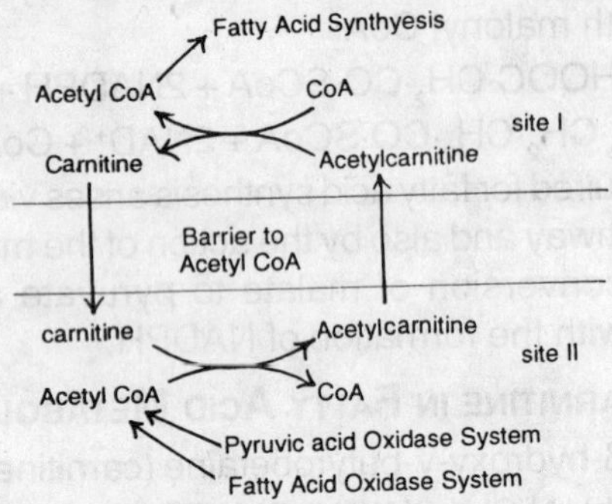

Fig. 7.2. Postulate role of the carnitine acetyltransferase system in shuttling of acetyl groups of acetyl CoA from intramitochondrial to extramitochondrial compartments.

in the movement of acetyl CoA and acetoacetyl CoA from the mitochondria to the nonmitochondrial sites of long-chain fatty acid synthesis.

Although the role of carnitine is as yet difficult to assess, it is important to note that other pathways are known by which mitochondrial acetyl CoA can gain access to the soluble portion of he cytoplasm—e.g., by conversion to citrate or acetate, which diffuse out of the mitochondrion and are then converted back again to acetyl CoA.

Unsaturated Fatty Acids

Burr showed many years ago that unsaturated fatty acids are essential components to the diet. That some unsaturated fatty acids could be formed from their saturated counterparts was shown by Schoenheimer and Rittenberg, who fed the methyl esters of fatty acids that had been obtained from lineseed oil after hydrogenation with deuterium (D_2). Unsaturated acids isolated some 7 to 12 days later contained deuterium. Stetten and Schoenheimer prepared deuteropalmitic acid biosynthetically and were able to isolate labeled palmitoleic acid. Later studies with improved techniques have shown that the polyunsaturated acids are not labeled in similar experiments.

The formation of oleic acid from stearic acid involves the formation of a double bond at some distance from the potentially activating carboxyl group, and without alteration in the certain length.

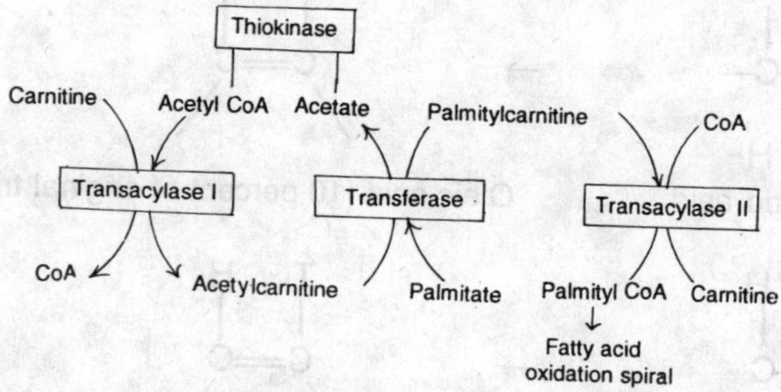

Fig. 7.3. Proposed scheme for the transfer of the methyl group from 5-CH_2-folate-H_2 to homocysteine.

An enzyme that carries out this reaction has been identified, and it requires oxygen, NADPH, and the fatty acyl CoA.

Just as the reaction of a dehydrogenase with the nicotinamide ring of NAD^+ involves a stereospecific approach of the substrate H to one or the other face of the ring, Bloch and Schroepfer have demonstrated that the enzymatic conversion of stearic acid to oleic acid is also stereospecific. For one thing, the unsaturated oleic acid has its double bond in the *cis* configuration. Secondly, the H atoms are removed specifically from the 9- and 10-carbon atoms. Finally, only the H atoms in the D-orientation on these two carbon atoms are removed during this desaturation reaction. Bloch and Schroepfer prepared the four tritium-labeled stearic acids—9D, 9L, 10D, and 10L—via a series of stereospecific chemical reactions and incubated each with *Corynebacterium diphtheriae.* Only tritium in the 9D or 10D configuration was removed during the formation of oleic acid. These results are illustrated by the following equation:

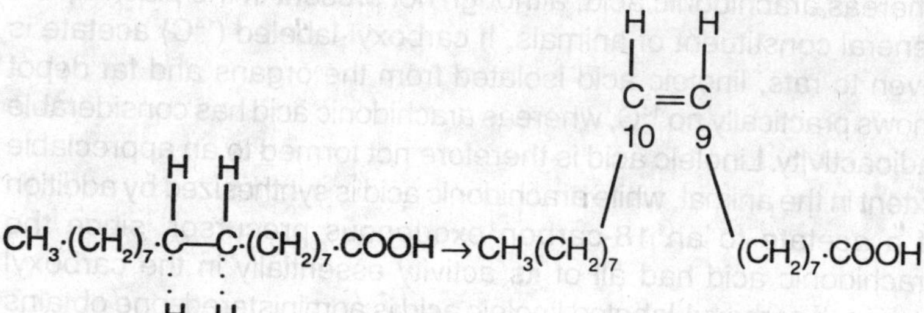

Stearic acid

Oleic acid (*cis*)

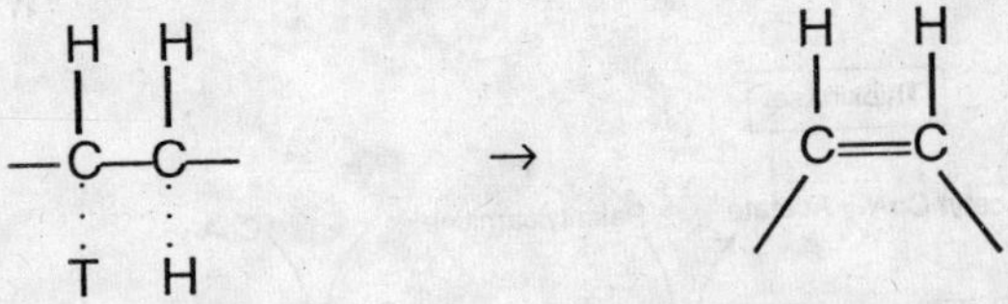

10D-Stearic acid → Oleic acid (10 percent of original tritium)

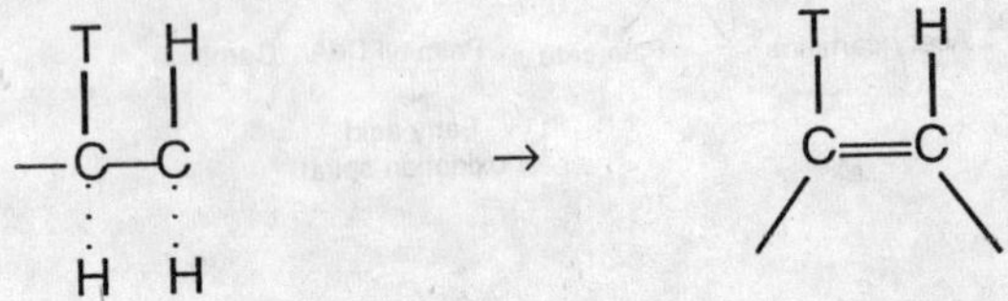

10L-Stearic acid → Oleic acid (87 percent of original tritium)

Polyunsaturated Fatty Acids

Linoleic acid α-linolenic acids are the major polyunsaturated fatty acids of higher plants, whereas arachidonic and similar C_{20}- and C_{22}-polyunsaturated acids of the γ-linolenic acid type are characteristic of higher animals. Bacteria lack polyunsaturated fatty acids. The formulas for several of the polyunsaturated acids are :

12 9
$CH_3{\cdot}(CH_2)_4{\cdot}CH{=}CH{\cdot}CH_2{\cdot}CH{=}CH{\cdot}(CH_2)_7{\cdot}COOH$ linoleic acid ($\Delta^{9,12}$-C_{18})

15 12 9
$CH_3{\cdot}CH_2{\cdot}CH{=}CH{\cdot}CH_2{\cdot}CH{=}CH{\cdot}CH_2{\cdot}CH{=}CH{\cdot}(CH_2)_7{\cdot}COOH$ α-Linolenic acid ($\Delta^{9,12,13}$-C_{18})

12 9 6
$CH_3{\cdot}CH_2)_4{\cdot}CH{=}CH{\cdot}CH_2{\cdot}CH{=}CH{\cdot}CH_2{\cdot}CH{=}CH{\cdot}(CH_2)_4{\cdot}COOH$ γ-Linolenic acid ($D^{6,9,12}$-C_{18})

14 11 8 5
$CH_3{\cdot}CH_2)_4{\cdot}CH{=}CH{\cdot}CH_2{\cdot}CH{=}CH{\cdot}CH_2{\cdot}CH{=}CH{\cdot}CH_2{\cdot}CH{=}CH{\cdot}(CH_2)_3{\cdot}COOH$ Arachidonic acid ($\Delta^{5,8,11,14}$-C_{20})

Linoleic and linolenic acids cannot be formed in the animal body, whereas arachidonic acid, although not present in the plasma, is a general constituent of animals. If carboxyl-labeled (^{14}C) acetate is given to rats, linoleic acid isolated from the organs and fat depot shows practically no ^{14}C, whereas arachidonic acid has considerable radioactivity. Linoleic acid is therefore not formed to an appreciable extent in the animal, while arachidonic acid is synthesized by addition of a acetate to an 18-carbon exogenous precursor, since the arachidonic acid had all of its activity essentially in the carboxyl carbon. If carboxyl-labeled linoleic acid is administered, one obtains the following distribution of ^{14}C (expressed as per cent of the total) in the isolated arachidonic acid

$$\underset{0}{\underset{\uparrow}{C_{17}H_{35}}}\cdot\underset{24.5}{\underset{\uparrow}{CH_2}}\cdot\underset{0.7}{\underset{\uparrow}{CH_2}}\cdot\underset{74.7}{\underset{\uparrow}{COOH}}$$

These results suggest that linoleic acid is incorporated into arachidonic acid unchanged. A possible mechanism for the conversion has been suggested :

$$\overset{12}{CH_3\cdot(CH_2)_4\cdot CH=CH}\cdot CH_2\cdot\overset{9}{CH=CH}\cdot(CH_2)_7\cdot COOH \quad \text{Linoleic acid}$$

$$\downarrow$$

$$\overset{12}{CH_3\cdot(CH_2)_4\cdot CH=CH}\cdot CH_2\cdot\overset{9}{CH=CH}\cdot CH_2\cdot\overset{6}{CH=CH}\cdot(CH_2)_4\cdot COOH \quad \gamma\text{-Linolenic acid}$$

$$\downarrow$$

$$\overset{14}{CH_3\cdot(CH_2)_4\cdot CH=CH}\cdot CH_2\cdot\overset{11}{CH=CH}\cdot CH_2\cdot\overset{8}{CH=CH}\cdot(CH_2)_6\cdot COOH \quad \text{Homo-g-linolenic acid}$$

$$\downarrow$$

$$\overset{14}{CH_3\cdot(CH_2)_4\cdot CH=CH}\cdot CH_2\cdot\overset{11}{CH=CH}\cdot CH_2\cdot\overset{8}{CH=CH}\cdot CH_2\cdot\overset{5}{CH=CH}\cdot(CH_2)_3\cdot COOH$$

Arachidonic acid

If one examines the oleic acid of organ and depot fat of an animal that had been given acetate-1-^{14}C, the oleic acid is found to be labeled. Also, another polyunsaturated acid—5,8,11-eicosatrienoic acid—is quite radioactive. Examination of the distribution of ^{14}C in these two compounds shows the following :

Oleic acid $\quad CH_3\cdot\ (CH_2)_7\cdot\ CH=CH\cdot\ (CH_2)_7\cdot\ COOH$

(upper brackets: $CH_3\cdot(CH_2)_7\cdot$ = 39.1; $CH=CH\cdot(CH_2)_7\cdot COOH$ = 60.9; lower brackets: CH_3 through $(CH_2)_7\cdot$ = 77.9; COOH = 22.1)

5,8,11-eicosatrienoic acid $\quad CH_3\cdot(CH_2)_7\cdot CH=CH\cdot CH_2\cdot CH=CH\cdot CH_2\cdot CH=CH\cdot CH_2\cdot CH_2\cdot CH_2\cdot COOH$

(upper brackets: $CH_3\cdot(CH_2)_7\cdot CH$ = 12.8; CH through CH_2 = 22.9; lower bracket: CH_3 through CH_2 = 27.5; arrows to the last CH_2, CH_2, COOH: 8.2, 1.8, 62.5)

It can be seen that the ratio of activity of the third through the eleventh carbon atoms of eicosatrienoic acid to that of the remaining nine carbon atoms at the methyl end of the chain (22.9/12.8=1.8) is approximately the same as that of the two halves of oleic acid (60.9/39.1=1.6). Similarly, if one compares the ratio of the 3-carbonofeicosatrienoic acid to the 4- through 20-carboxyl groups of oleic acid to the remainder of the chain (22.1/77.9 = 0.28). These results strongly suggest that oleic acid is the precursor for eicosatrienoic acid, possibly as follows:

$$\overset{9}{CH_3\cdot(CH_2)_7\cdot CH{=}CH\cdot(CH_2)_7\cdot COOH} \quad \text{Oleic acid}$$
$$\downarrow$$
$$CH_3\cdot(CH_2)_7\cdot \overset{9}{CH{=}CH}\cdot CH_2\cdot \overset{6}{CH{=}CH}\cdot(CH_2)_4\cdot COOH$$
$$\downarrow$$
$$CH_3\cdot(CH_2)_7\cdot \overset{11}{CH{=}CH}\cdot CH_2\cdot \overset{8}{CH{=}CH}\cdot(CH_2)_6\cdot COOH$$
$$\downarrow$$
$$CH_3\cdot(CH_2)_7\cdot \overset{11}{CH{=}CH}\cdot CH_2\cdot \overset{8}{CH{=}CH}\cdot CH_2\cdot \overset{5}{CH{=}CH}\cdot(CH_2)_3\cdot COOH$$

Eicosatrienoic acid

Fatty Acid Synthesis in E.coli

Figure summarizes the biosynthetic routes in *E.coli* for unsaturated and saturated fatty acids. The reactions are identical until they reach the stage of β-hydroxydecanoic acid where they diverge, one branching to α,β-decenoic acid and thence to palmitic

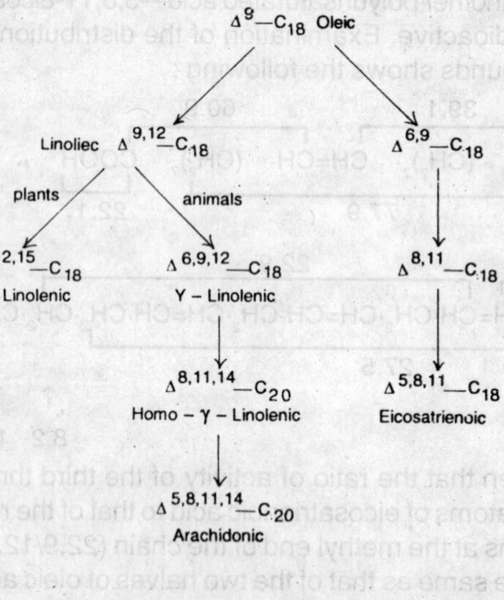

Fig. 7.4. Pathways of polyunsaturated acid synthesis.

and stearic acids, and the other to β,γ-decenoic acid, the precursor for palmitoleic and vaccenic acids.

The same enzyme, β-*hydroxydecanoyl thioester dehydrase,* catalyzes the formation of β,γ-decenoic and α,β-decenoic acids. It is found only in bacteria which produce long-chain fatty acids by the anaerobic pathway. Kinetic, chemical, and labeling studies lead to the conclusion that an enzyme-bound intermediate is involved :

$$\begin{array}{ccc} & \beta\text{-OH} & \\ & \downarrow\uparrow & \\ & (\alpha,\beta)\text{-Enz} & \\ \swarrow\nearrow & & \searrow\nwarrow \\ \beta,\gamma & & \alpha,\beta\ (\text{free}) \end{array}$$

Synthesis of Unsaturated Fatty Acids in Plants

A third mechanism for the synthesis of monounsaturated acids is found in green plants. Although the evidence for this pathway is not conclusive, it is postulated in the following scheme (Bloch) for *Chlorella*:

$CH_3 \cdot (CH_2)_{10} \cdot COOH$	Lauric acid
$\downarrow$	
$CH_3 \cdot (CH_2)_8 \cdot CH{=}CH \cdot COOH$	2-Dodecenoic acid
$\downarrow$	
$CH_3 \cdot (CH_2)_8 \cdot CHOH \cdot CH_2 \cdot COOH$	β-Hydroxydodecanoic acid
$\downarrow$	
$CH_3 \cdot (CH_2)_7 \cdot CH{=}CH \cdot CH_2 \cdot COOH$	3-Docenenoic acid
$\downarrow 3C_2$	
$CH_3 \cdot (CH_2)_7 \cdot CH{=}CH \cdot (CH_2)_7 \cdot COOH$	Oleic acid

The green plant system further desaturates oleic acid to linoleic and linolenic acids by the NADPH-linked pathway and molecular oxygen.

Role of Polyunsaturated Fatty Acids

The gross symptoms of essential fatty acid deficiency—increased metabolic rate, increased water consumption, and eventual growth failure—were originally described by Burr. The increased activity of cytochrome oxidase in the liver of rats maintained on a fat-deficient diet suggests the involvement of oxidative processes, and it has been demonstrated that such livers have a defect in

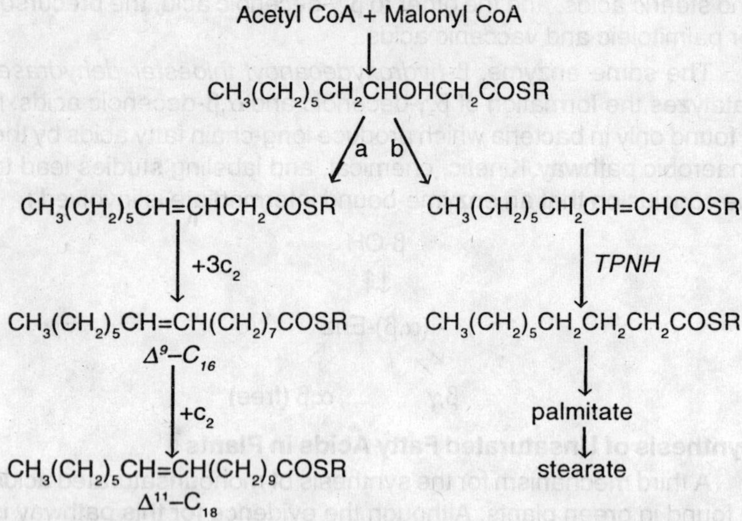

Fig. 7.5. Fatty acid biosynthesis in E. coli. Reaction a initiates the pathway to unsaturated acids and reaction b, the branch to saturated acids.

oxidative phosphorylation at or near the level of NAD^+ reduction and oxidation. These results are suggestive of a role for polyunsaturated fatty acids in oxidative phosphorylation in animal tissues.

An interesting aspect of comparative metabolism of the polyenoic acids has been reported by Bloch as a result of studies with the phytoflagellate *Euglena gracilis*, which has retained enzyme systems for production of fatty acids of the α-linolenate and the γ-linolenate types. It can grow in the light as a strict autotroph (making its own food by photosynthesis) or, after adaptation, as a heterotroph (obtaining its food from organic material) in the dark. This change is accompanied by a shift from the presence of α-linolenate, as the major unsaturated fatty acid to the appearance of the animal type C_{20}-, C_{22}-, and C_{24}- polyenoic acids. The galactolipids present in the organisms that grow in light are replaced, in the dark-adapted cells, by phospholipids, mostly lecithin. These results suggest some role for the α-linolenate-containing galactolipids during photosynthesis. Since photosynthesizing bacteria neither synthesize nor require α-linolenate and also do not evolve oxygen, it has been suggested that the fatty acid or its galactolipid may be necessary for one or more of the steps leading to oxygen evolution (Hill reaction) during photosynthesis.

BIOSYNTHESIS OF CHOLESTEROL

We consider the synthesis of cholesterol and its hormonal derivatives. Although cholesterol can be made in almost all tissues in humans, the principal sites of synthesis are the liver and intestinal mucosa; also important are the adrenal cortex, testes, and ovaries. The enzymes and cofactors required for synthesis are located in the endoplasmic reticulum and the cytosol.

One of the earliest demonstrations of the applicability of isotopic labelling in metabolic studies was the discovery that both carbons of acetate were incorporated into the cholesterol of liver. As these analyses were extended, it was found that the 2 carbons contributed equally to the cholesterol molecule when the labeled cholesterol was degraded systematically, the origin of each carbon was shown to be as follows:

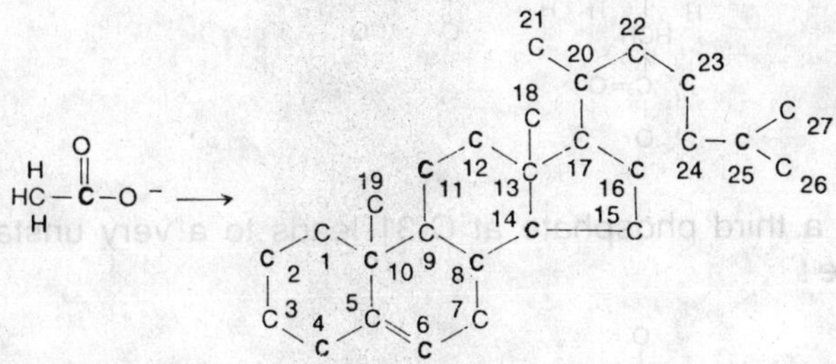

Cholesterol synthesis begins with two common intermediates in lipid metabolism acetyl-CoA and acetoacetyl-CoA.

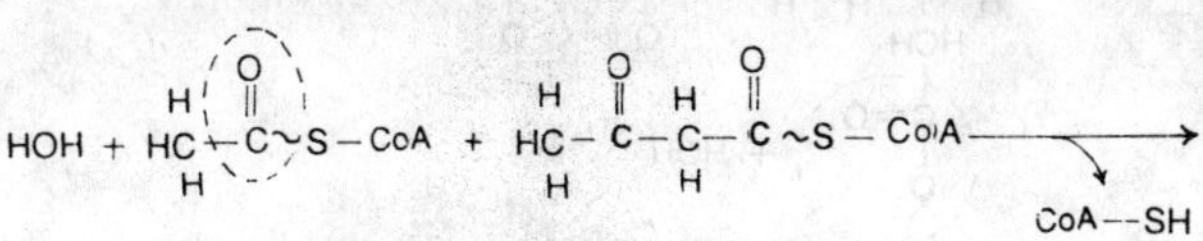

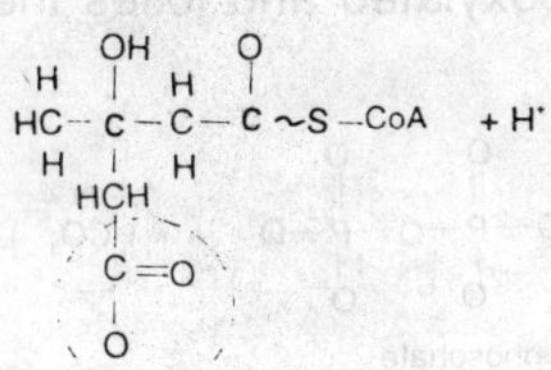

Hydroxymethylglutaryl – CoA

With the addition of four electrons from two NADPH, the thioester end of the glutaryl derivative is reduced to yield L-mevalonate:

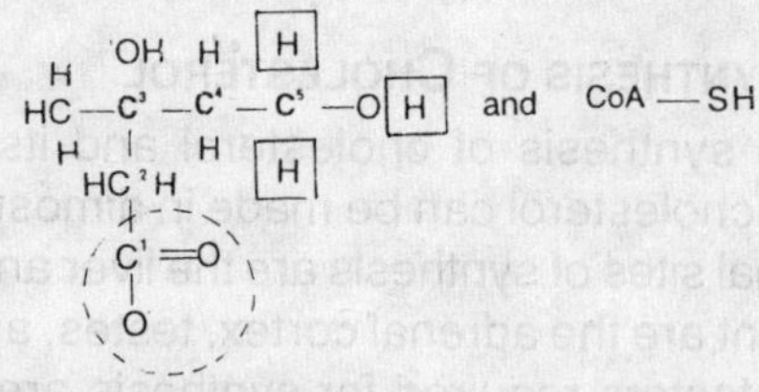

The irreversible reaction is the committed step in cholesterol synthesis and the enzyme that catalyzes the process, 3-hydroxy-3-methylglutaryl-CoA reductase, is an important control point in the biosynthetic pathway.

By two successive phosphorylation with ATP, the mevalonate is converted to its 5-pyrophosphate form :

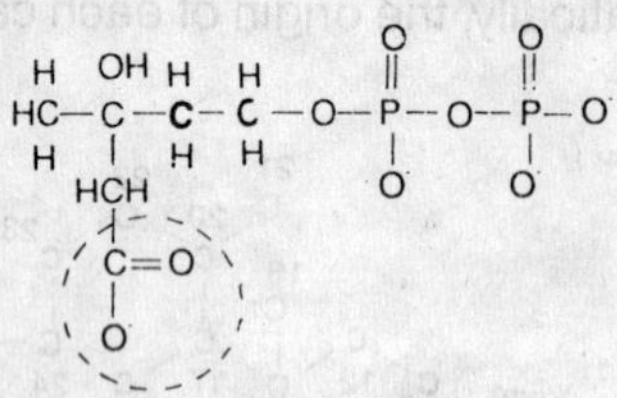

Addition of a third phosphate at C-31 leads to a very unstable intermediate :

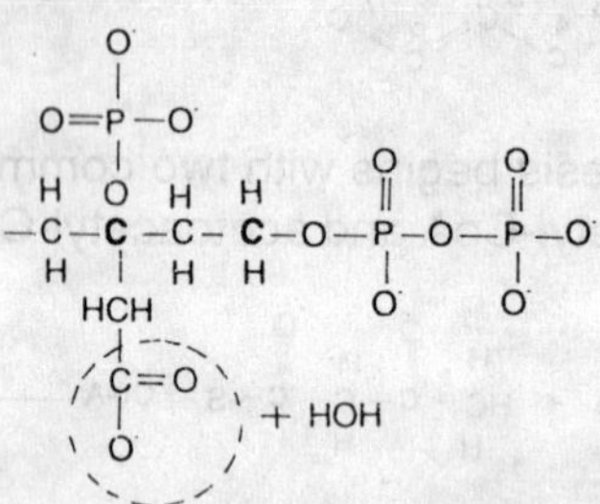

This compound is decarboxylated and loses the phosphate last added:

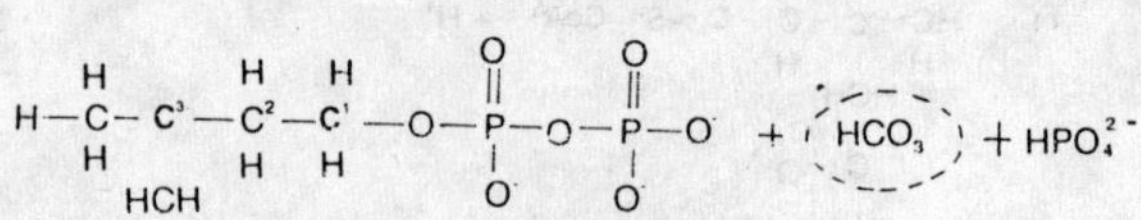

3- Isopentenylpyrophosphate

The isopentenyl unit is isomerized to yield 3,3-dimethylallylpyrophosphate:

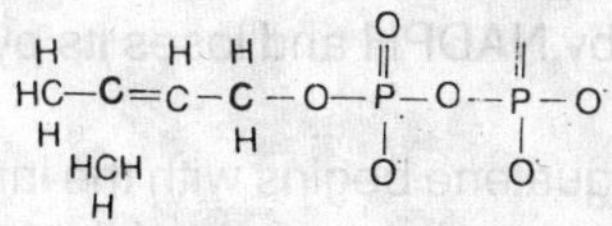

Both of these isoprenyl pyrophosphates are required for cholesterol synthesis and undergo condensation to form a 10-carbon pyrophosphate derivative, geranylpyrophosphate:

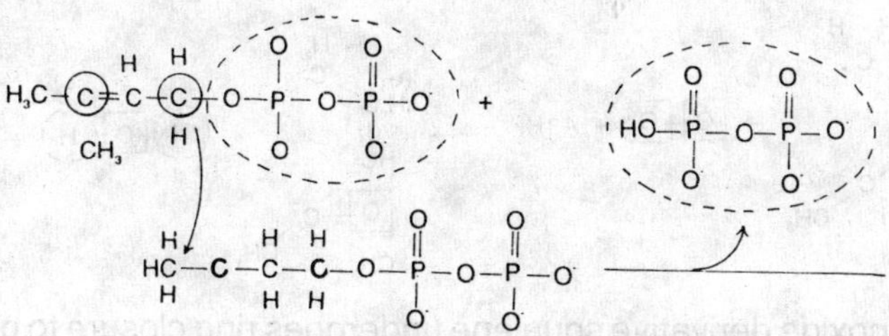

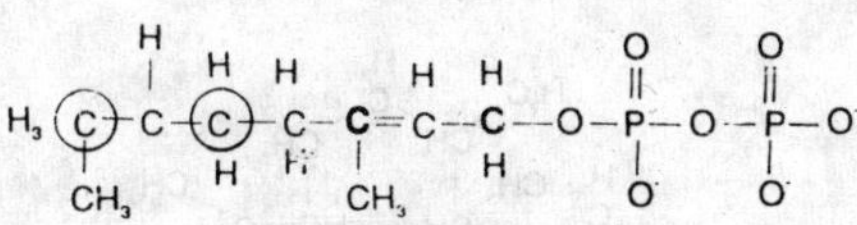

Geranylphyrophosphate

A second condensation reaction with an isopentenylpyro-phosphate yield the 15-carbon derivative, farnesylpyrophosphate:

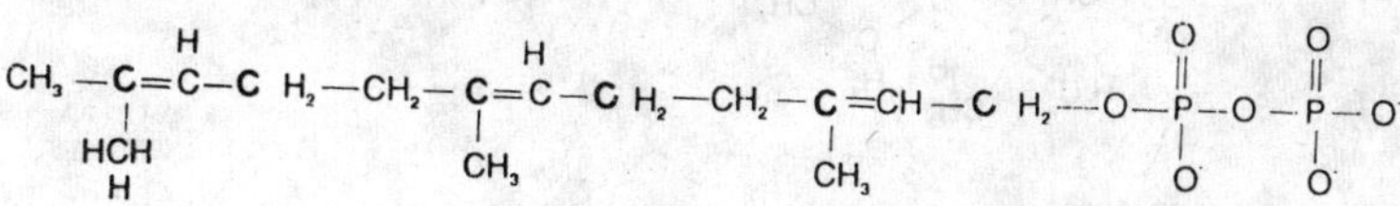

Two farnesylpyrophosphates condense to give a 30-carbon pyrophosphate derivative, presqualene pyrophosphate. This

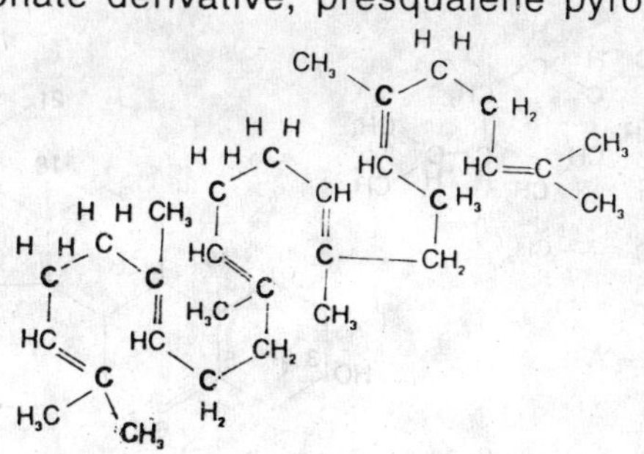

intermediate is reduced by NADPH and loses its pyrophosphate to become squalene.

The ring closure of squalene begins with the introduction of the oxygen atom which will ultimately be a part of the hydroxyl group at position 3 of cholesterol. This reaction utilizes molecular O_2 and, like other monooxygenase-catalyzed reactions, requires a reductant in addition to the oxygen:

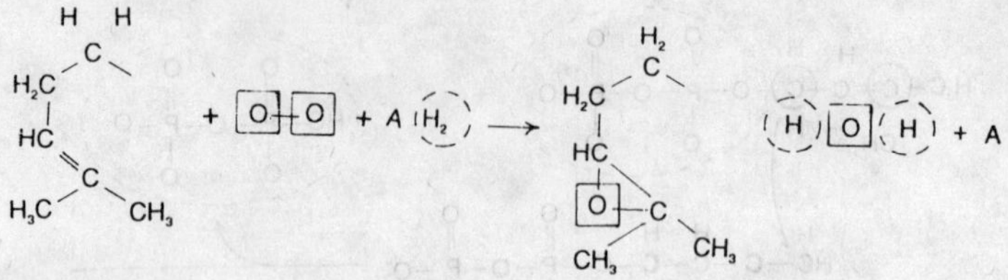

and the epoxide derivative squalene undergoes ring closure to give the lanosterol :

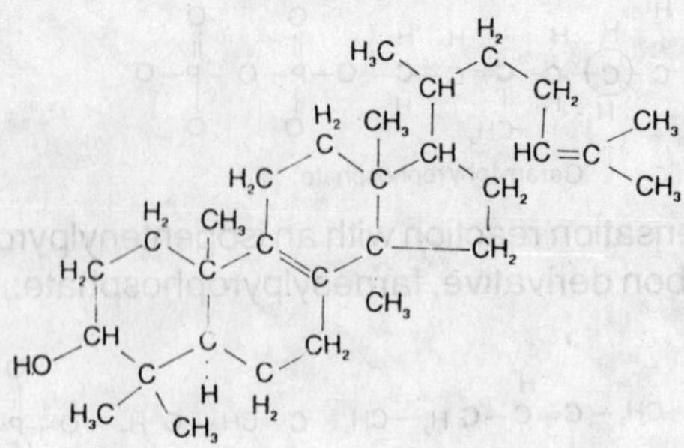

The remaining steps for the completion of cholesterol synthesis involves the oxidative removal of three methyl groups, hydrogenation of the side chain, and a double-bond shift from carbon 8,9 to 5,6 A

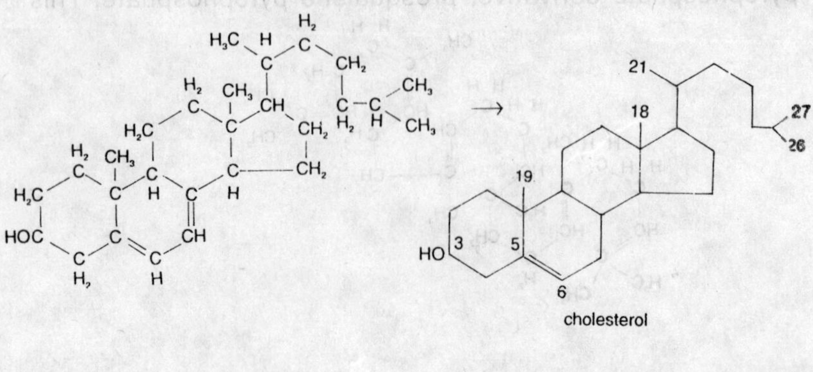

cholesterol

prominent intermediate in these final reactions is 7-dehydro-cholesterol.

Fatty Acid Esters of Cholesterol

The greater part of the cholesterol in the tissues and in transit in the blood is esterified at position 3 with long-chain fatty acids. In the liver, intestine, and adrenal cortex the esterification reaction begins with the ATP-dependent synthesis of the fatty acyl-CoA derivatives. The activated acyl group is enzymatically transferred to the 3-OH of cholesterol. Cholesterol, which may be released by the liver into the plasma as a component of the lipoprotein, can also be esterified in the plasma through the action of a phosphatidylcholine cholesterol acyl transferase :

Phosphatidylcholine + cholesterol = lysophosphatidylcholine + cholesterol fatty acid ester

Cholesterol as the Source of Steroid Hormones

The principal sites of biosynthesis of the steroid hormones are the adrenal gland, gonads, and placenta. Together, these tissues are responsible for the synthesis of four classes of steroids: the glucocorticoids, the mineralocorticoids, the androgens, and the estrogen. Irrespective of the tissue involved, the synthesis of all of these hormones, starting with cholesterol, proceeds through two intermediates: pregnenolone and progesterone. For a general orientation of the structural differences of the steroids, the overview of pathways and the data may be helpful. For sake of simplifying the nomenclature, stereochemical designations (i.e., α, β) are not included. This section on the steroid hormones describes only the pathways of their synthesis.

Table 7.2. Functional Groups of Steroid Hormones

Hormone	Functional Groups and Numbers				
	—OH	C=O	H–C=O	—CH_2OH	—CH_3
Pregnenolone	3	20			18, 19, 21
Progesterone		3, 20			18, 19, 21
Cortisol	11, 17	3, 30		21	18, 19
Cortisone		3, 11, 20		21	18, 19
Corticosterone	11	3, 20		21	18, 19
Aldosterone		3, 20	18	21	19
Testosterone	17	3			18, 19
Estradiol	17				18
Estrone		17			18

Conversion of Cholesterol to Pregnenolone

The rate-limiting process in the conversion of cholesterol to the steroid hormones is the initial oxidative removal of the side chain yielding a six-carbon fragment, isocaproic aldehyde, and pregnenolone. The oxidative cleavage is accomplished by two successive mixed-function oxidations at C-20 and C-22 :

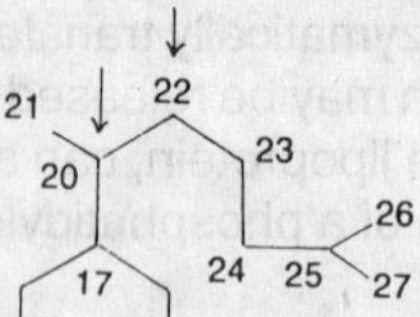

The mitochondrial enzymatic system responsible for the process, called *desmolase,* occurs in both the adrenal cortex and gonadal tissue. The overall reaction scheme outline below is typical of mixed-function oxidase or hydroxylation reactions with the exception that an additional electron carrier is interposed between the flavin and cytochrome P_{450}. This carrier, labeled *adrenodoxin* in the adrenal mitochondria and *testodoxin* in testicular mitochondria, is an iron-sulfur protein. The mechanism of hydroxylation at C-20 of cholesterol, designated below as RH, can be formalized as follows :

1. $R\mathbf{H} + \text{cyto } P_{450}\ (Fe^{3-}) \rightarrow R\mathbf{H} \cdot \text{Cyto } P_{450}\ (Fe^{3+})$
2. $R\mathbf{H} \cdot \text{cyto } P_{150} (Fe^{3+}) - (Fe^{2+}\text{–}5)_I \rightarrow R\mathbf{H} \cdot \text{Cyto } P_{450}\ (Fe^{2+}) + (Fe^{3+}\text{-}S)_I$
3. $R\mathbf{H} \cdot \text{cyto } P_{450}\ (Fe^{2+}) - (O_2) \rightarrow R\mathbf{H} \cdot \text{Cyto } P_{450}\ (Fe^{2+}) \cdot (O_2)$
4. $R\mathbf{H} \cdot \text{cyto } P_{450}\ (Fe^{2+}) \cdot (O_2) + (Fe^{2+}\text{-}S)_{II} \rightarrow$

$$R\mathbf{H} \cdot \text{Cyto } P_{450}\ (Fe^{2+}) \cdot (O_2) + (Fe^{3+}\text{-}S)_{II}$$

Next, the cytochrome P_{450} must be returned to its Fe^{3+} state and concurrently the hydroxylation of RH occurs:

5. $\text{NADP } \boxed{H} + \boxed{H^+} + \text{FAD} \rightarrow \text{NADP}^- + \text{FAD} \cdot \boxed{2H}$

$$2(Fe^{3+}\text{-}S)_{(\text{I and II})} + R\mathbf{H} \cdot \text{cyto } P_{450}\ (Fe^{2+})(O_2) + \text{FAD} \cdot 2H \rightarrow$$

$$\text{FAD} + 2(Fe^{2+}\text{-}S)_{(\text{I and II})}\ R\ (O\ \mathbf{H} + \text{cyto } P_{450}\ (Fe^{3+}) + \boxed{H}\ (O\ \boxed{H}$$

The net reaction becomes

$$R\mathbf{H} + \text{NADP } \boxed{H} + \boxed{H^+} + (O_2) \rightarrow RO\mathbf{H} + \text{NADP}^+ + \boxed{H}\ (O\ \boxed{H}$$

The hydroxylation of C-22 proceeds by the same mechanism and with an additional oxidation, cleavage occurs between C-20 and C-22 to yield pregnenolone and isocaproic aldehyde:

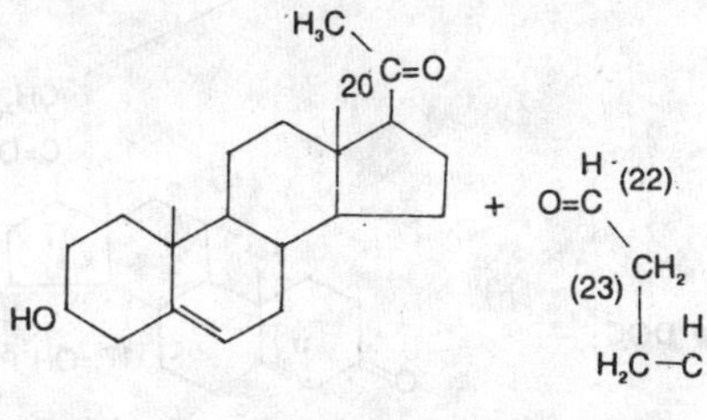

Conversion of Pregnenolone to Progesterone

Following its synthesis, pregnenolone leaves the mitochondria and is converted to progesterone by two microsomal enzyme, $\Delta^{4,5}$-isomerase and 3-β-ol-dehydrogenase:

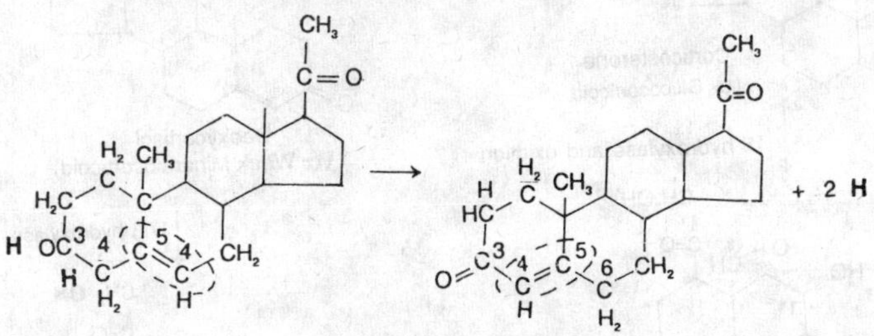

In the corpus luteum of the ovaries there are no reactions beyond this point and progesterone becomes the hormonal product of this tissue.

Conversion of Progesterone to Cortisol and Cortisone

In considering the synthesis of the mineralo- and glucocorticoids it is useful to recall that the site of their synthesis—the adrenal cortex—has two different types of cell layers. Just beneath the adrenal capsule lies the zona glomerulosa, which produces aldosterone. Deeper in the cortex is the zona fasciculata, the site of cortisol synthesis. The hydroxylation reactions to be described below are divided between the mitochondria and the endoplasmic reticulum and involve the mixed-function type of mechanisms. The sequence of the hydroxylation reactions and the physiologic activities of the intermediate may be formulated as follows:

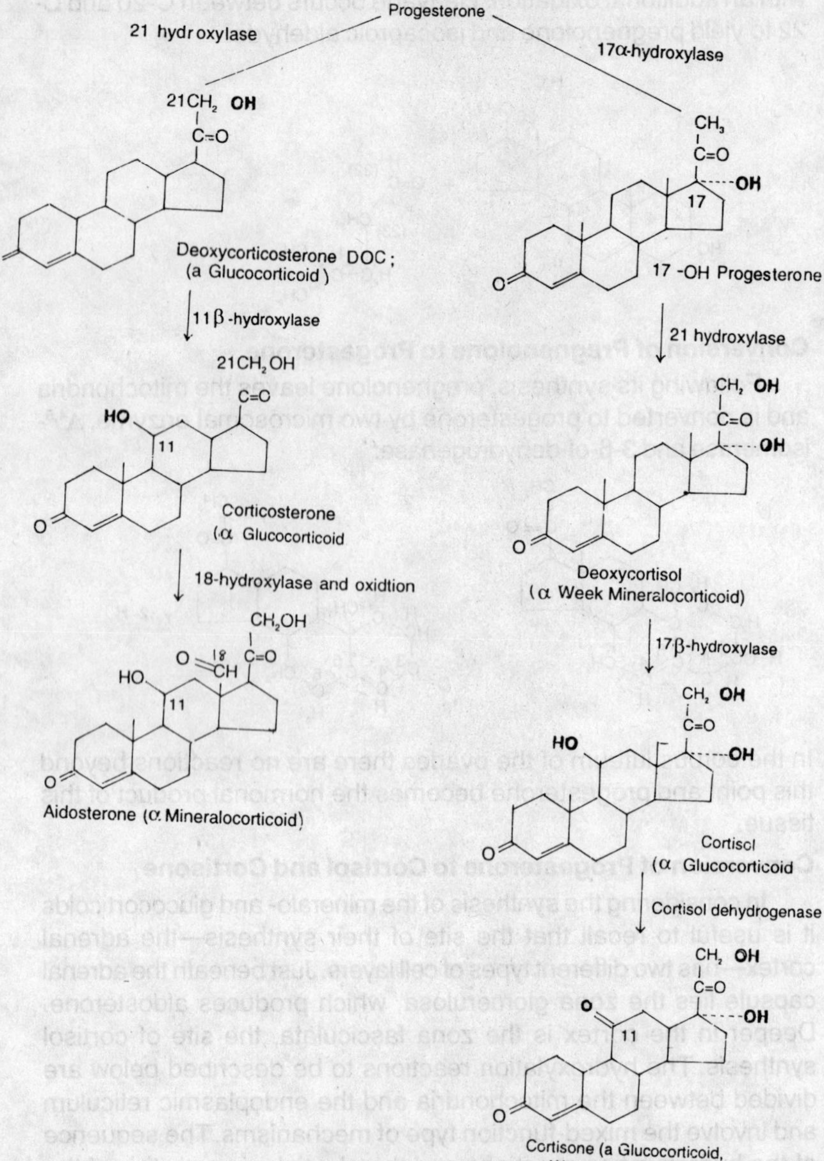
Progesterone
21 hydr oxylase
17α-hydroxylase
21CH₂ OH
C=O
Deoxycorticosterone DOC ;
(a Glucocorticoid)
CH₃
C=O
OH
17
O
17 -OH Progesterone
11β -hydroxylase
21CH₂OH
C=O
HO
11
O
Corticosterone
(α Glucocorticoid
21 hydroxylase
CH₂ OH
C=O
OH
O
Deoxycortisol
(α Week Mineralocorticoid)
18-hydroxylase and oxidtion
CH₂OH
O
18
CH
C=O
HO
11
O
Aidosterone (α Mineralocorticoid)
17β-hydroxylase
CH₂ OH
C=O
HO
OH
O
Cortisol
(α Glucocorticoid
Cortisol dehydrogenase
CH₂ OH
C=O
O
OH
11
O
Cortisone (a Glucocorticoid,
Weaker than cortisol)

The 17α-hydroxylase and the 21-hydroxylase are located in the endoplasmic reticulum and are faster acting than the 11β-hydroxylase, which is in the mitochondria. The bifurcation of the synthetic pathways from progesterone is due primarily to stereochemical restrictions related to C-21. When hydroxylation occurs first at C-21, the presence of the —OH at this position prevents hydroxylation at C-17. As a result, deoxycortisol cannot be formed and the subsequent hydroxylations are routed via DOC and corticosterone to aldosterone. The slow-acting 11β-hydroxylase can hydroxylate either DOC or deoxycortisol once they have entered the mitochondria.

Conversion of Progesterone to the Six Hormones in the Gonads

The synthetic sequence for converting progesterone to the androgens in the endoplasmic reticulum of Leydig cells in the testes begins with hydroxylation of C-17.

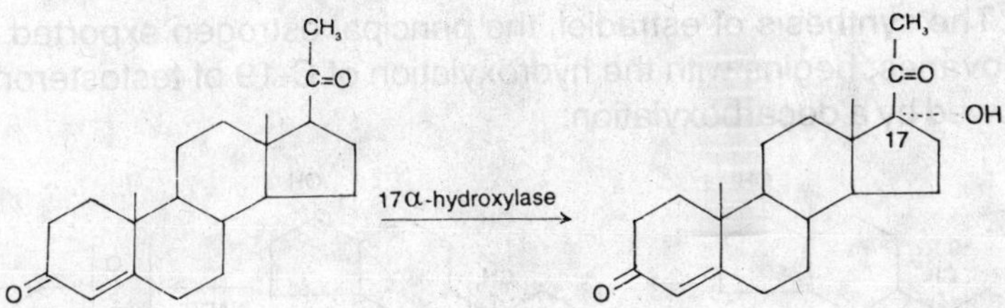

C-20 and C21 are now removed oxidatively to yield a C-17 ketone, Δ^4-androstenedione:

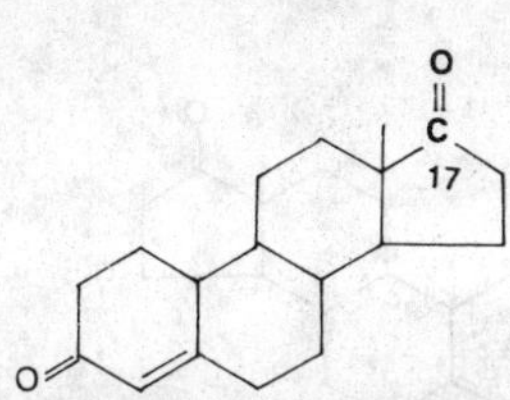

The 17-keto group is capable of a reversible oxidation-reduction and yields testosterone:

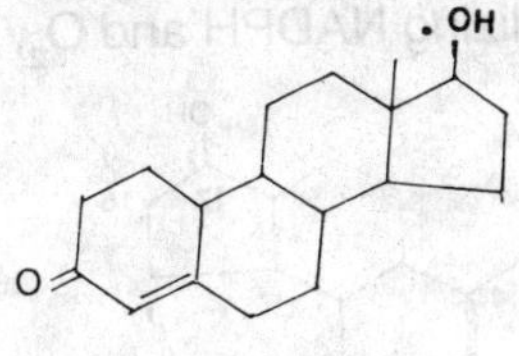

It may be recalled that testerone can also be formed in the adrenal cortex by the isomer of progesterone Δ^5-pregnenolone :

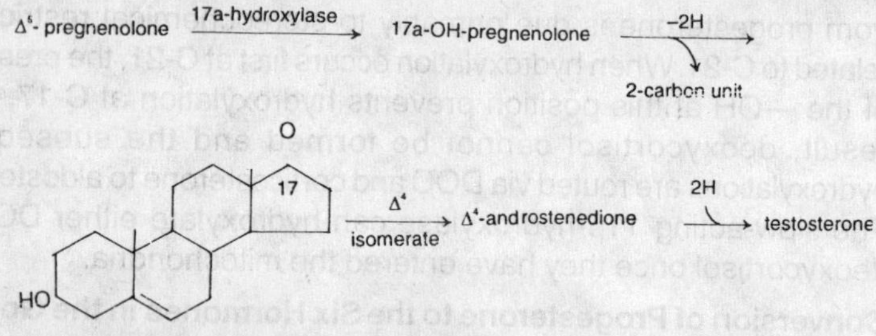

Dehydroepiandrosterone (DHA)

Formation of the Female Steroid Hormones from Testosterone

The synthesis of estradiol, the principal estrogen exported by the ovaries, begins with the hydroxylation of C-19 of testosterone, followed by a decarboxylation:

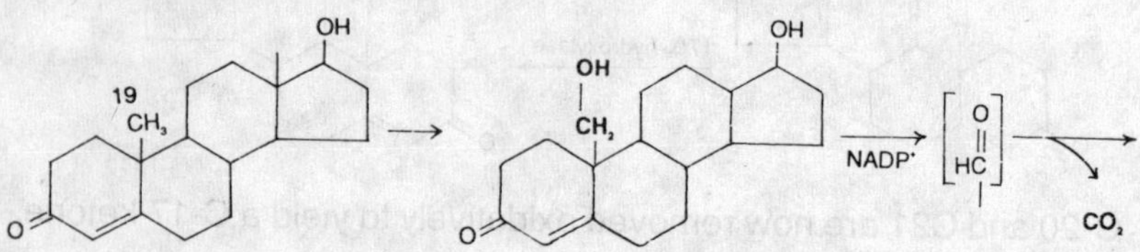

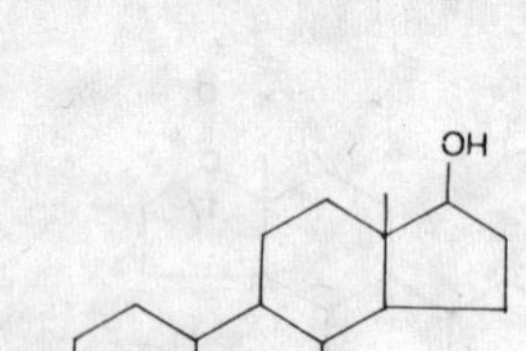

19-Nortestosterone

The A ring of the nortestosterone is now aromatized by a mixed-function oxidase system, utilizing NADPH and O_2, and the product is 17β-estradiol:

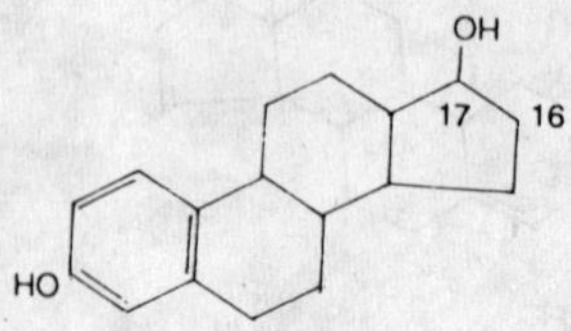

Estradiol can be converted into two less potent hormones either by oxidation of the C-17 hydroxyl or by hydroxylation of C-16:

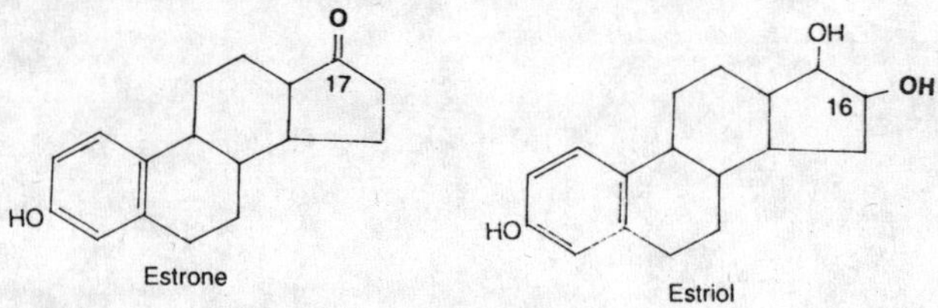

FORMATION OF DIHYDROTESTOSTERONE FROM TESTOSTERONE

As the principal circulating androgen, testosterone can either cells by passive diffusion and be reduced in the endoplasmic reticulum or the nuclear membrane :

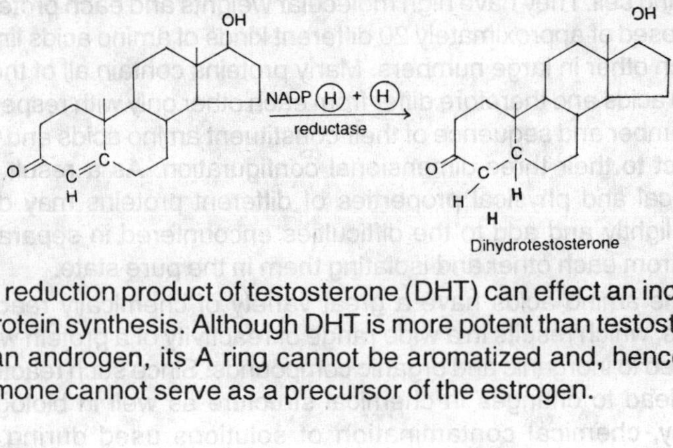

The reduction product of testosterone (DHT) can effect an increase in protein synthesis. Although DHT is more potent than testosterone as an androgen, its A ring cannot be aromatized and, hence, this hormone cannot serve as a precursor of the estrogen.

8

PROTEINS

Proteins are the most characteristic chemical compounds found in the living cell. They have high molecular weights and each protein is composed of approximately 20 different kinds of amino acids linked to each other in large numbers. Many proteins contain all of the 20 amino acids and therefore differ from each other only with respect to the number and sequence of their constituent amino acids and with respect to their three-dimensional configuration. As a result, the chemical and physical properties of different proteins may differ only slightly and add to the difficulties encountered in separating them from each other and isolating them in the pure state.

The amino acids have a great variety of chemically reactive groups, which results in a wide range of reactivity of a protein when exposed to inorganic and organic compounds. Since such reactions often lead to changes in chemical structure as well in biological activity, chemical contamination of solutions used during the procedure of protein purification must be rigorously prevented. A lack of appreciation of the sensitivity of proteins to such chemical reagents often leads to the isolation of a protein in a form different from that present in the cell.

In addition to covalent bonds, which bind amino acids to each other, proteins possess weaker but very important bonds that hold the macromolecule in a unique configuration. Such bonds are quite sensitive to environmental conditions—e.g., excessive stirring of a protein solution in air, exposure to ultraviolet light, elevated temperatures, marked changes in pH, and organic solvents. These procedures lead to alternation of protein structure characterized by loss of solubility and of any biological activity, even though covalent bonds may not have been broken. The protein is said to be *denatured* and frequently the change is irreversible; the *native state* has been destroyed.

Occasionally changes in environmental conditions lead to dissociations of a protein into molecules of smaller size, or of association into larger aggregates. Chemical as well as biological properties of the protein are affected by such changes.

Despite the difficulties encountered in protein purification, many proteins have been isolated in the crystalline state with a high degree of purity and biological activity. A variety of techniques have been developed that take into account the sensitivity of proteins to the usual laboratory manipulations. Possession by a protein molecule of some measurable biological activity, enzymatic or hormonal, simplifies the problem of purification, since we may use an increase in its *specific activity* as a measure of purification. Specific activity may be defined as the activity (enzymatic or hormonal) per unit weight of solids, or, since all enzymes are proteins, their specific activity is usually defined per unit weight of total protein. Specific activity of the protein enzyme or hormone will increase until no further purification occurs, either because the protein has reached maximum purity or because the procedure has not effected further purification. Since minor alterations in protein structure often lead to loss of activity, isolation of a protein with its activity intact is some assurance that the isolated molecule resembles that which is present in the cell in the native state.

Classification of Proteins

According to Structure

Fibrous proteins. These proteins are associated with cellular elements and often serve the function of supporting specific structures of the cell. Formerly they were called albuminoids or selerins. Examples are wool, silk fibroin, collagen (connective tissue), myosin (muscle), keratin (hair), and fibrin (blood clot). These proteins are largely insoluble in aqueous media and have high molecular weights that cannot be accurately estimated because of difficulties encountered in their purification. They appear as fibers made up of linear molecules that are arranged roughly parallel to the fiber axis. They are amorphous and some are capable of stretching and contracting. Human fibrin has the molecular dimensions of 38 × 700 Å.

Globular Proteins. In contrast to fibrous proteins, the globular proteins are soluble in aqueous media and can be isolated in the crystalline state. Although not necessarily spherical, they are less asymmetric than fibrous proteins and consist of polypeptides (chains of amino acid residues) that are held together by cross-linked groups. Such units are interwoven in a unique fashion and held together in a three-dimensional structure by relatively weak bonds.

According to Solubility

Albumins. These are water-soluble proteins that may be precipitated from solution at high salt concentrations, a process called "salting out." Decrease in solubility in the presence of high concentration of salts is due to competition between the salt and protein molecules for water, as well as to a decrease in charge on the protein molecules. Examples are egg albumin, serum albumin, lactalbumin (milk), and leucosin (wheat).

Globumins. Unlike albumins, these proteins are generally insoluble in salt-free water, and soluble in dilute salt solutions but insoluble in salt solutions at 30 to 50 percent saturation. With adequate care (low temperature and minimal stirring) the precipitation of globulins or albumins by salting-out procedures need not be accompanied by denaturation. Examples are serum globulins, ovoglobulin (egg), myosinogen (muscle), edestin (hemp seed), amandin (almonds), and excelsin (Brazil nuts).

Glutelins. There are insoluble in neutral aqueous solutions but soluble in dilute acid or alkali. Examples are glutenin (wheat) and oryzenin (rice).

Gliadins (*Prolamins*). These are soluble in 70 to 80 percent ethanol and insoluble in water or absolute ethanol. Examples are gliadin (wheat), hordein (barley), and zein (corn).

Histones. More basic than most proteins, the histones tend to form complexes with acidic compounds in the cell (nucleic acids). They are soluble in water and insoluble in dilute ammonia solution. Examples are globin (hemoglobin), thymus histone (also called nucleohistone since it is found combined with nucleic acids), and scombrone (mackerel).

Protamines. Compared with most proteins, these are relatively small molecules. The protamines are soluble in water and basic in character. They are found associated with nucleic acids in the sperm of fish, and these are often called nucleoprotamines. Examples are salmine (salmon), sturine (sturgeon), clupeine (herring), and cyprinine (carp).

According to Nonprotein Moiety (Conjugated Proteins)

Nucleoproteins. In addition to nucleohistones and nucleoprotamines, there is a third group of proteins capable of binding nucleic acids. These nucleoproteins are not basic and hold the nucleic acid by secondary valence bonds. They occur in microorganisms and are soluble in isotonic salt solution.

Glycoproteins. These contain carbohydrate moities, including hexoses, hexosamines, and hexuronic acids. Examples are mucin (salvia), osscomucoid (bone), and tendomucoid (tendon).

Phosphoproteins. Examples are casein (milk) and vitellin (egg yolk).

Chromoproteins. These proteins are pigmented owing to the prosthetic nonprotein group. They include hemoglobin (containing heme, an iron-protoporphyrin), cerulaplasmin (copper), hemocyanin (copper), ascorbic acid oxidase (copper), and ferritin (iron).

Dehydrogenases. These are conjugated protein enzymes that possess oxidizing activity because of the presence of prosthetic groups such as NAD (nicotinamide adenine dinucleotides), NADP (nicotinamide adenine dinucleotide phosphate), FMN (flavin mononucleotide), and FAD (flavin adenine dinucleotide).

According to State of Degradation

Native Protein. Protein in a state presumed to be identical with that present inside the living cell and possessing a unique structure required for biological activity are called native proteins.

Denatured Protein. A protein whose unique native state has been disorganized, so that the original three-dimensional configuration has been converted to that of a random coil is said to be denatured. No chemical groups have been lost from the original molecule.

Derived Proteins. These are products resulting from alterations in the original protein that involve the removal of amino acid residues or larger fragments (polypeptides). In some cases, this change does not lead to denaturation or loss of biological activity. Indeed, the conversion of pepsinogen to pepsin involves the change from an inactive protein to the active enzyme. Hydrolytic cleavage of proteins leads to ill defined products, listed in order of increasing state of degradation : primary protein derivatives, metaproteins, proteoses, peptones and peptides. These terms have little chemical significance.

Primary Structure of Proteins

Proteins are made of one or more polypeptide chains. Some proteins contain other kinds of molecules conjugated with the polypeptides, such as carbohydrates or fats, but a substance isn't a protein unless it contain a polypeptide.

The fundamental principle of protein chemistry is that one protein differs from another because each of its polypeptide chains consists of amino acids polymerized in a particular sequence. The amino acid sequences of the constituent polypeptides are the primary structure of the protein. It is this defined arrangement of amino acid side chains that ultimately gives a protein its properties. Any change in primary structure creates a different protein. Proteins that differ in only a single amino acid residue at some locations may have nearly identical properties, but there are other locations at which substitution of only one amino for another can cause changes in the nature of the protein that are drastic enough to be life-threatening if the protein has a critical function.

Number of different Polypeptides

Although there are only 20 different kinds of amino acids from which proteins are made, there is no danger of exhausting the possible primary structures for the creation of new proteins. A protein made by combining only 100 amino acids molecules is quite small as proteins go, and yet 20 different amino acids can be combined 100 at a time in 20^{100} different ways. This number is so large that every single protein molecule conceivable in our universe, existing now or at any time in the past, could be unique without using more than an infinitesimal portion of the possibilities.

In fact, however, each protein molecule is not unique. Informed guesses at the number of different kinds of proteins that a newborn infant makes range between 10,000 and 100,000. (These figures neglect differences caused by partial hydrolysis or other modifications of a polypeptide chain after it is formed). Assuming there are 40,000 the infant will have on the average about 1×10^{17} identical molecules of each kind of protein. In addition, every one of the $\sim 1.5 \times 10^8$ infants born each year will contain identical copies of many of these kinds of protein molecules. Their common humanity is a result of this fact, and it is a marvelous thing that something like 20,000 atoms, to use a typical number, can repeatedly be assembled with so few errors in exactly the same way in each and every human conceived.

However, some protein molecules will be different in two infants taken at random (but not in identical twins). That is, there are many proteins of which one infant will have 10^{17} molecules and the other infant will have none. The second infant is likely to have molecules of a similar, but not identical, protein in its place. The individuality of the two infants comes from these differences in protein composition.

Similarly, the fly that buzzes around the infant will contain many copies of protein molecules that are identical to molecules found in another fly and some that differ in detail. However, it is almost certain that the human and the fly have no protein molecules in common; their functions are too different to be served in exactly the same way. Proteins are made to satisfy these diverse needs by varying the way in which the 20 kinds of amino acids are put together, that is, by altering the primary structure, and specification of the variations is a major part of the heredity of the organism.

Amino Acid Composition of Proteins

In order to determine the quantity of each amino acid in a protein, the latter is subjected to hydrolysis, using acid, alkali, or enzymes, depending on the amino acid to be estimated. Acid hydrolysis, in the absence of air is usually preferred. Alkali hydrolyses the amide groups of asparagine and glutamine, destroys cystine and arginine, and leads to recemization. Enzymes are not only slow in their action out the hydrolysis may be incomplete, and the presence of the enzyme protein may interfere with the analysis. Some of the analytical methods that have been devised follow.

Chemical Methods

The earliest, and from the quantitative point of view, the crudest method involved the direct isolation of amino acids or their derivatives, based on solubility differences. The introduction of specific colour reactions for many amino acids, replaced the less quantitative isolation procedures but did not provide highly specific methods for all amino acids.

Microbiological Methods

Many microorganisms can be grown in media that are fairly simple and whose composition can be varied at will. These microorganism often require particular amino acids (essential amino acids) for growth. Within certain limitations a direct relationship can be demonstrated between the quantity of such an amino acid in the medium and the ability of the organism to grow (divide). Sometimes,

Table 8.1. Amino Acid Composition of Some Representative Proteins

Amino acid	Human Serum Albumin		Horse Hemoglobin		β-Lactoglobulin		Salmine	
	A	B	A	B	A	B	A	B
1	2	3	4	5	6	7	8	9
Alanine	...	...	7.40	54	7.07	30	1.12	1
Glycine	1.60	15	5.60	48	1.50	7	2.95	3
Valine	7.70	45	9.10	50	5.67	18	3.14	2
Leucine	11.00	58	15.40	75	15.48	44	0	0
Isoleucine	1.70	9	0	0	5.88	17	1.64	1
Proline	5.10	31	3.90	22	5.27	17	5.80	4
Phenylalanine	7.80	33	7.70	30	3.86	9	0	0
Tyrosine	4.70	18	3.03	11	3.69	8	0	0
Trytophan	0.20	1	1.70	5	1.92	3	0	0
Serine	3.34	22	5.80	35	4.07	14	9.1	7
Threonine	4.60	27	4.36	24	5.15	16	0	0
Cystine/2	5.60	32	0.45	2.5	2.29	7	0	0
Cysteine	0.70	4	0.56	3	1.10	3	0	0
Methionine	1.30	6	1.0	4.5	3.21	8	0	0
Arginine	6.20	25	3.65	14	2.88	6	85.2	40
Histidine	3.50	16	8.71	36	1.60	4	0	0
Lysine	12.30	58	8.51	38	11.30	29	0	0
Aspartic acid	8.95	46	10.60	51	11.46	32	0	0
Glutamic acid	17.0	80	8.50	38	19.10	48	0	0
Amide N	0.88	44	0.87	36	1.07	28	0	0

Table (Continue)

Insulin (Ox)		Ribonuclease		Pepsin		Keratin (Wool)		Gelatin	
A	B	A	B	A	B	A	B	A	B
10	11	12	13	14	15	16	17	18	19
4.5	6	...	...	...	...	4.14	46.4	9.3	104.6
4.3	7	1.3	3	6.4	29	6.53	87.0	26.9	359.0
7.75	8	7.3	9	7.1	21	4.64	39.7	3.3	28.2
13.2	12	0	0	10.4	27	11.3	86.3	3.4	26.2
2.77	3	3.1	4	10.8	28	...	...	1.8	13.8
2.9	2-3	3.6	5	5.0	15	9.5	82.6	14.8	129.0
8.14	6	3.6	3	6.4	13	3.65	22.1	2.55	15.5
12.5	8	7.93	7	8.5	16	4.65	25.7	1.0	5.5
0	0	0	0	2.4	4	1.8	8.8	0	0
5.23	6	12.0	17	12.2	40	10.01	95.4	3.18	30.0
2.08	2	9.0	11	9.6	28	6.42	53.9	2.2	18.5
12.5	12	6.51	8	1.64	4	11.9	98.9	0	0
0	0	0.6	0.7	0.5	2	...	...	0	0
0	0	4.43	5	1.7	4	0.7	4.7	0.9	6.1
3.07	2	5.16	5	1.0	2	10.4	59.7	8.55	49.2
5.21	4	4.22	4	0.9	2	1.1	6.83	0.73	4.71
2.51	2	10.4	11	0.9	2	2.76	18.9	4.60	31.5
6.80	6	14.2	16	16.0	41	7.2	54.1	6.7	50.6
18.60	15	13.0	13	11.9	28	14.1	96.0	11.2	76.1
1.39	12	2.05	22	1.32	32	1.17	83.2	0.07	5.3

A=grams of amino acid in 100 grams of protein; B=residues of amino acid per molecule; C=residues of amino acid per 10^5 grams of protein.

evidence of a specific metabolic activity—e.g., lactic acid production—is measured. A *standard* growth curve (or acid production curve) can be constructed to show this relationship for specific quantities of the amino acid. By comparison of this standard curve with the growth of the organism in a medium that lacks the amino acid under examination but to which has been added varying quantities of a neutralized protein hydrolysate, we can calculate the quantity of amino acid in the hydrolysate. This method has been quite helpful but suffers from several limitations. Amino acids that are available as standards are not always pure, microorganisms often substitute one amino acid for another, and the reaction of an organism to specific quantities of an amino acid often differs in the presence of other amino acids.

Isotope Dilution Method

This method can be extremely accurate, but it requires pure isotopically-labeled amino acids, it is time consuming, and sometimes as is the case with stable isotopes, it requires costly assay equipment (mass spectrometer). In principle, an amino acid containing an isotope of known specific activity is dissolved in a known quantity of a protein hydrolysate, and the same amino acid is then isolated and purified, either as such or as a derivative. It is not necessary to isolate the amino acid quantitatively, but rather to be certain of its purity. If we assume that the tagged amino acid that was added has the same solubility as the nonisotopic amino acid in the hydrolysate, a comparison of the specific isotope activities of the amino acid that was added and the one that was isolated will reveal the extent of dilution of the added acid by that present in the hydrolysate. For example, the amount of glycine originally present in a hydrolysate (y) can be calculated from the amount of glycine added (x), and it content of N^{15} $(C_0)_9$ as well as the N^{15} content of the isolated glycine (C):

$$y = \left(\frac{C_0}{C} - 1\right) x$$

Chromatographic Analysis

By far the most useful and precise method for determining the quantity of each amino acid in a protein hydrolysate is the separation of the amino acids on a column of adsorbent, followed by the consecutive elution of each acid. As little as 2 mg. of pure protein is required for a complete analysis. The amino acids are eluted separately by buffer solutions of varying composition. The solutions,

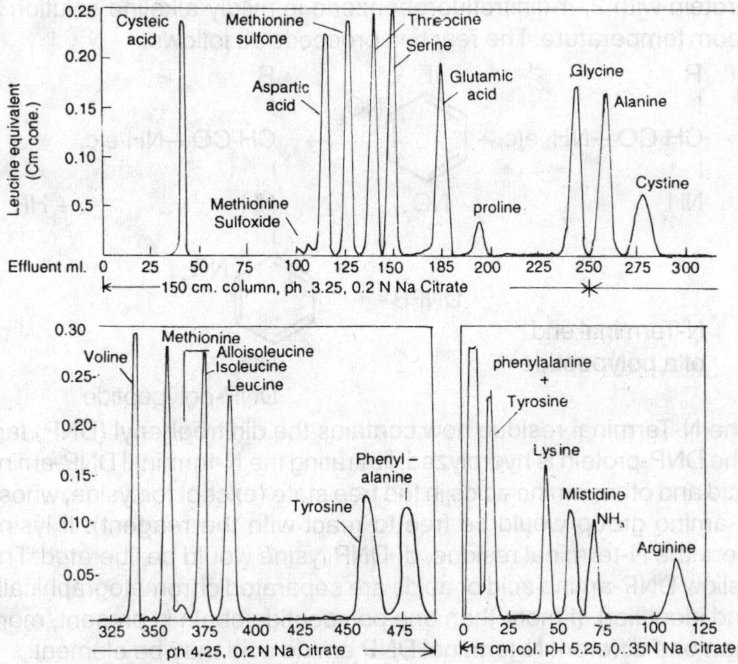

Fig. 8.1. Chromatographic fractionation of a synthetic mixture of amino acids on columns of Amberlite IR-120.

as they leave the column, are separated into small fractions and ninhydrin reagent is added to each for colorimetric estimation of amino acid content. By the use of known amino acids and standardization of conditions, the position of each amino acid in the effluent can be determined. Figure illustrates the use of this technique. Automatic devices are now incorporated into an apparatus that simplifies the determination. The quantity of each amino acid is calculated from the area under each curve.

Sequence Determination

A knowledge of the number of amino acid residues of each kind per molecule of protein does not reveal the sequence in which these amino acid residues appear in the polypeptide chain. Techniques for the determination of amino acid sequences will now be discussed.

N-Terminal Amino Acids

A single polypeptide chain contains one N-terminal amino acid residue whose amino group is usually free. Sanger devised a technique for identifying this residue based on the reaction of the protein with 2, 4-dinitrofluorobenzene in mildly alkaline solution at room temperature. The reaction proceeds as follows:

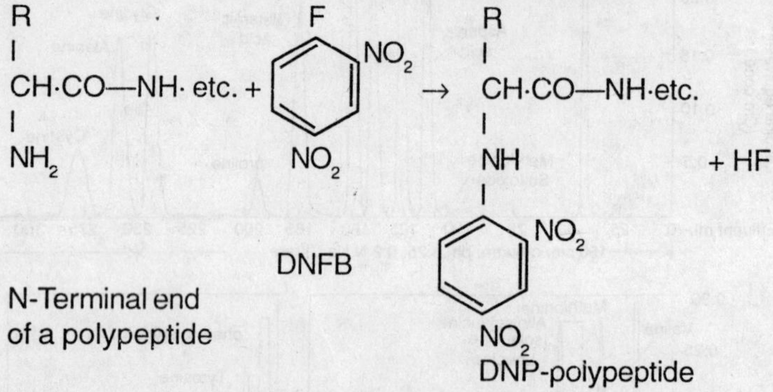

The N-Terminal residue now contains the dinitrophenyl (DNP) tag. The DNP-protein is hydrolyzed, liberating the N-terminal DNP-amino acid and other amino acids in the free state (except for lysine, whose ϵ-amino group would be free to react with the reagent). If lysine were the N-terminal residue, di-DNP lysine would be liberated. The yellow DNP-amino acid or acids are separated chromatographically and identified. If more than one polypeptide chain is present, more than one different N-terminal DNP amino acid may be element.

Another method for identifying the N-terminal residue, and one that is more helpful for the analysis of small peptides because amino and residues can be removed from the N-terminal position one at a time. The peptide is reacted with phenylisothiocyanate and then treated with acid, liberating the N-terminal residue as the phenylthiohydantoin, together with the peptide, which now has a new N-terminal residue. The reaction is illustrated as follows:

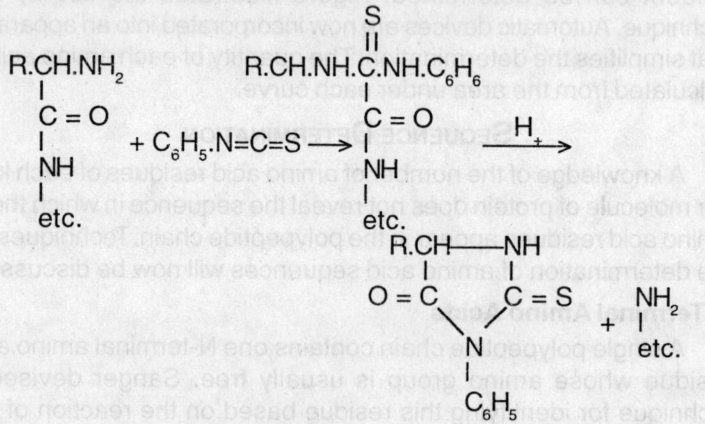

C-Terminal Amino Acids

Several methods for determining the nature of the C-terminal residue are available. One of these is based on the specificity of the enzyme carboxypeptidase, which preferentially hydrolyzes the peptide linkage involving the C-terminal residue. It is necessary to examine the free amino acids liberated at various intervals in order to determine which amino acid is first liberated, because as thus is amino residue is removed, another C terminal residue appears.

Another method for identifying the C-terminal residue involves the conversion of the terminal carboxyl group to the primary alcohol by means of $LiAlH_4$ and $NaBH_4$, followed by hydrolysis and isolation of the amino alcohol and its identification.

Polypeptide Cross-linkages

Although the basic unit of some proteins contains but one polypeptide chain, other basic units contain two or more. In eitner case the polypeptide chains may be held together by disulfide groups of two half-cystine residues. Treatment of the protein with performic acid results in oxidation of the disulfide to two sulfonic acid groups that are then part of different cysteic acid residues. The modified polypeptide chains can then be separated. Disulfide bridges can also connect portions of the same polypeptide chain, but their cleavage by performic acid oxidation does not give rise to several polypeptides:

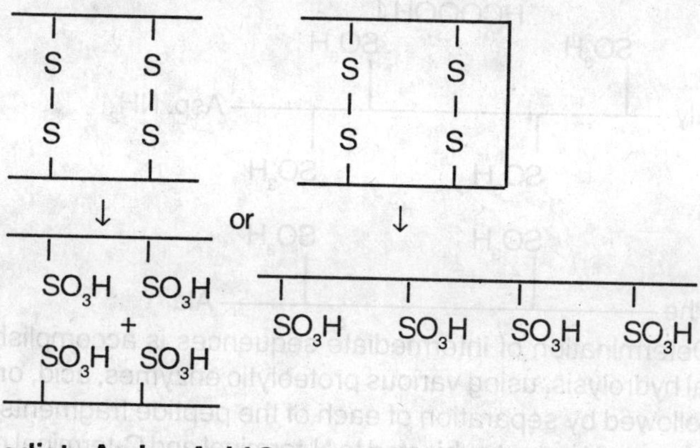

Intermediate Amino Acid Sequences

Completion of the primary structural analysis of a protein requires and elucidation of the sequence of residues between the N-terminal and C-terminal residues of each polypeptide chain. We can illustrate this technique with insulin, whose complete primary structure was the first to be determined (Sanger).

The unit monomer of insulin has a molecular weight of approximately 6000. Application of the Sanger method of N-terminal amino acids analysis results in the isolation of two N-terminal amino acids—glycine and phenylalanine. The C-terminal amino acids are asparagine and alanine. The insulin molecule contains three disulfide bridges. Treatment of insulin with performic acid results in the production of two altered polypeptides—A and B—which can be separated. The A chain has glycine and asparagine as N- and C-terminal residues respectively, whereas the B chain has phenylalanine and alanine as N- and C-terminal residues, respectively. The altered A chain contains for cysteic acid residues, whereas the B chain has two such residues. Reconstruction of the original molecule from these data suggests the following preliminary structure:

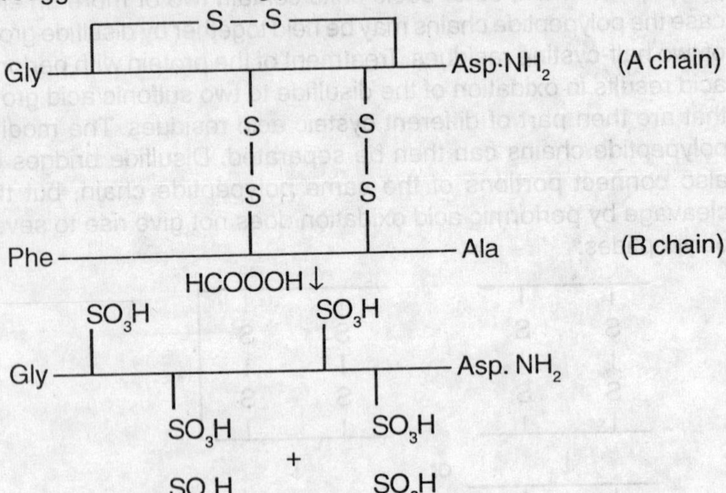

Determination of intermediate sequences is accomplished by partial hydrolysis, using various proteolytic enzymes, acid, or alkali, and followed by séparation of each of the peptide fragments. Each small fragment is next subjected to N-terminal and C-terminal residue analyses and further degraded until the complete sequence of each fragment is determined. As examination of the sequences in all of the fragments show enough overlapping of residues to enable one to reconstruct the sequences in the original polypeptide.

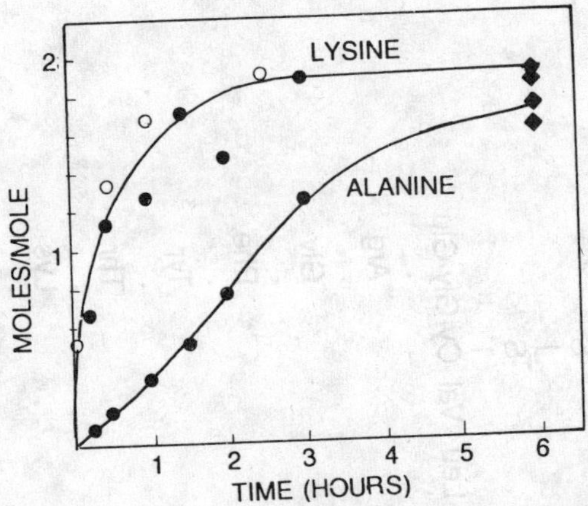

Fig. 8.2. The kinetics of amino acid release from rabbit muscle enolase during digestion with carboxypeptidase.

Table 3.2. Deduction of Peptide Sequence from Peptide Fragments

---

Dipeptides

ser—his leu—val glu—ala

his—leu val—glu

Tripeptides

ser—his—leu

leu—val—glu

val—glu—ala

ala—leu—tyr

tyr—leu—val

Higher peptides

ser—his—leu—val—glu

val—glu—ala—leu leu—val—$CySO_3H$—gly—

Deduced sequence

—ser—his—leu—val—glu—ala—leu—tyr—leu—val—$CySO_3H$—gly—

Figure shows the complete sequence of residues in the insulin molecule. It consists of two polypeptide chains held together by disulfide bridges.

Figure shows the complete amino acid sequence for the first enzyme molecule whose primary structure was clarified—ribonuclease. This molecule, which contains 124 residues, consists

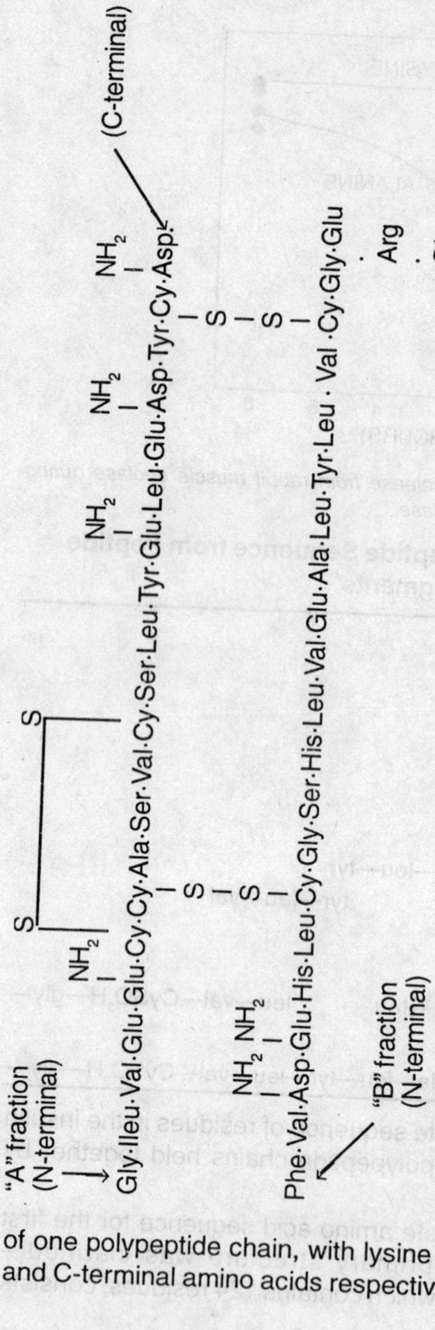

Ala = alanine
Arg = arginine
Asp-NH_2 = asparagine
Cy-S-S-Cy = cystine
Glu = glutamic acid
Glu-NH_2 = glutamine
Gly = glycine
His = histidine
Ileu = isoleucine

Leu = leucine
Lys = lysine
Phe = phenylalanine
Pro = proline
Ser = serine
Thr = threonine
Tyr = tyrosine
Val = valine

Fig. 8.3. The structure of beef insulin.

of one polypeptide chain, with lysine and valine as the N-terminal and C-terminal amino acids respectively. The disulphide bridges of eight half-cystine residues connect different portions of the chain resulting in a structure that has several restrictions for the three-dimensional folding of the molecule.

Figure shows the complete amino acid sequence in cytochrome c from horse heart. It is of interest for two reasons—the N-terminal amino acid glycine is acetylated (also true for tobacco mosaic virus protein, whose N-terminal residue is acetyl serine), and the two cysteine residues are attached to the vinyl side chains of the porphyrin by thioether bonds.

5 10 15
Acetyl-gly-asp-val-glu lys-gly-lys-ileu-phe-*val*-*gln*-lys-cys-*ala*-gln-cys-
└ heme ┘

20 25 30 35
his-thr-val-glu-lys-gly-gly-lys-his-lys-thr-gly-pro-asn-leu-his-gly-leu-phe-

40 45 50 55
gly-arg-lys-thr-gly-gln-ala-pro-gly-*phe* -*thr*-tyr-thr-*asp*-ala-asn-lys-asn-lys-

60 65 70
gly-ileu-*thr*-try-*lys*-glu-*glu*-thr-leu-met-glu-tyr-leu-glu-asn-pro-lys-lys-

75 80 85 90
tyr-ileu-pro-gly-thr--lys-met-ileu-phe-*ala*-gly-ileu-lys-lys-lys-*thr*-glu-agr-

95 100
glu-asp-leu-ileu-ala-try-leu-lys-lys-ala-thr-asu-glu-COOH

Fig. 8.4. The amino acid sequence of horse heart cytochrome C. Residues that differ in human heart protein are indicated by italics.

Amino Acids Replacement in Hemoglobin

The hemoglobins have served as an excellent source of a pure protein that can be obtained from an almost unlimited number of individuals from various around the world and that can be analyzed for differences in amino acid sequences.

Normal adult human hemoglobin consists of four polypeptide chains—two α- and β-chains. It may therefore be designated by the symbols $\alpha_2{}^A\beta_2{}^A$, where A identifies the chains as belonging to adult hemoglobin (HbA). Each α-chain consists of 141 amino acid residues, with valine and arginine as the N-terminal and C-terminal residues respectively. Each of the β-chains contains 146 residues, with valine and histidine as N-terminal and C-terminal residues. The intermediate sequences for both chains have been identified.

Sickle cell anemia is a disease characterized by the "sickling" or red cells induced by lowered oxygen tension. The hemoglobin in such cell appears to aggregate, its solubility is decreased, and the cell is removed from the circulation at an abnormally rapid rate, leading to severe anemia. The disease is hereditary, and severe sickle cell anemia occurs in individuals homozygous for an abnormal gene. It also occurs as a sickle cell trait, in the absence of an anemia, in carriers who are heterozygous, having one normal and one abnormal gene. The demonstration of an abnormal hemoglobin (Hb S) in red cells from individuals with the disease was made by Pauling, Itano, Singer, and Wells. It was shown that the hemoglobin prepared from individuals with this disease possessed an electro-phoretic mobility, when measured at a pH alkaline to its IEP, lower than that of normal adult hemoglobin. Pauling called sickle cell anemia a "molecular disease", to replace the earlier designation "inborn error of metabolism" suggested by Garrod for the inherited disease alcaptonuria.

As information about the amino acid sequence in hemoglobin developed, it became apparent that the differences between Hb A and Hb S was slight. The exact differences was narrowed to be replacement of a glutamic acid residue at position 6 of the β-chain in Hb A by a valine residue in Hb S. The reason for the lower mobility of Hb S was clear, since the negatively charged glutamic acid residue had been replaced by an uncharged glutamic acid residue had been replaced by an uncharged valine. Detection of such a minor difference is somewhat easier than might be anticipated, because of the use of a "mapping" or "finger print" technique. The protein (or polypeptide chain) is treated with a proteolytic enzyme, such as trypsin, which hydrolyzes the chain at specific positions. The resulting peptide mixture is separated by the use of paper electrophoresis and chromatography. When trypsin acts on Hb A and Hb S, the results in shown figure are obtained. The peptide fragments appear as sports produced by reaction with ninhydrin. Close examination of the two "maps" reveals that Hb S has one peptide that occupies a new and different position from the corresponding peptide in Hb A. In a duplicate chromatogram (not treated with ninhydrin), the spot is cut from the paper, and the peptide is eluted and subjected to further analyses to reveal its exact amino acid sequence. Peptide No. 4 is an octapeptide having the following sequences:

Hb A val-his-leu-thero-pro-glu-glu-lys

Hb S val-his-leu-threo-pro-val-glu-lys

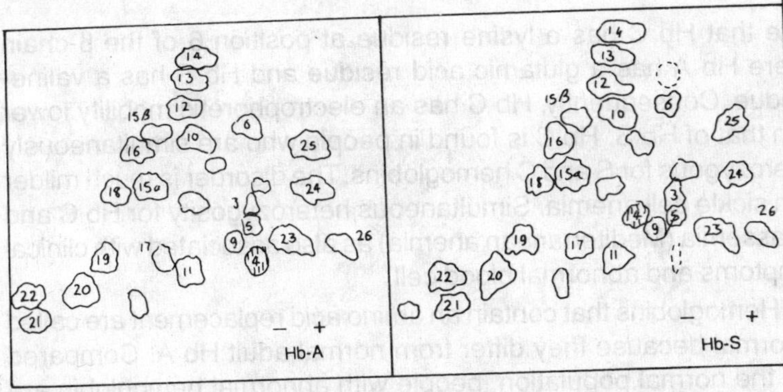

Fig. 8.5. Peptide fingerprints of hemoglobins A and S.

This technique has made it possible to examine the hemoglobin of large numbers of individuals, and many instances of "abnormal" hemoglobins have been discovered. These all appear to have a single amino acid replacement, either in the α-chain or the β-chain. They are summarized in Table 8.3.

Table 8.3. Amino Acid Replacement in the Hemoglobin of Humans.

	α-Chain						β-Chain					
	16	30	57	58	68	116	6	7	26	63	67	125
Hb A	lys	glu	gly	his	asn	his	glu	glu	glu	his	val	glu
Hb I	asp											
Hb GHonolulu	glu											
HbNorfolk			asp									
Hb M Boston				tyr								
Hb GPhiladelphia					lys							
Hb OIndonesia						lys						
Hb S							val					
Hb C							lys					
Hb CSan Jos								gly				
Hb E									lys			
Hb MSaskatoon										tyr		
Hb MMilwaukee											glu	
Hb DPunjab												gln
HbZurich										arg		
Hb OArabia												lys

Note that Hb C has a lysine residue at position 6 of the β-chain where Hb A has a glutamic acid residue and Hb S has a valine-residue. Consequently, Hb C has an electrophoretic mobility lower than that of Hb S. Hb C is found in people who are simultaneously heterozygous for S and C hemoglobins. The disorder is much milder than sickle cell anemia. Simultaneous heterozygosity for Hb C and thalassemia (Mediterranean anemia) as also associated with clinical symptoms and abnormal blood cell.

Hemoglobins that contain an amino acid replacement are called abnormal because they differ from normal adult Hb A. Compared with the normal population, people with abnormal hemoglobin are rare, and it is quite certain that the alteration in structure represents a gene mutation. However, not all abnormal hemoglobins are associated with clinical manifestations of disease. Since the oxygen-carrying power of hemoglobin is crucial to life in animals that possess hemoglobin, it is obvious that mutations would be lethal if they resulted in alterations greatly affecting this oxygen-carrying power. Later in this chapter we illustrate how an understanding of the molecular structure of hemoglobin, in primary sequence as well as in three-dimensional conformation, can be used to develop a concept of the function in this molecule.

Amino Acid Sequences in Normal Hemoglobin during Development

Examination of hemoglobin isolated from the blood of the developing fetus has revealed the presence of several polypeptides that differ from the α- and β-chains found in adult hemoglobin. Illustrates the appearance of these polypeptide chains in fetal hemoglobin. Soon after the fetus begins to develop, there is a rapid production of α- and ε-chains. The α-chains persists, whereas the ε-chain disappears, and new polypeptide, the γ-chain makes its appearance. During the late stage of fetal growth the β-chain increases in quantity as the γ-chain decreases correspondingly. The composition of fetal and adult hemoglobins is illustrated by the following symbols:

Hb A (Adult)	$\alpha_2^A\beta_2^A$
Hb F (Fetal)	$\alpha_2^A\gamma_2^F$
Hb A_2	$\alpha_2^A\delta_2^A$

Our knowledge about the amino acid sequence in the ∈-chain is still incomplete. In all of the fetal hemoglobins, the α-chain remains a constant constituent. The β-, γ-, and δ-chains each contain 146 amino acid residues and closely resemble each other with respect to their amino acid sequences. There are only 10 differences between the β- and δ- chains and approximately 39 differences between the β- and γ-chains. Figure shows the entire sequence of the γ-chain, as well as those amino acids residues that are replaced in the β-chain.

We see that the three-dimensional configurations of the α- and β-chains of normal adult hemoglobin are very similar although the amino acid residues in these two chains differ in number. If the amino acid sequences of the two chains are compared, it is possible to identify 65 residues that are the same, providing one is permitted to allow spaces between certain residues (deletions) and one assumes that insertions have been made during their development

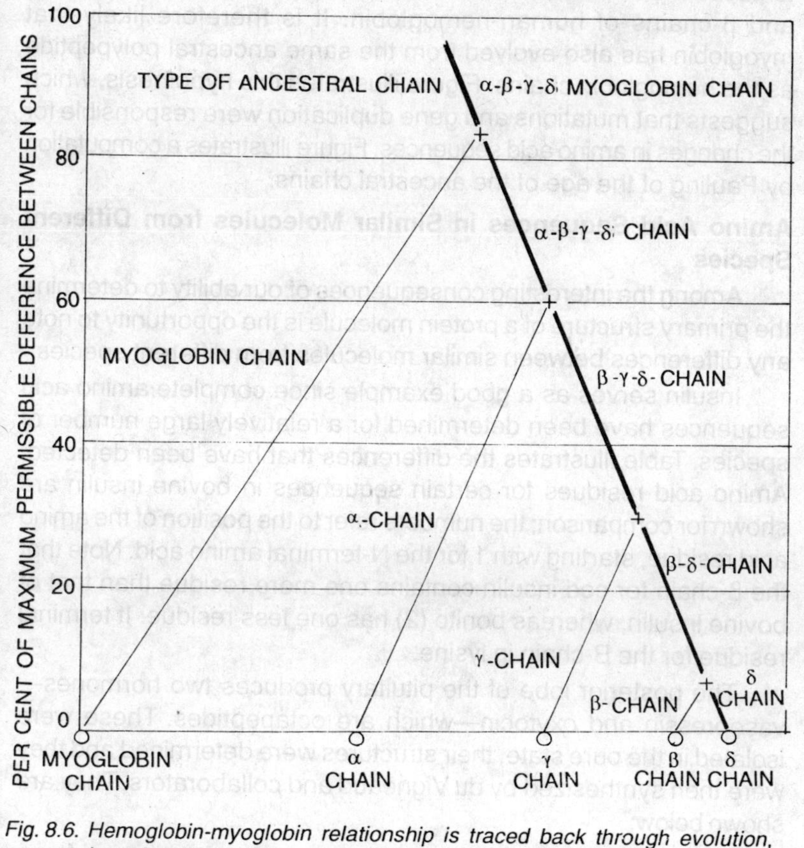

Fig. 8.6. Hemoglobin-myoglobin relationship is traced back through evolution, based on the number of differences in the various chains.

from some common ancestral polypeptide. This is considered likely since all the chains serve the same biological function and possess similar conformations, in addition to the similarities in their amino acid sequences.

In considering the possibility of the evolution of present polypeptide chains of hemoglobin from a common ancestral polypeptide, we must take into account the protein myoglobin of muscle, which also binds molecular oxygen reversibly. This protein has a molecular weight roughly one-quarter the molecular weight of hemoglobin, and it consists of but one polypeptide chain with 153 amino acid residues. Although the sequences in human myoglobin are not yet completely determined, while myoglobin has been shown to possess 37 residues that are the same as those in the α-chain and β-chains of human-hemoglobin. It is therefore likely that myoglobin has also evolved from the same ancestral polypeptide as the hemoglobin chains. Figure illustrates this hypothesis, which suggests that mutations and gene duplication were responsible for the changes in amino acid sequences. Figure illustrates a computation by Pauling of the age of the ancestral chains.

Amino Acid Sequences in Similar Molecules from Different Species

Among the interesting consequences of our ability to determine the primary structure of a protein molecule is the opportunity to note any differences between similar molecules from different species.

Insulin serves as a good example since complete amino acid sequences have been determined for a relatively large number of species. Table illustrates the differences that have been detected. Amino acid residues for certain sequences in bovine insulin are shown for comparison; the numbers refer to the position of the amino acid residue, starting with 1 for the N-terminal amino acid. Note that the β-chain for cod insulin contains one more residue than that of bovine insulin, whereas bonito (2) has one less residue. It terminal residue for the B-chain in lysine.

The posterior lobe of the pituitary produces two hormones—vasopressin and oxytocin—which are octapeptides. These were isolated in the pure state, their structures were determined and they were then synthesized by du Vigneaud and collaborators. They are shown below:

Table 8.4. Amino Acid Replacements Among Insulins of Various Species.

	A-Chain				B-Chain				
	4	8	9	10	1	3	29	30	31
Bovino	Glu	Ala	Ser	Val	Pho	Asp	Lys	Ala	—
Sheep			Gly						
Horse		Thr	Gly	Ile					
Pig		Thr		Ile					
Dog		Thr		Ile					
Man		Thr		Ile				Thr	
Rabbit		Thr		Ile				Ser	
Rat (1)	Asp	Thr		Ile		Lys		Ser	
Rat (2)	Asp	Thr		Ile		Lys	Met	Ser	
Whale (Sel)				Thr					
Whale (Sperm)		Thr		Ile					
Cod					Ser				Lys
Bonito (1)					Leu				
Bonito (2)					Ala			—	—

Oxytocin (bovine)

```
┌────── S—S ──────┐
Cy—try—ile—gln—asn—Cy—pro—leu—gly
```

Vasopressin (bovine)

```
┌────── S—S ──────┐
Cy—tyr—phe—gln—asn—Cy—pro—arg—gly
```

Vasopressin (hog)

```
┌────── S—S ──────┐
Cy—tyr—phe—gln—asn—Cy—pro—lys—gly
```

Amino acid differences in hemoglobins of several different species are shown in Table 8.5.

Table 8.5. Amino Acid Differences in Hemoglobins of Different Species.

	56	57	58	59	60	61	(α-chain)
Human A	lys	gly	his	gly	lys	lys	
Bovine					ala		
Sheep					glu		
Goat					glu		
Horse		ala					
Orangutan		asp					

Melancyte Stimulating Hormones (MTH)

Colour differences among different species are due to differences in the quantity and distribution of the pigment *melanin.* This compound is derived from the amino acid tyrosine. Melanin is part of the colour in skin and hair, and in the "ink" released by the squid. The pituitary glands of many animals have yielded two forms of the polypeptide hormone MSH that promote rapid skin colour changes by affecting the melanocytes—cells that are derived embryonically from nerve tissue. One of these, α-MSH appears to be same in the pig, cow, horse, monkey, and probably man, whereas β-MSH varies somewhat from one species to another. Their amino acid sequences are shown in figure.

Amino Acid Sequence and Biological Activity

The relation of enzyme activity to the amino acid sequence around the "active site" is discussed under enzymes.

An interesting study of biological activity of a polypeptide hormone and its amino acid sequence has been reported for the adrenocorticotropic hormone (ACTH), a straight-chain polypeptide containing 39 residues. Synthesis of large units of this chain have revealed that : (1) The smallest unit that has biological activity is the N-terminal tridecapeptide, corresponding to residues 1—13, (2) Further elongation of the chain result in increased activity, with maximum potency attained by a chain of 20 residues.

1 2 3 4 5 6 7 8 9 10 11 12 13 14 15 16 17 18
H·Ser·Tyr·Ser·Met·Glu·His·Phe·Arg·Try·Gly·Lys·Pro·Val·Gly·Lys·Lys·Arg·Arg·
HO·Phe..Pro·Tyr·Val·Lys·Val·Pro
39 24 23 22 21 20 19

Fig. 8.7. Primary sequence of amino acid residues in ACTH.

The "active site" of the polypeptide lies in the N-terminal tridecapeptide, whereas residues 14 to 20 appear to increase the affinity of the effective subunit for the receptor of the cell.

SECONDARY STRUCTURE OF PROTEINS

The spatial arrangement of polypeptide chains is now known in many proteins through the use of x-ray crystallography. Even casual inspection shows that the chains are not bunched up in random masses; most proteins contain many segments with recognizable patterns. These segments of regular geometry are called secondary structures.

Three types of secondary structures are common in protein : *helixes, reverse turns,* and *pleated sheets.* It is the sequence of amino acid side chains that determines whether a particular portion of he polymer will fold into one of these patterns, but there are some general considerations that apply to all.

Structure will be favoured in which there is a minimum of unfilled space. *Dispersion forces,* also called *London forces,* stabilize structures in which atoms are tightly packed. Either the structure itself must be snughly together r it must have crevices large enough to be crammed with solvent molecules.

A structure will be favoured if it brings all of the C=O and N—H groups of the polypeptide back one into positions where they can form *hydrogen bonds* with each other or with side chain groups. Otherwise, the polypeptide would tend to open up so that these groups could form hydrogen bonds with water. Put another way, the protein tends to fold so that the energy released by internal hydrogen bonding of the back bone groups is as great as the energy that could be released by bonding with water.

BONDS

The forces that are invclved in folding a protein molecules are briefly summarized in the following sections.

Covalent Bonds

The interaction between two atoms arising from the sharing of an electron pair produces a strong interatomic force—the covalent bond. Such forces may be involved in holding two polypeptide chains together or two section of a single polypeptide chain in a folded configuration—for example, the cross-linking disulfide bond contributed by two half-cystine residues. Such bonds are not easily broken, but they may be converted to sulfhydryl groups by reductions, resulting in a separation of the two polypeptide chains or a unfolding of a single chain.

Ionic Bonds

These are due to the charged groups present in the various amino acid side chains, the extent and the kind of charge varying with pH. These are illustrated in Figure. They interact by mutual attraction, giving rise to specific conformations of the protein.

Table 8.6. Dissociation Energies to Typical Bonds.

Bond	Dissociation Energy (ev.)
H—H	4.40
—C—C—	2.55
—C—H	3.80
—C—N	2.13
—C—N (peptide)	3.03
C = C	4.35
C = O	6.30
—C≡C—	5.35

Van der Waals Forces

These forces are due to the mutual attraction of nonpolar hydrocarbon side chains of amino acids, like valine and leucine. These forces attract when the atoms are at a certain distance from each other, but repel when the atoms approach each other more closely. The forces are determined by the distribution of electrons around the two atoms. By means of x-ray diffraction analysis of a crystal, one can measure the distance separating the atoms and assign a van der Waals radius to the atom. This distance is the closest approach that an atom can make to a neighbouring atom without forming a chemical bond. The energy of such forces is about 1 kcal./mole.

Steric Effects

These effects are caused by the presence in amino acid residues of bulky side chains, which can prevent the close approximation of groups or chains.

Hydrogen Bonds

These bonds are the result of the affinity of hydrogen atoms for electronegative atoms such as N or O. Electronegative atoms have a tendency to share electrons with H.

$$—\overset{-}{N}—\overset{+}{H}......\overset{-}{O}=\overset{+}{C}—$$

In this illustration the attraction of the H atom for the O atom results in a *hydrogen bond* between the amide and carbonyl groups that are constituents of the peptide bond in proteins. Weak hydrogen bonds are electrostatic in nature, whereas strong hydrogen bonds are shorter and have a greater covalent character. The energy of such bonds is less than 10 kcal./mole, but because there are a large number of peptide bonds in a protein molecule, hydrogen bonds assume great significance in stabilization of the three-dimensional structure of the molecule. We may visualize two ways in which such bonds might serve to restrict protein structure : (a) by holding two polypeptide chains together, or (b) by holding sections of a single polypeptide chain in specific orientation in space.

Hydrogen Bonds

1. Between peptide groups:

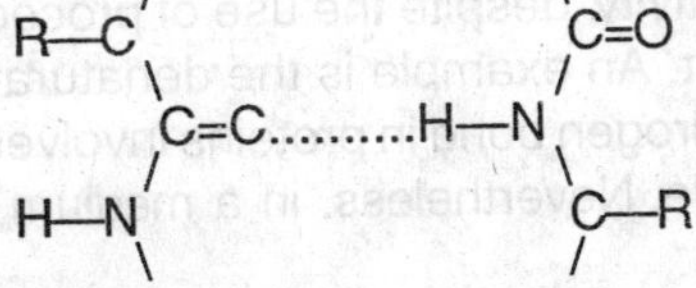

2. Between neutral groups of side chains :

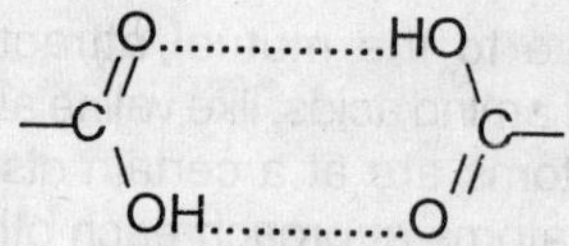

3. Between charged and neutral groups of side chains :

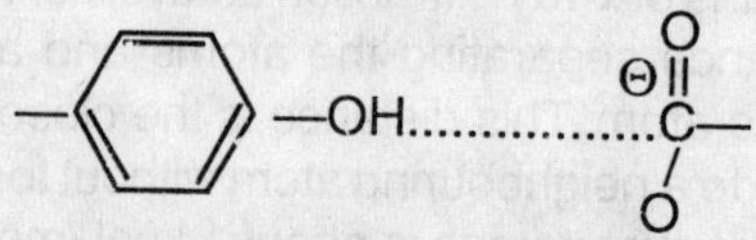

4. Between charged groups of side chains :

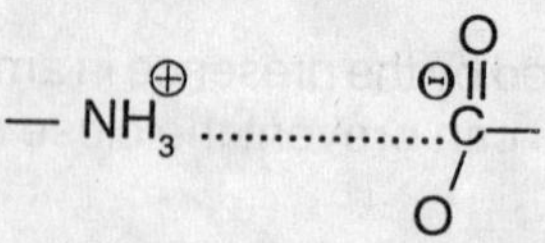

5. Between side chains and peptide groups :

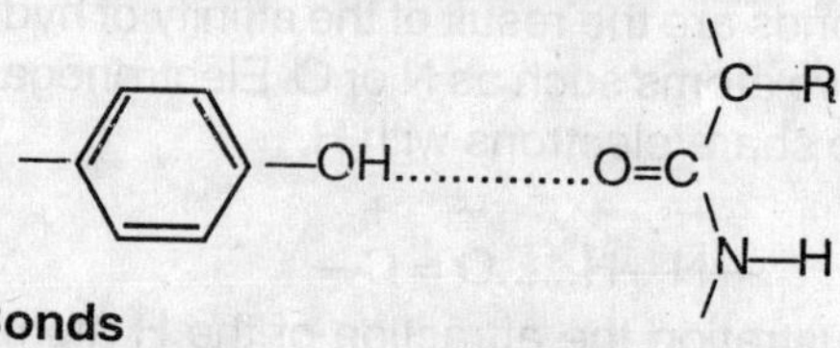

Hydrophobic Bonds

1. Between two nonpolar side chains group.

$$-CH_3......H_2C-$$

2. Between nonpolar side chain and backbone groups :

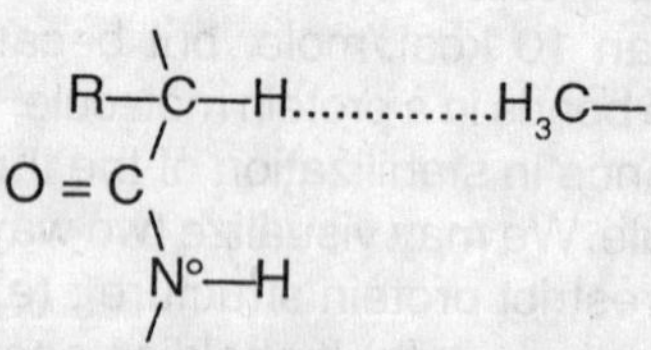

Fig. 8.8. Examples of hydrogen bonds and hydrophobic bonds in proteins.

The concept of the hydrogen bond, when applied to protein structure, helps to explain why there is often an alteration of the three-dimensional structure, and in the case of biologically active proteins, a loss of activity, despite the use of procedures that leave covalent bonds intact. An example is the denaturation of proteins. The breaking of a hydrogen bond in proteins involves the absorption of some 10 kcal./mole. Nevertheless, in a medium that has a high

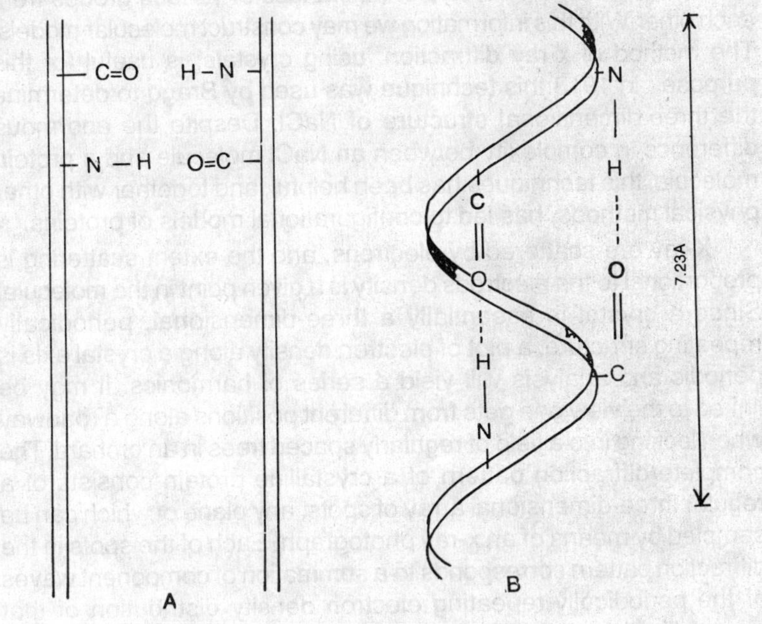

Fig. 8.9.

dielectric, whose molecules themselves have a marked tendency to form hydrogen bonds, the solvent molecule e.g., water causes disruption of the hydrogen bond with the absorption of only 1.5 kcal./mole.

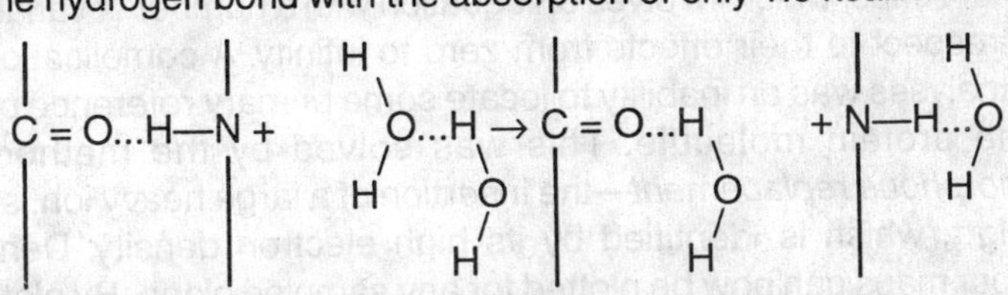

Effective denaturing agents for proteins are urea and guanidine. The action of urea as a hydrogen bond breaker has been questioned, but it has been supposed that the effectiveness of such compounds of denaturing agents is associated with their ability to disrupt hydrogen bonds, causing an unfolding of the polypeptide chains.

Spatial Arrangement of the Polypeptide Chain; X-ray Diffraction Analysis

In order to determine the spatial arrangements possible for structures like a polypeptide chain, it is necessary to know bond distances, bond angles, and the distances of various groups from each other. With this information we may construct molecular models. The method of x-ray diffraction, using crystals, is useful for this purpose : in 1913 this technique was used by Bragg to determine the three-dimensional structure of NaCl. Despite the enormous difference in complexity between an NaCl molecule and a protein molecule, this techniques has been helpful, and together with other physical methods, has led to configurational models of proteins.

X-ray are scattered by electrons, and the extent scattering is proportional to the electrods density is a given point in the molecule. Since a crystal is essentially a three-dimensional, periodically repeating structure, a plot of electron density along a crystal axis is periodic and analysis will yield a series of harmonics. It may be linked to the view one gets from different positions along a roadway when looking into a field of regularly spaced trees in an orchard. The complete diffraction pattern of a crystalline protein consists of a regular three-dimensional array of spots, any plane of which can be sampled by means of an x-ray photograph. Each of the spots in the diffraction pattern corresponds to a summation of component waves of the periodically-repeating electron density distribution of that crystal. Each wave has a characteristic amplitude wave length, and phase.

With relatively simple molecules, the solution to the problem can be accomplished by use of a mathematical technique known as *Fourier series,* which involves an equation with a number of variables with respect to their effects from zero to infinity. A complication in the analyses was an inability to locate some primary reference point in the protein molecule. This was solved by the method of *isomorphous replacement*—the insertion of a large heavy ion, such as Hg^{++}, which is identified by its high electron density. Density contour maps can now be plotted for any sampled plane. By plotting these maps on clear plastic and stacking them, a three-dimensional picture showing regions of high and low density is obtained.

Application of this method to proteins requires an enormous number of calculations in order to get the fine resolution required to identify features of the molecule. High-speed computers are required to get a resolution of 2Å, which involves the treatment of 10,000 reflections. The importance of resolving power in an analysis by this method lies in the fact the distance between covalently bonded atoms is 1 to 1.5Å, whereas atoms of neighbouring groups brought together by forces such as hydrogen bonding or van der Waals forces lie 2.8 to 4Å apart. A resolution of 2Å would separate polypeptide chains or groups of atoms, whereas 1Å would allow a separation of individual atoms. Actually, 1.4Å is just distinguishable by this method; below this the diffraction pattern fades away.

Helixes

Polypeptides, like any long repeating polymer, tend to form coils. Why is this so? Because any repetitious displacement in space travels a spiral path. To eliminate the effects of differing side chains, imagine a polypeptide made from one kind of amino acid. Except for those residues near the ends of the chain, each will be exposed to much the same environment as its neighbours. Therefore, they will tend to form the same Ramachandran angles relative to each other; the polypeptide backbone will regularly bend in the same way. It so happens that there are particular spiral arrangements with sterically favourable Ramachandran angles that are reasonably compact and also enable all possible hydrogen bonds to be formed with the polypeptide backbone.

The α-Helix. The most common helix is protein occurs when the polypeptide chain twists into a right-handed screw with the N—H group of each amino acid residue hydrogen-bonded to the C=O group of the fourth following residue, which is in the adjacent turn of the helix. This screw is the α-helix. It is a rod-like structure coated with amino acid side chains extending outward from its axis. While it is obviously suited for creating long fibre-like molecules, we shell see that it also commonly occurs in globular protein molecules.

Other helixes can be formed by single polypeptide chains so as to hydrogen bond all possible backbone group. All that must be done in principle is to tighten or lossen the coils of the helix to align the bonding groups at other residues in the chain. For example, the chain can be tightly twisted so that each residue forms hydrogen bonds with the third successive residues, rather than the fourth. Such a helix has exactly three is called a 3_{10} helix because each hydrogen bond creates a 10-membered ring and there are three residues per turn. (If the twisting is somewhat less tight, an N—H group can be bonded to the C=O groups of both the third and the fourth following residues, forming a true of α_{11}-helix).

Similarly, untwisting the α-helix so as to align groups on the fifth following residues will create a π-helix.

Of all of these possibilities, the α-helix is the most common. Why? Because it is the one in which the twist is just tight enough to pack the backbone snugly without straining the bonds or leaving a hollow core. Even so, forces occasionally favour localized formation of these other helixes.

Reverse Turns

Polypeptide chains often back upon themselves so as to change, or even reverse, direction. Important forms of these bonds involve hydrogen bonding form one residue to the third following residue. One of six ways in which this can be done. There are other specific geometries that do not involve hydrogen bonding within the bend, but in either case these "reverse turns" or "chain reversals" occur in specific favoured configurations.

Pleated Sheets

Compact sheet like structures form when segments of polypeptide chains lie side by side at nearly maximum extension. These β-structures, like the α-helix, are made with all possible hydrogen bonds between the segments of backbone. They are also known as pleated sheets because of the zigzag appearance they have when viewed on edge. The sheets may be *antiparallel,* with the chain running in opposite direction in adjacent segments. This is the form taken when a single chain is successively folded back upon itself in the most compact possible way. The sheets also may be *parallel*; this requires that the polypeptide chain be looped back upon itself in some way so that adjacent segments of the sheet run in the same direction, creating a β_P structure.

Sheets in proteins are often formed with both parallel and antiparallel segments, and frequently are somewhat twisted. These configurations are represented by a convention in which the general direction of the backbone is shown by a flat arrow, pointing from the amino terminal of the chain toward the carboxyl terminal. The figure also illustrates the resemblance of a commonly occurring pleated sheet, when depicted in this way, to a design motif often found on ancient pottery—the Greek key.

The striking characteristic of the pleated sheet is that side chains of successive amino acid residues appear on opposite sides of the sheet; the sheets therefore tend to form when every other side chain can interact in some effective way. The intervening side chains are segregated on the opposite side of the sheet.

A small irregularity known as the β-bulge is sometime introduced into pleated sheets by insertion of an extra amino acid residue into one segment. The result is a greater distortion of the sheet enabling the formation of some structures to be discussed shortly.

Nonordered Arrangement

Part of a polypeptide chain may lack any recognizable geometrical order of the kinds just described and is sometimes said to be in a random coil. This is misleading term because the arrangement of a polypeptide chain is seldom random, even if it appears as tangled as a crushed coathanger. Lack of the repetitive elements of design that we call secondary structure simply reflects

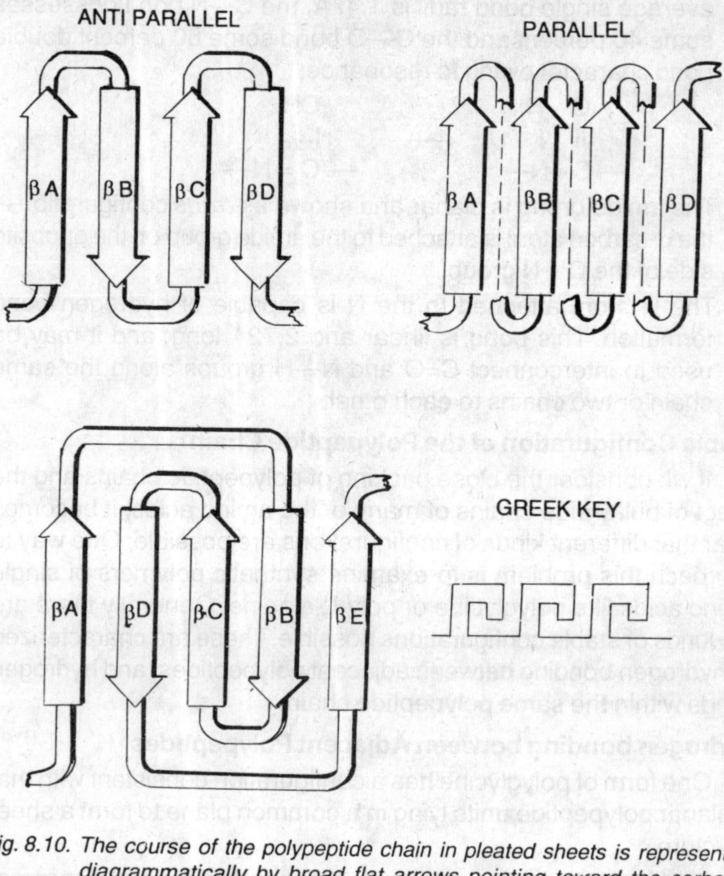

Fig. 8.10. The course of the polypeptide chain in pleated sheets is represented diagrammatically by broad flat arrows pointing toward the carboxyl terminal of the chain.

response to nonrepetitive forces—nonrepetitive because they involve the side chains of the amino acid residues.

The Polypeptide Chain

The polypeptide chain in its fully extended form, together with important dimensions, is illustrated. The distance of 7.23Å refers to the repeat distance between one side chain (R) and the next one pointing in the same direction. Several features of this structure are:

1. Regardless of the particular amino acid involved in peptide linkage, a similarity of peptide group dimension is observed—1.24Å for the C—O distance, and 1.32Å for the C—N distance. Since the average double bond covalent radii is 1.21Å, and the average single bond radii is 1.47Å, the C—N bond possesses some 40 percent and the C—O bond some 60 percent double bond character owing to resonance:

```
   O  H            OH
   ‖  |            |
 —C—N—    ⟷     — C = N—
```

2. The amide group is planar and shown as *trans* configuration—the α-carbon atom is attached to the amide group on the opposite side of the C—N group.
3. The H atom attached to the N is capable of hydrogen bond formation. This bond is linear and 2.72Å long; and it may be used to interconnect C=O and N—H groups along the same chain or two chains to each other.

Stable Configuration of the Polypeptide Chain

If we consider the close packing of polypeptide chains and the effect of bulky side chains of many of the amino acids, it becomes clear that different kinds of configurations are possible. One way to approach this problem is to examine synthetic polymers of single amino acids like polyglycine or poly-L-alanine. Generally there are two kinds of stable configurations possible. These are characterized by hydrogen bonding between adjacent polypeptides, and hydrogen bonds within the same polypeptide chain.

Hydrogen bonding between Adjacent Polypeptides

One form of polyglycine has a configuration consistent with that of planar polypeptide units lying in a common plane to form a sheet structure.

Silk fibroin, a protein containing 44 percent glycine and 40 percent alanine plus serine, has a structure similar in many respects to

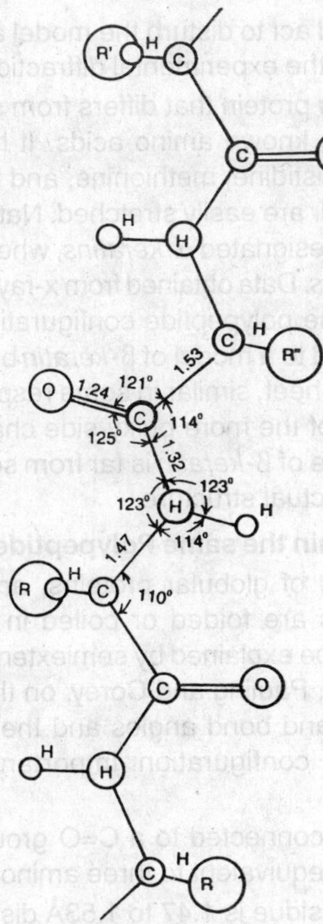

Fig. 8.11. Dimensions of the fully extended polypeptide chain.

polyglycine, but with the polypeptide chain less extended, in order to make room for the side chains of amino acids other than glycine. The chains are arrayed alongside in an antiparallel fashion and bent to give the appearance of a pleated sheet. The polypeptide backbone has s zigzag rather than a straight form, and the C=O and N—H groups are oriented so that all H bonds can be formed by aligning a series of these polypeptides with alternate chains moving in opposite directions. The successive carbonyl and amid groups are hydrogen bonded to chains on either side of the original chain.

Although most of the properties of silk fibroin can be understood from this simplified structure, this model does not account for all the amino acids found in the protein, including many with extensive side chains, that would act to disturb the model and produce minor complications seen in the experimental diffraction pattern.

Keratin is a fibrous protein that differs from silk fibroin in that it contains nearly all the known amino acids. It has a high cystine content and is low in histidine, methionine, and tryptophan. Unlike silk fibers, wool and hair are easily stretched. Natural keratins in the unstretched form are designated α-*keratins,* whereas the stretched variety are the β-*keratins.* Data obtained from x-ray diffraction studies, together with one of the polypeptide configurations suggested by Pauling and Corey, lead to a model of β-*keratin* best described as a parallel chain pleated sheet, similar in some respects to silk fibroin, but differing because of the more bulky side chains. Although the problem of the structure of β-*keratin* is far from solved, the working model is close to the actual structure.

Hydrogen Bonds within the same Polypeptide Chain

α-keratin, sections of globular proteins, and some forms of synthetic polypeptides are folded or coiled in a more compact configuration than can be explained by semiextended polypeptides. To explain such coiling, Pauling and Corey, on the basis of known interatomic distances and bond angles and the study of models, suggested the α-helix configuration. Important features of this structure are :

1. Each NH group is connected to a C=O group by a hydrogen bond at a distance equivalent to three amino acid residues.
2. Each amino acid residue is 1.47 to 1.53Å distant from the next residue, and the helix makes a complete turn for each 3.6 to 3.7 residues. The α-helix is only one of many different helices possible and is determined by the number of residues per turn. For example, the π-helix would have 4.4 amino acid residues per turn.
3. The vertical distance along the axis from any point on the helix to a corresponding point directly above it is approximately 5.4Å (3.6 × 1.5) and is referred to as the *pitch of the helix.*
4. With 3.6 residues per turn, five complete turns correspond to 18 residues, yielding a repeating interval of approximately 27A (1.5 × 18), which has been found in diffraction patterns made of fibers of polyamino acids of certain kinds. There are two ways of coiling a helix—a right handed and left-handed screw. The right-handed helix has been shown to be more stable.

In the case of α-keratin, agreement of experimental x-ray diffraction reflections with those that have been calculated is less clear than those obtained from synthetic polyamino acids, although the 1.5Å reflection is strong. Instead of a spacing of 5.4Å, corresponding to the pitch of the α-helix just described, a spacing of 5.1Å is obtained for α-keratin and in a slightly different position. These differences can be reconciled if we assume that the α-helices of α-keratin are coiled into bundles or cables of "coiled coils". Similar structures may be used to picture the configuration of other members of the keratin group—myosin of muscle, epidermin of skin, and fibrinogen of plasma.

The α-helix is a closely packed structure with no open spaces in the center. Inside this structure is the repeating part of the polypeptide chain, exposing all amino acid side chains to the surface. When proline residues are incorporated into the right-handed α-helix, little disturbance in structure need occur. However, if these residues are inserted into the left-handed α-helix, the sequence is disrupted, and they may serve as a turning point for the helix—it will reverse direction. In collagen—the principal fibrous protein of mammalian connective tissue, tendon and bones—there is one proline or hydroxyproline per four residues, and this protein represents a special problem in configuration. It is suggested that collagen is constructed of left-handed helical structures for each of three polypeptide chains. These three left-handed structures twist around each other forming a large right-handed helix.

Tertiary Structure

At this level of organization of the molecule, a protein has size and shapes as well as a unique arrangement of its polypeptide chains. To gather such structural information requires the application of a variety of physiochemical methods that can be used with protein solutions, as well as the x-ray diffraction method, which can only be applied to the crystal. A brief account of these methods is given.

Two Examples

The tertiary structure of a polypeptide chain is the complete form assumed by the chain—the combination of various secondary structures and nonordered segments into an ordered whole. We shall concentrate for now on globular proteins and consider the more specialized fibrous proteins as we come to them. Outlines the polypeptide backbone in each of two proteins. The one on top is part of liver alcohol dehydrogenase, a protein that is responsible for the oxidation of ethanol to acetaldehyde. The one on the bottom is the polypeptide of myoglobin (muscle hemoglobin), a protein responsible for binding and transporting oxygen in muscle fibers.

Although both of these proteins are relatively globular, their overall shapes give on clue to the quite different arrangements of the polypeptide chains creating the proteins. The liver alcohol dehydrogenase contains a central core of a bent parallel pleated sheet made from six segments of chain enclosed by four recognizable segments of α-helix. (The segments are lettered according to the order in which they appear from the amino terminal of the chain. βA is the first segment used to form a pleated sheet, αA, is the first segment appearing in an α-helix, and so on). Between these segments are various twists and bends.

The myoglobin polypeptide, on the other hand, contains no pleated sheet and is made up almost entirely of segments of α-helix, with what appears to be a minimal amount of connecting chain between some segments.

Patterns of Tertiary Structure

The two examples of tertiary structure we used are not at all unusual. The general arrangement shown for liver alcohol dehydrogenase is similar to one found in several other proteins, some of quite different function. The same arrangement of stacked helixes seen in myoglobin is also present in other oxygen-binding proteins, ranging from a hemoglobin found in the nitrogen-fixing nodules of legumes to circulating human hemoglobins.

Indeed, is now possible to recognize several ways in which groups of pleated sheet and helix segments are organized to make globular proteins. Some polypeptide chains fold into stacks of pleated sheets surrounded by bends and nonordered segments. In others, a large pleated sheet is folded into a cylinder (the β-barrel). Many others have pleated sheet cores surrounded by helixes, much like alcohol dehydrogenase, but this can be accomplished in many ways. (The term *super secondary structure* has been applied to these recurring patterns, but they do not have the fixed geometry of the fundamental secondary structures.)

Domains

One polypeptide chain need not be folded into one compact unit. For example, in alcohol dehydrogenase part of the chain is folded to make the structure. However, another part of the chain is folded into a similar but not identical structure with a separate function even though it is covalently liked to the first structure by the connecting region of the polypeptide chain. Domains, then, are recognizable units of tertiary structure resembling separate proteins but formed from one polypeptide chain. They often fold independently of each other and retain their character if the chain linking them is broken.

Origin of the Tertiary Structure

Why do liver alcohol dehydrogenase and myoglobin differ so much in the conformation of lowest energy content? We know the difference in amino acid composition of the two proteins is the root cause, exactly how do the forces developed by the varying side chains generate the structure? This is a question we cannot answer in detail : it is not yet possible to deduce exactly how a polypeptide chain of known primary structure will fold, but great progress has been made toward the development of the general principles.

Secondary Structures are Preferred

Since the pleated sheets and helixes saturate the backbone with hydrogen bonds, permitting it to be buried away from water, it is not surprising that many polypeptides preferentially fold into a combination of these shapes, often connected by specific kinds of reverse turns. Even those globular proteins without recognizable sheets or helixes, and they are not common, are mainly made from specific turns.

The Structures are Compact

The same forces that cause secondary structures to be compact also cause them to combine into snug tertiary structures. Openings are usually more than one atom wide and are intended for access of solvent or for combination with other molecules. The protein may have bizarre forms, but it has little unoccupied space. It is not porous.

The Surface is Polar and the Interior is Nonpolar

One of the major forces affecting the shape of a protein comes from the tendency of polar side chains to associate with water and of nonpolar side chains to agglomerate in the interior of the molecule. These effects stabilize contortions of the polypeptide that bring the hydrophilic residues to the surface and bury the hydrophobic residues.

Hydrophilic Side Chains

The most hydrophilic amino acid residues are those with net charges, such as aspartate and glutamate with their negatively charged carboxylate groups, and lysine and arginine with their positively charged ammonium and guanidinium groups. The chain must nearly always fold so as to keep them in contact with water. It is only possible to bury these groups when the charge in neutralized by forming the uncharged forms, combining positive and negative groups, or by dissipating the charge through multiple hydrogen bonds. Even when this is done, the folding must be strongly favoured elsewhere to remove these polar groups from contact with water. The carboxamide groups of glutamine and asparagine are also polar enough to strongly favour conformations that keep them in contact with water.

Hydrophobic Bonds

The force causing nonpolar side chains to huddle together out of contact with water comes not from any specific attraction of the groups for each other but from the propensity of surrounding water to resist being forced into ordered structures by the presence of the nonpolar groups. The so-called hydrophobic bond is really an association created by spurned water, and it is the same kind of force that causes insoluble oils to form compact spheres upon suspension in water. How does it arise?

Molecules of water have a strong tendency to form hydrogen bonds with each other as is shown by the release of heat when water freezes into the fixed lattice of ice. There is an opposing tendency of all molecules to stay dispersed, to be random and unordered. This tendency, measured as entropy, is increased by a rise in temperature. The two effects just balance at the melting point. At higher temperature, hydrogen bonds persist for shorter and shorter times; each molecule becomes more and more promiscuous.

A nonpolar molecule introduced into liquid water represents a eunuch at an orgy. Since it cannot form hydrogen bonds or other bonds created by polarity, neighbouring water molecules lose some of their possibilities for random union, and associate with each other longer than they would in the absence of the nonpolar intruder. However, this longer association represents greater order; a step toward freezing into ice, and we have already said that creation of such order at temperatures above the melting point can be accomplished only by the introduction of energy. It follows that dispersing nonpolar molecules in water requires energy; if they are left to themselves, they diminish their exposed surfaces by clustering, thus impeding fewer random liaisons among water molecules. They become a less effective moral minority, as it were.

Some Nonpolar Chains are on the Surface

If the degree of polarity of the side chains were the only determinant of structure, polypeptide would fold so that every group not interacting with water was buried in the interior. The tendency to make this separation is very strong and can aid in shaping complex surfaces, but the separation is not complete. There are usually more nonpolar side chains than can be accommodated in the interior of a protein, and some are exposed on the surface of most proteins. This is an important part of the development of specific function, as well as shape. Having more reactive groups separated by the relatively unreactive nonpolar side chains give individual character to the surface geometry—the stars the bright because the sky is black.

GENERATION OF SPECIFIC STRUCTURES

The general nature of the forces acting on a polypeptide chain seem clear, but it is more difficult to understand exactly what influence a particular amino acid sequence may have on protein conformation. However, we have some clear-cut examples with which to make a start.

Proline is a Helix-Breaker

This is the one apparently infallible general rule for prediction of structure. A residue of proline cannot occur in the middle of a straight α-helix :

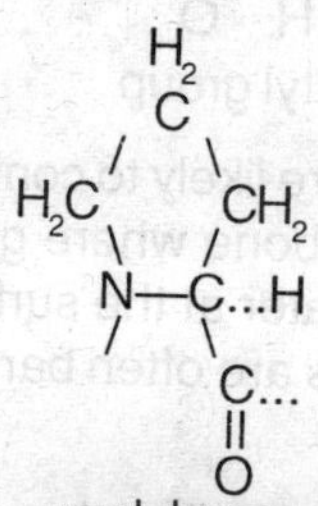

prolyl group

The nitrogen atom of a prolyl group has no attached hydrogen atom with which to bond a preceding carbonyl group, and the fixed orientation of the bulky side chain gets in the way of any preceding turn. Therefore, a prolyl group can only be in the first turn in an α-helix; the preceding amino acid residue must occur in some other configuration, at least bending, and usually breaking, the helix.

Reverse Turns are often Exposed

When causes chain reversals? The chain can be bent upon itself at any residue if the bend enables strong bonding elsewhere, but there are some amino acids whose side chains favour reverse turns. Proline, with its odd geometry, is an example. Glycine is another; its lack of any side chain permits some tight folds that are not possible with other amino acids, and it is more likely to occur in a reverse turn than any other :

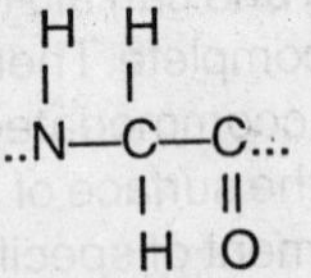

glycyl group

Other amino acids favour the formation of reverse turns because they have hydrogen-bonding groups only a short distance removed from the peptide backbone. These include asparagine, serine, and aspartate :

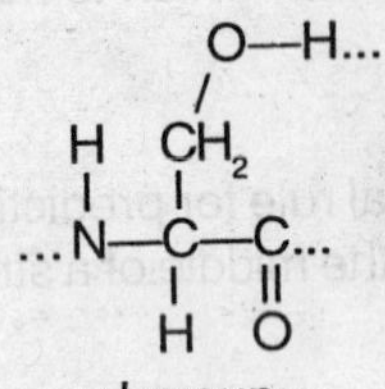

seryl group

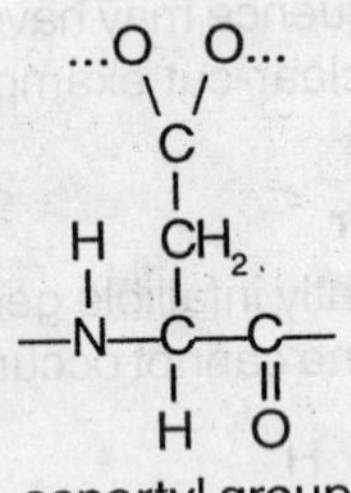

aspartyl group

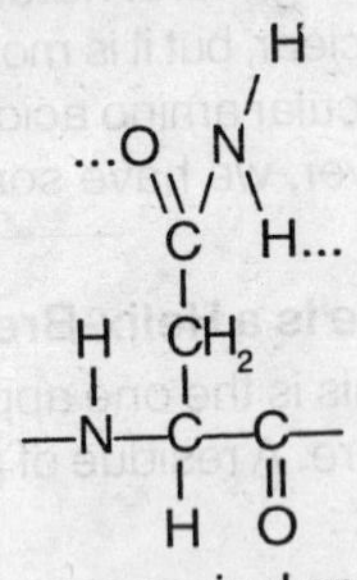

asparaginyl group

Here we have it : reverse turns are likely to contain polar side chains or to have a nearly naked backbone where glycine occurs. Either event will favour exposure to water at the surface of the molecule. The elbows of protein molecules are often bared.

Pleated Sheet or α-Helix?

Since the pleated sheets and helixes have complete hydrogen bonding of the backbones, the forces favouring one over the other must be developed by the differing dispositions of the side chains. For example, generation of a secondary structure with a hydrophobic surface that can be faced toward the interior may facilitate construction of a compact molecule. We can sometimes predict if formation of a pleated sheet or α-helix is more likely to occur by picturing what happens in the two cases.

Alternate amino acid side chains are exposed on the same face of a pleated sheet. Therefore, if we plot the amino acid sequence as

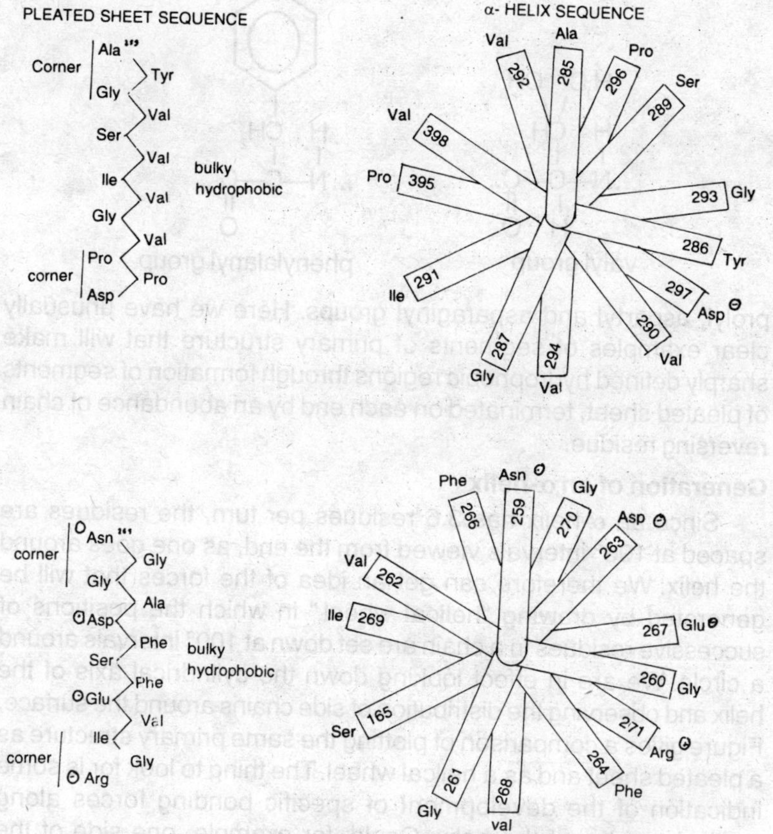

Fig. 8.12. The forces creating segments of secondary structure can sometimes be visualized by plotting the amino acid sequence in the order it would assume in the structure.

a zigzag, those residues to one side will be on one face of a sheet, those on the other side on the other face. The left part of figure shows that happens when this is done with the sequences of the two longest segments of pleated sheet from liver alcohol dehydrogenase, together with the preceding and following residues. Folding these sequences as a sheet creates a massive aggregation of hydrophobic groups from the side chains of valine and phenylalanine :

The other face of the sheet has smaller or charged side chains. Also notice the nature of the residues at the two ends of the sequences. The bends at these positions are loaded with glycyl,

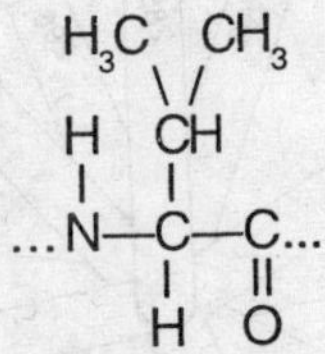

valyl group

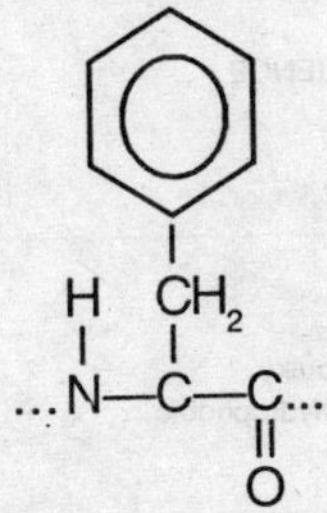

phenylalanyl group

prolyl, aspartyl and asparaginyl groups. Here we have unusually clear examples of segments of primary structure that will make sharply defined hydrophobic regions through formation of segments of pleated sheet, terminated on each end by an abundance of chain reversing residue.

Generation of an α-helix

Since an α-helix has 3.6 residues per turn, the residues are spaced at 100° intervals viewed from the end, as one goes around the helix. We therefore can get an idea of the forces that will be generated by drawing "helical wheel," in which the positions of successive residues in a chain are set down at 100° intervals around a circle. We are in effect looking down the cylindrical axis of the helix and observing the distribution of side chains around the surface. Figure gives a comparison of plotting the same primary structure as a pleated sheet and as a helical wheel. The thing to look for is some indication of the development of specific bonding forces along particular sides of the helix. Could, for example, one side of the helix be facing the hydrophobic interior and another side be facing the aqueous phase? If so, a helix becomes more likely.

When we look at the segments in those terms, they appear rather nondescript. The segments from 12 o'clock are relatively polar; the remaining portion is nonpolar, but the groups are too widely scattered to be definitive. The first segment couldn't be a continuous α-helix in any case because of the prolyl residues that end it. The second segment possibly might be, but the apparent bonding forces are nothing like those seen with a pleated sheet configuration.

Now let us look at a sequence from myoglobin. Here the pleated sheet configuration results in a rather random distribution of groups, while the wheel shows that an α-helix will create a well-defined region of bulky hydrophobic groups. In fact, this segment of the chain is a

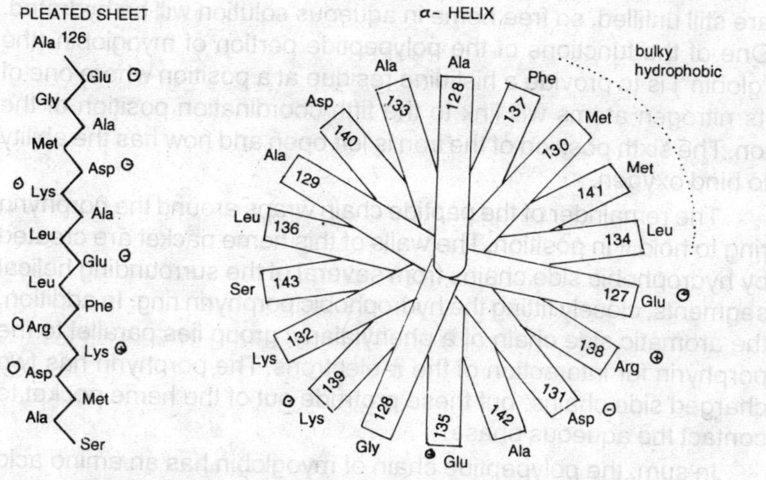

Fig. 8.13. Plots of an amino acid sequence from human myoglobin as a pleated sheet and an α-helix.

portion of an α-helix, with the hydrophobic segment in contact with other parts of the chain in the interior, and the remainder of the surface being exposed to surrounding water. It is the potential for creating this distribution of groups that makes myoglobin fold in the way it does.

The Heme Pocket

There is another important aspect to the structure of myoglobin. It is a heme protein, meaning that it belongs to a class of conjugated proteins that contain heme groups in addition to polypeptide chains, and the primary structure must be arranged so as to hold the heme in position.

Heme is combination of iron (II) and protoporphyrin IX. We shall say great deal more about the porphyrins later. The point of importance now is that the porphyrins contain a highly resonant planar ring system in which four nitrogen atoms are fixed in the centre at a spacing that is ideal for bonding of their unshared electrons with a metal ion. The porphyrins therefore act a tetradentate (four-toothed) ligands, and protoporphyrin IX has an especial affinity for iron.

Iron (II) ions have a coordination number of six, which means that they can associate with six electron pairs atom. When a ferrous ion forms a chelate with a porphyrin, two of its coordinations positions are still unfilled, so free heme in aqueous solution will be hydrated. One of the functions of the polypeptide portion of myoglobin (the "globin") is to provide a histidine residue at a position where one of its nitrogen atoms will link to the fifth coordination position of the ion. The sixth position of the iron is left open and now has the ability to bind oxygen.

The remainder of the peptide chain wraps around the porphyrin ring to hold it in position. The walls of this heme packet are created by hydrophobic side chains from several of the surrounding helical segments, closely fitting the hydrophobic porphyrin ring. In addition, the aromatic side chain of a phenylalanyl group lies parallel to the porphyrin for interaction of the π-electrons. The porphyrin has two charged side chains, but these protrude out of the heme pocket to contact the aqueous phase.

In sum, the polypeptide chain of myoglobin has an amino acid sequence that favours the formation of a particular set of α-helices that will snugly nestle around a heme group and hold it in four ways : chelation of the iron, hydrophobic exclusion of water, interaction of aromatic rings, and close packing of the atoms. Indeed, the polypeptide is built to fold into a particular tertiary structure in the presence of heme. (Otherwise, the hydrophobic heme pocket would be exposed to water).

Sedimentation in the Centrifugal Field

Development of the ultracentrifuge by Svedberg in 1925 marked a milestone in modern biochemistry. Its subsequent modification has led to its general use in the biochemistry laboratory. Since the rate of movement of a protein molecule is a function, among other things, of its size, one can determine the apparent molecular weight of a protein in solution under a variety of conditions. Two methods used are sedimentation velocity and sedimentation equilibrium.

Sedimentation Velocity

By means of a suitable optical system it is possible to measure the rate of movement of protein molecules in a centrifugal field of high intensity in a solvent (e.g., aqueous buffer) less dense than that of the protein solution. If a single molecular species is present, a single boundary between the moving protein molecules and the buffer will be observed. If several species of protein molecules are present, these will be observed as a series of boundaries, each boundary moving at a different rate, depending on its molecular weight.

Calculation of the molecular weight involves a knowledge of the sedimentation constant *s*, the diffusion constant D, density of the medium d_m, and the partial specific volume V : R and T are the gas constant and absolute temperature respectively. This relationship follows :

$$M = \frac{RT}{D(1 - Vd_m)} . (s)$$

where

$$s = \frac{dx / dt}{x\omega^2}$$

and *dx*/*dt* is the rate at which the particle travels at distance *x* from the center of rotation, and ω is the velocity of revolutions in radians per second (angular velocity). The diffusion constant is obtained by observing the rate of motion of the protein boundary in the absence of a centrifugal field. The partial specific volume is determined by observing the density of solutions of the protein as a function of concentration of the anhydrous protein. The sedimentation constant *s* is characteristic of the molecule under stated conditions and varies from 1×10^{-13} second to 200×10^{-13} second for proteins. Because of the negative exponents, another units, the Svedberg (S), is used. This is a unit of sedimentation constant equaling 1×10^{-13}, so that proteins may be said to have value of S to 200 S.

Sedimentation Equilibrium

In this method, we observe the distribution of protein molecules in cell in which the gravitational field is less than that used for the sedimentation velocity method. The molecules distribute themselves at equilibrium according to size, and average molecular weights can be calculated over the range force 1000 to 10,000.

Light Scattering

The scattering of light is proportional to the weight-average molecular weight, and the proportionality factor can be experimentally determined. The sensitivity of this method increases with the molecular weight. In addition, this method can yield information regarding the shape of molecules. Molecules of dimensions of 200 to 300Å show interference because of the light scattered from different parts of the same molecule. As a consequence, the scattering is diminished, the effect increasing with the scattering angle. For example, cellulose nitrate has a rather broad distribution of molecular weights, and the size indicative of an extended random coil. On the other hand, tobacco mosaic virus shows rods of 300Å in length for a molecular weight of 39.5 million, with an effective diameter of 150Å.

Osmotic Pressure

The pressure required to prevent the diffusion of solvent through a membrane into a solution of a protein dissolved in that solvent is termed osmotic pressure. The osmotic pressure of a protein solution is a property that gives a macroscopic response with dilute macromolecular solutions, measuring the number of molecules of solute per unit volume of solution. Knowing the weight of material per unit volume of solution, we can determine the number of molecules and thus get a number-average molecular weight.

Table lists the molecular weights of several proteins determined by various methods.

Optical Rotation

As was noted earlier proteins have specific optical rotations that, on denaturation, become more negative. It is believed that this

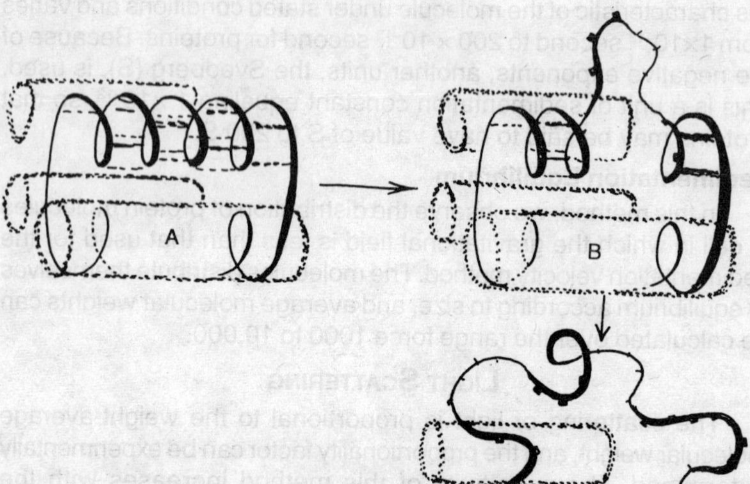

A : Native Protein
B : Initial Stage of Unfolding
C : Partially Denatured Protein

Fig. 8.14. Denaturation.

Table 8.7. Molecular Weights of Crystalline Proteins Determined by Various Methods

Method	Ovalbumin	Serum Albumin	Hemoglobin	Lysozyme	β-Lactoglobulin	Insulin
Containing weight	43,000	70,000	66,700	14,000	42,000	6000
Osmotic Pressure	45,000—46,000	66,000—76,000	67,000	17,000	35,050—38,000	12,000—48,000
Diffusion	34,500		68,600	13,000	40,000	40,900
Sedimentation	43,800—45,000	65,360—70,200	68,000	14,000—17,000	38,000—41,500	12,000—48,000
Light scattering	45,700—46,000	72,500		14,700	33,700—35,500	12,000—48,000
X-radiation		65,600—66,200	66,700	13,900	33,000—35,600	36,000—52,400

is due to the disorganization of the helical configuration, which lacks reflection symmetry and contributes, in the native protein, to the positive rotation, which keeps the specific rotation of the most native proteins at values of –30° to –60°. On denaturation, the degree of rotation generally increases to values of –80° to –120°. Doty has used such measurements to estimate the helical content of proteins by assigning a value of $[\alpha]_D = -90°$ for a 100 percent random coil, and $[\alpha]_D = -10°$ for a 100 percent helix. The method of *optical rotatory* dispersion—the variation of rotation with wave length of light—has also been used to approximate the helical content of globular proteins. On changing from the native to the denatured state, the shape of the curve is changed markedly. Using this method it has been estimated that ribonuclease contains 15 percent, whereas ovalbumin and serum albumin have a 45 percent helical configuration. Tropomyosin has been estimated to exit in the helical form to an extent greater than 90 percent.

Deuterium Exchange

Insight into protein structure may be obtained by determining the rate and extent of exchange of deuterium between the protein and the surrounding medium (water). Whereas the hydrogen in various groups of amino acids—e.g., amino carboxyl, amide, and hydroxyl—is instantaneously exchangeable with deuterium, the presence of strong internal hydrogen bonds in a protein may decrease this rate or eliminate it entirely. For example, at a pH of 4.7, the enzyme ribonuclease has some 243 exchangeable hydrogen atoms, of which 123 are peptide hydrogens. In the native state, 185 atoms of hydrogen exchange promptly, whereas if the disulfide bonds are broken by oxidation or if the native enzyme is treated with a denaturing agent, all 243 atoms exchange instantaneously. It appears, therefore, that the hydrogen-bonded folded structure of the ribonuclease molecule shields some of the hydrogen atoms from instantaneous exchange.

Tertiary Structure From X-ray Diffraction Analysis : Myoglobin and Hemoglobin

Just as Sanger was able to give us the first structural formula of a protein—insulin—Kendrew and his collaborators have presented us with first three-dimensional picture of a protein molecule reconstructed entirely from x-ray diffraction data. Myoglobin was the protein chosen for this study, since it has a molecular weight of only 17,000 and crystallizes easily from muscle extracts of the sperm whale. It consists of a protein, globin (not the same as that found in hemoglobin), which is a single polypeptide of 153 amino acid residues, and one heme group. Heme (in oxymyoglobin) is composed of a substituted glutamic acid residue in Hb $M_{Milweukes-1}$ leads to a stable complex. Again, this glutamic acid residues is four residues removed from his-63 and might therefore be in close approximation to the iron atom, since one turn of the helix accommodates the equivalent of 3.6 amino acid residues.

Denaturation

Some denaturation phenomena suggest that tertiary structure is determined by amino acid sequence. Transformation from the native to the denatured state is often irreversible, although Anson and Mirsky, as early as 1930, demonstrated the reversibility of hemoglobin denaturation. Yet, if the native configuration is the most stable, denaturation should, under certain conditions, be reversible.

This reversibility has been demonstrated with the enzyme ribonuclease, which has four cystine disulfide bridges stabilizing its tertiary structure. If ribonuclease is reduced to the —SH form, in the presence of the powerful denaturing agent urea, the enzyme loses all activity. Reoxidation of the reduced enzyme by air leads to a completely active enzyme indistinguishable from the native protein. With eight sulphydryl groups, the reduced enzyme might reoxidize in many ways. It would appear, therefore, that the amino acid sequence by itself is sufficient to determine the most stable configuration. It should be noted, however, that ribonuclease contains but one polypeptide chain. In the case of insulin which contains two chains, attempted oxidation of the reduced protein leads to a recovery of only 1 or 2 percent of the original activity. It is possible that permanent damage is done or that bonds are formed during denaturation, which stabilizes the unfolded or denatured, state. Many other enzymes have been reversibly denatured—e.g., catalase—to the extent of a 50 percent yield.

We may summarize the various effects of denaturation :

1. Decreased solubility at the isoelectric point.
2. Loss of biological activity.
3. Increased reactivity of chemical groups that had been masked by folding in the native protein—e.g., —SH, —S—S—, and phenolic hydroxyl of tyrosine.
4. Increased number of ionizable groups available for titration.

5. Increase in levorotation.
6. Increase in asymmetry of the molecule.
7. Increase in susceptibility to hydrolysis by proteolytic enzymes.
8. Loss of crystallizability.
9. Cleavage of hydrogen bonds.

Quaternary Structure

This level of organization of the protein molecule results from the association of folded units into relatively stable aggregates. Many examples are available from molecular weight studies of the effect or change in environmental conditions on the degree of association or dissociation of proteins. For example, a 1 percent solution of insulin yields a sedimentation constant of 3.7 S for the protein, corresponding to a molecular weight of 36,000. At acid pH a value of 2.0 S is obtained, which corresponds to a molecular weight of 12,000.

Some protein molecules are complexes of more than one polypeptide chain; each chain with its tertiary structure then constitutes a subunit of a larger molecule. The geometry of the combination of subunits into the complete molecule is said to be its quaternary structure.

The force available for the formation of quaternary structure are the same as those that create secondary and tertiary structure. One of the most important is the tendency for hydrophobic groups to combine so as to exclude water. Suppose that a polypeptide chain folds in such a way as to create a face with exposed hydrophobic side chains. That face would tend to stick to any other hydrophobic surface. If there are two polypeptide chain with areas of hydrophobic surface that can fit closely together, then the two chains will bond into a large molecule, and this is the way in which many proteins containing multiple subunits are constructed. (Of course, tertiary structure with an exposed hydrophobic face wouldn't be very stable as an isolated entity, but when two or more chains are present simultaneously that can form matching faces, otherwise transient areas of hydrophobic surface will be stabilized by combining with each other).

Liver alcohol dehydrogenase, for example, is a dimer that contains two identical subunits. The domains with pleated sheet cores that we discussed earlier comprise the tertiary structure of each subunit. However, these subunits have one segment of pleated sheet exposed at the surface in such a way that if two subunits about, a continuous pleated sheet will be formed from one subunit to the other. The subunits are then joined both by association of hydrophobic side chains and by hydrogen bonding.

Hemoglobin

Perhaps the most thoroughly studied example of quaternary structure is that of hemoglobin. The molecules of hemoglobin that carry oxygen in human blood are tetramers made from a pair of each of two kinds of polypeptide chains. These chains are designated α and β in the most abundant hemoglobin in adults, and this hemoglobin A therefore has the chain formula $\alpha_2\beta_2$. Each of these chains has a tertiary structure much like that of myoglobin, which is made from only one polypeptide chain. However, they differ from myoglobin in that they are built to associate in particular ways. That is, α chains tend to bind β chains, and *vice versa*. The result is a tetrahedral molecule with the four subunits in close contact, except for a hole in the center that is freely accessible to water.

The hemoglobin chains associate and the myoglobin chains do not because some of the amino acid residues on the surfaces are different. The hemoglobin subunits have many additional hydrophobic residues exposed; they can only be buried by bringing the subunits together. They also have different polar side chains at locations where they form hydrogen bonds when the subunits about. It is the additional surface charge and the lack of hydrogen bonding that keep the myoglobin molecules separated. (The hemoglobin subunit surfaces are also contoured for closer contact).

On the other hand, the surface remaining exposed after formation of the tetramer are designed to minimize any tendency for further association. Red blood cells are stuffed with hemoglobin almost to the limit of its solubility, with adjacent molecules not much more distant than they are in a crystal. Even a moderate tendency to aggregate would not permit such a high concentration, and therefore would diminish the capacity to carry oxygen. We shall changes in single amino acid residues sometimes cause problems in this way.

What is the point of building hemoglobin, or any other protein, out of multiple polypeptide chains when many respectable protein have only one? Sometimes proteins are built this way simply to become large enough so that they won't leak through membranes, but we shall see when we turn to the biological functions of proteins that those composed of more than one subunit can respond to changes in environmental concentrations in a more complex way than can those made from only one polypeptide chain. But before we examine these functions we should consider how proteins are made, and this process involves another class of macromolecular polymers, the nucleic acids.

FUNCTION OF PROTEINS

Protein play crucial roles in virtually all biological processes. The significance and remarkable scope of their functions are exemplified in :

1. Enzymatic Catalysis

Nearly all chemical reactions in biological systems are catalyzed by specific macromolecules called enzymes. Some of these reactions, such as the hydration of carbon dioxide, are quite simple. Others, such as the replication of an entire chromosome, are highly intricate. Nearly all enzymes exhibit enormous catalytic power. They usually enhance reaction rates by at least a million fold. Indeed, chemical transformations rarely occur at perceptible rates *in vivo* in the absence of enzymes Several thousand enzymes have been characterized, and many of them have been crystallized. The striking fact is that all known enzymes are proteins. Thus, striking fact is that all known enzymes are proteins. Thus, proteins play the unique role of determining the pattern of chemical transformations in biological systems.

2. Transport and Storage

Many small molecules and ions are transported by specific proteins. For example, hemoglobin transports oxygen in erythrocytes, whereas myoglobin, a related protein, transports oxygen in muscle. Iron is carried in the plasma of blood by transferring and is stored in the liver as a complex with ferritin, a different protein.

3. Co-ordinated Motion

Proteins are the major component of muscl^ Muscle contraction is accomplished by the sliding motion of two kinds of protein filaments. On the microscopic scale, such co-ordinated motions as the movement of chromosomes in mitosis and the propulsion of sperm by their flagella also are produced by contractile assemblies consisting of proteins.

4. Mechanical Support

The high tensile strength of skin and bond is due to the presence of collagen, a fibrous protein.

5. Immune Protection

Antibodies are highly specific proteins that recognize and combine with such foreign substances as viruses, bacteria, and cells from other organisms. Proteins thus play a vital role in distinguishing between self and nonself.

6. Generation and Transmission of Nerve Impulses

The response of nerve cells to specific stimuli is mediated by receptor proteins. For example, rhodopsin is the photoreceptor protein in retinal rod cells. Receptor molecules that can be triggered by specific small molecules, such as acetylcholine, are responsible for transmitting nerve impulses at synapses—that is, at junctions between nerve cells.

7. Control of growth and Differentiation

Controlled sequential expression of genetic information is essential for the orderly growth and differentiation of cells. Only a small fraction of the genome of a cell is expressed at any one time. In bacteria, repressor proteins are important control elements that silence specific segments of the DNA of a cell. A quite different way in which proteins act in differentiation is exemplified by nerve growth factor, a protein complex that guides the formation of neutral networks in higher organisms.

9

METABOLISM OF AMINO ACIDS

The amino acids are relatively simple molecules and serve several functions in metabolism. They serve as (1) the substrates for protein synthesis, an anabolic process that also includes such diverse events as tissue repair, hormone synthesis, etc.; (2) precursors of other amino acids; and (3) precursors of other compounds of biologic importance that are of a nonprotein character (e.g., heme, purine, pyrimidines).

We should be reminded that, even though they become vital components in metabolism, amino acids per se are found in only minor quantities in the common human diets. We consume amino acids as they occur in proteins of both animal and plant origin. The proteins we ingest must, of course, be hydrolyzed before their constituent amino acids can be made available for the many metabolic reactions in which they are capable of participating.

If one accounts for the totality of oxidative metabolism of the carbon chains of the amino acids found in humans, it would be necessary to consider at least 20 different multienzyme sequences responsible for the oxidation of the 20 different amino acids. However, these seemingly diverse pathways ultimately converge into a relatively few terminal pathways which all lead to the tricarboxylic acid (TCA) cycle. For example, the carbon chains of 10 of the amino acids are converted to active acetate, five to α-ketoglutarate, three to succinyl-CoA, and two to oxaloacetate.

Before the carbon chains of the amino acids are metabolized, the amino group is removed, either by (1) a nonoxidative process called *transamination,* or (2) by oxidative deamination. Although the nitrogen atoms can become part of a great number of different intermediary metabolites in humans, they share a common terminal fate : excretion as urea, which accounts for about 90 percent of the total nitrogen excreted per day, and as NH_4^+ (or NH_3), which contributes about 4 percent of the total nitrogen excreted per day. Let us examine first the metabolic consequences of transamination.

TRANSAMINATION

In humans, the α-NH_2 groups of 12 amino acids can be transferred to the α-C of the following keto acids:

$$CH_3-\overset{\overset{\Large O}{\|}}{C}-COOH \qquad \text{Pyruvic Acid}$$

$$HOOC-CH_2-\overset{\overset{\Large O}{\|}}{C}-COOH \qquad \text{Oxaloacetric Acid}$$

$$HOOC-CH_2-\overset{\overset{\Large O}{\|}}{C}-COOH \qquad \alpha\text{-Ketoglutaric Acid}$$

In the process the amino acids lost their α-NH_2 groups to become α-keto acids, and alanine, aspartate, or glutamate are formed.

$$R'-\underset{H}{\overset{NH_2}{C}}-COOH + R''-\overset{\overset{O}{\|}}{C}-COOH = R'-\overset{\overset{O}{\|}}{C}-COOH \text{ and } R''-\underset{H}{\overset{NH_2}{C}}-COOH$$

These reactions are freely reversible with a *K'* of approximately l. Compared with a process such as reductive amination, transamination provides an energetically economical means to furnish the nitrogen for amino acid synthesis.

In human metabolism α-ketoglutarate is quantitatively the most important acceptor of α-NH_2 groups by the transamination process. Alanine, arginine, aspartate, asparagine, cysteine, isoleucine, phenylalanine, tryptophan, tyrosine, and valine can have their α-NH_2 groups transferred to α-ketoglutarate to form glutamate. The significance of this convergence of amino groups from so many amino acids in the synthesis of glutamate will become apparent when we discuss the process of deamination.

Mechanism of Transamination

The enzymes responsible for the transfer of α-NH_2 groups to α-keto acids are called transaminases or aminotransferases. No matter which α-NH_2 acid/α-keto acid partners are involved, all transamination reactions share a common reaction mechanism involving a coenzymic derivative of vitamin B_6, or pyridoxal phosphate:

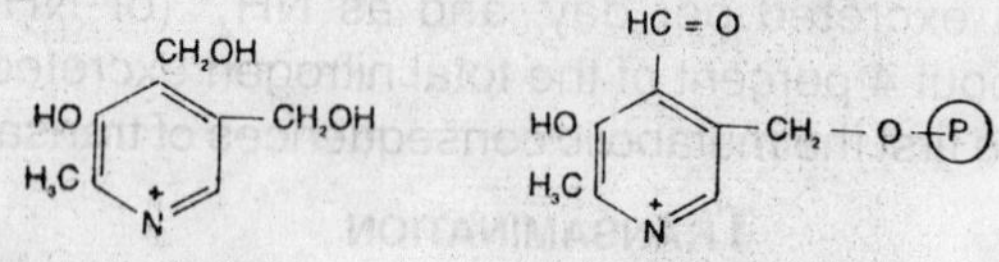

Thus, a transaminase consists of a specific protein plus the pyridoxal phosphate as a coenzyme. The pyridoxal phosphate is bound noncovalently, via its nitrogen, to the enzyme protein. With the exception of lysine and threonine, specific transaminases for all of the amino acids have been described and all of them utilize pyridoxal phosphate as the coenzyme. The role of the coenzyme in the transamination process can be summarized as follows:

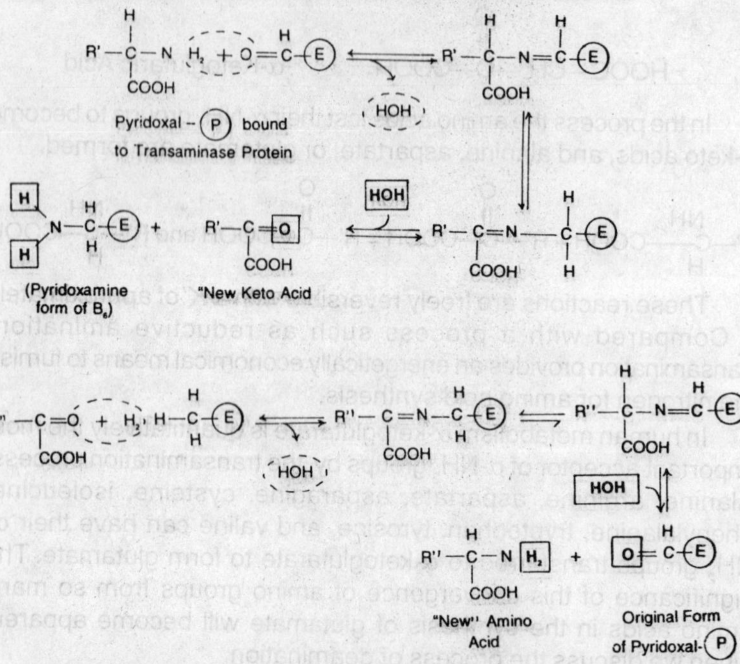

$$\text{Net}: R'\text{—}\underset{COOH}{\overset{H}{C}}\text{—}NH_2 + R''\text{—}\underset{COOH}{C}{=}O = R'\text{—}\underset{COOH}{C}{=}O + R''\text{—}\underset{COOH}{\overset{H}{C}}\text{—}NH_2$$

This mechanism of transamination can be confirmed by labeling the amino group of the amino acid substrate,

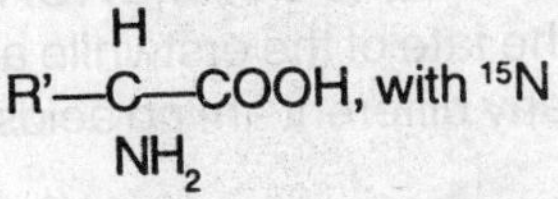

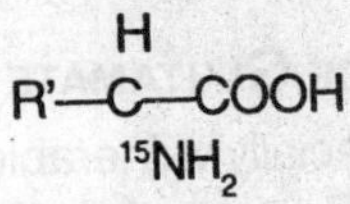

The label is found to be completely recovered as

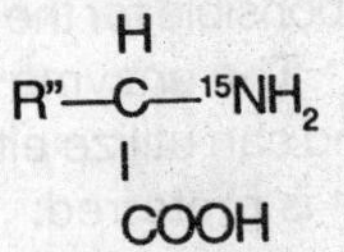

It may be noted that, although pyridoxal phosphate is the obligatory and only coenzyme for the transaminases, it is versatile coenzyme and functions in other types of metabolic reactions such as decarboxylations and recemizations.

Location of Transaminases in Cells and Tissues

The transaminases are found in both the mitochondria and cytosol of cells of a variety of tissues and organs. As expected, if there is damage to a tissue and/or organ, the transaminases of the injured or diseased cells will appear in the extracellular fluid—the serum. Two transaminases that are commonly analyzed in differential diagnoses are serum glutamate-pyruvate transaminase (SGPT) and serum glutamate-oxaloacetate transaminase (SGOT). For example, after a myocardial infarction, only SGOT is elevated; SGPT is not changed significantly. In liver disease both SGOT and SGPT are elevated.

Key Role of α-Ketoglutarate in Transamination

As mentioned earlier, at least 12 of the amino acids can transaminate very readily and in most cases α-ketoglutarate is the keto acid acceptor and "collector" of their amino groups:

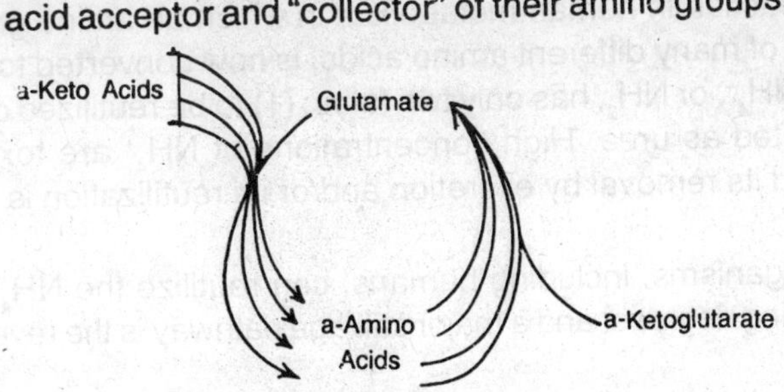

As will be described in a later reaction, the keto acids derived from the various amino acids are subject to oxidative degradation to other intermediary metabolites and ultimately to CO_2 and HOH. The present discussion will be confined to the fate of the erstwhile amino groups that were originally a part of many different amino acids but are now the α-N of glutamate.

Oxidative Deamination of Glutamate

The α-NH_2 group of glutamate is especially vulnerable to oxidative removal by the process of deamination. Particularly in the liver, there is a very active dehydrogenase that is responsible for the deamination of glutamate to α-ketoglutarate and NH_4^+. The enzyme is present in both the cytosol and the mitochondria and can utilize either NAD^+ or $NADP^+$ as its coenzyme, although NAD^+ is preferred:

$$^{+}Glu^{2-} + HOH + \underset{\text{or } NADP^+}{NAD^+} \xrightarrow[\text{Dehydrogenase}]{\text{glutamate}} NH_4^+ + \alpha\text{-}KG^{2-} + \underset{\text{or NADPH}}{NADH + H^+}$$

Like the NADH from many other dehydrogenations, the NADH (or NADPH) generated by this reaction can be oxidized by the electron-transport chain of the mitochondria with the formation of ATP. The deamination of glutamate is a process, therefore, that can initiate the formation of three ATP molecules.

With regard to its macromolecular structure, glutamate dehydrogenase consists of six identical subunits. As would be surmised, this subunit character of the enzyme gives it the possibility of being subject to allosteric control. Indeed, the dehydrogenase is inhibited by NADH, ATP, and GTP, and is stimulated by ADP, GDP, and some amino acids. The thyroid hormone, thyroxine, can also influence the activity of glutamate dehydrogenase but the significance of this action is not yet established.

Glutamate dehydrogenation may be underscored as a major oxidative reaction in human metabolism in which an amino group, once a part of many different amino acids, is now converted to free NH_4^+/NH_3. NH_4^+, or NH_3, has only two fates: (1) to be reutilized or (2) to be excreted as urea. High concentrations of NH_4^+ are toxic to humans and its removal by excretion and/or its reutilization is very important.

Most organisms, including humans, can reutilize the NH_4^+ (or NH_3) to varying degrees, and a major salvage pathway is the reversal of the glutamate dehydrogenase reaction: namely, the reconversion of NH_4^+ to the α-NH_2 group of glutamate :

$$\alpha\text{-KG}^{2-} + NH_4^+ + NADH + H^+ \rightarrow {}^+\text{Glu}^{2-} + NAD^+$$

The equilibrium point of the glutamate dehydrogenase reaction is, in fact, far in the direction of glutamate synthesis. For this reason, this reaction can have some interesting metabolic consequences. For example, in an organ such as the brain, the removal of α-ketoglutarate by NH_4^+ and the resulting inhibition of the TCA cycle could have untoward effects. In the liver the reversal of the glutamate dehydrogenase reaction and inhibition of the TCA cycle could lead to excess ketone-body formation from acetyl-CoA.

Conversion of NH_4^+ to the Amide Group of Glutamine

Another pathway of reutilization of the NH_4^+ is its conversion to an amide group as in the case of glutamine :

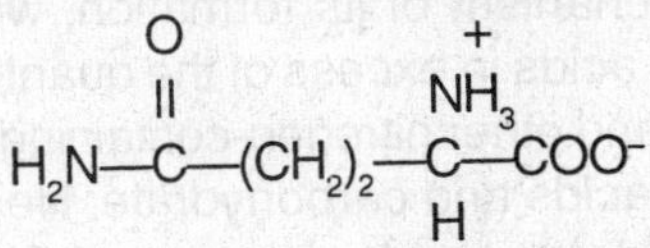

In this form not only is the erstwhile NH_4^+ made innocuous as a neutral amide group, but it can also function as a nitrogen donor in important biosynthetic reactions.

Glutamine is synthesized primarily in the liver by a complex enzyme system, glutamine synthetase.

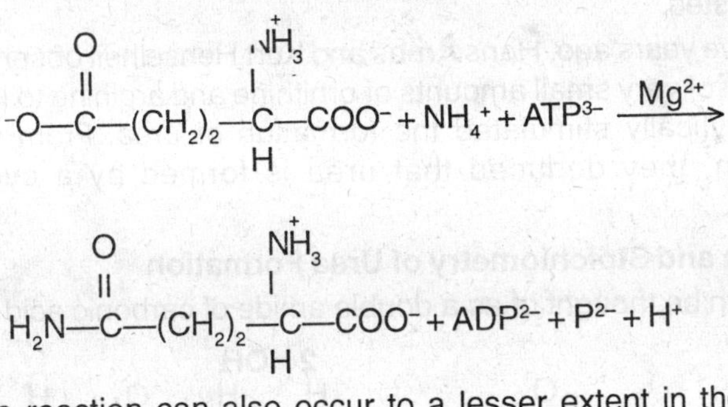

This reaction can also occur to a lesser extent in the brain and kidneys.

Hydrolysis of Glutamine to Glutamate and NH_4^+

A major metabolic fate of glutamine is its hydrolysis, regenerating glutamate and NH_4^+. The hydrolytic enzyme responsible for this reaction is located exclusively in the kidney and plays an important role in renal regulation of acid-base balance as well as in nitrogen excretion:

$$Glu\text{—}NH_2 + HOH \xrightarrow{\text{glutaminase}} {}^{+}Glu^{2-} + NH_4^+$$

The NH_4^+ so formed enters the urine directly. As much as 5 percent of the nitrogen excreted per day is in the form of NH_4^+ and almost all of this is produced by the glutaminase reaction.

SYNTHESIS OF UREA

The dominant end product of nitrogen metabolism in humans is not NH_4^+ but is instead a NH_4^+ derivative : urea, $NH_2\text{—}\underset{\underset{O}{\|}}{C}\text{—}NH_2$.

Before discussing the mechanism of its formation, we should be reminded again that amino acids in excess of the quantities needed for the synthesis of protein and other nitrogen-containing metabolites are not stored as are fatty acids and carbohydrate. Neither are the excess amino acids excreted. Any surplus is used as fuel for oxidative metabolism and in the process, virtually all of the nitrogen is excreted in the form of urea and NH_4^+. On an average diet we excrete about 25 to 30 g of urea per day and this represents about 90 percent of the total nitrogenous substances in our urine. As would be expected, the amount of urea excreted is directly proportional to the amount of protein ingested.

Forty-five years ago, Hans Krebs and Kurt Henselheit observed that addition of very small amounts of ornithine and arginine to liver slices catalytically stimulated the formation of urea. From this observation, they deduced that urea is formed by a cyclic mechanism.

Mechanism and Stoichiometry of Urea Formation

Urea can be thought of as a double amide of carbonic acid :

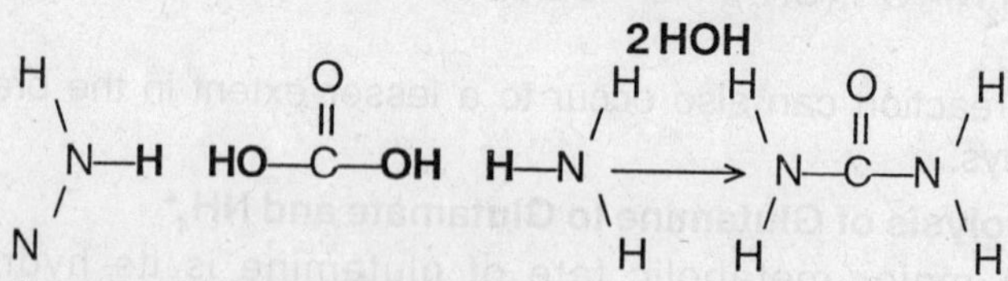

Obviously, the ultimate substrates for urea synthesis must be CO_2 (or HCO_3^-) and two amino groups or their equivalents.

The overall stoichiometry of the urea cycle becomes

$$2\,NH_4^+ + HCO_3^- + 3\,ATP^{3-} + 2\,HOH \rightarrow$$
(from Glu, Asp)

$$NH_2\text{—}\underset{\underset{O}{\|}}{C}\text{—}NH_2 + 2\,ADP^{2-} + 2\,P^{2-} + AMP^{-} + PP^{2-} + 3\,H^{-}$$

Since pyrophosphate is converted to orthophosphate, the energy requirement for the production of one molecule of urea is equivalent to four high-energy phosphates. The equivalent of four high-energy phosphates would be needed to restore all of them ADP and AMP to ATP.

Relation of the Urea Cycle to the Tricarboxylic Acid Cycle

Since fumarate is a product of the urea cycle, the process of urea synthesis is linked to the tricarboxylic acid cycle :

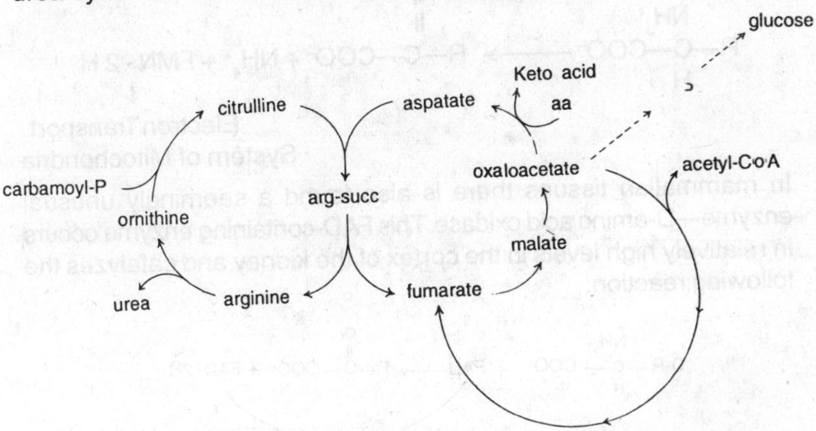

Fumarate is hydrated to malate, which in turn can be oxidized to oxaloacetate. This key metabolite has several fates, which include transamination to aspartate, conversion to glucose, and condensation with acetyl-CoA as the first step in the TCA cycle.

Inherited Defects of the Urea Cycle

Inherited enzymatic defects of the urea cycle has been identified and, predictably, the end point of all of these diseases is hyperammonemia. A complete block of any of the steps in the urea cycle is lethal since there is no known alternative pathway for the synthesis of urea in humans. Patients suffering from these genetic defects have an intolerance to dietary protein and have excessive amounts of NH_3/NH_4^+ in the blood. Some patients can be given a replacement of dietary protein by a mixture of the α-keto acids corresponding to the essential amino acids, the rationale of this treatment being that these keto acids are converted to the amino acids with the concomitant removal of NH_3/NH_4^+ from the blood.

To recapitulate, when an α-NH_2 group has been removed from the carbon chain of an amino acid, it has two fates: (1) to be excreted—as NH_4^+ and urea or (2) to be utilized in the biosynthesis of other amino acids or similar compounds.

L- AND - D-AMINO ACID OXIDASES

To complete the story on deamination, it should be mentioned that there are other oxidative enzymes that play a lesser role in amino acid oxidation—the L-amino acid oxidases. These enzymes use FMN as the coenzyme :

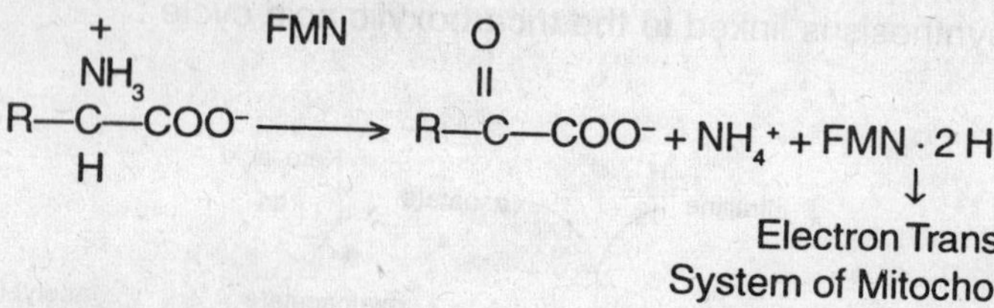

In mammalian tissues there is also found a seemingly unusual enzyme—D-amino acid oxidase. This FAD-containing enzyme occurs in relatively high levels in the cortex of the kidney and catalyzes the following reaction:

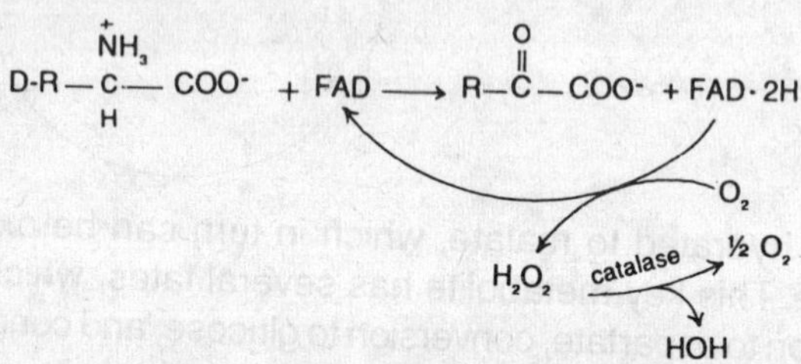

This interesting flavoenzyme—obtainable in crystalline form—has been used widely in mechanistic studies. Since D-amino acids do not occur naturally in mammalian tissues, of what use is the enzyme? Is its function to facilitate oxidation of D-amino acids derived from bacterial cell walls? Is it one of nature's mistakes or dead ends that was never erased through eons of evolution?

Overview of Nitrogen Balance and Self-Regulation of Nitrogen Metabolism

Having sketched the metabolic fate of the nitrogen of the amino acids, we have also arrived at an overview of nitrogen balance. Clearly, unlike the carbon of carbohydrate and of lipid, nitrogen is not stored in humans. In a 70-kg man who is eating an average diet, about 400 g of protein turns over per day. About three-fourths of this may be recycled. Depending on the circumstances and the state of the individual, 6 to 20g of N, derived from the α-NH_2 groups of the amino acids is excreted each day, mostly as urea. Even when protein is not ingested, as in nitrogen starvation, a person can still excrete as much as 5 g of nitrogen per day, and amount corresponding to the loss of some 30 g of protein. The rate of utilization and the proportions of the loss of some 30 g of protein. The rate of utilization and the proportions of the different amino acids used by humans depends on many factors, as will become more evident when we examine the fate of the carbon chains of the amino acids.

The pathways of nitrogen metabolism described so far are self-regulated. As will be more apparent later, the syntheses of many nitrogen compounds that are not amino acids, but may be derived from them, are regulated at their "committed steps." Therefore, they operate at rates that disregard the supply of amino acids. The remarkable constancy of the urinary excretion of these compounds support this conclusion.

The maintenance of nitrogen balance depends, of course, on the control of urea synthesis. Nitrogen balance is maintained if the amount of urea excreted equals the difference between the total nitrogen ingested and the total of all of the other forms of nitrogen in the urine. To illustrate, assume that there is a transient accumulation of amino acids. They will transaminate with α-ketoglutarate. The glutamate formed will yield NH_3/NH_4^+ by oxidation, which by conversion to carbamoyl phosphate and aspartate tends to force the formation of arginine and, thence, urea. This is the law of mass action, and the results are predictable from the equilibrium constants and the concentrations of the compounds involved. This is the simplest kind of control of understand.

However, the kinetics of the processes involved are also implicated in the control mechanism. There is a higher order of controls in an organism as complicated as a human that makes certain that these processes will go at rates commensurate with need. Consider the following illustrations. On a high-protein diet, the liver concentrations of many, if not all, of the transaminases go up significantly. This is in response to the adrenal cortical steroids that are secreted in higher amounts on high-protein diets. The quantity of dietary protein will also influence the enzymes needed for making arginine. On a high-protein diet, all five enzymes increase in quantity. On a low-protein diet, all five decline in amount and the rates of the reactions leading to urea change accordingly.

Metabolic Relationships of Amino Acids, Carbohydrates, and Lipids

Having examined the major pathways for the removal and metabolic disposition of α-NH_2 groups from amino acids, we may now turn our attention to the fate of the carbon chains of the deaminated amino acids. Although the oxidative metabolism of the 20 different amino acids requires at least a like number of different multienzyme reactions, these seemingly diverse pathways converge into a relatively few terminal pathways. Equally important, these pathways become tangental to those responsible for the metabolism of carbohydrates and lipids. From these metabolic interrelationships, one can also appreciate the biochemical basis for characterizing the amino acids as essential, nonessential, glycogenic, or ketogenic.

Glucogenic, Ketogenic, Essential, and Nonessential Amino Acids

Amino acids whose carbon chains can be oxidized more or less directly to acetyl-CoA can also yield acetoacetate, a *ketone body*. Formation of acetoacetate from such amino acids is particularly evident in fasting animals and, accordingly, they are labeled *ketogenic*. These amino acids are leucine, lysine, isoleucine, tryptophan, phenylalanine, and tyrosine. If we define a ketogenic amino acid as one whose carbon chain can give rise only to acetoacetate, there are only two ketogenic amino acids—leucine and lysine. The other four—isoleucine, tryptophan, phenylalanine, and tyrosine—can also give rise to glucose precursors and are, therefore, both ketogenic and glucogenic.

Amino acids that can serve as precursors of phosphoenolpyruvate (therefore, of glucose) are called *glycogenic* or *glucogenic*. These amino acids are :

alanine	cysteine(ine)	histidine	serine
arginine	glutamate	methionine	threonine
aspartate	glutamine	proline	tryptophan
asparagine	glycine	hydroxyproline	valine

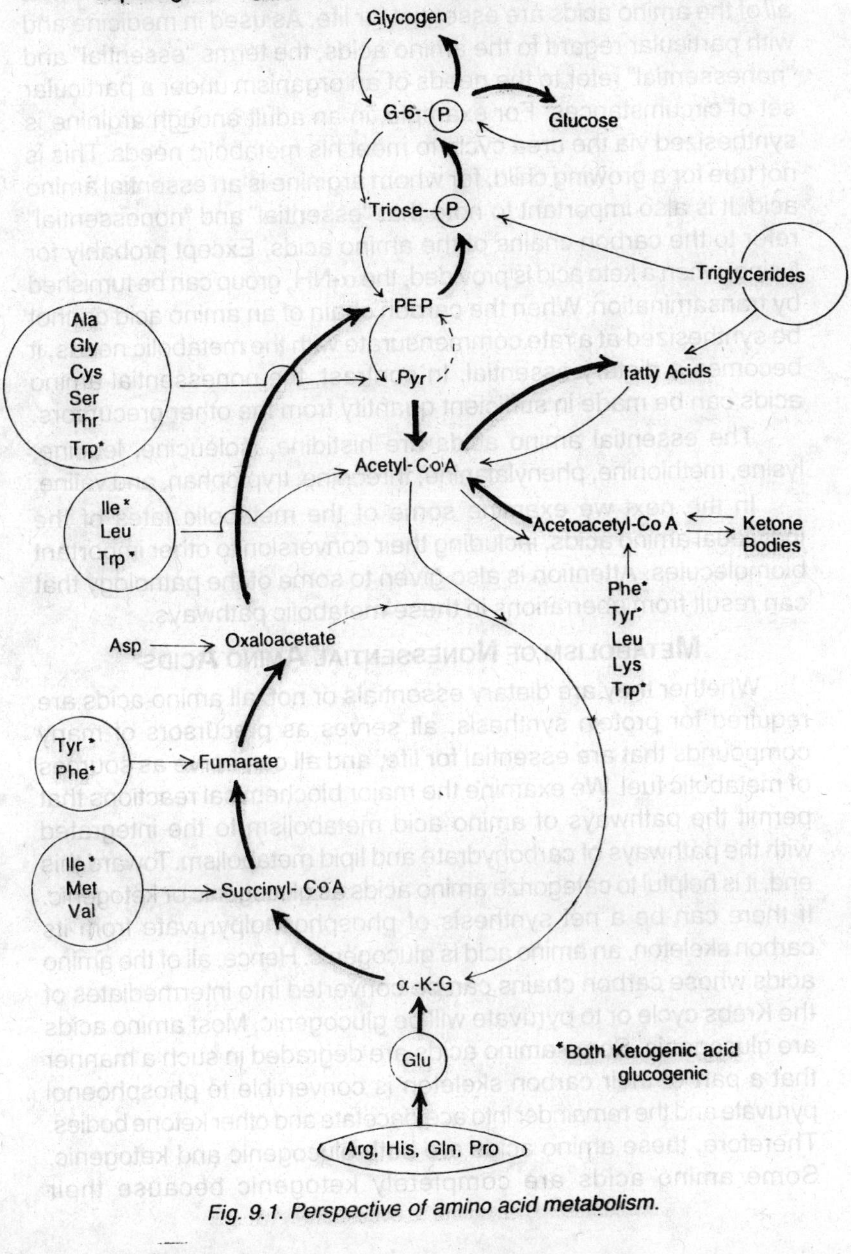

Fig. 9.1. Perspective of amino acid metabolism.

It is important to emphasize that "essential" and "nonessential" are strictly dietary terms and are not germane to the relative importance of the amino acids in intermediary metabolism. That is, *all* of the amino acids are essential for life. As used in medicine and with particular regard to the amino acids, the terms "essential" and "nonessential" refer to the needs of an organism under a particular set of circumstances. For example, in an adult enough arginine is synthesized via the urea cycle to meet his metabolic needs. This is not true for a growing child, for whom arginine is an essential amino acid. It is also important to note that "essential" and "nonessential" refer to the carbon chains of the amino acids. Except probably for lysine, when a keto acid is provided, the α-NH_2 group can be furnished by transamination. When the carbon chain of an amino acid cannot be synthesized at a rate commensurate with the metabolic needs, it becomes a dietary essential. In contrast, the nonessential amino acids can be made in sufficient quantity from the other precursors.

The essential amino acids are histidine, isoleucine, leucine, lysine, methionine, phenylalanine, threonine, tryptophan, and valine.

In the next we examine some of the metabolic fates of the individual amino acids, including their conversion to other important biomolecules. Attention is also given to some of the pathology that can result from aberrations in these metabolic pathways.

Metabolism of Nonessential Amino Acids

Whether they are dietary essentials or not, all amino acids are required for protein synthesis, all serves as precursors of many compounds that are essential for life, and all can serve as sources of metabolic fuel. We examine the major biochemical reactions that permit the pathways of amino acid metabolism to the integrated with the pathways of carbohydrate and lipid metabolism. Toward this end, it is helpful to categorize amino acids as glucogenic or ketogenic. If there can be a net synthesis of phosphoenolpyruvate from its carbon skeleton, an amino acid is glucogenic. Hence, all of the amino acids whose carbon chains can be converted into intermediates of the Krebs cycle or to pyruvate will be glucogenic. Most amino acids are glucogenic. Some amino acids are degraded in such a manner that a part of their carbon skeleton is convertible to phosphoenol pyruvate and the remainder into acetoacetate and other ketone bodies. Therefore, these amino acids are both glucogenic and ketogenic. Some amino acids are completely ketogenic because their intermediate oxidation products can be accounted for entirely as acetyl-CoA and its ketone body derivatives.

Let us now examine the amino acids that are not dietary essentials for humans, accounting for their biosynthesis, their glucogenic or ketogenic character, and their roles as intermediates in the biosynthesis of metabolically essential compounds.

Glycine

As the simplest of the amino acids, glycine also participates in more metabolic reactions than any of the other amino acids. As a nonessential amino acid, the most direct route for its synthesis is from serine:

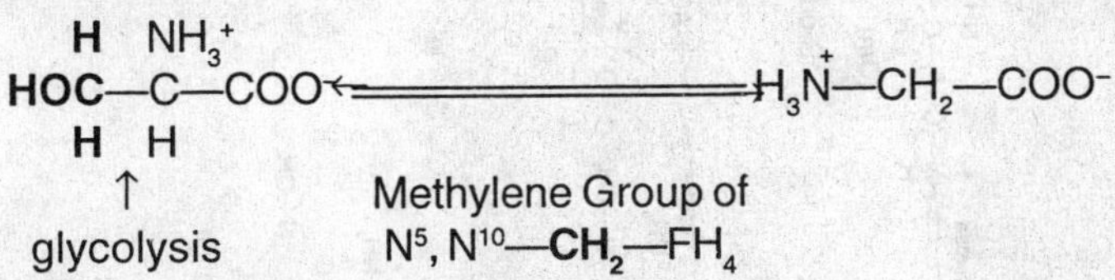

Glycine is glucogenic because it can be reconverted into serine, which in turn can form pyruvate :

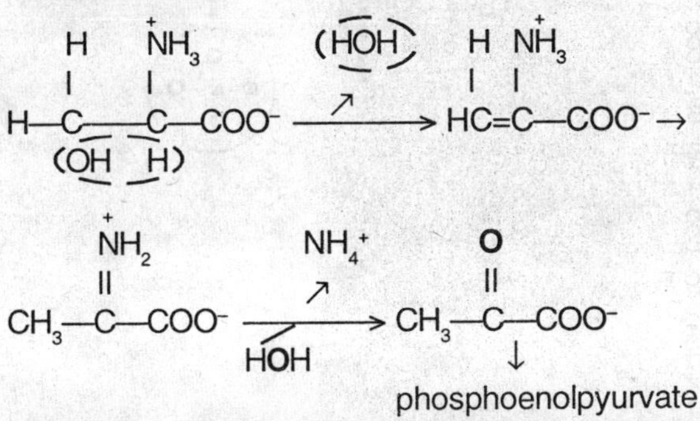

It should be recalled that pyruvate can be converted to phosphoenolpyruvate (and glucose) by the following reactions:

Pyruvate + CO_2 + ATP → oxaloacetate + ADP + P

Oxaloacetate + GTP → PEP + CO_2 + GDP

Pyruvate + ATP + GTP → phosphoenolpyruvate + ADP + GDP + P

↓↑

glucose

Creatine, Creatinine, and Creatine Kinase

Glycine is essential for the synthesis of creatine, an important constituent of muscle, but whose formation can also involve the kidneys and liver :

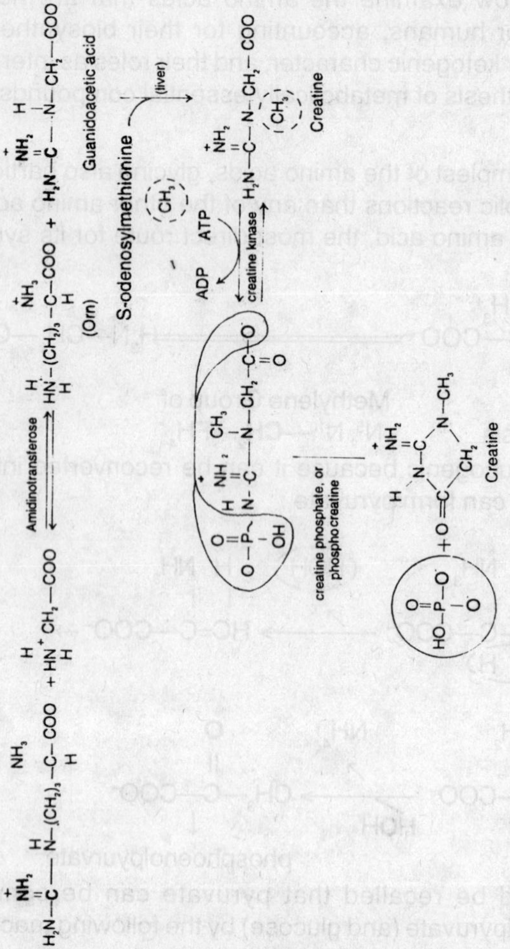

The assay of creatine kinase is a common and sensitive procedure in the diagnostic laboratory and has particular application in the diagnosis of myocardial infraction. This enzyme has three isozymic forms that are dimers of two polypeptide chains, designated as B and M. The MM forms is present in higher concentration in skeletal muscle and heart. In principle the analysis is as follows :

$$\text{Creatine} + \text{ATP} \xrightleftharpoons{\text{CPK kinase}} \text{creatine} \sim \text{phosphate} + \text{ADP}$$

ADP is then measured by the coupling of the following reactions, using PEP, pyruvate kinase, and NADH as reagents :

$$\text{ADP} + \text{phosphoenolpyruvate} \xrightarrow[\text{Pyruvate kinase}]{} \text{ATP} + \text{pyruvate}$$

$$\text{Pyruvate} + \text{NADH} + \text{H}^{+} \longrightarrow \text{lactate} + \text{NAD}^{+}$$

The rate of disappearance of NADH is a measure of CPK activity in the original sample.

In muscle creatine ~P is converted to creatinine spontaneously and irreversibly :

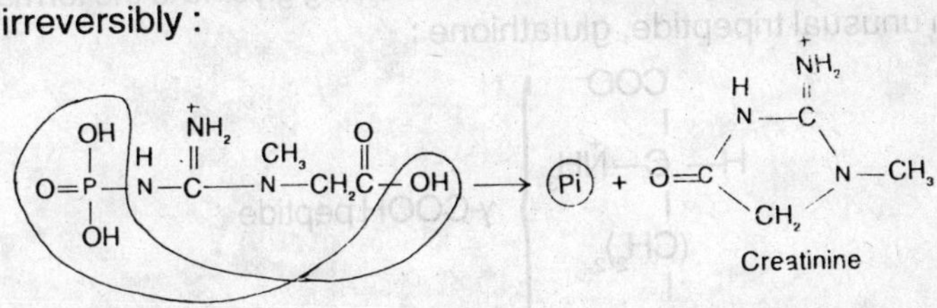

Creatinine

By this irreversible process, creatinine become the major excretory form of creatine. Creatinine has no known function in humans and is metabolically inert. When related to body size or, more particularly, to muscular mass, the 24-hour excretion of creatinine by an individual is remarkably constant from day to day, accounting for sex and age. Measurement of creatinine excretion is also employed to determine glomerular filtration rates.

Hippuric Acid

The liver has a remarkable capacity to *detoxify* a wide variety of compounds particularly those that may be foreign to the body and/ or that may offer problems related to their solubility. One example of such a detoxication reaction and one in which glycine plays a role is that involving benzoic acid (benzoate) :

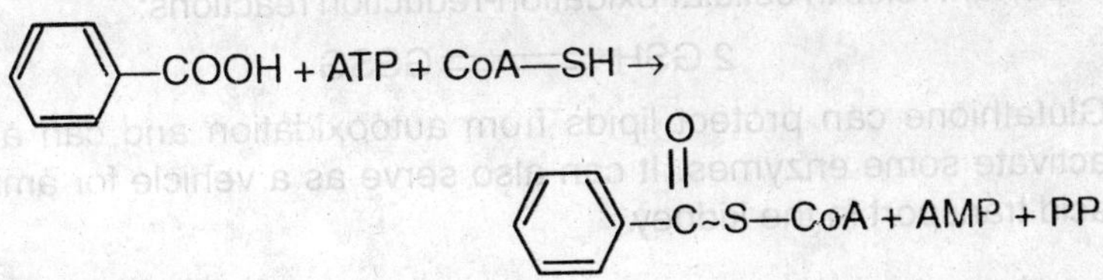

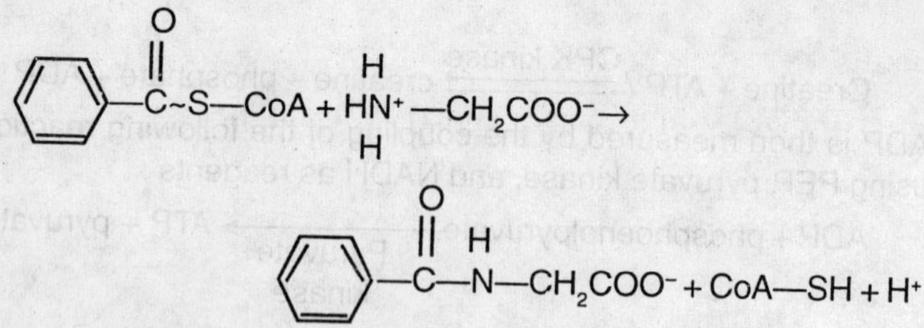

In this reaction, a slightly soluble compound has been converted into a soluble derivative of glycine that can be easily eliminated.

Glutathione

An important synthetic reaction involving glycine is the formation of an unusual tripeptide, glutathione :

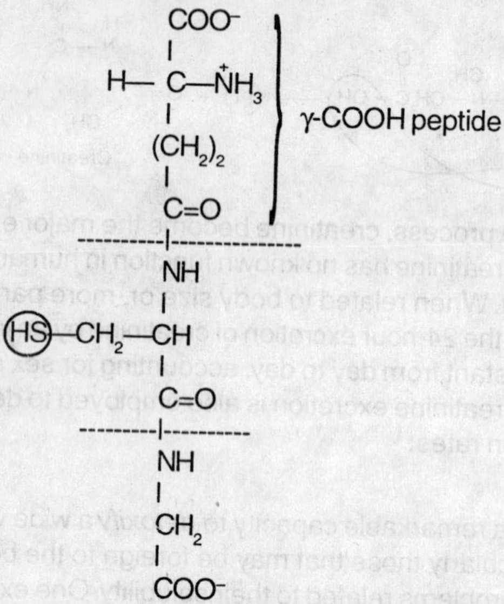

It should be noted that the peptide linkage with the glutamyl residue is at the γ-COOH. This widely distributed sulfhydryl compound plays important roles in cellular oxidation-reduction reactions:

$$2\,GSH \rightleftharpoons GSSG$$

Glutathione can protect lipids from autooxidation and can also activate some enzymes. It can also serve as a vehicle for amino acid transport in the kidney.

Glycine as a Precursor of Purines and Porphyrins

The metabolic versatility of glycine is underscored by its role as a precursor in two other biosynthetic reactions. One of these that will be described in more detail later is the synthesis of purines:

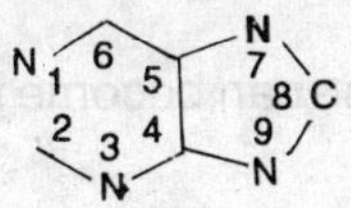

Another important reaction requiring glycine is the synthesis of porphyrins :

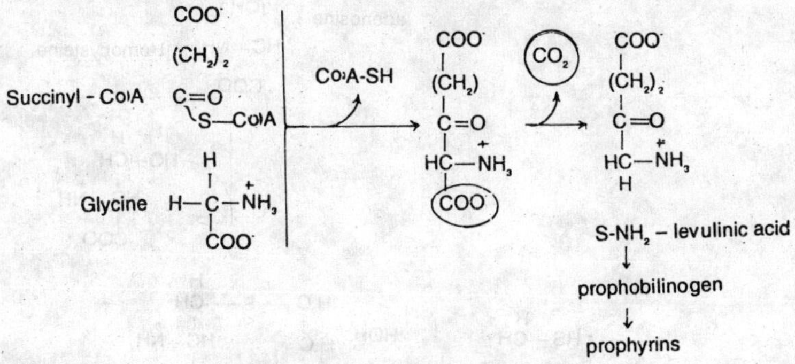

Formation of Oxalic Acid

A major degradative reaction of glycine begins with its oxidative deamination to form glyoxylic acid, which in turn can be further oxidized to oxalic acid, formaldehyde, and carbon dioxide:

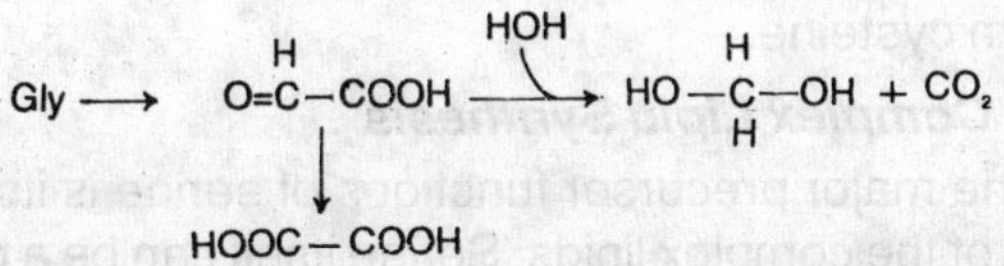

A genetic disorder has been identified in which there are high urinary levels of oxalate, accompanied by calcium oxalate calculi in the kidney.

Serine

The overall reaction of the conversion of serine to pyruvate, the key step in the series of reactions that accounts for the glucogenic character of this amino acid, has been outlined in figure. Serine is a nonessential amino acid because it can be synthesized from glycine and also from 3-phosphoglyceric acid, an intermediate of glycolysis.

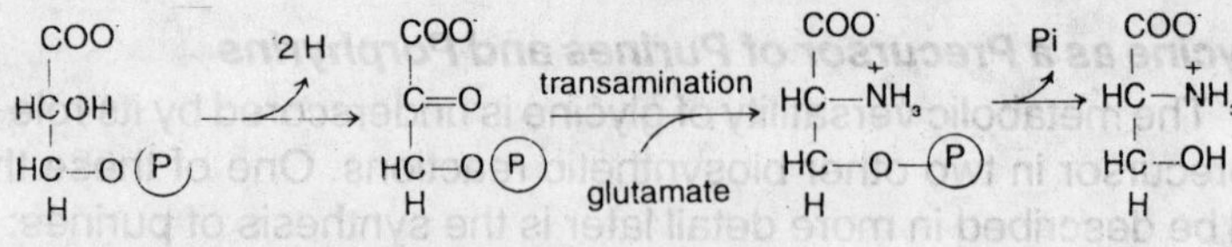

Role in Cysteine Synthesis

The carbon chain of serine can become the carbon chain of the amino acid cysteine:

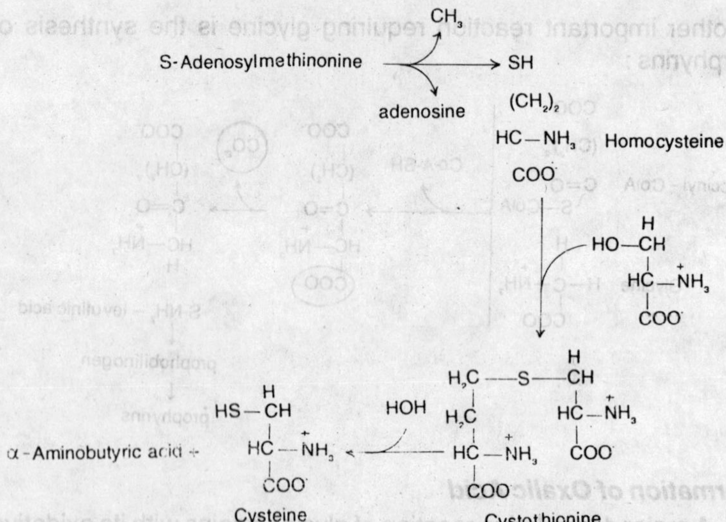

In humans the formation and splitting of cystathionine are irreversible reactions. Methionine is an essential amino acid. It cannot be made from cysteine.

Function in Complex Lipid Synthesis

One of the major precursor functions of serine is its role in the biosynthesis of the complex lipids. Serine itself can be a component of the complex lipids such as following :

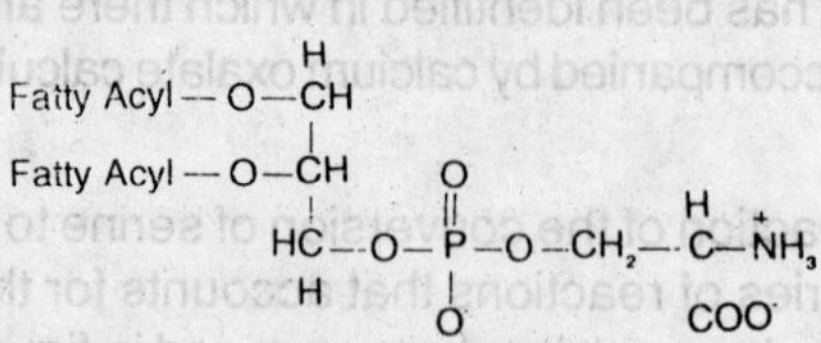

Phosphatidylserine

In addition, serine can lose its COOH group to form aminoethanol:

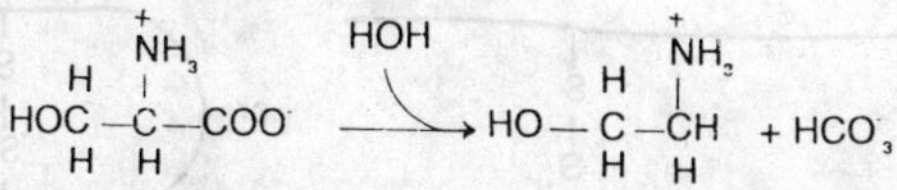

By successive additions of methyl groups to the nitrogen by S-adenosylmethionine, the ethanolamine is converted to choline:

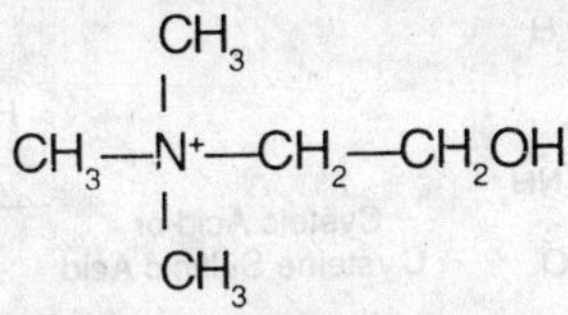

All of these reactions can occur with phosphatidyl derivatives.

$$\text{Phosphatidylserine} \xrightarrow[CO_2]{} \text{phosphatidylethanolamine}$$

$$\xrightarrow{\cdot CH_3} \xrightarrow{\cdot CH_3} \xrightarrow{\cdot CH_3} \text{phosphatidylcholine}$$

Cysteine

It will be recalled that this amino acid can be synthesized from a reaction between serine and homocysteine.

Cysteine is glucogenic because it can be converted to phosphoenolpyruvate:

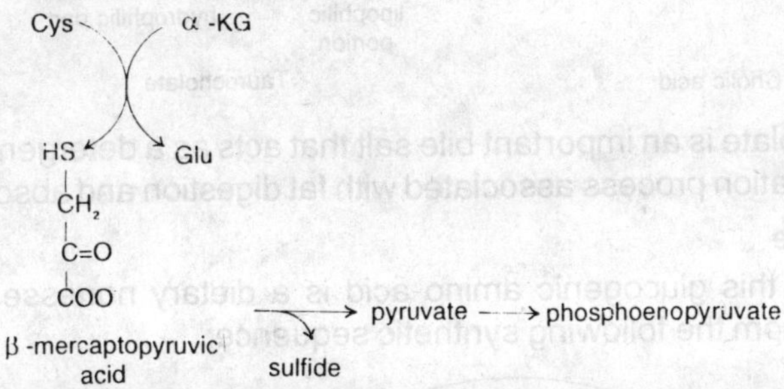

Oxidation of the —SH Group

The oxidizability of the cysteine sulfhydryl group gives the amino acid considerable metabolic versatility and also makes it an important determinant of protein structure :

$$2\ \text{Cys—SH} \rightarrow \text{Cys—S—S—Cys} + 2\ \text{H}.$$

Thus, both inter- and intrapeptide chain —S—S— linages can be formed:

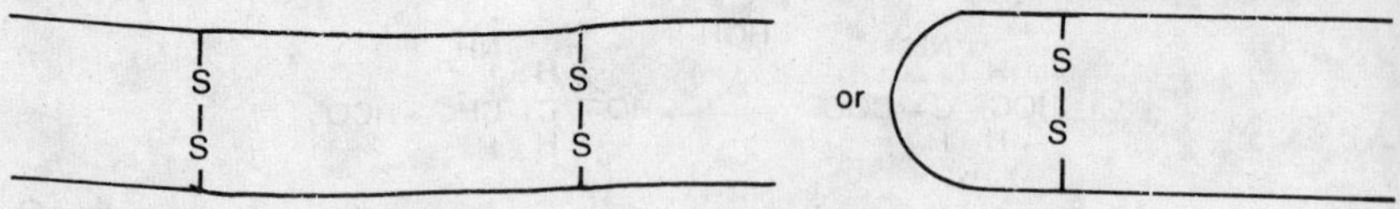

The S of cysteine can also be oxidized beyond the S—S level :

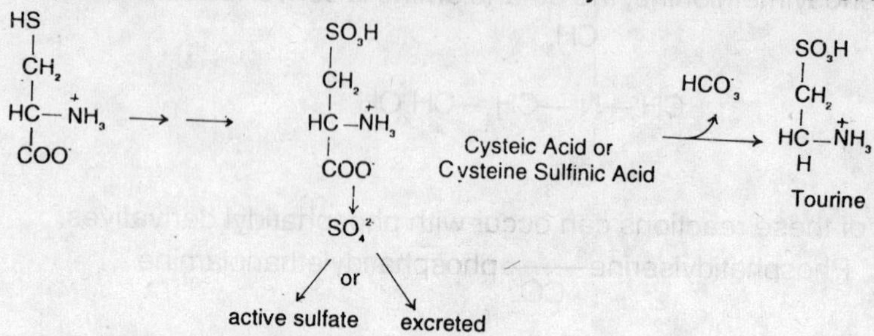

Formation of Bile Salts

Bile salts are formed as follows:

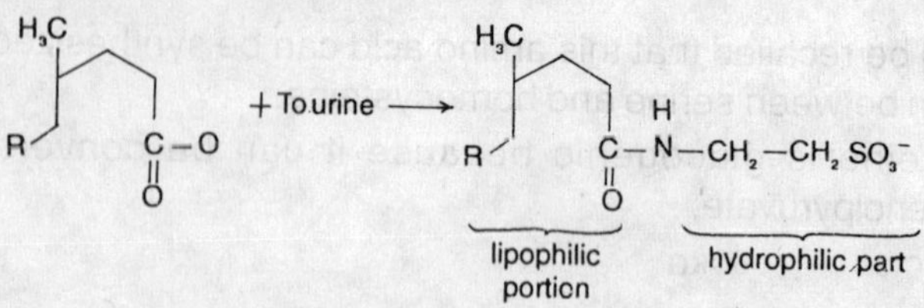

Cholic acid Taurocholate

Taurocholate is an important bile salt that acts as a detergent in the emulsification process associated with fat digestion and absorption.

Aspartate

That this glucogenic amino acid is a dietary nonessential is evident from the following synthetic sequence:

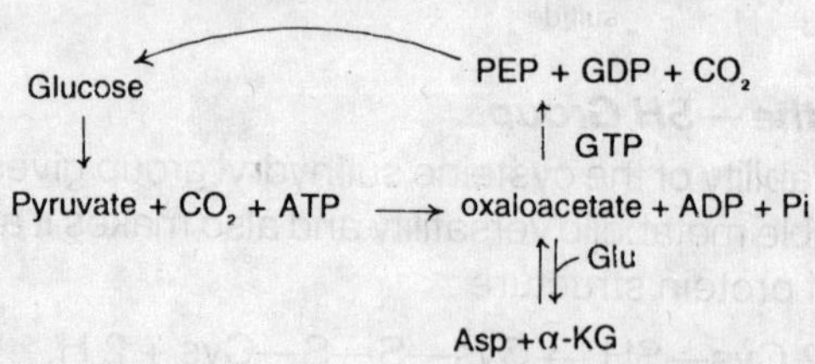

It should be remembered that the oxaloacetate also serves as the "trigger" for the tricarboxylic acid cycle.

Participation in Synthesis of Urea, Purines, and Pyrimidines

Aspartate participates in three vital synthetic reactions. It contributes one of the nitrogens in the synthesis of urea, serving as a substrate for the synthesis of arginosuccinate:

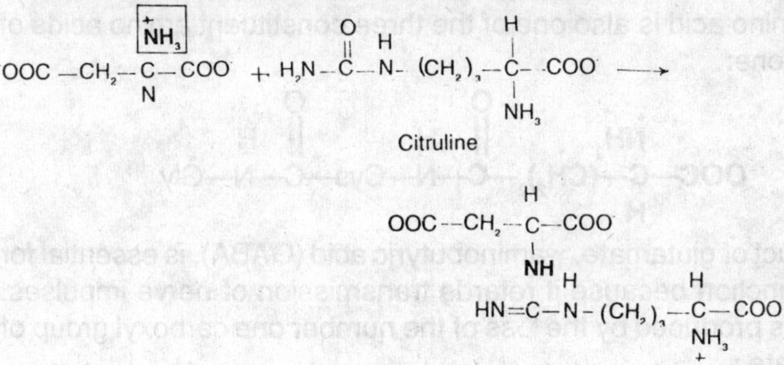

Aspartate is also one of the nitrogen donors in purine synthesis.

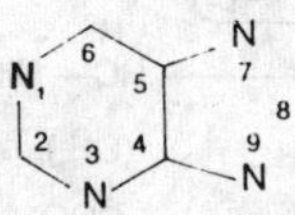

Finally, both its carbon chain and nitrogen become a major part of the pyrimidine ring system.

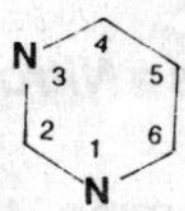

Glutamate

As pointed out previously, this amino acid is formed from α-ketoglutarate by either transamination or reductive amination. Glutamate is glucogenic because it can be converted to phosphoenolpyruvate:

Glutamate $\rightarrow$ α-ketoglutarate $\rightarrow$ succinyl-CoA $\rightarrow$ succinate $\rightarrow$ fumarate $\rightarrow$ malate $\rightarrow$ oxaloacetate $\rightarrow$ phosphoenolpyruvate (PEP) $\rightarrow \rightarrow \rightarrow$ glucose

Role in Biosynthesis of Urea, Glutathione, and γ-Aminobutyric Acid

Glutamate plays a significant role in biosynthetic reactions and as a precursor or other biological compounds. As pointed out earlier, it is one of the prime amino group donors in transamination. One of the nitrogens of urea is provided by glutamate:

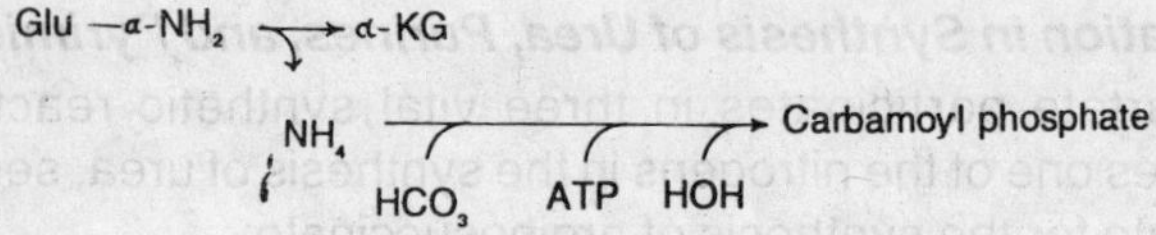

This amino acid is also one of the three constituent amino acids of glutathione:

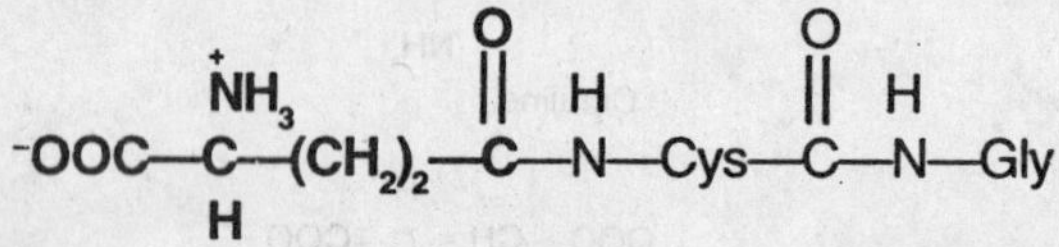

A product of glutamate, γ-aminobutyric acid (GABA), is essential for brain function because it retards transmission of nerve impulses. GABA is produced by the loss of the number one carboxyl group of glutamate :

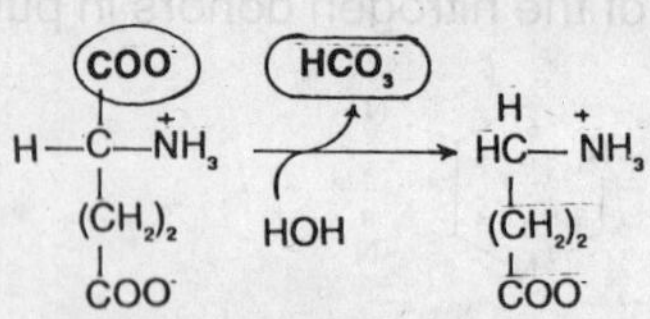

It may be noted that 75 percent of the total free amino acids of brain can be accounted for by aspartate, glutamate, and their derivatives.

Amide Group of Glutamine as a Nitrogen Source in Biosynthetic Reactions

When converted to glutamine, the amide group on C-5 of glutamate becomes a prime source of nitrogen in many biosynthetic reactions. N^3 and N^9 of the purine ring system and the amino nitrogen at position 2 of guanylic acid are derived from glutamine:

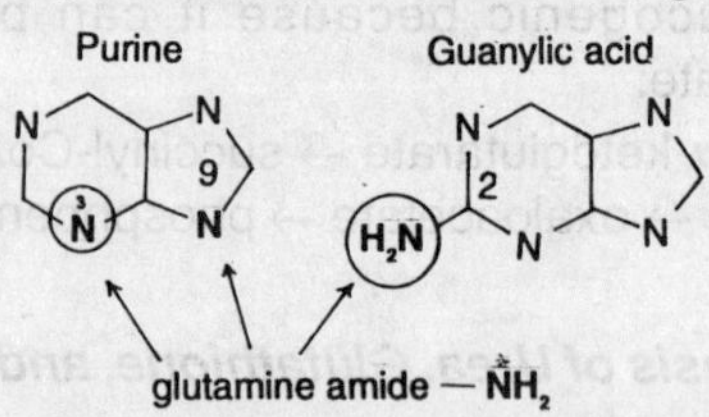

In the cytoplasm of eukaryotic cells is a carbamoyl phosphate synthetase that needs glutamine as a nitrogen donor. This carbamoyl phosphate synthetase provides a carbamoyl group for pyrimidine synthesis.

$$\text{Glutamine} + HCO_3^- + 2ATP + HOH \rightarrow$$

$$(-NH_2) \quad H_2N\text{—}\overset{\displaystyle O}{\overset{\|}{C}}\text{—}O\text{Ⓟ} + 2\,ADP + P + \text{Glutamate}$$

Glutamine also donates its amide nitrogen in the synthesis of amino sugars:

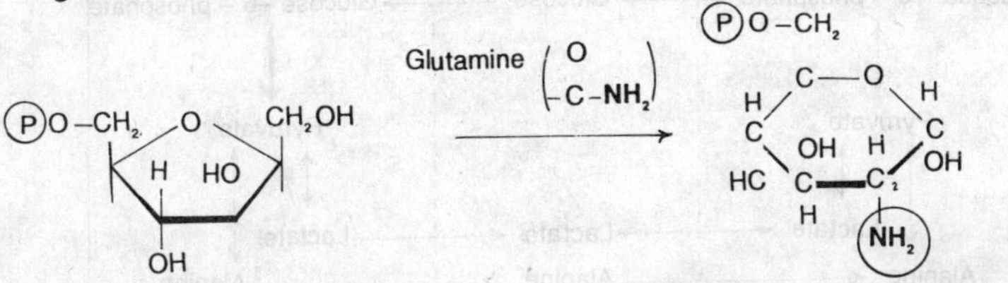

Fructose – 6 – phosphate   Glucosamine – 6 – phosphate

Glutamine participates in a cross-linking reaction of fibrin in the blood clotting process, resulting in the final insoluble polymeric form, or *Hard clot* :

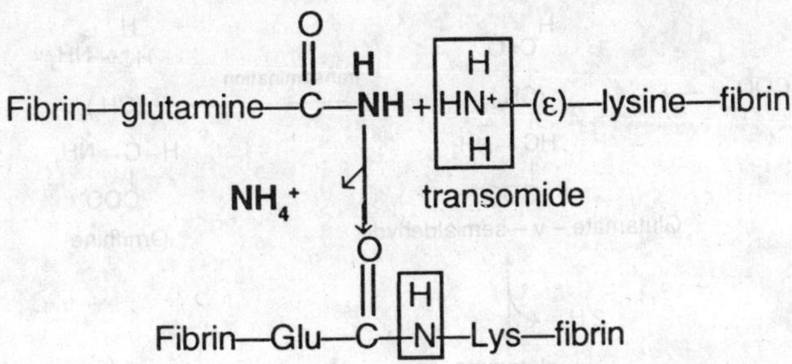

Alanine

The synthesis and metabolism of the carbon chain of alanine is apparent from the following metabolic sequence :

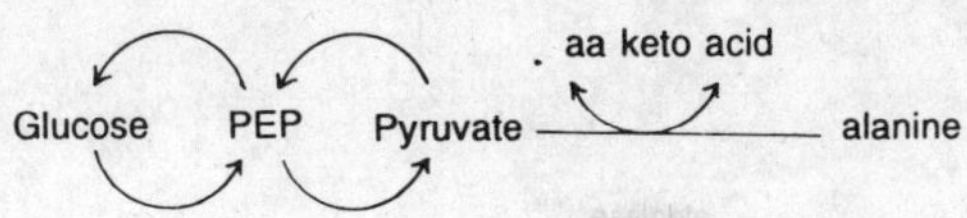

Alanine-Glucose Cycle

The Cori cycle represents a metabolic link between liver and muscle. The liver furnishes glucose, produced from lactate by gluconeogenesis, to skeletal muscle. In the muscle lactate is regenerated from glucose by glycolysis. Just as lactate is formed reductively from pyruvate is skeletal muscle, so can alanine be produced from pyruvate by transamination from a number of amino acids. Because of its interconvertibility with pyruvate by transamination, alanine can serve in the same manner as lactate in a recycling mechanism between skeletal muscle and the liver:

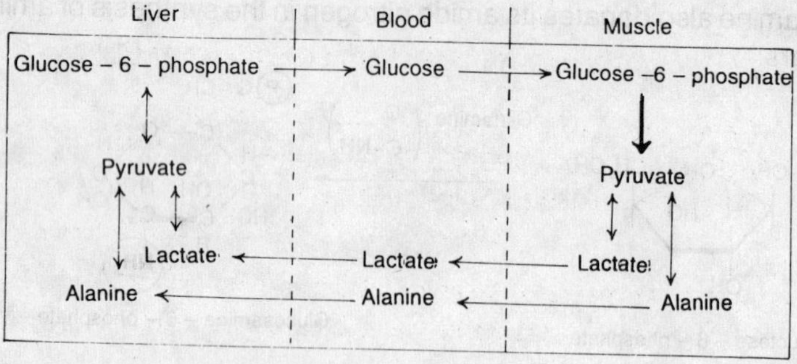

Proline and 3-Hydroxyproline

The nonessentiality of these amino acids and their glucogenic nature is evident from the following reactions:

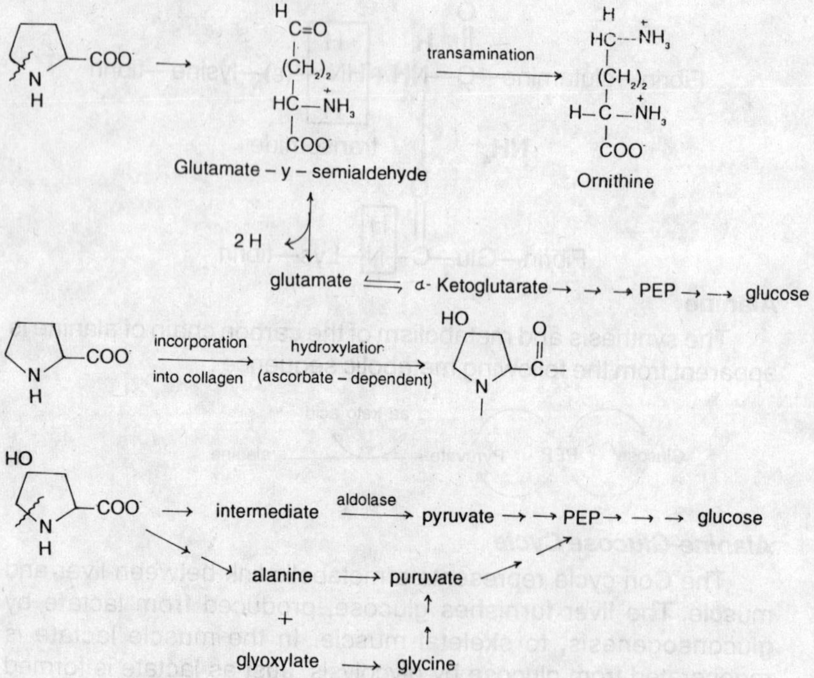

Arginine

Arginine is formed in the urea cycle:

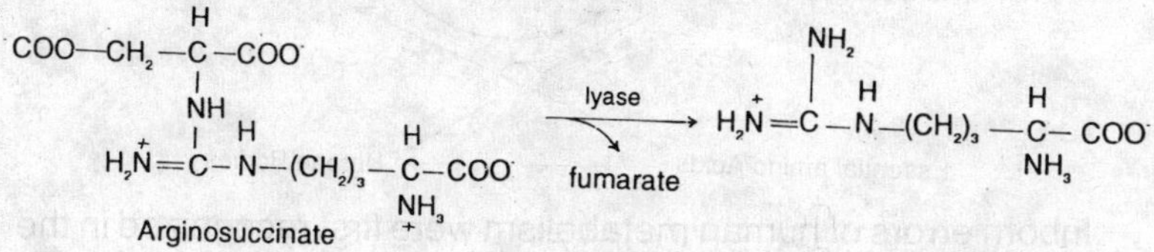

and is synthesized at a rate sufficient for the metabolic needs of an adult but not rapidly enough to support growth in the young animal.

The metabolism of arginine proceeds mainly via ornithine. However, not all of the ornithine continues in the cycle. Some is converted to glutamate:

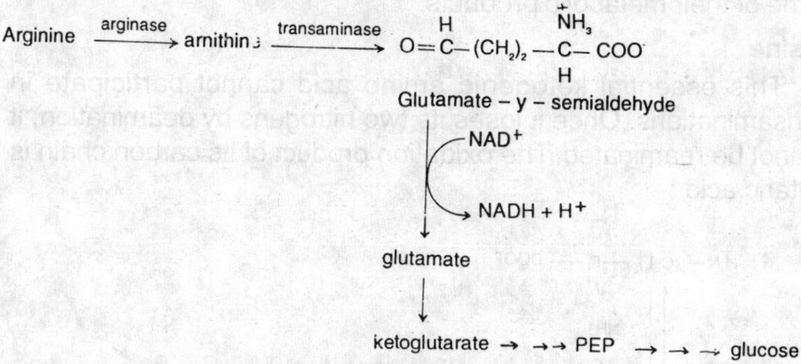

Hence, arginine is glucogenic.

METABOLISM OF ESSENTIAL AMINO ACIDS

An amino acid is classed as essential if it must be supplied in the diet for optimal growth of an immature animal or for the maintenance of nitrogen balance in the mature animal. In the context of this definition, histidine may be considered borderline as a dietary essential for human beings because it can be synthesized and is also salvaged from protein breakdown to a considerable degree. However, since it is essential for infants, this amino acid will be included in the group of nine amino acids to be described here.

As is true for the nonessential amino acids, the biochemical reactions of the essential amino acids are integrated with the pathways of carbohydrate and lipid metabolism. The interplay of bidirectional and unidirectional processes in these metabolic relationships may be viewed as follows:

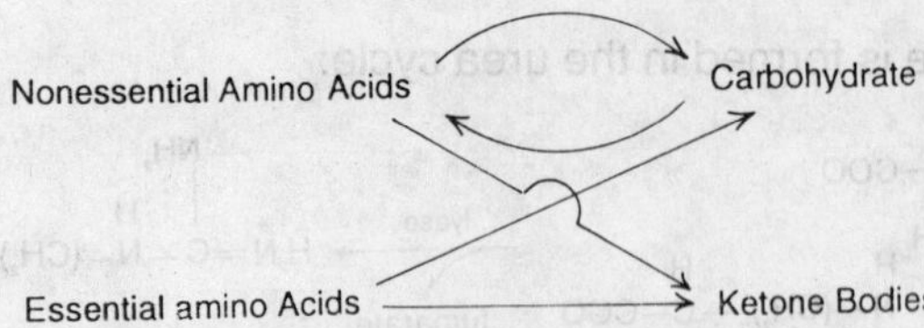

Inborn errors of human metabolism were first recognized in the area of amino acid metabolism. Although catabolic reactions of both the essential and nonessential amino acids are affected by such genetic defects, the inherited errors in metabolism of essential amino acids such as phenylalanine, tryptophan, and the branched chain acids are particularly prominent. We now examine the major catabolic pathways of the essential amino acids and the precursor roles of some of their metabolic products.

Lysine

This essential ketogenic amino acid cannot participate in transaminations. Once it loses its two nitrogens by deamination, it cannot be reaminated. The oxidation product of its carbon chain is glutaric acid :

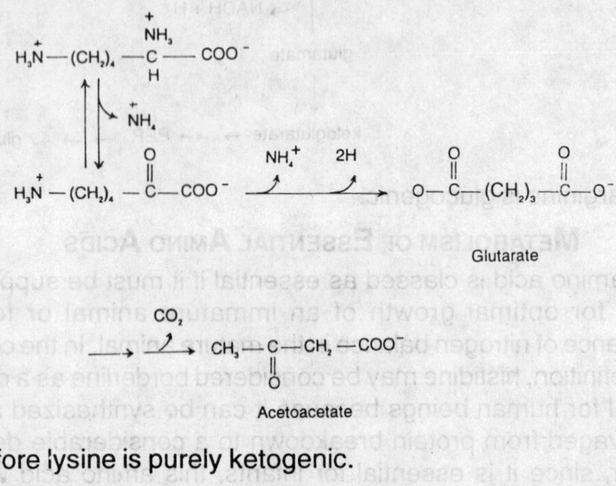

Therefore lysine is purely ketogenic.

Threonine

This was the most recent essential amino acid to be discovered. Since it resembles serine, it is not unexpected that it is metabolized analogously to serine:

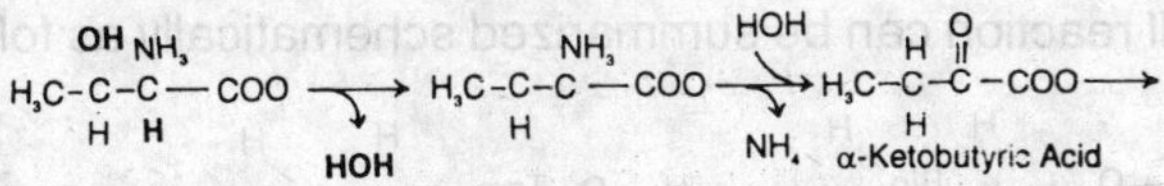

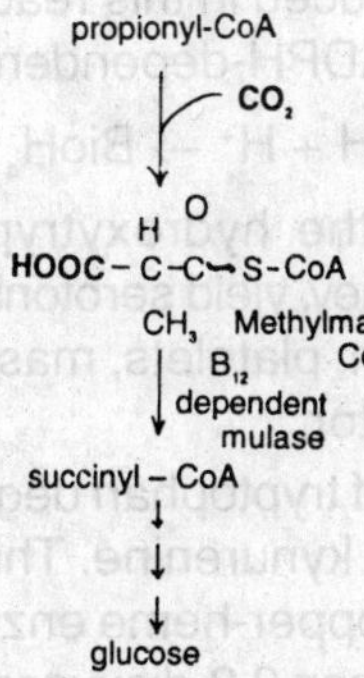

Tryptophan

This essential, glucogenic amino acid has two major pathways of degradation and gives rise to two biologically active compounds, serotonin and NAD.

The oxygenase system responsible for the introduction of an oxygen atom into the tryptophan ring requires the pteridine derivative, tetrahydrobiopterin:

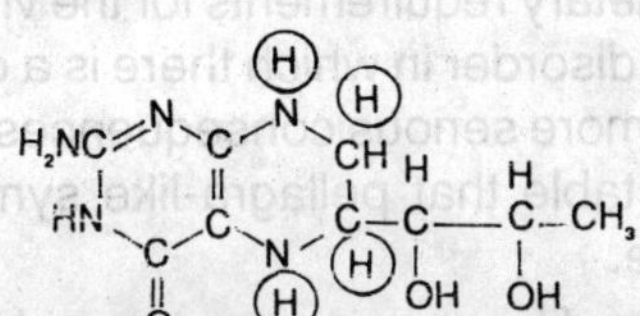

In the oxygenation of the tryptophan the tetrahydrobiopterin is converted to the dihydroform ($BioH_2$) :

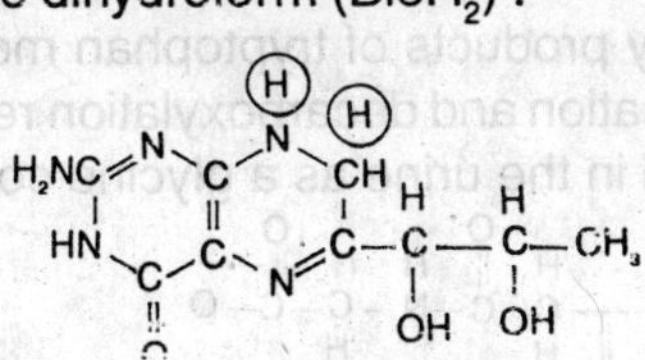

The overall reaction can be summarized schematically as follows:

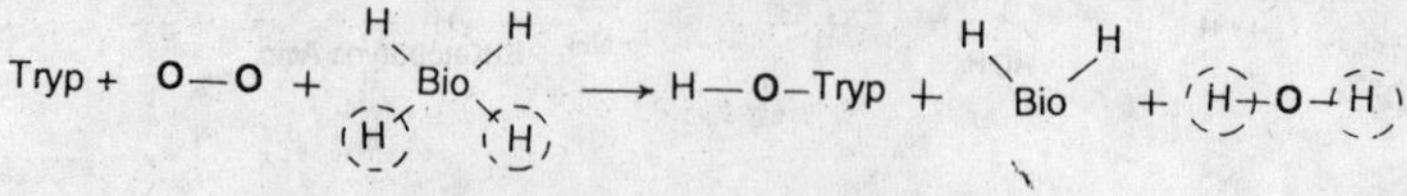

The dihydrobiopterin produced in this reaction can be restored to its tetrahydro form by an NADPH-dependent reduction :

$$BioH_2 + NADPH + H^+ \rightarrow BioH_4 + NADP^+$$

The decarboxylation of the hydroxytryptophan, a reaction prominent in both brain and kidney, yield serotonin. This neurohumoral agent, which is present in brain, platelets, mast cells, and intestinal cells, is a potent vasoconstrictor.

The other major pathway of tryptophan degradation results in its oxidation and ring opening to kynurenine. This pathway, which is prominent in liver, involves a copper-heme enzyme system, termed tryptophan pyrrolase or tryptophan 2,3-dioxygenase. Kynurenine can be converted in the liver to quinolinic acid,

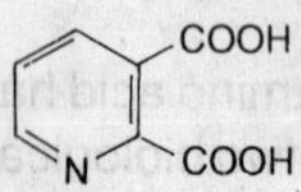

which is an intermediate in the synthesis of nicotinic acid ribonucleotide and NAD. Although tryptophan cannot replace niacin obtained from exogenous sources, it can have a demonstrable "sparing" effect on the dietary requirements for the vitamin. *Hartnup's Disease* is a hereditary disorder in which there is a deficiency of the pyrolase. In addition to more serious consequences such as mental retardation, it is predictable that pellagra-like symptoms may be manifest in this disease.

The carbon skeleton of kynurenine can be oxidatively degraded to yield both acetyl-CoA and alanine. Hence, tryptophan is both ketogenic and glucogenic.

The major excretory products of tryptophan metabolism arise from successive deamination and decarboxylation reactions. Indole acetic acid is eliminated in the urine as a glycine conjugate.

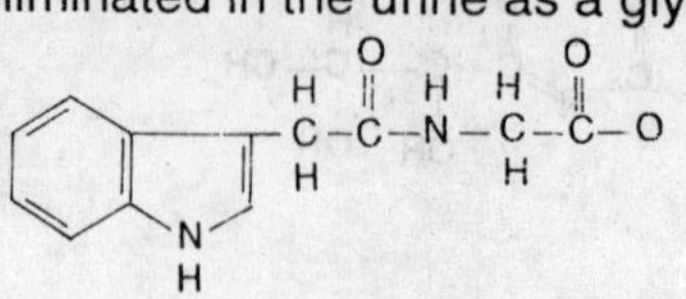

More extensive degradation, resulting in skatole, occurs by bacterial action in the lower bowel.

Histidine

Like arginine, histidine can be considered as a "semiessential" amino acid. It is a dietary requirement for the young animal but can be synthesized at a rate adequate to maintain nitrogen balance in the adult. In *E.coli* and *Salmonella*, histidine is synthesized from phosphoribosyl-1-pyrophosphate (PRPP), ATP, and glutamine. Its glycogenic character is evident from the following reactions:

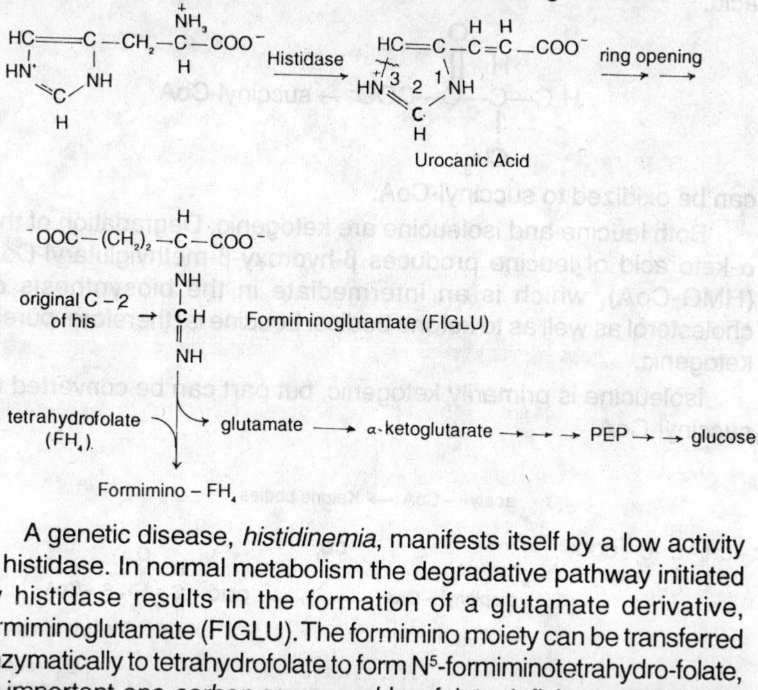

A genetic disease, *histidinemia*, manifests itself by a low activity of histidase. In normal metabolism the degradative pathway initiated by histidase results in the formation of a glutamate derivative, formiminoglutamate (FIGLU). The formimino moiety can be transferred enzymatically to tetrahydrofolate to form N[5]-formiminotetrahydro-folate, an important *one-carbon compound.* In a folate deficiency, FIGLU will be excreted because there is inadequate tetrahydrofolate to "trap" the formimino group.

In most cells of lung, liver, and gastric mucosa, histidine undergoes an enzymatic decarboxylation :

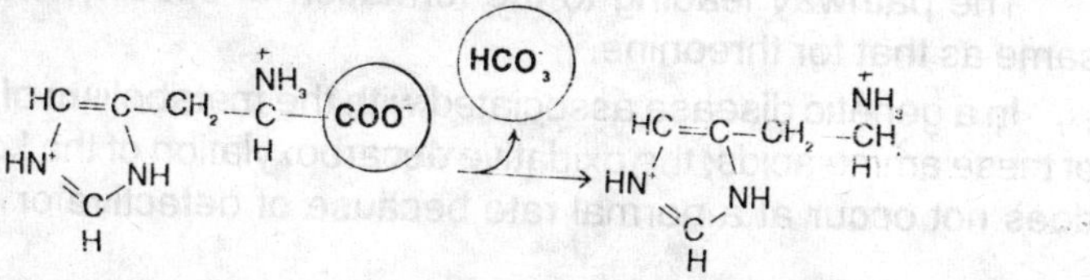

Histamine

to form histamine, a powerful vasodilator. This amine is liberated in traumatic shock and is localized in areas of inflammation.

The five remaining amino acids to be considered are essential, ketogenic, or in some cases, also glucogenic.

Valine, Leucine, and Isoleucine

As is most often the case, the metabolic degradation of these three essential amino acids begins with a transamination to produce the corresponding keto acids. The keto acid of valine, α-ketoisovaleric acid:

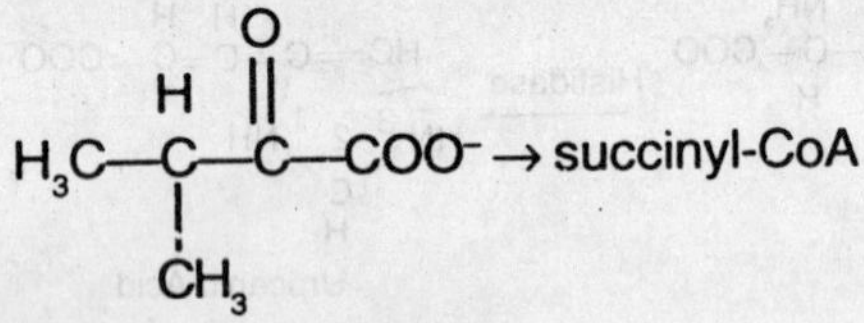

can be oxidized to succinyl-CoA.

Both leucine and isoleucine are ketogenic. Degradation of the α-keto acid of leucine produces β-hydroxy-β-methylglutaryl-CoA (HMG-CoA), which is an intermediate in the biosynthesis of cholesterol as well as to ketone bodies. Leucine is, therefore, purely ketogenic.

Isoleucine is primarily ketogenic, but part can be converted to succinyl-CoA:

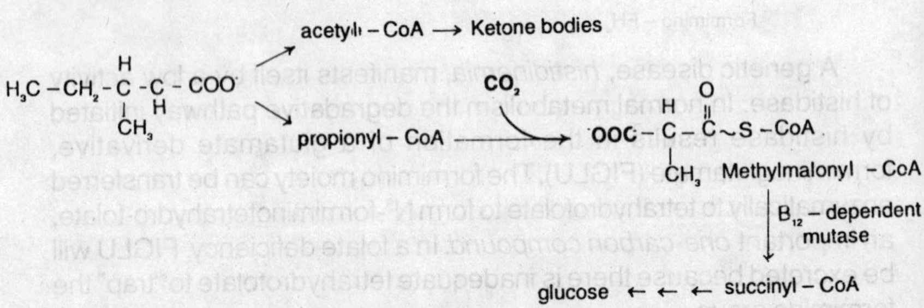

The pathway leading to the formation of succinyl-CoA is the same as that for threonine.

In a genetic disease associated with the metabolism of all three of these amino acids, the oxidative decarboxylation of the keto acids does not occur at a normal rate because of defective or deficient oxidative enzymes. Therefore, the keto acids accumulate and appear in the urine, causing the odor of maple sugar. This serious condition, called *maple syrup disease*, appears in early infancy with a high mortality.

Phenylalanine and Tyrosine

Humans and most all other animal organisms cannot synthesize the benzene ring. Hence, phenylalanine is a dietary essential. Tyrosine is synthesized from phenylalanine. Other than to be utilized for protein synthesis in normal metabolism, the only known fate of phenylalanine is its conversion to tyrosine. Much of the dietary requirement for phenylalanine is due, therefore, to the need for tyrosine.

Synthesis of Tyrosine

The synthesis of tyrosine from phenylalanine again illustrates a type of synthetic reaction in which molecular oxygen is a direct reactant. The hydroxylation of phenylalanine, like that of tryptophan, requires tetrahydrobiopterin:

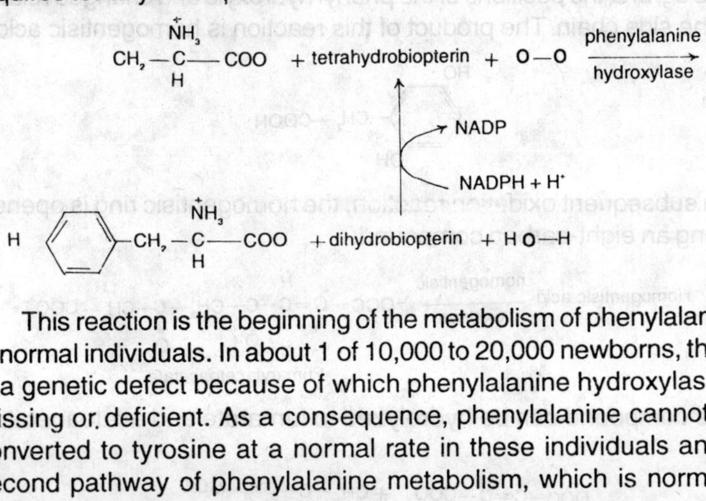

This reaction is the beginning of the metabolism of phenylalanine in normal individuals. In about 1 of 10,000 to 20,000 newborns, there is a genetic defect because of which phenylalanine hydroxylase is missing or deficient. As a consequence, phenylalanine cannot be converted to tyrosine at a normal rate in these individuals and a second pathway of phenylalanine metabolism, which is normally insignificant, comes into play:

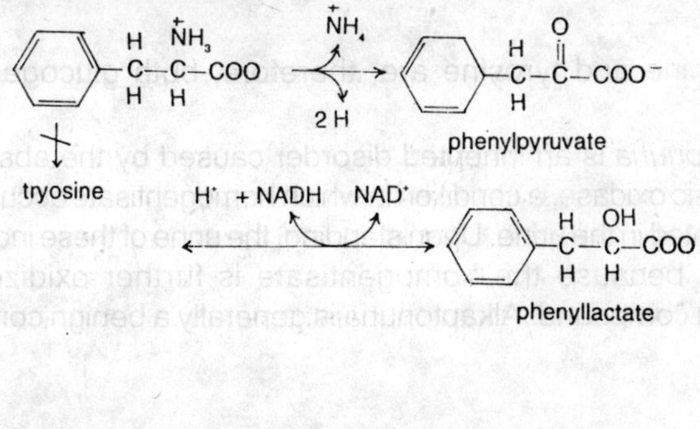

There is now an increased concentration of phenylalanine in all body fluids and phenylpyruvate and phenyllactate are excreted in high levels. This disease is called phenylketonuria. Early diagnosis of this tragic condition is absolutely essential. A common therapy is a diet low in phenylalanine.

Alkaptonuria

The first step in the catabolism of tyrosine is the loss of its α-amino group to form the keto acid.

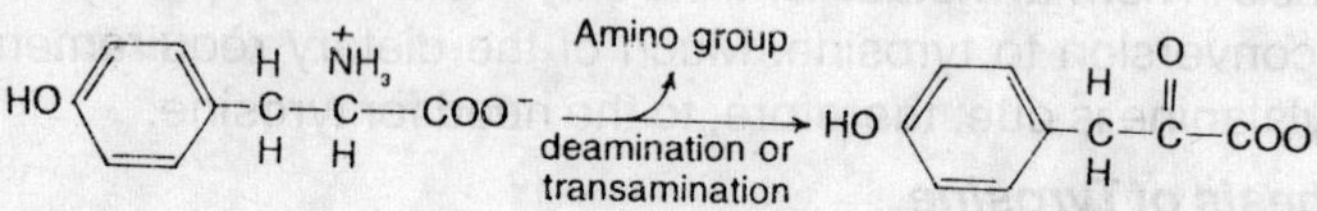

4 – Hydroxyphenylpyruvate

The next reaction is a complicated oxidation rearrangement in which another hydroxyl group is added to the ring, the carboxyl group is lost as CO_2, and the positions of the phenyl hydroxyls and changed relative to the side chain. The product of this reaction is homogentisic acid:

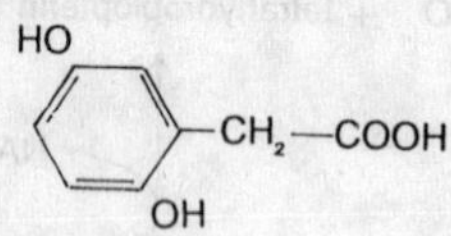

In a subsequent oxidation reaction, the homogentisic ring is opened, giving an eight-carbon compound:

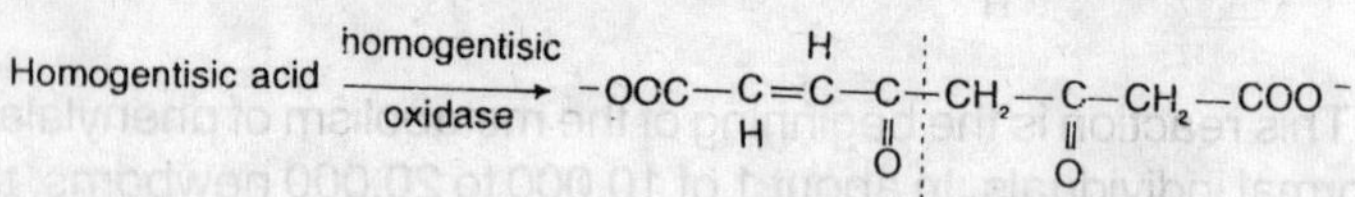

Fumarylacetoacetate

This compound can be hydrolyzed to fumarate and acetoacetate:

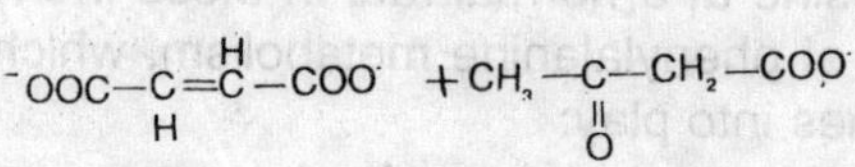

Phenylalanine and tyrosine are, therefore, both glucogenic and ketogenic.

Alkaptonuria is an inherited disorder caused by the absence of homogentisic oxidase, a condition in which homogentisate accumulates and is excreted in the urine. Upon standing, the urine of these individuals turns dark because the homogentisate is further oxidized to a melaninlike compound. Alkaptonuria is generally a benign condition.

Tyrosine as a Precursor of Melanin, Catecholamines and Thyroxine

Tyrosine can be hydroxylated to yield a dihydroxy derivative that is an important intermediate in mammalian metabolism. In common with other hydroxylation reactions, molecular oxygen and tetrahydrobiopterin ($BioH_4$) are required :

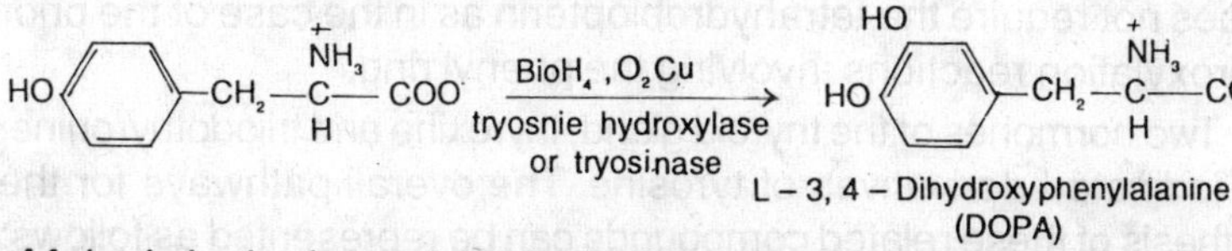

Melanin is the brownish-black pigment of skin that is produced in the melanocytes in the basal layer of the epidermis. Melanin is a group of high molecular weight polyhydroxyphenyl polymers produced by the action of tyrosinase on DOPA. The oxidation involves two quinone intermediates:

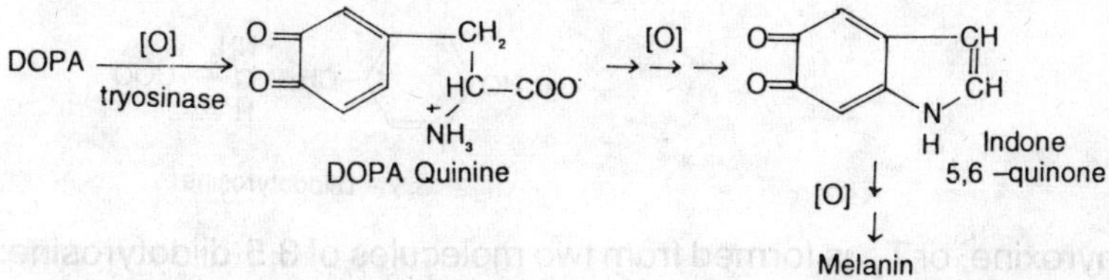

A deficiency in the tyrosinase system of the melanocytes results in the condition called albinism.

Another important precursor function of DOPA is its role in the synthesis of the catecholamines in the adrenal medulla and sympathetic nerve terminals.

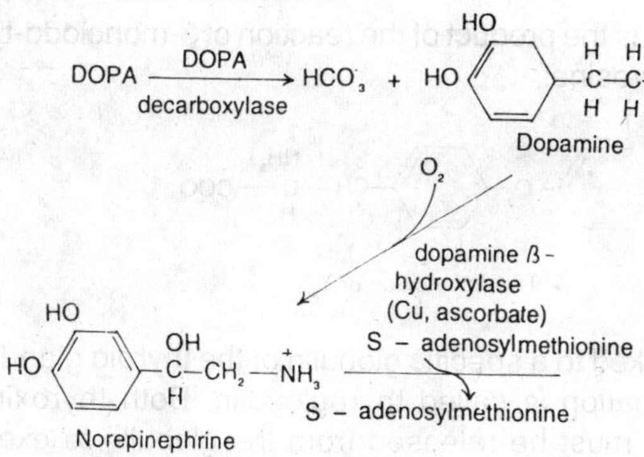

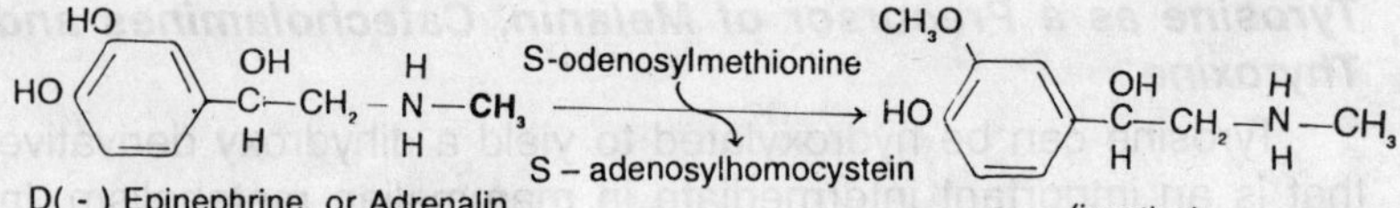

D(-) Epinephrine, or Adrenalin — (inactive)

Dopamine β-hydroxylase is a Cu-containing mixed-function oxidase. It does not require the tetrahydrobiopterin as in the case of the prior hydroxylation reactions involving the phenyl ring.

Two hormones of the thyroid gland, thyroxine and triiodothyronine, are iodinated derivatives of tyrosine. The overall pathways for the synthesis of these related compounds can be represented as follows:

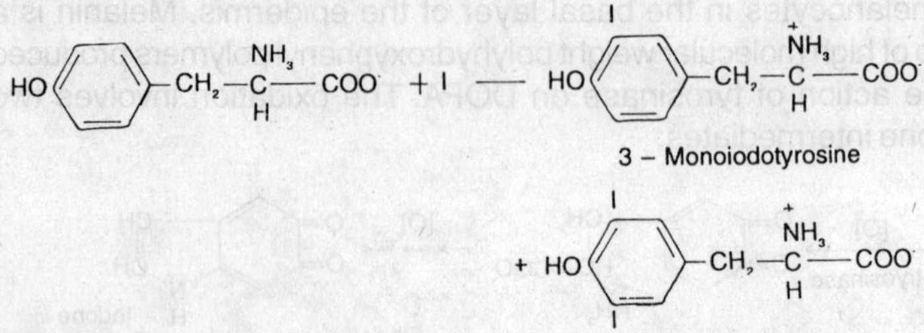

3 – Monoiodotyrosine

3,5 – Diiodotyrosine

Thyroxine, or T_4, is formed from two molecules of 3,5-diidotyrosine:

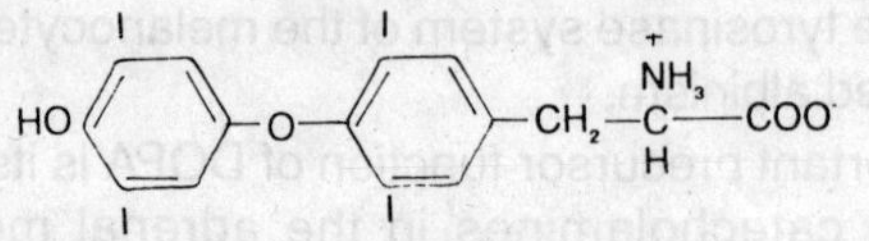

Thyroxin

Triiodothyronine is the product of the reaction of 3-monoiodo-tyrosine and 3,5-diiodotyrosine:

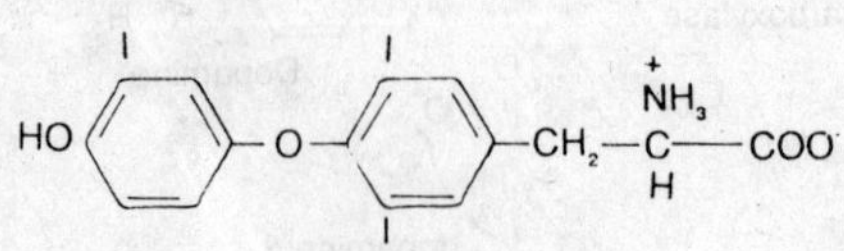

Triiodothyronine or T_3

T_3 and T_4 are linked to a specific globulin of the thyroid gland colloid and the combination is called thyroglobulin. Both thyroxine and triiodothyronine must be released from the globulin to exert their hormonal action.

Methionine

This essential amino acid is glucogenic. Following removal of its methyl group the compound that remains, homocysteine, can be converted oxidatively to succinyl-CoA by the same pathway involved in the oxidation of the carbon chains of threonine and isoleucine.

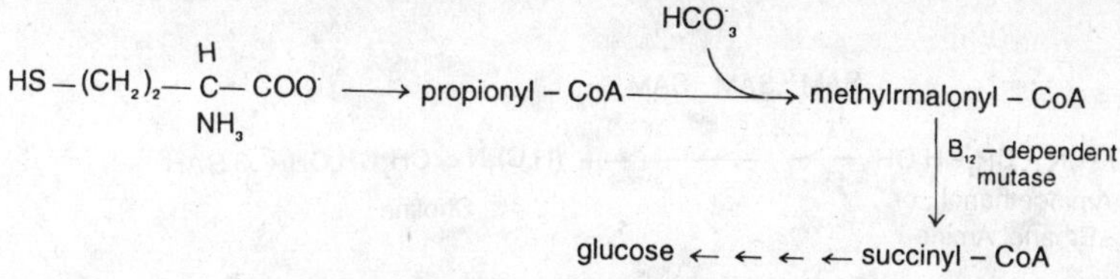

The S of homocysteine can become the S of cysteine (and cystine) via the intermediate cystathionine.

S-Adenosylmethionine as a Methyl Group Donor

An important biochemical property of methionine is its ability to serve as a methylating agent. Methionine can donate its methyl group to other compounds in toto, without the loss of any of the H atoms of the methyl group. Evidence for such a mechanism was obtained in experiments employing isotopes of C and H :

Methionine—$^{14}CD_3$ + methyl acceptor → homocysteine + acceptor—$^{14}CD_3$

To accomplish this group transfer, called *transmethylation,* methionine must first be converted into a sulfonium derivative, S-adenosylmethionine. ATP is the partner in this transformation. The reaction is unusual in that all of the phosphates in the ATP are converted to pyrophosphate and orthophosphate :

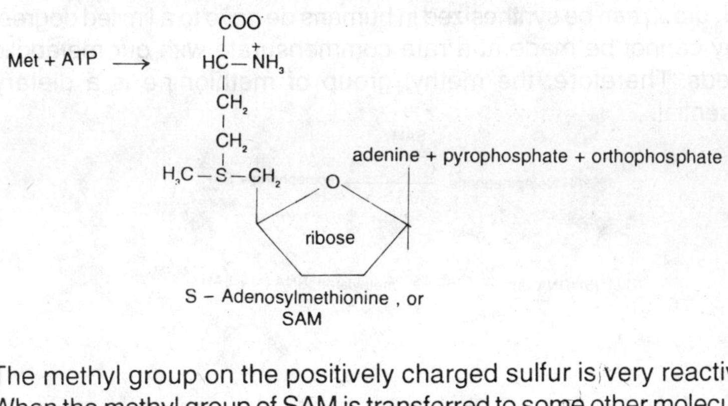

S – Adenosylmethionine , or SAM

The methyl group on the positively charged sulfur is very reactive. When the methyl group of SAM is transferred to some other molecule, the homocysteine remains intact as S-adenosylhomocysteine, SAH. Representative transmethylation reactions involving SAM in humans include the following:

(1) $H_2N - CH_2CH_2OH \xrightarrow{\text{SAM SAM SAM}} (H_3C)_3N - CH_2CH_2OH + 3\ SAH$

Aminoethanol, or Ethanol Amine → Choline

(2) Guanidoacetate $\xrightarrow{\text{SAM}}$ creatine + SAH

(3) Nicotinamide $\xrightarrow{\text{SAM}}$ N – methylnicotinamide + SAH

↑ NAD

Each of these methylation requires a specific enzyme protein (i.e., a methyltransferase) and all can be considered to be unidirectional reactions.

The reasons for the dietary essentiality of methionine are twofold: humans cannot make the C—S chain of the acid; and, although —CH_3 group can be synthesized in humans de novo to a limited degree, they cannot be made at a rate commensurate with our metabolic needs. Therefore, the methyl group of methionine is a dietary essential.

(4) Norepinephrine $\xrightarrow{\text{SAM}}$ epinephrine + SAH

(5) tRNA (s) $\xrightarrow{\text{SAM}}$ methylated tRNA (s) + SAH

10

Enzymes and Biological Catalysts

A living organism is a magnificent assembly of chemical reactions, an entity rather than chaos because most of its components make an orderly contribution to existence of the whole. Many of these reactions occur at significant rates only because specific proteins—the enzymes—are present to catalyze them. The enzymes are constructed to control both the kinds of reaction and the rates at which they occur, and it is the existence of these proteins that creates a complex harmony of chemical function.

Enzymes contain the same amino acids found in other proteins, and they have a three-dimensional structure that represents the conformation of least energy content, as do other proteins. All of our present information supports the view that enzymatic catalysis involves the same kinds of bonds that appear during ordinary organic reactions, using structures supplied by amino acid residues and by prosthetic groups. The binding of prosthetic groups is itself determined by conformation of amino acid residues, so the study of the nature of enzymes is in many ways only an extension of the principles of protein structure that we have already developed. There are thousands and thousands of different kinds of enzymes, each designed to enhance the rate of a specific reaction. As a result of a century of intensive laboratory investigation, much is now known about what enzymes are, how they work, and how they are regulated. Because of the ubiquitous role of enzymes in the cell, it is important for us to consider these general issues at this early stage in the book.

Discovery of Biological Catalysts

In the early nineteenth century the Swedish chemist Jon Berzelius discovered that an extract of potatoes is more effective than concentrated sulphuric acid in promoting the breakdown of starch. Berzelius concluded that this effect is due to the presence of biological catalysts and with remarkable insight went on to predict that all materials found in living organisms are synthesized under the influence of such catalysts. This observation stimulated interest in the biochemical identity of these catalysts, and Louis Pasteur soon postulated that the catalytic effect is intimately associated with cell structure and so cannot be separated from living cells. If true, this meant that the catalysts could not be isolated and studied in purified form. It was therefore a great milestone when, in 1897, Edward Buchner demonstrated that an extract of yeast, from which all intact cells had been removed, is capable of catalyzing the breakdown of glucose. This was an especially significant observation because glucose degradation is a major metabolic pathway in cells. This finding firmly established that biological catalysts, in the absence of any cell organization, are capable of enhancing the rates of metabolic reactions.

The increasing importance of this class of molecules soon led to suggestions for a uniform nomenclature system to facilitate communication among biochemists. The term *enzyme* was adopted as a general designation for biological catalysts, and it was decided that individual enzymes would be named by adding the suffix "ase" to the name of the *substrate* (the substance upon which the enzyme acts). For example, the enzyme catalyzing the degradation of urea was termed urease, enzymes catalyzing hydrolysis of nucleic acids were designated nucleases, the enzyme catalyzing glucose oxidation was called glucose oxidase, and so forth. This basic approach to enzyme nomenclature is still in use today, although some older names introduced prior to the adoption of this system (e.g., trypsin pepsin) have been retained.

In the years immediately following the initial demonstration of cell-free enzyme activity by Buchner, a large number of other enzymes were identified by biochemists. But little progress was made in understanding the structure or mechanism of these cellular catalysts. An example of the difficulties experienced during this period is the classic story of Richard Willstatter's encounters with the enzyme *peroxidase,* one of the first enzymes to be obtained in a highly purified form. When Willstatter chemically analyzed his purified enzyme preparation, he found no lipid, no sugar (and therefore no nucleic acid), and no protein. These results led to the prevailing opinion in the early twentieth century that enzymes belong to none of the commonly recognized classes of biological molecules.

In 1926, however, James Summer reported the first crystallization of an enzyme, in this case urease. He found that the crystals consisted of pure protein, even after repeated cycles of recrystallization. Since crystallization is generally recognized as the ultimate criterion of purity, these results suggested to Summer that the enzyme urease is a protein. Although this conclusion initially met with skepticism, similar experiments by John Northrop and his colleagues on other enzymes soon led to general acceptance of the idea that enzymes are made of protein. Willstatter, it turns out, had been misled by the fact that enzymes are active in extremely small quantities. Thus his peroxidase preparation exhibited measurable enzyme activity even though the protein content of the sample was too low to be detected by the analytical techniques then available.

In the years since the pioneering work of Summer and Northrop, over a thousand enzymes have been purified and identified as proteins. Though the number of different reactions catalyzed by these enzymes is very large, one can subdivide this group into a relatively small number of categories. All cellular reactions are catalyzed by enzymes that fit into one of these categories.

As Berzelius predicted many years ago, enzymes act as extraordinarily efficient chemical catalysts. Enzymes therefore exhibit the four defining characteristics shared by all catalysts: (1) They modify the rates of chemical reactions. (2) They are not consumed in the process. (3) They are effective in extremely small quantities. (4) They do not alter the equilibrium of the reaction being catalyzed.

In addition to these four attributes common to all catalysts, enzymes exhibit several unique features that distinguish them from the catalysts routinely encountered in organic or inorganic chemistry. First, enzymes are exceedingly efficient compared to other catalysts. A typical enzymatic reaction may proceed as much as 10^8—10^{11} times faster than the same reaction catalyzed by a nonenzymatic catalyst. Also, unlike the nonbiological catalysts routinely encountered in chemistry, the catalytic activity of enzymes can be regulated to suit specific cellular needs. Finally, one of the most striking properties of enzymes is their selectivity. Unlike other catalysts enzymes are highly specific, both in the types of reactions they catalyze and in the structures of the substances upon which they act. For example, the enzyme glucose oxidase is specific not just for the oxidation of glucose versus other six-carbon sugars but also for a particular three-dimensional form of the sugar, namely α-glucose.

Table 10.1. Current nomenclature of six types of biochemical reactions catalyzed by enzymes.

Enzyme Class	Type of Reaction Catalyzed	Example
Oxidoreductases	Oxidation-reduction	Dehydrogenases, oxidases, perioxidases, hydroxylases, reductase, oxygenase.
Transferases	Transfer of functional group from one molecule to another (AX+B → A + BX)	Transaminases, transmethylases
Hydrolases	Hydrolytic cleavage	Esterases, amidases glycosidases, peptidases, phosphatases
Lyases	Cleavage of C—C, C—O, C—N, etc by elimination; leaving a double bond; or addition to double bonds.	Aldolases, synthases, deaminases hydrases, decarboxylases, nucleotide cyclases
Isomerases	Intramolecular rearrangements	Isomerases; racemases, epimerases, mutases
Ligases	Joining together of five molecules coupled to the cleavage of a high-energy bond.	Synthetases

The unique properties of enzymes (efficiency, regulation, and specificity) all stem from the fact that enzymes are protein molecules and therefore exhibit all the remarkable characteristics of proteins. They study of how enzymes work is therefore intimately tied to the study of protein chemistry.

Catalytic Site

Why then do someproteins act as very effective catalysts—more effective than any other known compounds in making some reactions go faster? The answer lies in the ability of proteins to be molded into a variety of shapes. Proteins acting as enzymes are built with clefts that function as catalytic sites. Many compounds can wander in and out, but when one with a proper fit diffuses in, the protein closes around it and brings to bear chemical groups in a particular geometric conformation. Here is the basis for enzyme function : The catalytic effectiveness of amino, carboxyl, and other bonding groups is increased by several order of magnitude placing them in fixed spatial arrangements where they can lock onto the affected compound; yet it is possible to stretch or otherwise deform particular bonds through relatively small changes in conformation of the protein.

We shall see that the use of proteins as catalysts conveys other advantages. They can be constructed with separate domains, making rapid multimolecular reactions feasible in some cases, or providing binding sites for allosteric effectors with which to regulate the rates of reactions.

Specificity

The *catalytic site* of enzymes is a part of a *binding site.* Enzymes, like antibodies, are frequently highly specific in binding particular compounds, because the amino acid residues in the binding site closely fit only a few compounds, stronger force with "good" substrates than with other compounds. The existence of this stronger force is an important factor in easy deformation of the substrate to break old bonds and create new ones. Enzymes differ from antibodies in that tight binding may change the bound compound.

Catalytic Groups

Another clue to the catalytic mechanism comes from the presence in proteins of groups known to be effective catalysts for ordinary organic reactions. For example, many amino acid side chains are capable of donating or accepting protons so as to act as *general acid* or *general base catalysts.* Other groups, such as hydroxyl or amino groups, may act as *nucleophiles,* and surrounding structures in an enzyme are often designed to reinforce their catalytic properties. (Nucleophiles is effect donate electrons to bond other groups lacking filled electron orbitals).

General Catalytic Mechanism

Enzymes owe their peculiar catalytic effectiveness to a combination of specific binding and the presence of catalytic groups. In an ordinary chemical reaction, a compound must comfort into an unfavourable configuration before it changes into something else the formation of the less probable intermediate state represents an activation energy barrier that must be overcome before reaction can occur.

Enzymes change this situation by being built to that their combination with the reacting compound is most stable when the compound is in a "high-energy" intermediate form. An enzymatic reaction proceeds through at least three stages—the formation of a complex between enzyme and substrate, the conversion of this complex to an enzyme-intermediate complex, and the further conversion to a complex between enzyme and product that can dissociate:

$$E + S = ES = EI = EP = E + P$$

The groups on the enzyme are arranged to make the total energy of the enzyme intermediate complex less than the total energy in the enzyme-substrate or enzyme-product complexes. The activation barriers in converting enzyme-substrate to enzyme-intermediate, or enzyme-intermediate to enzyme-product are substantially lower than the barrier without enzyme.

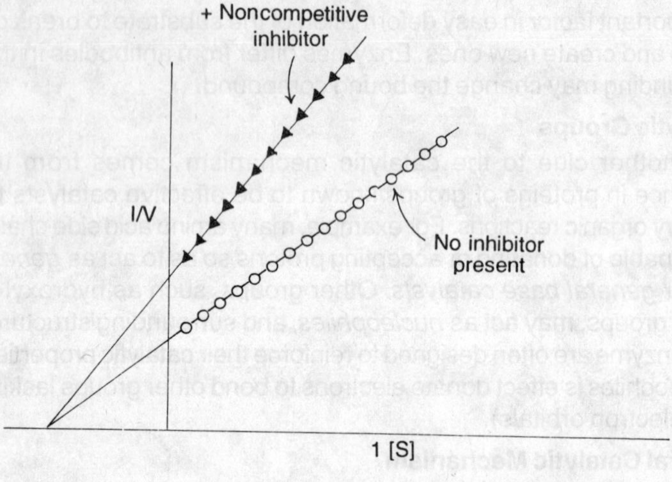

Fig. 10.1. A double-reciprocal plot of enzyme kinetics in the presence (▲-▲-▲) and absence (O-O-O) of a noncompetitive inhibitor; K_M is unaltered by the noncompetitive inhibitor, where V_{max} is decreased.

The specific binding of substrate to enzyme therefore serves not only to confine particular reactions to a few compounds but also to create a complex that spontaneously favours the formation of a more reactive configuration. The enzyme does several things to achieve this end. It provides catalytic groups in a single molecule, eliminating the need for collision with two or more molecules during the course of the reaction (*entropy effect*); it aligns the substrate at favourable angles relative to the catalytic groups so that the interaction with these groups is facilitated (*orbital steering*), and it provides binding groups at positions that stabilize a reactive intermediate (*propinquity effect*).

Perhaps the easiest way to get a feel for what enzymes do and how they go about it is to examine a few of them in detail, not because these details are something that ought to be committed to memory, but so we can develop some general principles by example. Enzyme mechanisms are rarely "seen" in the sense that they can be frozen for inspection by x-ray crystallography; they are deduced through the use of techniques such as molecular probes that react with active sites, destruction of selected amino acids, spectral changes, and more recently by neutron diffraction; hence our concentration on the principal features rather than the more conjectural fine points of mechanism.

Enzymes are Highly Specific

Enzymes are highly specific both in the reaction catalyzed and in their choice of reactants, which are called *substrates*. An enzyme usually catalyzes a single chemical reaction or a set of closely related reactions. The degree of specificity for substrate is usually high and sometimes virtually absolute.

Let us consider *proteolytic enzymes* as an example. The reaction catalyzed by those enzymes is the hydrolysis of a peptide bond.

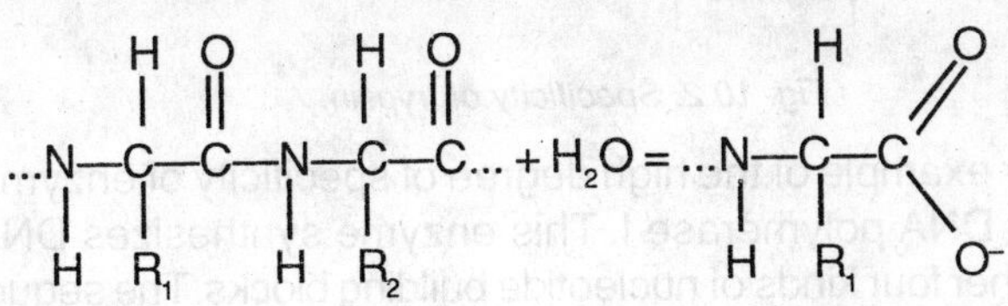

Peptide

Carboxyl component

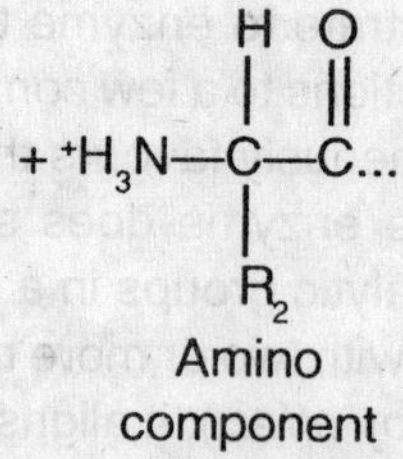

Most proteolytic enzymes also catalyze a different but related reaction, namely the hydrolysis of an ester bond.

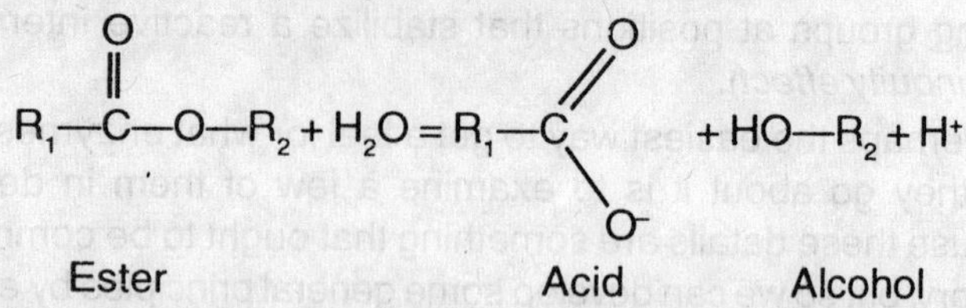

Proteolytic enzymes vary markedly in their degree of substrate specificity. Subtilisin, which comes from certain bacteria, is quite undiscriminating about the nature of the side chains adjacent to the peptide bond to be cleaved. Trypsin, as was mentioned in previously is quite specific in that it splits peptides bond on the carboxyl side of lysine and arginine residues only. Thrombin, an enzyme participating in blood clotting, is even more specific than trypsin. The side chain on the carboxyl side of the susceptible peptide bond must be arginine, whereas the one on the amino side must be glycine.

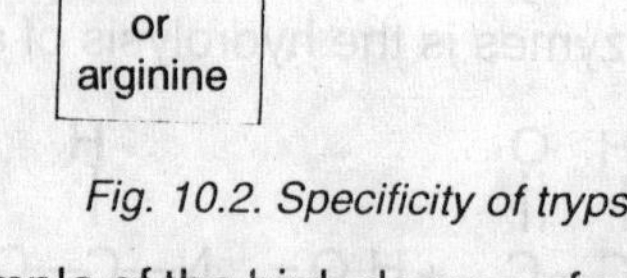

Fig. 10.2. Specificity of trypsin.

Another example of the high degree of specificity of enzymes is provided by DNA polymerase I. This enzyme synthesizes DNA by linking together four kinds of nucleotide building blocks. The sequence of nucleotides in the DNA strand that is being synthesized is determined by the sequence of nucleotides in another DNA strand that serves as a template DNA polymerase I is remarkably precise in carrying out the instructions given by the template. The wrong nucleotide is inserted into a new DNA strand less than once in a million times.

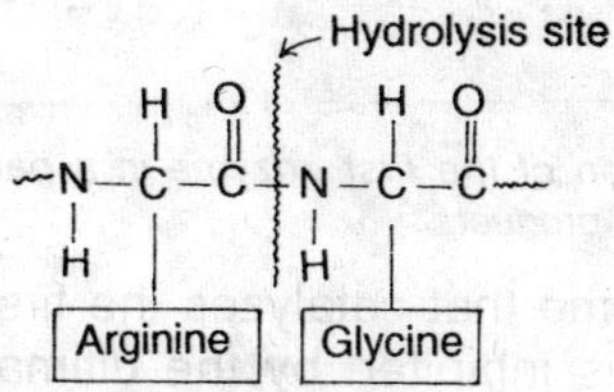

Fig. 10.3. Specificity of thrombin, a clotting factor.

Some enzymes are synthesized in an *inactive precursor form* and are activated at a physiologically appropriate time and place. The digestive enzymes exemplify this kind of control. For example, trypsinogen is synthesized in the pancreas and is activated by peptide-bond cleavage in the small intestine to form the active enzyme trypsin. This type of control is also repeatedly used in the sequence of enzymatic reactions leading to the clotting of blood. The enzymatically inactive precursors of proteolytic enzymes are called *zymogens.*

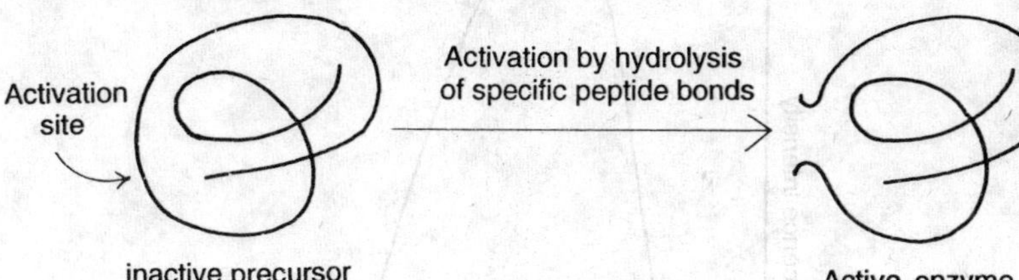

Fig. 10.4. Zymogen activation by hydrolysis of specific peptide bonds.

Another mechanism that controls activity is the covalent insertion of a small group on an enzyme. The control mechanism is called *covalent modification.* For example, the activities of the enzymes that synthesize and degrade glycogen are regulated by the attachment of a phosphoryl group to a specific serine residue on these enzymes. This modification can be reversed by hydrolysis. Specific enzymes catalyze the insertion and removal of phosphoryl and other modifying groups.

A different kind of regulatory mechanism affects many reaction sequences resulting in the synthesis of small molecules such as

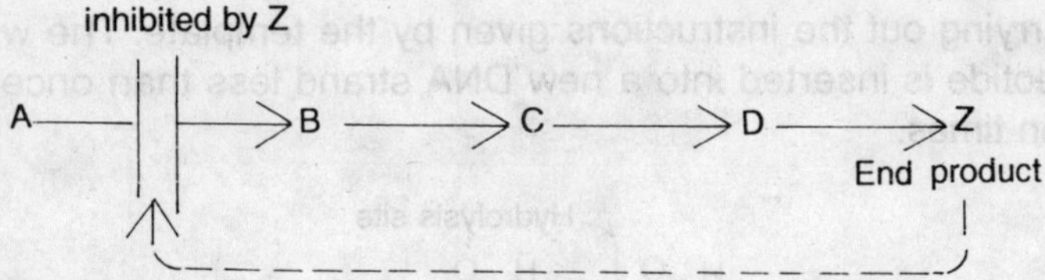

Fig. 10.5. Feedback inhibition of the first enzyme in a pathway by reversible binding of the final product.

amino acids. The enzyme that catalyzes the first step in such a biosynthetic pathway is inhibited by the ultimate product. The biosynthesis of isoleucine in bacteria illustrates this type of control, which is called *feedback inhibition.* Threonine is converted into isoleucine in five steps, the first of which is catalyzed by threcnine deaminase. This enzyme is inhibited when the concentration of isoleucine reaches a sufficiently high level. Isoleucine binds to a regulatory site on the enzyme, which is distinct from its catalytic site. The inhibition of threonine deaminase is mediated by an *allosteric*

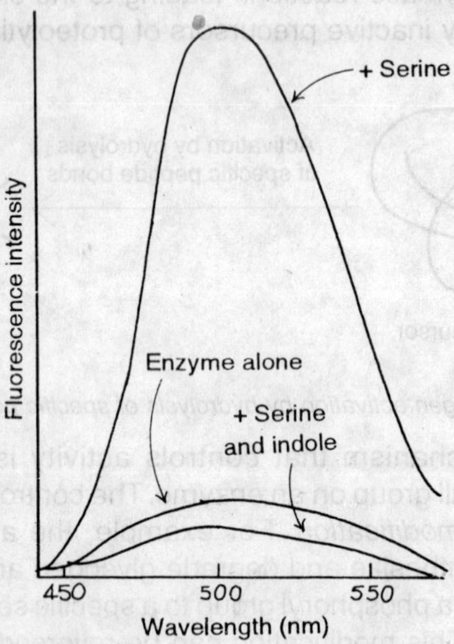

Fig. 10.6. Florescence intensity of the pyridoxal phosphate group at the active site of tryptophan synthetase changes upon addition of serine an dindole, the sub-strates.

interaction, which is reversible. When the level of isoleucine drops sufficiently, threonine deaminase becomes active again, and consequently isoleucine is again synthesized.

The specificity of some enzymes is under physiological control. The synthesis of lactose by the mammary glands is a particularly striking example. Lactose, synthetase, the enzyme that catalyzes the synthesis of lactose, consists of a catalytic subunit and a modifier subunit. The catalytic subunit by itself cannot synthesize lactose. It has a different role, which is to catalyze the attachment of galactose to a protein that contains a covalently linked carbohydrate chain. *The modifier subunit alters the specificity of the catalytic subunit* so that it links galactose to glucose to form lactose. The level of the modifier subunit under hormonal control. During pregnancy, the catalytic subunit is formed in the mammary gland but little modifier subunit is formed. At the time of birth, hormonal levels change drastically, and the modifier subunit is synthesized in large amounts. The modifier subunit then binds to the catalytic subunits to form an active lactose synthetase complex that produces large amounts of lactose. This system clearly shows that hormones can exert their physiological effects by altering the specificity of enzymes.

Enzymes Transform Different Kinds of Energy

In many biochemical reactions, *the energy of the reactants is converted into a different form with high efficiency.* For example, in photosynthesis, light energy is converted into chemical-bond energy. In mitochondria, the free energy contained in small molecules derived from foods is converted into a different currency, that of adenosine triphosphate (ATP). The chemical-bond energy of ATP is then utilized in many different ways. In muscular contraction, the energy of ATP is converted into mechanical energy. Cells and organelles have pumps that utilize ATP to transport molecules and ions against chemical and electrical gradients. These transformations of energy are carried out by enzyme molecules that are integral parts of highly organized assemblies.

Enzymes do not alter Reaction Equilibria

An enzyme is a catalyst and consequently is cannot alter the equilibrium of a chemical reaction. This means that an enzyme accelerates the forward and reverse reaction by precisely the same factor.

Consider the interconversion of A and B. Suppose that in the absence of enzyme the forward rate (k_F) is 10^{-4} sec^{-1} and the reverse rate (k_R) is 10^{-6} sec^{-1}. The equilibrium constant K is given by the ratio of these rates :

$$A \underset{10^{-6}\,\text{sec}^{-1}}{\overset{10^{-4}\,\text{sec}^{-1}}{\rightleftharpoons}} B$$

$$K = \frac{[B]}{[A]} = \frac{k_F}{k_R} = \frac{10^{-4}}{10^{-6}} = 100$$

The equilibrium concentration of B is 100 times that of A, whether or not enzyme is present. However, it would take several hours to approach this equilibrium without enzyme, whereas equilibrium would be attained within a second when enzyme is present. Thus, enzymes accelerate the attainment of equilibria but do not shift their positions.

Enzymes Decrease the Activation Energies of Reactions Catalyzed by Them

A chemical reaction, A=B, goes through a *transition state* that has a higher energy than either A or B. The rate of the forward reaction depends on the temperature and on the difference in free energy between that of A and the transition state, which is called *Gibbs free energy of activation* and symbolized by $\Delta G^{\ddagger}$.

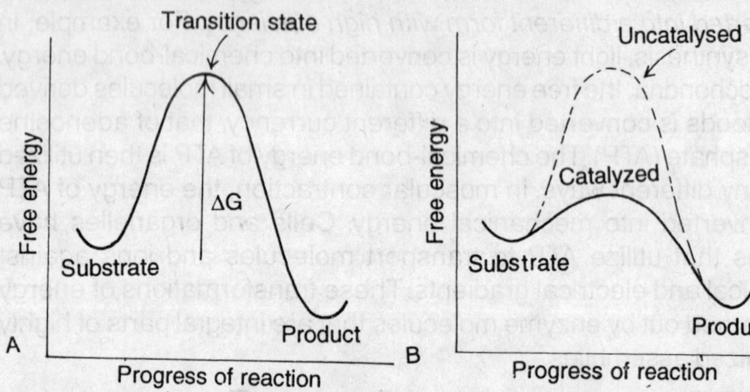

Fig. 10.7. A—Definition of $\Delta G^{\ddagger}$, the free energy of activation. B—Enzymes accelerate catalysis by reducing $\Delta G^{\ddagger}$

$$\Delta G^{\ddagger} = G_{\text{transition state}} - G_{\text{substrate}}$$

The reaction rate is proportional to the fraction of molecules that have a free energy equal to or greater than $\Delta G^{\ddagger}$. The proportion of molecules that have an energy equal to or greater than $\Delta G\ddagger$ increases with temperature.

Enzymes accelerate reactions by decreasing $\Delta G\ddagger$, the activation barrier. The combination of substrate and enzyme creates a new reaction pathway whose transition-state energy is lower than it would be if the reaction were taking place in the absence of enzyme.

Formation of an Enzyme-Substrate Complex is the First Step in Enzymatic Catalysis

The making and breaking of chemical bonds by an enzyme are preceded by the formation of an *enzyme-substrate* (ES) complex. The substrate is bound to a specific region of the enzyme called the *active site.* Most enzymes are highly selective in their binding of substrate. Indeed, the catalytic specificity of enzymes depends in large part on the specificity of the binding process. Furthermore, the control of enzymatic activity may also take place at this stage.

The existence of ES complexes has been shown in a variety of ways :

1. ES complexes have been *directly visualized by electron microscopy and X-ray crystallography.* Complexes of nucleic acids and their polymerase enzymes are evident in electron micrographs. Detailed information concerning the location and interactions of glycyl-L-tyrosine, a substrate of carboxypeptidase A, has been obtained from X-ray studies of that ES complex.
2. The *physical properties* of an enzyme, such as its solubility or heat stability, frequently change upon formation of an ES complex.
3. The *spectroscopic characteristics* of many enzymes and substrates change upon formation of an ES complex just as the absorption spectrum of deoxyhemoglobin changes markedly when it binds oxygen or when it is oxidized to the ferric state, as described previously. These changes are particularly striking if the enzyme contains a coloured prosthetic group. Tryptophan synthetase, a bacterial enzyme that contains a pyridoxal phosphate prosthetic group, affords a nine illustration. This enzyme catalyzes the synthesis of L-tryptophans from L-serine and indole. The addition of L-serine to the enzyme produces a marked increase in the fluorescence of the pyridoxal phosphate group. The subsequent addition of indole, the second substrate, quenches this fluorescence to level lower than that of the enzyme

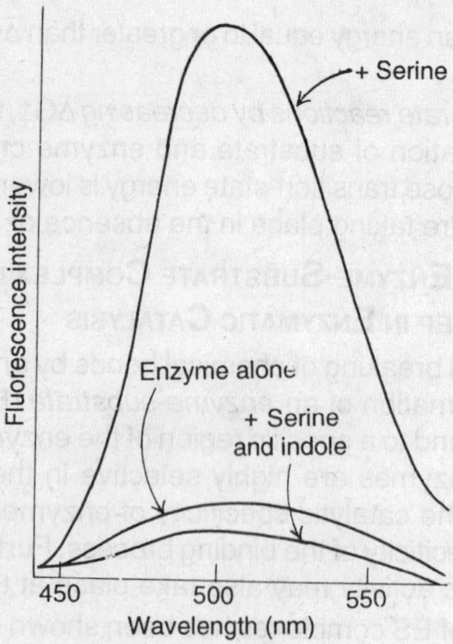

Fig. 10.8. Fluorescence intensity of the pyridoxal phosphate group at the active site of tryptophan synthetase changes upon addition of serine and indole, the substrates.

alone. Thus, fluorescence spectroscopy reveals the existence of an enzyme-serine complex and of an enzyme-serine indole complex. Other spectroscopic techniques, such as nuclear and electron magnetic resonance, also are highly information about ES interactions.

4. A high degree of *stereospecificity* is displayed in the formation of ES complexes. For example, D-serine is *not* a substrate of tryptophan synthetase. Indeed, the D-isomer does not even bind to the enzyme. This implies that the substrate-binding site has a very well define shape.
5. ES complexes can sometimes be *isolated in pure form.* For an enzyme that catalyzes the reaction A + B = C, it is sometimes possible to isolate an EA complex. This can be done if the enzyme has a sufficiently high affinity for A and B is absent from the mixture.
6. At a constant concentration of enzyme, the reaction rate increases with increasing substrate concentration until a maximal velocity is reached. In contrast, uncatalyzed reactions do not show this saturation effect. In 1913, Leonor Michaelis interpreted

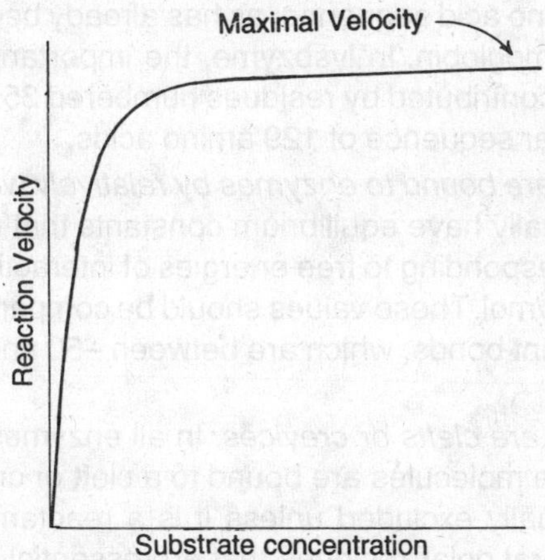

Fig. 10.9. Velocity of an enzyme-catalyzed reaction as a function of the substrate concentration.

the *maximal velocity of an enzyme-catalyzed reaction* in terms of the formation of a discrete ES complex. At a sufficiently high substrate concentration, the catalytic sites are filied and so the reaction rate reaches a maximum. This is the oldest and most general evidence for the existence of ES complexes.

Some Features of Active Sites

The active site of an enzyme is the region that binds the substrates (and the prosthetic group, if any) and contributes the residues that directly participate in the making and breaking of bonds. These residues are called the *catalytic groups*. Although enzymes differ widely in structure, specificity, and mode of catalysis, a number of generalizations concerning their active sites can be stated :

1. *The active site takes up a relatively small part of the total volume of an enzyme.* Most of the amino acid residues in an enzyme are not in contact with the substrate. This raises the intriguing question of why enzymes are so big. Nearly all enzymes are made up of more than 100 amino acid residues, which gives them a mass greater than 10 kdal and a diameter of more than 25Å.

2. *The active site is a three-dimensional entity.* The active site of an enzyme is not a point, a line, or even a plane. It is an intricate three-dimensional form made up of groups that come from different parts of the linear amino acid sequence-indeed, residues far apart in the linear sequence may interact more strongly than adjacent

residues in the amino acid sequence, as has already been seen for myoglobin and hemoglobin. In lysozyme, the important groups in the active site are contributed by residues numbered 35, 52, 62, 63 and 101 in the linear sequence of 129 amino acids.

3. *Substrates are bound to enzymes by relatively weak forces.* ES complexes usually have equilibrium constants that range from 10^{-2} to 10^{-8}M, corresponding to free energies of interaction ranging from –3 to –12 kcal/mol. These values should be compared with the strengths of covalent bonds, which are between –50 and 110 kcal/mol.

4. *Active sites are clefts or crevices.* In all enzymes of known structure, substrate molecules are bound to a cleft or crevice from which water is usually excluded unless it is a reactant. The cleft also contains several polar residues that are essential for binding and catalysis. The nonpolar character of the cleft enhances the binding of substrate. In addition, the cleft creates a microenvironment in which certain polar residues acquire special properties essential for then catalytic role.

5. *The specificity of binding depends on the precisely defined arrangement of atoms in on active site.* A substrate must have a matching shape to fit into the site. Emil Fischer's metaphor of the lock and key, stated in 1890, has proved to be an essentially correct and highly fruitful way of looking at the stereospecificity of catalysis.

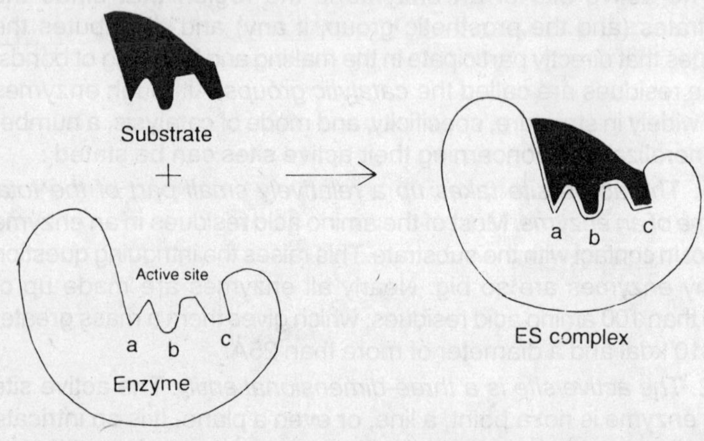

Fig. 10.10. Lock-and-key model of the interaction of substrates and enzymes. The active site of the enzyme by itself is complementary in shape to that of the substrate.

However, recent work suggests that the active sites of some enzymes are not rigid. In such an enzyme, the shape of the active site is modified by the binding of substrate. The active has a shape complementary to that of the substrate only *after* the substrate is bound. This process of dynamic recognition is called *induced fit.* Further, some enzymes preferentially bind a strained form of the substrate corresponding to the transition state.

Thermodynamics of Chemical Reactions

In order to discuss the question of how enzymes work, one must first have a basic familiarity with the concepts of thermodynamic and kinetics. *Thermodynamics* refers to the study of the changes in energy that occur during chemical reactions. Such information determines the extent to which equilibrium conditions favour the formation of product, but tells us nothing about reaction rates.

Energy can be transferred into and out of systems in various forms, such as heat, light, electricity, mechanical energy and chemical energy. However, the total energy content of any system plus its surroundings must remain constant, a principle known as the *first law of thermodynamics.* In addition to this overall conservation energy, there are limitations on the types of energy transformations that can occur. The *second law of thermodynamics* states that every even proceeds in such a way that a system plus its surrounding exhibit a net increase in disorder, or *entropy.* This propensity to maximize entropy in the universe is the ultimate driving force of all chemical reactions. The portion of the total energy used to increase the entropy is not available to do work. This amount of energy can be calculated by multiplying the change in entropy (ΔS) times the absolute temperature (T), yielding $T\Delta S$. The energy available to do work, on the other hand, is known as the *free energy* (G). The changes in free energy and entropy occurring in a system undergoing transformation are related by the equation

$$\Delta G + T\Delta S = \Delta H$$

where H equal the *enthalpy* or heat content. In biological systems, where pressure and volume are constant, enthalpy is equivalent to the total energy. Hence the above equation, which combines both the first and second laws of thermodynamics, tells us that in any system undergoing transformation, the total change in energy is equal to the changes in (1) free energy and (2) the energy involved in altering the entropy.

Although the second law of thermodynamics tells us that the entropy of a system plus its surroundings always proceeds toward a maximum, this does not necessarily mean that the entropy of an individual system always increases. It may either increase, decrease, or stay the same. If the system decreases in entropy, however, the entropy of the surroundings must increase by a sufficient amount so that the total entropy of system plus surroundings increases. This is exactly what happens during the development of living organisms. As highly organized living cells are produced, there is a decrease in entropy of the system, but only at the expense of an even greater increase in entropy of the surroundings.

The fact that a net increase in entropy accompanies all events proceeding toward equilibrium means that there must be a comparable decrease in free energy (due in essence to the loss of the free energy used to increase the entropy). Because of this tendency to free energy to decrease to a minimum, free-energy calculations are useful indicators of the direction in which reactions tend to proceed. They are especially useful in this regard because changes in free energy are much easier to determine than changes in entropy.

Determination of Free-Energy Changes

For the generalized chemical reaction

$$\text{Reactants} \rightarrow \text{products}$$

the change in free energy (ΔG) can be calculated from the equation

$$\Delta G = \Delta G^\circ + 2.303\,RT\log_{10}\frac{[\text{products}]}{[\text{reactants}]} \qquad (1)$$

where R is the gas contants, T is the absolute temperature, [reactants] is the mathematical product obtained by multiplying together the initial molar concentrations of each of the reactants, [products] is the mathematical product obtained by multiplying together the initial molar concentrations of each of the products of the reaction, and ΔG° is the standard free-energy change. This *standard free-energy change* (ΔG°) is a measure of the amount of free energy released during conversion of reactants to products under "standard conditions" (defined as all reactants and products present at an initial concentration of 1.0M). It is critical that the distinction between ΔG and ΔG° be clearly understood. On the one hand ΔG is a measure of the *actual* free-energy change that occurs given a mixture of reactants and products at any particular concentrations; the value of ΔG thus varies, depending on the concentrations involved. On the other hand ΔG° is a constant for any given reaction at a given temperature. It is the free energy change that occurs under standard conditions of reactant and product concentration.

When equilibrium is achieved, it means by definition that no further net change in free energy can occur, that is ΔG = 0. Substituting ΔG=0 into Equation 2-1, we obtain the equilibrium expression

$$0 = \Delta G^\circ + 2.303RT \log_{10} \frac{[\text{products}_{eq}]}{[\text{reactants}_{eq}]}$$

where $[\text{reactants}_{eq}]$ and $[\text{products}_{eq}]$ are the molar concentrations of reactants and products at equilibrium, respectively. Rearranging terms we obtain

$$\Delta G^\circ = -2.303RT \log_{10} \frac{[\text{products}]}{[\text{reactants}]} \qquad (2)$$

Because the *equilibrium constant* (K'_{eq}) for any chemical reaction is defined as

$$K'_{eq} = \frac{[\text{products}]}{[\text{reactants}]} \qquad (3)$$

K'_{eq} can be substituted into Equation 2-2 to obtain the general relationship

$$\Delta G^\circ = -2.303\ RT \log_{10} K'_{eq} \qquad (4)$$

This final equation provides us with a relatively easy way of determining the standard free-energy change for any particular chemical reaction. One simply measures the concentrations of reactants and products after equilibrium has been achieved, uses this information to calculate K'_{eq} (Equation 2-3), and then substitutes this value of K'_{eq} into Equation 2-4 obtain ΔG°.

It can be readily calculated from Equation 2-4 that reactions with an equilibrium constant of 1.0 (equal concentrations of reactants and products at equilibrium) exhibit aΔG° of 0. This means that no free-energy change occurs during conversion of reactants to products under standard conditions. For reactions with an equilibrium constant greater than 1.0 (equilibrium favouring products). ΔG° will be negative and free energy will therefore be released during conversion of reactants of products. When the equilibrium constant is less than

1.0 (equilibrium favouring reactants), ΔG° will be positive and free energy will therefore be consumed during conversion of reactants to products.

Reactions with a negative ΔG°, termed *exergonic* reactions, can produced spontaneously in the direction written. However, it is important to emphasize that in this particular context, the term *spontaneous* does not signify anything about the speed at which the reaction will occur. A spontaneous reaction may take seconds or years to achieve equilibrium; its spontaneity refers only to the fact that no net input of energy is required in order to make the reaction proceed. Reactions in which ΔG° is positive are termed *endergonic,* and cannot proceed spontaneously in the direction written. Instead, they will proceed in the reverse direction.

In a living cell, of course, one does not generally start with 1.0M concentrations of all ingredients, the standard conditions that apply to the above discussion of ΔG°. Under other conditions it is ΔG, the actual free-energy change, that determines the direction in which a reaction proceeds. Depending on the particular conditions involved, ΔG may be smaller, larger, or the same as ΔG°.

As an example to help summarize the above points, let us consider the following reaction that occurs as part of the pathway for metabolizing glucose :

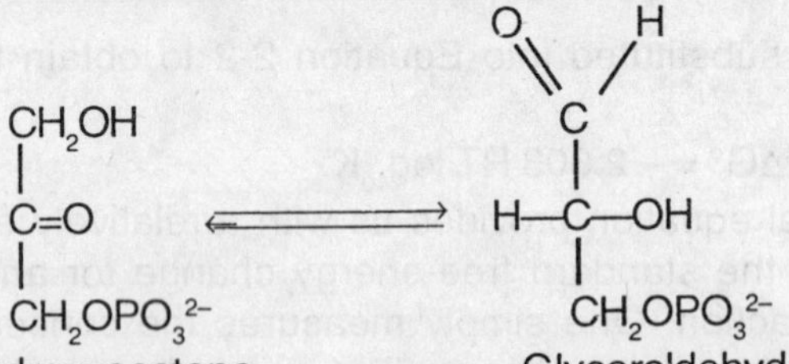

Dihydroxyacetone phosphate ⇌ Glyceraldehyde 3-phosphate

No matter what the starting conditions, the ratio of glyceraldehyde 3-phosphate to dihydroxyacetone phosphate is always found to be 0.0475 once equilibrium has been attained. By substituting this value of K'_{eq} into Equation 2-4 we obtain

$$\Delta G^\circ = -2.303RT \log_{10} (0.0475)$$

Substituting appropriate values for the gas constant R (0.00198 kcal/mol/K) and absolute temperature T (25°C = 298K), we can calculate

$$\Delta G^\circ = -2.303 \times 0.00198 \times 298 \times \log_{10} (0.0475)$$
$$= +1.8 \text{ kcal/mol}$$

Hence for every mole of dihydroxyacetone phosphate converted to glyceraldehyde 3-phosphate *under standard conditions,* 1.8 kcal of energy is consumed.

Now let us observe what would happen to this reaction under an arbitrary set of initial conditions such as might exist in the cell, for example, a dihydroxyacetone phosphate concentration of 10^{-4}M and a glyceraldehyde 3-phosphate concentration of 10^{-6}M. The actual free-energy change ΔG can be calculated for these conditions by substituting into equation 2-1 as follows :

$$\Delta G = \Delta G^\circ + 2.303\,RT\log_{10}\frac{[\text{products}]}{[\text{reactants}]}$$

$$= 1.8\text{ kcal / mol} + 2.303\,RT\log_{10}\frac{10^{-6}}{10^{-4}}$$

$$= 1.8\text{ kcal/mol} + (2.303 \times 0.00198 \times 298 \times -2)$$

$$= 1.8\text{ kcal/mol} - 2.7\text{ kcal/mol}$$

$$= -0.9\text{ kcal/mol}$$

Thus in spite of the positive value of ΔG° previously calculated for the reaction, ΔG turns out to be negative under these particular conditions. This negative value of ΔG means that under the above specified conditions, the reaction will proceed spontaneously with a release of free energy, even though such would not be the case under standard conditions. Since ΔG is calculated on the basis of the prevailing initial concentrations of reactants and products rather than arbitrary standard conditions, ΔG rather than ΔG° determines the direction in which reactions proceed in the cell.

Kinetics of Chemical Reactions

The thermodynamic approach is useful in defining the directions in which reactions tend to proceed, but it tells us nothing about the rates at which they occur. Analysis of the factors involved in determining how fast a reaction will occur is referred to as the study of *kinetics.* Reaction kinetics are ultimately based on the fact that the conversion of reactants to products always involves an intermediate stage, called the *transition state,* whose energy level is higher than that of either the reactants or the products. The term *free energy of activation,* or ΔG‡, is used to refer to the difference between the free-energy level of the reactants and the free-energy level of the transition state. The rate at which a chemical reaction proceeds is directly proportional to the number of molecules attaining the high energy transition state. As a result, any factor that increases the number of molecules having sufficient energy to attain the transition state will increase the rate of the reaction.

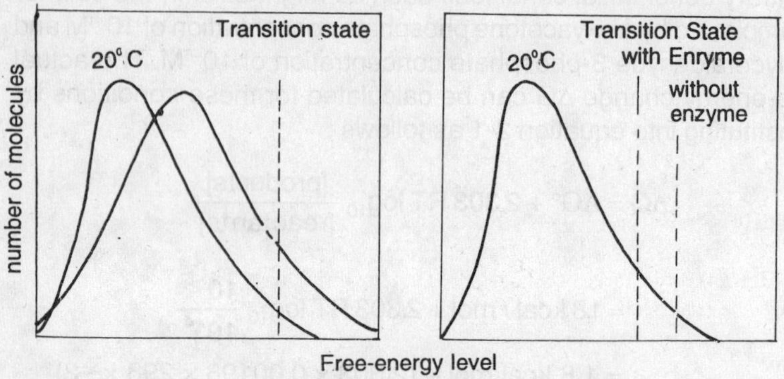

Fig. 10.11. Energy distribution diagrams illustrating two ways in which reaction rates can be increased. (a) Raising the temperature increases the proportion of the molecules in a population having sufficient free energy to achieve the transition state. (b) The presence of a catalyst creates a transition state with a lower free energy, so at any given temperature more molecules will possess sufficient free energy to attain this state.

There are two general ways in which this can be accomplished. The first involves temperature. At moderate temperatures a typical reaction mixture consists of a population of molecules with varying energy levels. Although the *average* free-energy level of the mixture is lower than that of the transition state, a small fraction of the molecules has sufficient energy to achieve this activated state. The relative number of molecules in this latter category determines how fact the reaction will proceed. Raising the temperature increases the overall thermal energy of the population as a whole, thereby increasing the number of molecules with free-energy levels high enough to enter the transition state. This higher proportion of molecules attaining the transition state in turn increases the rate of product formation.

The second general method for increasing reaction rates involves the use of catalysts. A catalyst combines with one or more reactants to generate a new transition state whose $\Delta G\ddagger$ is lower them the $\Delta G\ddagger$ of the transition state for the uncatalyzed reaction. In other words, the catalyst provides an alternative reaction pathway with a transition state of lower free energy. Hence at any given temperature, the number of molecules possessing sufficient free energy to attain the transition state will be higher for the catalyzed reaction that for the uncatalyzed reaction.

Because the new transition state created by a catalyst is part of both the forward and reverse reactions, catalysts increase reaction rates in both directions rather than selectively enhancing the rates of either the forward or backward reaction. A catalyst therefore can only speed up the rate at which equilibrium is achieved. The equilibrium concentration of reactants and products are solely determined by the thermodynamic calculation of ΔG, which is not influenced by the presence or absence of a catalyst.

Enzyme-catalyzed reactions obey all the basic principles of thermodynamics and kinetics described above. Like other catalysts, enzymes provide reactions with a new transition state of lower free energy, thereby speeding up the rate at which equilibrium is attained. And like other catalysts, enzymes do not affect the relative concentrations of reactants and products that prevail once equilibrium is attained. These equilibrium conditions are determined solely by the difference in free-energy levels (ΔG) between reactants and products, a factor upon which enzymes have no effect. Thus in the presence of enzymes, reactants need less of an energy input to be able to react to form products. This is paramount biological significance for it permits metabolic reactions to occur at the relatively low temperatures compatible with living cells.

In addition to the above properties shared by all catalyzed reaction, enzymatic reactions exhibits several unusual characteristics not commonly encountered elsewhere. In the following section we shall elaborate upon some of these characteristics and discuss their implications for the mechanism of enzyme action.

The Michaelis-Menten Concept of Enzyme Action

What is the relationship between *v* and [S] when the rate of conversion of S to P is increased by adding a catalyst—an enzyme? The plot of *v* versus [S] now takes the form. In contrast to the nonenzymetic reactions we now find three zones in the relationship between *v* and [S] : (1) at low [S], *v* is virtually proportional to [S] as in any first order reaction, except that the rate may be magnitudes greater than observed in either the unheated or heated nonenzymatic reactions; (2) beyond a certain [S], *v* increases less, is no longer linearly related to [S], and instead shows the characteristics of a mixed -order reaction; and (3) finally a value of [S] is reached at which *v* becomes constant, being essentially independent of [S].

It is typical of enzyme reactions that the reaction rate becomes independent of [S] when [S] is increased to a sufficiently high level. This observation, from which it can be deduced that the enzyme can be saturated with substrate, led to the proposal by Michaelis and Menten that the initial and essential step in an enzyme-catalyzed reaction is the reversible formation of an enzyme-substrate complex. Accounting for the kinetics of formation and breakdown of such a complex, one can derive a useful mathematical relationship between the reaction observed at any particular [S] and the maximal rate which is achieved at a high [S]. The *Michaelis-Menten equation,* which is developed below, describes the kinetics of any enzyme reaction in which there is only one substrate.

The Michaelis-Menten theory assumes the following sequence of events:

$$E + S \underset{k_2}{\overset{k_1}{\rightleftharpoons}} E\,.\,S \underset{k_4}{\overset{k_3}{\rightleftharpoons}} E + P \tag{5}$$

It shall also be assumed that k_4 is very small or, in other words, that E . S → P + E is a one-way reaction.

Let us examine first the change in the concentration of E.S. The enzyme-substrate complex, with time. The rate of formation of E.S. will be

$$\frac{d(ES)}{dt} = k_1[E_{free}]\,[S] \tag{6}$$

Since

$$[E_{total}] = [E_{free} + ES] \text{ or } [E_{free}] = [E_{total} - ES] \tag{7}$$

$$\frac{d[ES]}{dt} = k_1\,[_{total} - ES]\,[S] \tag{8}$$

The rate of breakdown of ES will be

$$\frac{-d[ES]}{dt} = k_2[ES] + k_3[ES] \tag{9}$$

When the rate of formation of ES equals its rate of disappearance, the reaction system has achieved the steady-state condition. [ES] now remains constant and equations (8) and (9) can be equated:

$$k_1[E_{total}] - [ES]\,[S] = k_2\,[ES] + k_3\,[ES] \tag{10}$$

Solving for [ES], we obtain

$$k_1[E_{total}][S] - k_1[ES][S] = k_2[ES] + k_3[ES]$$
$$k_1[E_{total}][S] = k_1[ES][S] + k_2[ES] + k_3[ES]$$
$$= ES[k_1[S] + k_2 + k_3]$$

and $$[ES] = \frac{k_1[E_{total}][S]}{k_1[S] + k_2 + k_3} = \frac{[E_{total}][S]}{[S] + (k_2 + k_3)/k_1}$$

$(k_2 + k_3)/k_1$ will be defined as K_M. Therefore

$$[ES] = \frac{[E_{total}][S]}{[S] + k_M} \quad (11)$$

[ES] is not a useful term because very often it cannot be determined directly. It will be noted from equation (5), however, that the breakdown of ES to product and E is a first-order reaction if k_4 is neglected. The rate of this reaction is termed the *initial velocity* v_0 of the enzyme reaction and is equal to

$$K_3[ES]$$

If both sides of equation (11) are multiplied by k_3, the resulting equation will be

$$k_3[ES] = \frac{K_3[E_{total}][S]}{[S] + K_M} v_0 \quad (12)$$

The maximal initial velocity of the reaction will be achieved when the total E in the system is present as ES. Therefore, it follows that $k_3[E_{total}]$ must equal the maximal initial velocity $V_{max.}$ Hence,

$$v_0 = \frac{V_{max}[S]}{[S] + K_M} \quad (13)$$

This is the Michaelis-Menten equation which describes the plot of v_0 versus [S].

To illustrate the utility of the equation, consider first the meaning of the constant K_M. Assume that the initial reaction rate v_0 is equal to one-half the value of V_{max}. Under this condition

$$\frac{V_{max}}{2} = \frac{V_{max}[S]}{[S] + K_M} \text{ or } \frac{1}{2} = \frac{[S]}{[S] + K_M}$$

$$2[S] = [S] + K_M \text{ or } K_M = [S] \quad (14)$$

Therefore, operationally defined, K_M is equivalent to the [S] at which the initial velocity of the enzyme reaction is half of the V_{max}. As defined in equation 6-16, K_M is a composite dissociation constant for the ES complex, accounting for both its reversible formation and dissociation as well as for the reaction in which ES breaks down to form P and regenerate E. Whenever it can be established for any enzyme reaction that k_3 is negligible compared with k_2, the equation for K_M becomes

$$K_M = \frac{k_2}{k_1}$$

and K_M in this approximation is equivalent to the dissociation constant for the ES complex and is designated as K_3 :

$$\frac{[E][S]}{[ES]} = K_5 \qquad (15)$$

When properly defined, K_M and K_5 are useful constants for describing enzyme activities both in tissues and purified preparations. For example, a determination of K_5 values permits comparison of affinities of a substrate for an enzyme isolated from different sources or a comparison of affinities of structurally related substrates with a given enzyme. Knowing the value of K_M is useful when one needs to determine how much S to employ to assure maximal velocity. Obviously, when [S] is very small compared with K_M, v_0 will directly with [S] as a simple first-order function :

$$v_0 = \frac{V_{max}}{K_M}[S]$$

On the other hand, if [S] is many times larger than K_M,

$$v_0 = V_{max}$$

When [S] is saturating and all of the E is present as ES, $v_0 = k_3$ [ES] = V_{max}. The plot of v_0 versus [E] becomes V_{max} versus [E] :

These are the experimental conditions required for measuring the enzyme level in a particular preparation or for comparing the activities of an enzyme from different sources.

Alternative Algebraic forms of the Michaelis-Menten Equation

The rectangular hyperbolic plot that relates the initial rate of a one-substrate enzyme reaction with the initial substrate concentration is not particularly convenient for determining V_{max} accurately. Two other algebraic forms of the equation are now employed widely in kinetic analyses of enzymatic reactions. One of these transformations is the equation obtained by taking the reciprocal of each side of the Michaelis-Menten equation :

$$\frac{1}{v_0} = \frac{K_M + [S]}{V_{max}\,[S]}$$

or
$$\frac{1}{v_0} = \frac{K_M}{V_{max}\,[S]} + \frac{[S]}{V_{max}\,[S]}$$

or
$$\frac{1}{v_0} = \frac{K_M}{V_{max}}\frac{1}{[S]} + \frac{1}{V_{max}} \qquad (16)$$

This form of the Michaelis-Menten relationship, referred to as the *Lineweaver-Burk Equation,* is the same equation that describes the straight-line plot of y versus x :

$$y = mx + b$$

in which m is the slope of the line and b is the intercept on the y axis. Hence, plotting $1/v_0$ versus $1/[S]$, the graphical representative of the Lineweaver-Burk equation is

The stop of the line is K_M/V_{max} and the intercept is $1/V_{max}$. The intercept of the extrapolation of this plot to the $1/[S]$ axis will have the value of $-1/K_M$ because

$$\frac{1}{v_0} = 0 = \frac{K_M}{V_{max}}\frac{1}{[S]} + \frac{1}{V_{max}}$$

and
$$-K_M\frac{1}{[S]} = 1$$

and
$$\frac{1}{[S]} = -\frac{1}{K_M} \qquad (17)$$

It is apparent that the linear double-reciprocal plot of the Line-weaver-Burk equation for determining V_{max} circumvents the problems of estimating this velocity as a limiting value in the curvilinear plot of v_0 versus [S].

Another commonly employed permutation of the Michaelis-Menten equation is that resulting when both sides of the Line-weaver-Burk expression are multiplied by v_0 :

$$\frac{1}{v_0} = \frac{K_M}{V_{max}}\frac{1}{[S]} + \frac{1}{V_{max}}$$

$$\frac{v_0}{v_0} = \frac{K_M v_0}{V_{max}\,[S]} + \frac{V_0}{V_{max}}$$

$$1 = \frac{R_M v_0}{V_{max}\,[S]} + \frac{V_0}{V_{max}}$$

$$\frac{v_0}{V_{max}} = 1 - \frac{K_M v_0}{V_{max}\,[S]}$$

$$v_0 = V_{max} - \frac{K_M v_0}{[S]} \tag{18}$$

The straight-line function described by this equation is
The intercept on the v_0 axis is V_{max} and that on the $v_0/[S]$ axis is V_{max}/K_M. The slope of the line is $-K_M$.

Mechanism of Enzyme Actions

There is a very close structural relationship between the molecular surface of an enzyme and its substrates. Enzymes are protein molecules with definite surface geometry. The functional groups of the enzyme are exactly complementary to those of the substrate. Only particular types of substrate molecules will fit with a given enzyme molecule. For example, substrates A and B will fit into the enzyme E but not substrate C. This is referred to as a 'lock-and-key' mechanism. Thus reactions involving A and B will be speeded up but, not reactions involving C.

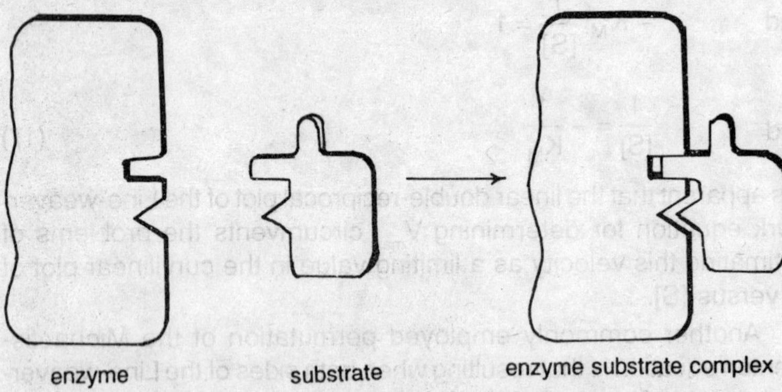

Fig. 10.12. Diagram illustrating how only particular substrates react with an enzyme.

The *enzyme* enters into a chemical combination with the *substrate* to form an *enzyme-substrate complex* (Michaelis-Menten hypothesis).

$$E + S \rightarrow ES$$

The enzyme substrate complex then breaks down to give the products of reaction. The enzyme is released and can be used over and over again.

$$ES \rightarrow E + \text{Products}$$

Active Site

An enzyme has a distinct cavity or cleft in which the substrate is bound. The cleft contains an *active centre* in which the amino acids are grouped together in such a way as to enable them to combine with the substrate. The reactive amino acids may lie widely separated in the polypeptide chain. The chain, however, undergoes folding in such a manner that the reactive amino acids come together in the active site.

It is believed that when the substrate molecule binds to the active site, its parts are hold together in such a way as to cause distortion of the chemical bonds, i.e., the bonds are weakened. This distortion of the chemical bonds of the substrate increases its reactivity, and thus speeds up the rate of the reaction. The products of the reaction are released because they are less firmly bound. The mechanism suggested above has been called the *strain model* of enzyme catalysis.

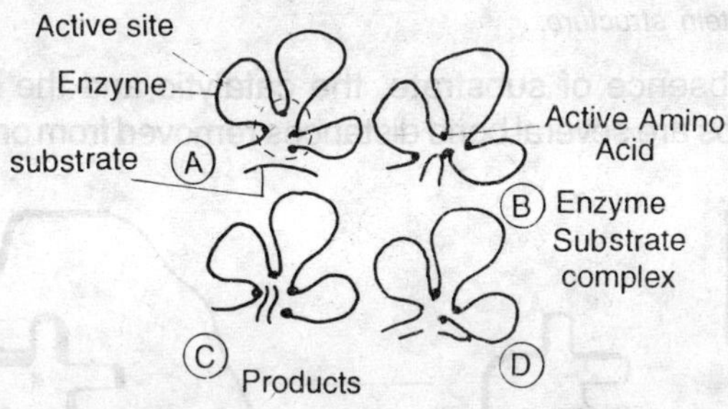

Fig. 10.13. Diagramatic representation of an enzyme and its active site. A—Enzyme with its reactive site consisting of four reactive amino acid. B—Substrate binds to enzyme to form enzyme-substrate complex. C—Reaction occurs in which the substrate is broken down into its products. D—Products removed from active site.

Another model, called the *rack model*, suppose that the conformational change in the substrate after binding leads to *increased* distortion of the substrate molecule. This bending after binding makes the substrate more reactive.

Another way in which the reactive site is believed to speed up a reaction is by *excluding water molecules* of the solvent from the site of reaction. The tight fit of the substrate in the active site of the enzyme molecule does not permit water molecules at the site of reaction. It is known that in the case of certain compounds exclusion of water molecules greatly affects the rate of reaction.

Induced Fit Model

This model was given by Koshland (1966). In the Fischer model, i.e., lock-and-key mechanism, the catalytic site is presumed to be preshaped to fit the substrate. In the induced fit theory, the substrate induces a conformational change in the enzyme. This aligns amino acid residues or the other groups on the enzyme in the correct spatial orientation for substrate binding, catalysis or both. At the same time, the other amino acid residues may get buried in the interior of the enzyme.

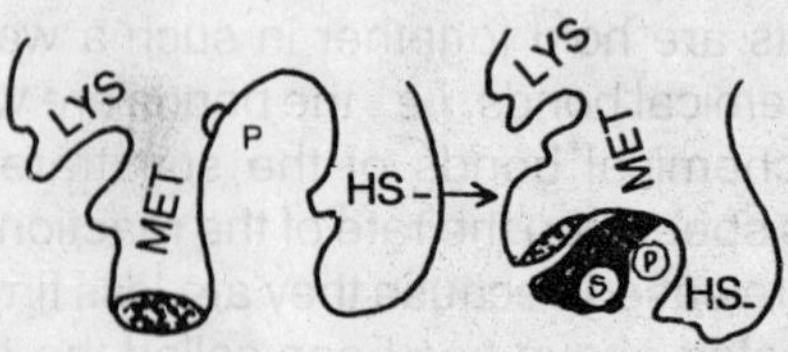

Fig. 10.14. Representation of an induced fit by a conformational change in the protein structure.

In the absence of substrate, the catalytic and the substrate-binding groups are several bond distances removed from one another.

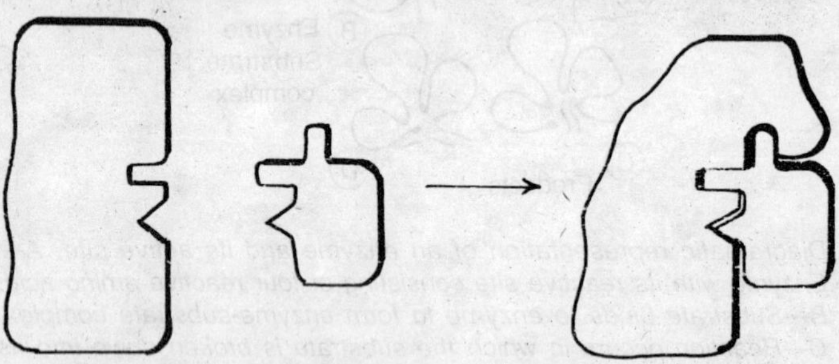

Fig. 10.15. Representation of alternative reaction paths for a substrate induced conformational change.

When the substrate approaches, there occurs a conformational change in the enzyme protein, aligning the groups correctly for substrate binding and for catalysis. At the same time there also occurs a change in the spatial orientations of the other regions.

The main evidence in favour of induced fit model comes from demonstration of conformational changes during substrate binding and catalysis with creative kinase, phosphoglucomutase, and several other enzymes. Upon this time, it could not be established about the exact sequence of events in a substrate-induced conformational change. There may be several possibilities which are depicted.

Even if one knows the complete primary structure of an enzyme, it is not very easy to decide exactly which residues exactly constitute the catalytic site.

Factors Affecting Enzyme Activity

1. Substrate Concentration

Enzyme activity has been found to first increase with increase in concentration of the substrate to a maximum and then it levels off. At this concentration of substrate, amount of enzyme may probably become limiting.

Two types of graphs are obtained from the plotting of substrate concentration and enzyme activity. Straight line or Michaelis type of curve is obtained with the enzymes if they are made up of one unit.

If the enzyme are made up of the several subunits and they exhibit interaction amongst subunits (allosteric enzymes) then sigmoidal type of curve is obtained.

Table 10.2. Amino acid sequence of active sites of various enzymes

Enzyme	Sequence of Active Site
Trypsin	Gly—Asp—Ser—Gly—Gly—Pro
Chymotrypsin	Gly—Asp—Ser—Gly—Gly—Pro
Elastase	Gly—Asp—Ser—Gly—Gly—Pro
Thrombin	Asp—Ser—Gly
Alkaline phosphate	Asp—Ser—Ala
Futyrylcholine esterase	Gly—Glu—Ser—Ala—Gly
Acetylcholine esterase	Glu—Ser—Ala
Aliesterase	Glu—Ser—Ala

* Note that Asp and Gly are similar amino acids (both acidic), as are Gly and Ala (both aliphatic)

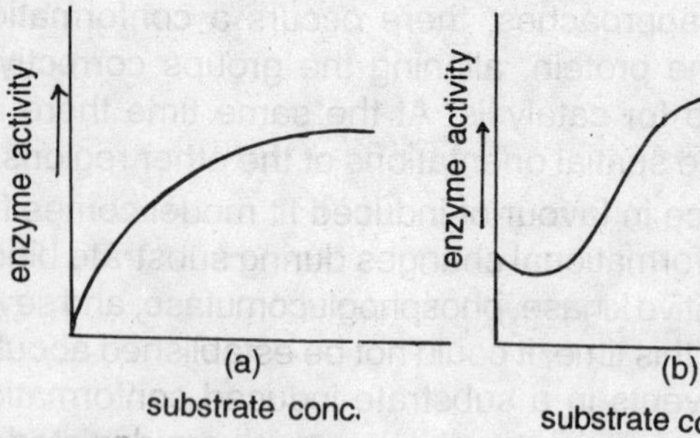

Fig. 10.16. Effect of substrate concentration on enzyme activity, a—Michaelis-Menten type curve, b—Sigmoidal curve.

2. Concentration of Enzymes

Increase in enzyme concentration has been found to increase enzyme activity. When an excess of substrate is present, increasing the enzyme concentration by two times generally doubles the rate of formation of the product. With the given concentration of the enzyme also, a point may be reached where all the substrate molecules get bound to the enzyme and further increase in enzyme does not have any effect on the formation of the product.

3. Effect of pH

As enzymes are made up of proteins, they are very sensitive to change in pH. A graph of enzyme activity against pH is generally bell shaped. There is a optimum pH, which depends upon the nature of the enzyme. However, some enzymes may be having two optima of pH or no sharp pH but a range of pH.

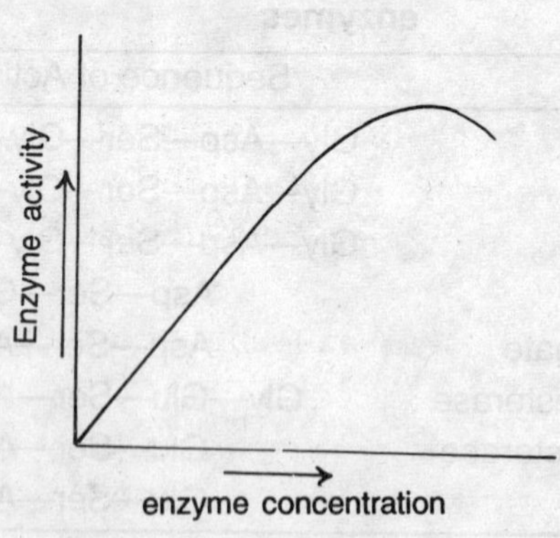

Fig. 10.17. Effect of enzyme concentration on enzyme activity.

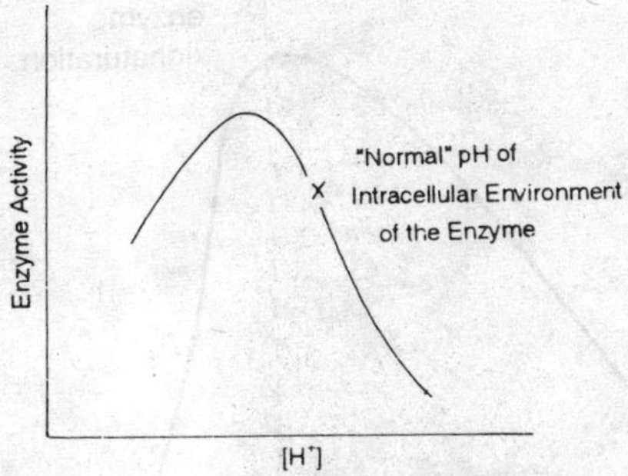

Fig. 10.18.

At optimum pH the activity of enzymes is maximum. For most enzymes the effective pH range is 4-9, Beyond these limits denaturation of enzymes takes place. The optimum pH for pepsin is 2.0 and for trypsin 8.0.

4. Effect of ions

Many enzymes only become active in the presence of a cation such as Mg^{2+}, Ca^{2+}, Mn^{2+}, Zn^{2+}, Na^{+} or K^{+}. In some cases, the cations get loosely bound to the enzyme, while in other cases they get bound to the substrate. Anions have also been found to increase the enzyme activity. For example, chloride enhances the activity of salivary amylase. Concentration of ions also influences the enzyme activity.

5. Effect of Temperature

Enzyme action is greatly affected by temperature. If the temperature is increased by 10°C the rate of most chemical reactions is doubled. However, at 40°C—60°C there is loss of enzyme activity because denaturation of proteins occurs at this temperature. There are few enzymes which continue to be active above 60°C also. The enzymes of dry tissues such as seeds and spores, are more resistant to high temperature.

6. Redox Potential

Enzymes have been found to be sensitive to the redox potential of the cell also. Oxidising and reducing enzymes also take part in changing the redox component of a cell and hence they get affected

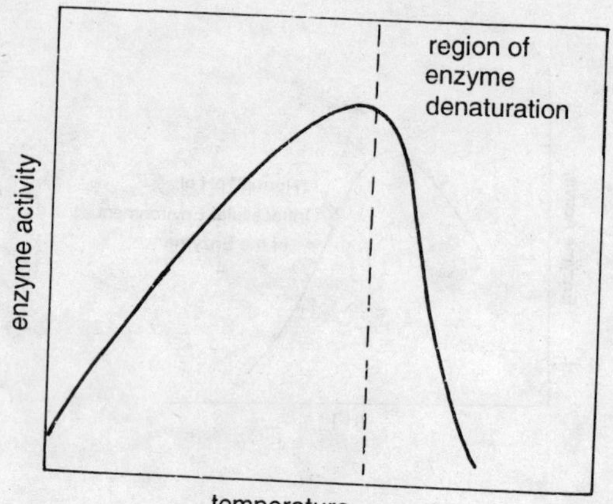

Fig. 10.19. Effect of increasing temperature on the rate of an enzyme-catalyzed reaction.

by the redox potential. Some enzymes are affected by redox potential due to the presence of readily oxidisable—SH group in their molecule.

Enzyme Inhibition

Just as medicine has gained a knowledge of normal metabolic pathways by observing the consequences of abnormal processes, so the biochemist has come to understand the chemistry of biological catalysis through analyses of enzyme inhibition. Directly or indirectly, all inhibitors affect the union of enzyme and substrate and/or the catalytic consequences thereof. The resulting inhibition of catalytic function can be categorized as competitive, uncompetitive, or noncompetitive. In all cases the effects of the inhibitor will be reflected as changes in the kinetics of enzyme-substrate interactions:

$$E + S = ES$$

To characterize the mechanism by which a particular inhibitor is adversely affecting an enzyme reaction, questions such as the following must be answered. Is the degree of inhibition at a fixed enzyme concentration and fixed inhibitor concentration dependent on [S]? Is the degree of inhibition time dependent, suggesting that the inhibitor may be combining irreversibly with the enzyme?

Consider first the simplest type of inhibition—that due to a reversible competition between the inhibitor and substrate for binding at the substrate-binding site.

SUBSTRATE-COMPETITIVE INHIBITION

When a competitive inhibitor binds reversibly with the enzyme, the complex formation can be described in the same manner as the enzyme-substrate interaction :

$$E + I = EI$$

The dissociation constant of the complex, K_i, can be defined as

$$K_i = \frac{[E][I]}{[EI]} \qquad (19)$$

Since the inhibitor is not chemically altered, the only fate of the EI complex is to dissociate into E and I. K_i is, therefore, a true dissociation constant and is comparable to K_s. The simplest experimental test of substrate-competitive inhibition is to determine whether the degree of inhibition at a fixed inhibitor concentration can be decreased by increasing the substrate concentration. Referring again to the relationship between v_0 and [S], the presence of a fixed amount of inhibitor with varying amounts of S will give the following plots:

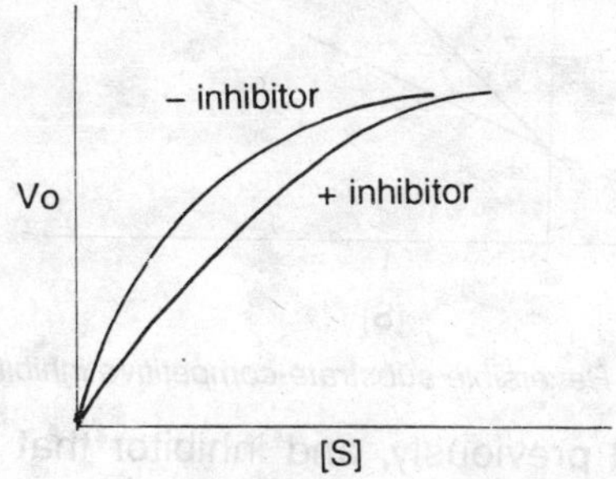

As would be predicted from the Michaelis-Menten model, a higher [S] is required to reach half-maximal velocity for the inhibited reaction than for the uninhibited reaction. Theoretically, the inhibition can be completely reversed by making the [S] [i] ratio very high; the two curves will merge asymptotically as V_{max} is approached.

Just as the kinetic parameters of the Michaelis-Menten relationship are conveniently evaluated by a double-reciprocal plot (equation 6-22), the kinetics of an inhibited enzyme reaction can also be analyzed by plots of 1/[S] versus $1/v_0$. Thus, when the [I] is kept constant and the [S] is varied, the uninhibited and competitively inhibited reactions will be related. The extrapolated intercepts will be

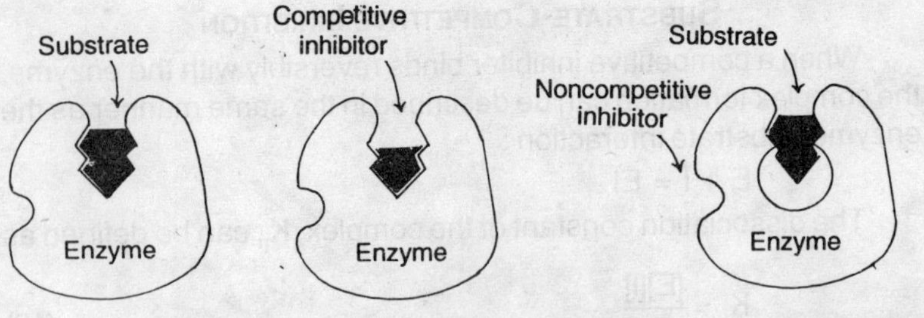

Fig. 10.20. Distinction between a competitive inhibitor and a noncompetitive inhibitor: (left) enzyme-substrate complex; (middle) a competitive inhibitor prevents the sub-strate from binding; (right) a noncompetitive inhibitor does not prevent the substrate from binding.

$$(a) -\frac{1}{K_M} \text{ and } (b) -\frac{1}{K_M(1+[I]/K_I)}$$

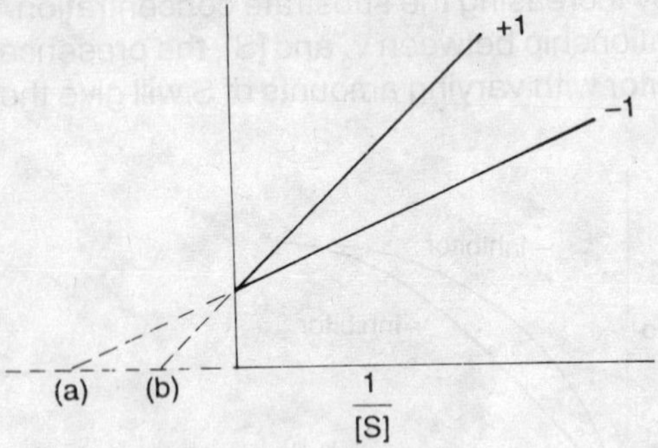

Fig. 10.21. Reversible substrate-competitive inhibiton.

As emphasized previously, and inhibitor that is competing reversibly with the substrate for a common binding site will not affect

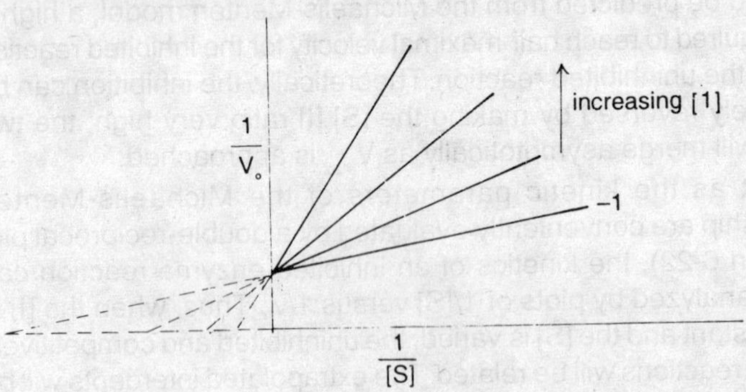

the V_{max} of the system. This type of inhibitor does not affect the rate of breakdown of the enzyme-substrate complex to E+S or to E+P. Accordingly, one criterion of substrate-competitive inhibition is the common intercept. $1/V_{max}$, in plots of $1/v_0$ versus 1/[S]. Corroborative evidence for concluding that an inhibition is of the single substrate-competitive type and also reversible is obtained by employing a range of fixed levels of inhibitor in kinetics analyses such as those shown in figure.

The mathematical formulation describing the reversible competitive between inhibitor and substrate must account for the following processes :

$$E + S = ES = E + P$$

$$E + I = EI \text{ (inactive)}$$

The final form of the Michaelis-Menten relationship in this case is

$$\frac{1}{v_0} = \frac{K_M}{V_{max}}\left(1 + \frac{[I]}{K_i}\right)\frac{1}{[S]} + \frac{1}{V_{max}} \tag{20}$$

In the reciprocal plot of v_0 versus [S] of the inhibited reaction, the intercept on the ordinate, I/V_{max}, remains the same as for the uninhibited reaction as pointed out above. The slope of the line now becomes

$$\frac{K_M}{V_{max}}\left(1 + \frac{[I]}{K_I}\right)$$

Compared to the uninhibited reaction the slope has been increased by a factor of $1 + [I]/K_i$. Equation 6-23 can also be used to calculate K_i, the dissociation constant of the inhibitor-enzyme complex. As will be discussed in later section, a knowledge of the values of K_M and K_I is not only relevant to our understanding of enzyme mechanisms but also has application in the design and clinical use of chemotherapeutic agents.

It was noted in comparing the hyperbolic plots of v_0 versus [S] in the substrate-competitive inhibitor system that to attain the same v_0 in both the uninhibited and inhibited reactions, the [S] in the inhibited reaction must be higher than in the uninhibited system. This is a reasonable consequence of the two competing reactions :

$$E + S = ES = E + P$$

and $$E + I = EI$$

At any given moment some of the enzyme, being in the form of EI, is unavailable for ES formation. It is important to realize, however, that the presence of I affects neither the association of E and S nor the dissociation of the ES complex. The value of K_M is the same in the presence of I as in its absence. Just as the I does not alter the K_M, neither does it affect V_{max}. As seen in the plot on previous page and in Figure. When the [S]/I ratio is very large, the V_{max} of the inhibited reaction is the same as the V_{max} of the uninhibited system.

The biochemical literature is replete with examples of substrate-competitive inhibition of enzyme systems and their application to medicine. Almost without exception, the inhibitors and substrates have striking similarities with regard to functional groups and/or molecular conformation.

NONCOMPETITIVE INHIBITION

Alteration of the conformation of an enzyme can be expected to affect its catalytic activity. An reagent that can combine reversibly with the enzyme in some region or with some functional group outside the substrate-binding site may cause a structural change in the protein if the affected region or site is needed to maintain the native shape of the molecule. Such a conformational change in turn results in a modified substrate-binding site. The kinetic consequences of such structural changes can be both a decreased rate of ES formation as well as decreased rate of ES breakdown to E and reaction product(s). The inhibitor can combine with either E or ES :

$$E + I = EI$$

and $$ES + I = ESI$$

The dissociation constants of these inhibitor complexes are

$$K_{EI} = \frac{[E][I]}{[EI]} \quad (21)$$

and $$K_{ESI} = \frac{[ES][I]}{[ESI]} \quad (22)$$

It may also be assumed that EI can react with S :

$$EI + S = ESI$$

The K for this reaction, K'_{ESI}, may not have the same value as the K_{ESI} for reaction (22). Thus, the "composite kinetics" of a noncompetitively inhibited enzyme reaction is complicated. Assuming that the three K values cited above are identical and that EI and ESI are inactive, the reciprocal form of the Michaelis-Menten relationship is

$$\frac{1}{v_0} = \frac{K_M}{V_{max}}\left(1+\frac{[I]}{K_I}\right)\frac{1}{[S]} + \frac{1}{V_{max}}\left(1+\frac{[I]}{K_I}\right) \tag{23}$$

The plot of this equation is shown in figure. The common intercept obtained by extrapolation of the inhibited and noninhibited plots is —1/K_M. Although the K, for the two reactions

E + S = ES and EI + S = ESI

can be expected to be different because the active site has been deformed by the noncompetitive inhibitor, the K_M for the two system can be the same, Why" Assuming that both the EI and ESI complexes are inactive, it is as if the noncompetitive inhibitor has merely removed some of the enzyme. That is, the v_0 and V_{max} for the inhibited enzyme are the velocities that would be obtained if the kinetic analysis of the uninhibited control had been repeated using less enzyme.

One test for reversible, noncompetitive inhibition is that the effects of the inhibitor cannot be reversed no matter how much the [S] is increased. The value of V_{max} cannot be increased to that of the uninhibited reaction. The inhibition can only be reversed by removing the inhibitor.

Uncompetitive Inhibition

Uncompetitive is the term used to describe an inhibition of enzyme activity resulting from the reversible combination of the inhibitor with the ES complex

ES + I = ESI (24)

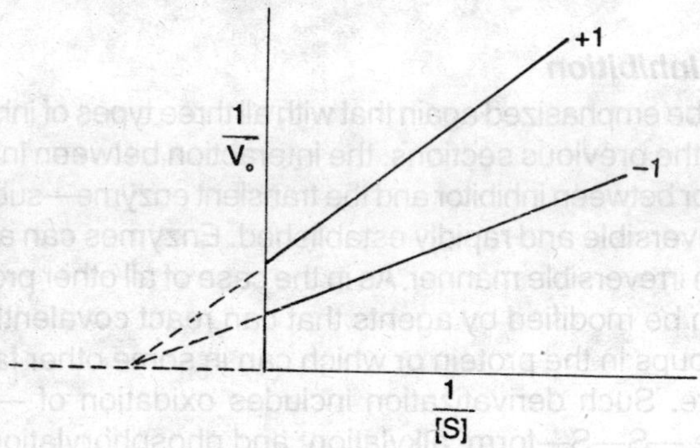

Fig. 10.22. Noncompetitive inhibition.

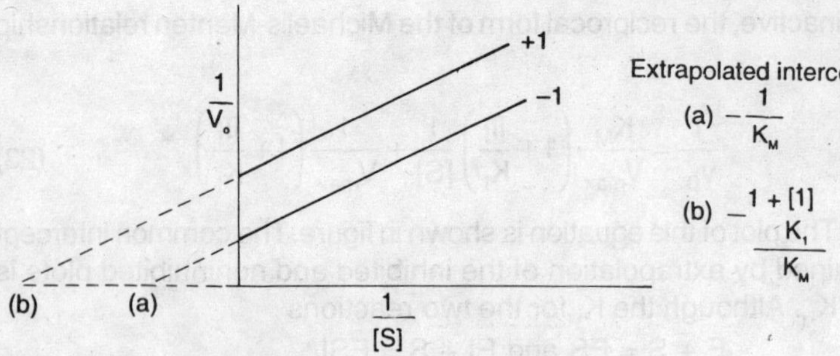

Fig. 10.23. Uncompetitive inhibition.

The ESI is inactive and the dissociation constant is :

$$K_I = \frac{[ES][I]}{[ESI]} \tag{25}$$

In the reciprocal form, the relationship between the v_0, [S], and dissociation constants for this type of inhibition is :

$$\frac{1}{v_0} = \frac{K_M}{V_{max}}\frac{1}{[S]} + \frac{1}{V_{max}}\left(1 + \frac{[I]}{K_I}\right) \tag{26}$$

The plot of initial velocities versus [S] in a system containing a fixed concentration of uncompetitive inhibitor is shown in Figure. Although the V_{max} of the inhibited reaction will be less than for the uninhibited system, the K_M of the substrate-enzyme complex itself is not affected by an uncompetitive inhibitor. From equation 6-29 it will be seen that the degree of inhibition by a fixed concentration of uncompetitive inhibitor can actually be increased by increasing the [S].

Irreversible Inhibition

It should be emphasized again that with all three types of inhibition discussed in the previous sections, the interaction between inhibitor and enzyme or between inhibitor and the transient enzyme—substrate complex is reversible and rapidly established. Enzymes can also be inhibited in an irreversible manner. As in the case of all other proteins, enzymes can be modified by agents that can react covalently with functional groups in the protein or which can in some other fashion alter structure. Such derivatization includes oxidation of —S—H groups to the —S—S—form, alkylation, and phosphorylation. The irreversible inhibition of an enzyme elicited by such agents is most often slow and time related, usually increasing with time. The kinetics of this type of inhibition do not fit the Michaelis-Menten model.

Enzyme Reactions Involving More Than One Substrate

Most metabolic reactions are two-substrate enzymatic reactions. For example, *both* the α-amino acid and the α-keto acid are substrates for the transaminase that catalyzes the transfer of the amino group of an amino acid to an α-keto acid :

α-keto $acid_A$+α-amino $acid_b$ $\xleftrightarrow{\text{transaminase}}$ α-$amino_A$+ α-keto $acid_B$. Although the overall kinetics of such a system are more complicated than for the process E + S = ES → E + P, the K_M for each substrate in a two-substrate system can be determined if the reaction is made pseudo first order. This is accomplished by keeping the concentration of one substrate constant, at its saturation level, and then measuring initial rates of reaction at various concentrations of the second substrate. The analysis is repeated, with the conditions being reversed, to obtain the second K_M.

Having determined the K_M for each of the two substrates, one can then carry out kinetic analyses of the reaction under conditions in which the concentration of one substrate (e.g., amino $acid_A$) is varied while the concentration of the other (e.g., keto $acid_B$), is kept constant and lower than its saturating level. The analysis is repeated with several concentrations of the second substrate (i.e., in this case, keto $acid_B$). Plots of $1/v_0$ versus 1/[amino $acid_A$] give a series of parallel lines :

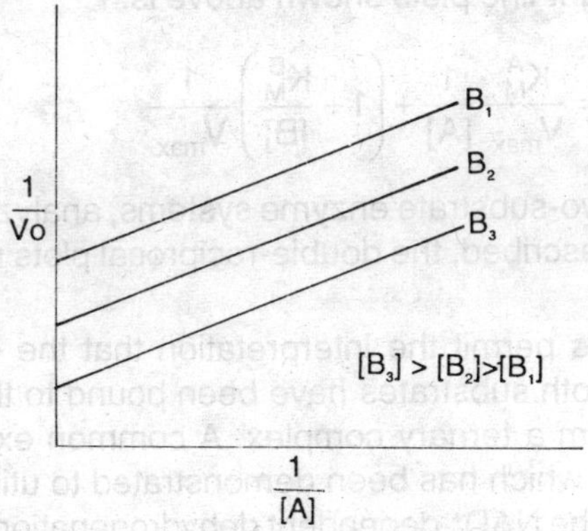

One substrate must be bound to the enzyme and one of the reaction products must be released before the second substrate can enter and the second product can be released. The enzyme, and/or its prosthetic group or coenzyme, is transiently and reversibly modified during the total reaction. The sequence of reactions can be summarized as follows [the transaminase bound to the aldehydic and amino forms of its coenzyme are designated as E—C(H)=O and E—C(H)(H)—NH$_2$]:

$$\text{amino acid}_A\text{—NH}_2 + \text{E—CH(=O)} = \text{amino acid}_A\text{—NH}_2\text{.E—CH(=O)}$$

$$\text{amino acid}_A\text{—NH}_2\text{.E—CH(=O)} = \text{keto acid}_A + \text{E—CH}_2\text{—NH}_2$$

$$\text{E—CH}_2\text{—NH}_2 + \alpha\text{-keto acid}_B = \text{E—CH}_2\text{—NH}_2\text{.}\alpha\text{-keto acid}_B$$

$$\text{E—CH}_2\text{—NH}_2\text{.}\alpha\text{-keto acid}_B = \text{E—CH(=O)} + \text{amino acid}_B\text{—NH}_2$$

Note that in this mechanism only binary complexes are involved.

The double-reciprocal form of the rate equation that describes each of the straight-line plots shown above is:

$$\frac{1}{v_0} = \frac{K_M^A}{V_{max}}\frac{1}{[A]} + \left(1 + \frac{K_M^B}{[B]}\right)\frac{1}{V_{max}} \qquad (27)$$

With some two-substrate enzyme systems, analyzed under the conditions just described, the double-reciprocal plots may take the following form :

These results permit the interpretation that the reaction can proceed only if both substrates have been bound to the enzyme's active site, to form a ternary complex. A common example of an enzyme reaction which has been demonstrated to utilize this type of mechanism is the NAD$^+$-dependent dehydrogenation of malate to oxaloacetate, one of the reactions of the Krebs cycle. In this system,

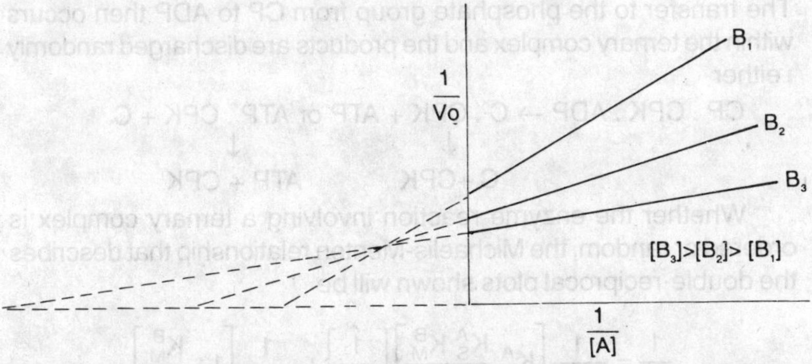

the combination of NAD^+ and malate with the dehydrogenase protein does not occur randomly; the NAD^+ is bound prior to the malate :

$$NAD^+ + \text{dehydrogenase} \rightarrow NA\overset{+}{D}-\text{dehydrogenase}$$

$$NA\overset{+}{D}\text{-dehydrogenase} + \text{malate} \rightarrow NA\overset{+}{D}\text{-dehydrogenase-malate.}$$

Following the transfer of electrons from the malate to the NAD^+ within the ternary complex, the release of product is similarly ordered; the NADH is the last reaction product to leave the complex. This sequence of substrate binding and product release is consistent with the observation that the dehydrogenation of malate is found to be inhibited in the presence of excess NADH. NADH can complete with NAD^+ and thereby hinders the normal rate of formation of the active ternary complex.

In contrast to the mechanism involving an ordered synthesis and breakdown of a ternary complex, some two-substrate enzyme reactions proceed via a complex in which substrate addition to the enzyme, and product discharge from the complex, occur randomly. For example, the transfer of a phosphate group from phosphocreatine to ADP, by the action of creatine kinase (CPK), occurs in such a nonordered manner. The overall reaction,

Phosphocreatine (CP + ADP $\overset{CPK}{\rightleftharpoons}$ creatine (C) + ATP requires formation of a ternary complex : either

CP + CPK = CP.CPK

CP.CPK + ADP = CP.CPK.ADP

or ADP + CPK = ADP.CPK

ADP.CPK + CP = ADP.CPK.CP

The transfer to the phosphate group from CP to ADP then occurs within the ternary complex and the products are discharged randomly : either

$$\begin{array}{ccc} \text{CP . CPK . ADP} \rightarrow & \text{C . CPK + ATP} & \text{or ATP . CPK + C} \\ & \downarrow & \downarrow \\ & \text{C + CPK} & \text{ATP + CPK} \end{array}$$

Whether the enzyme reaction involving a ternary complex is ordered or random, the Michaelis-Menten relationship that describes the double-reciprocal plots shown will be

$$\frac{1}{v_0} = \frac{1}{V_{max}}\left[K_M^A \frac{K_S^A K_M^B}{[B]}\right]\left[\frac{1}{[A]}\right] + \frac{1}{V_{max}}\left[1 + \frac{K_M^B}{[B]}\right]$$

In this equation, K_A^M is the composite constant for the dissociation of EA and EAB :

EA and EAB = E + EB + A

The extrapolated point of the plots of the figure represents— $1K_S^A$.

Coenzymes

Definition

All enzymes are protein molecules. Some enzymes are simple proteins whereas others are conjugated proteins. The non-protein part of a conjugated protein or enzyme is known as prosthetic group. It is observed that prosthetic group is necessary for protein to act as an enzyme and is generally termed as *coenzyme,* the protein part of a conjugated or protein as *apoenzyme* and the intact molecule or conjugated protein or *holoenzyme.* Thus,

Holoenzyme = Apoenzyme + Coenzyme

Thus, the *coenzyme may be defined as a non-proteinous substance necessary for the activity of the enzyme.*

Only when enzyme and coenzyme are present together, catalysis will occur.

It is often difficult to distinguish between what is meant by the terms "coenzymes", "prosthetic group" and "activators". Perhaps it is best to reserve the "coenzyme" for the organic fragments that are not normally isolated with the apoenzyme. A "prosthetic group" would then represent a firmly attached coenzyme such as iron protoporphyrin and "activator" might be reserved for metals, non-specific reducing agents, and so forth. At any rate, all these substances are generally of low molecular weights and are dialyzable and heat stable.

Coenzymes are organic molecules of a size intermediate between the small-molecule intermediary metabolites, which serve as the substrates of enzymatic reactions, and the macromolecular proteins. A coenzyme is an easily dissociable portion (sometimes called the prosthetic portion or group) attached to the protein component (apoenzyme) to form the complete, enzymatically active, conjugated protein (holoenzyme). Each coenzyme (or cofactor) acts usually as acceptor or donor of some specific type of atom or group of atoms to be removed from or added to a small molecule substrate in a reaction catalysed by the hole-enzyme. For example, the folic acid coenzyme accept or donate single carbon units (at various states of oxidation) in a considerable number of enzymatic reactions grouped together as single carbon unit metabolism.

It is often helpful to regard the coenzyme a second substrate i.e., a cosubstrate. This is attributed to two reasons :

(i) The chemical changes in the coenzyme exactly counter-balance those taking place in the substrate. An example is in transphosphorylation reactions involved in the metabolism of sugars, for every molecule of sugar phosphorylated, one molecule of ATP is dephosphorylated and converted to ADP.

(ii) The second reason for the importance of co-enzymes is that their reactions may be of physiological importance. For example, the importance of ability of muscle to convert anaerobically pyruvate to lactate resides neither in pyruvate nor in lactate themselves but the reaction merely serves to convert NADH to NAD^+.

Without NAD^+, glycolysis cannot occur and anaerobic ATP synthesis ceases and hence muscular work ceases.

Many coenzymes are closely related to *vitamins* and are the derivatives of vitamins. The B group vitamins (except biotin and lipoic acid) function as parts of some coenzyme, e.g., CoA is a derivative of *pantothenic acid.* When there is vitamin deficiency the coenzyme concentration decreases. Consequently enzyme function is depressed. Nucleotides may also function as coenzymes in certain metabolic reactions.

In some cases enzymes may be activated by simple substances like metal ions which are then called *activators.* The substrate forms a complex with the metal ion and then reacts with the enzyme examples of metal ions which function as cofactors are Na^+, K^+, Ca^{2+}, Mg^{2+}, Mn^{2+}, Cd^{2+}, Fe^{2+}, Cr^{3+} and Al^{3+}.

Structure of Coenzymes

Metallic ions, vitamins and other types of organic compounds have been found to function as prosthetic groups or coenzymes. As essential component of most of coenzymes is generally phosphate in the form of nucleotides. For example, nicotinamide adenine dinucleotides, also known as coenzyme I (NAD) or coenzyme II (NADP) act as coenzymes for dehydrogenases. Iron containing cytochromes make the electrons to get transferred from flavorproteins to cytochrome oxidase. Pyruvate decarboxylase is an enzyme that splits pyruvate into CO_2 and acetaldehyde in the yeast cells. It contains thiamine (vitamin B_1) as a part of its coenzyme.

Classification of Coenzymes

The coenzymes have been classified in accordance to the reactions they participate in. Such a classification with their corresponding vitamins has been summarised in table.

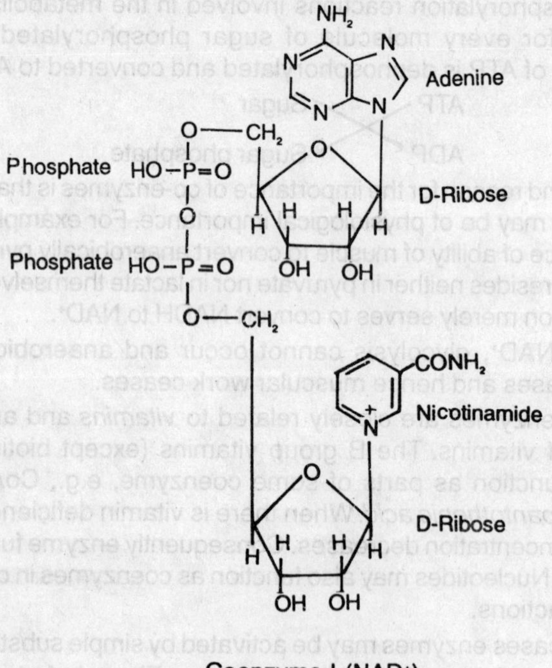

Coenzyme I (NAD^+)

Table 10.3 Classification of Major Coenzymes

Class	Abbrevation	Corresponding vitamin	Group transferred
1. *Hydrogen transferring coenzymes*			
(i) Nicotinamide adenine dinucleotide	$NAD^+/NADP$	Nicotinamide	Hydrogen, electrons
(ii) Flavin mononucleotide	FMN	Riboflavin	Hydrogen, electrons
(iii) Flavin adenine dinucleotide	FAD	Riboflavin	Hydrogen, electrons
(iv) Ubiquinone	Q	Riboflavin	Hydrogen, electrons
(v) Lipoic acid	$Lip(S_2)$	—	Hydrogen, electrons and acyl group
(vi) Cytochromes	—	—	Electrons
2. *Group transferring coenzymes*			
(i) Adenosine triphosphate	ATP	—	Phosphate, AMP
(ii) Adenosyl methionine	—	Methionine	Methyl group
(iii) Phospho adenyl sulphate	PAPs	—	Sulphate group
(iv) Pyridoxal phosphate	PALP	Pyridoxine	Amino group
(v) Cytidine diphosphate	CDP	—	Phosphoryl choline
(vi) Uridine diphosphate	UDP	—	Sugar uronic acid
(vii) Tetrahydrofolate	COF	Folate	Formyl group
(viii) Biotin	—	Biotin	Carboxyl group
(xi) Coenzyme A	CoA	Pantothenate	Acyl group
(x) Thiamine pyrophosphate	TPP	Thiamine	C_2 aldehyde group
(xi) Glutathione	—	—	Addition of thiol group
3. *isomerase and lyase coenzymes*			
(i) Uridinediphosphate	UDP	—	Sugar isomerisation
(ii) Pyridoxal phosphate	PALP	—	Decarboxylation
(iii) Thiamine pyrophosphate	TPP	Pyridoxine	Decarboxylation
(iv) B_{12} coenzyme	—	Thiamine	Carboxyl displacement
(v) Glutathione	—	Cobalamin	Isomerization of olefins

Mechanism of Action of Coenzymes

Now we will describe the chemical structure and mechanism of action of some important coenzymes.

1. Coenzyme I

It is also known as *coenzymase,* DPN (diphosphopyridine phosphate nucleotide), codehydrogenase I or NAD^+ (nicotinamide adenine dinucleotide).

Coenzyme I is mainly involved in many hydrogen transferring reactions. In these reactions, H is added to a substrate from C-4 atom of the pyridine nucleus. The pyridine gets reduced thereby retaining only two double bonds.

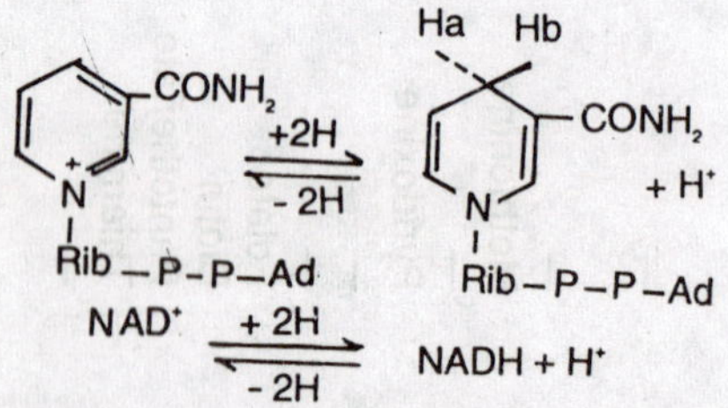

From the experimental work, it could be concluded that NAD^+ enzyme complex is usually stereospecific, only one hydrogen (H_a or H_b) is reacting exclusively. Which face of NAD^+ is attacked and which hydride ion from NADH is transferred depends upon the nature of the enzyme.

NADH absorbs light at 340 nm whereas NAD^+ shows no absorption at that wavelength. Thus, when NADH is formed during a reaction, absorption of light at 340 nm increase. The increase in absorption has been found to be proportional to the enzyme activity. This property finds use extensively in the bioassay of dehydrogenase and other oxidoreductases which employ NADH as coenzymes.

2. Coenzyme II

It is also known as *phosphocoenzymase,* TPN (triphosphopyridine nucleotide), *codehydrogenase II* or $NADP^+$ (nicotanamide adenine dinculeotide phosphate). The coenzyme has one additional phosphate group than coenzyme I molecule in position 2' of ribose molecule of adenosine.

The role of coenzyme II is similar to that of coenzyme I.

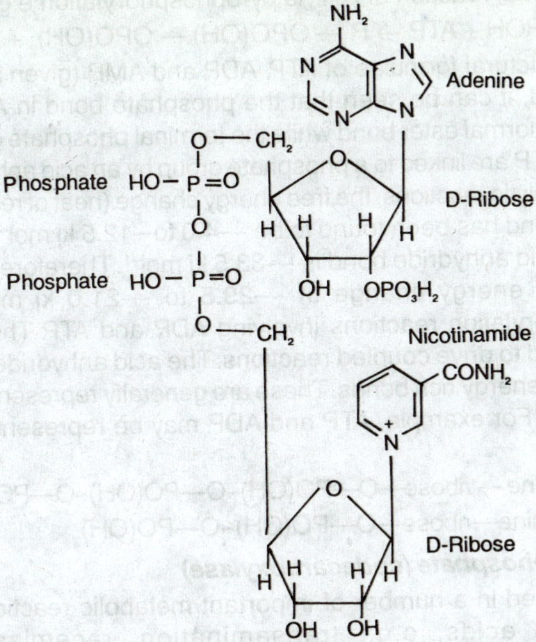

Coenzyme II

3. Adenosine monophosphate, adenosine diphosphate and adenosine triphosphate

These coenzymes are involved in transphosphorylation.

Adenosine triphosphate is a polyelectrolyte molecule having four negative charges. Since it has low molecular weight, it moves freely in cells. It acts as a mediator to receive energy from one reaction and transfers this energy to drive another reaction. ATP is synthesised in several metabolic reactions of the cell (cyclic, noncyclic phosphorylation and oxidative phosphorylation).

In order to overcome the energy barriers, the energy must be supplied. In biosynthetic processes, ATP supplies this energy in transphosphorylation reactions in the presence of a suitable enzyme and it is converted into adenosine diphosphate (ADP).

$$ROH + ATP \rightarrow R — OPO(OH)_2 + ADP$$

ADP also behaves as a phosphorylating agent and it is converted into adenosine monophosphate (AMP).

$$ROH + ADP \rightarrow R — OPO(OH)_2 + AMP$$

A less usual reaction of ATP is pyrophosphorylation e.g.,

$$ROH + ATP \rightarrow R — OPO(OH) — OPO(OH)_2 + AMP$$

If the structural formulae of ATP, ADP and AMP (given above) are inspected, it can be seen that the phosphate bond in AMP is linked by the normal ester bond while the terminal phosphate groups in ADP and ATP are linked to a phosphate group by an acid anhydride bond. In hydrolytic reactions, the free energy change (heat of reaction) of an ester bond has been found to be ~–4.0 to –12.5 kj mol^{-1} while that for the acid anhydride bond is ~–33.5 kj mol^{-1}. Therefore, there is a net free energy change of ~–29.5 to ~–21.0 kj mol^{-1} in transphosphorylation reactions involving ADP and ATP. This free energy is used to drive coupled reactions. The acid anhydride bond are known as energy rich bonds. These are generally represented by the symbol ~. For example, ATP and ADP may be represented as follows:

ATP:Adenine — ribose—O—PO(OH)~O—PO(OH)~O—PO$(OH)_2$

ADP:Adenine—ribose—O—PO(OH)~O—PO$(OH)_2$

4. *Pyridoxal phosphate (codecarboxylase)*

It is involved in a number of important metabolic reactions of the α-amino acids, e.g., transamination, racemisation, decarboxylation and elimination reactions.

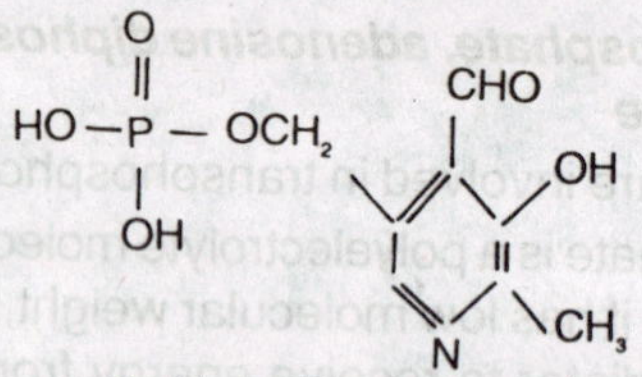

Pyridoxal Phosphate (PALP)

Role

Nearly all pyridoxal enzymes catalyse reactions of amino acids recemisation, decarboxylation, transamination, elimination of substituents and others. All the reactions can be described by the same mechanism.

Pyridoxal phosphate is attached to the apoenzyme tightly; the aldehyde group of pyridoxal reacts with amino acids to form a Schiff's base. For racemisation, there occurs a shifting of electrons in Schiff's base, thereby leading to labilisation of all bonds around the α-carbon atom of the amino acid.

Pyridoxal phosphate also acts as a coenzyme reaction between amino acids and keto acids.

$$\begin{array}{c}R\\|\\HCNH_2\\|\\COOH\end{array} + \begin{array}{c}R'\\|\\CO\\|\\COOH\end{array} \rightleftharpoons \begin{array}{c}R\\|\\CO\\|\\COOH\end{array} + \begin{array}{c}R'\\|\\CHNH_2\\|\\COOH\end{array}$$

This reaction, of great importance biologically is catalysed by pyridoxal phosphate and its mechanism is similar to racemisation.

5. Thiamine pyrophosphate (cocarboxylase)

It is involved in the transfer of acyl and carboxyl groups. It is derived from the vitamin, thiamine (vitamin B_1).

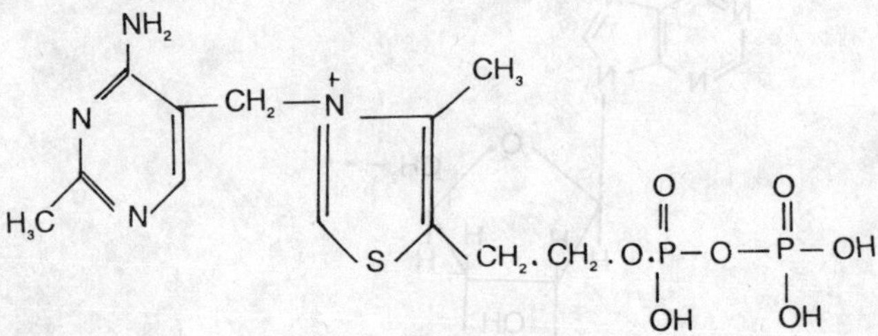

Cocarboxylase

Action

The enzyme *carboxylase* breaks down pyruvic acid into acetaldehyde in fermentation and carbohydrate metabolism. This reaction takes place in the presence of coenzyme, *cocarboxylase.*

$$\underset{\text{Pyruvic acid}}{CH_3COCOOH} \xrightarrow[\text{Cocarboxylase}]{\text{Carboxylase}} CH_3CHO + CO_2$$

The mechanism of the cocarboxylase depends upon the ionisation of the proton at C—2 in the thiazolium ring. The liability of this hydrogen atom is by its ready displacement by deuterium when thiazolium salts are dissolved in acidified deuterium oxide.

The most important reaction catalysed by TPP is the decarboxylation of pyruvic acid to acetaldehyde.

The decarboxylation of pyruvic acid by this mechanism involves a non-enzymatic process and in it there occurs transfer of electron pairs. In all other thiamine catalysed enzymic reactions also, the mechanism has been found to be almost similar. Thiamine cleaves the carbonyl-x bond, where x is generally —H or —COO^- group. A substrate-thiamine complex is later on cleaved at α-carbon atom to liberate the coenzyme.

6. Coenzyme A

This coenzyme is a complex thiol derivative and is usually written as CoA—SH but CoA is also in common usage. In contrast to coenzyme I and coenzyme II, it is not an oxidising-reducing coenzyme but it is acylating; it is an acyl transfer coenzyme and a cofactor for a wide variety of biologically acylations, e.g., the formation of acetoacetate from acetate in pigeon liver.

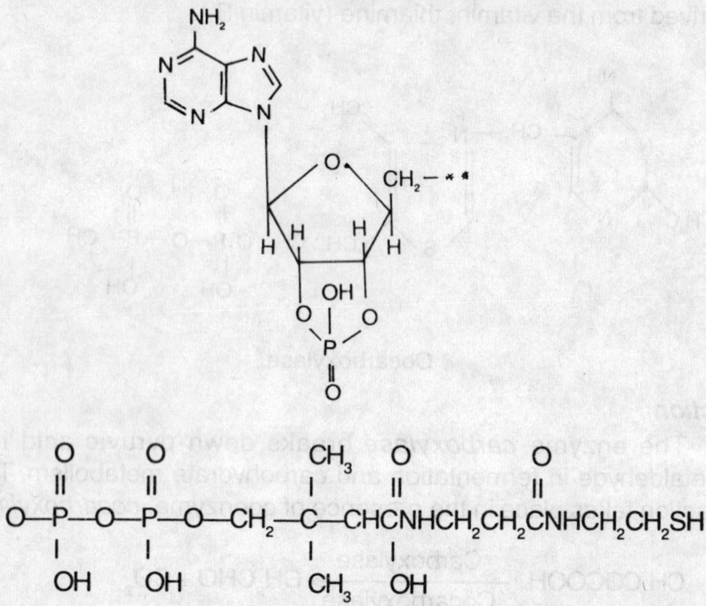

Role

Coenzyme A is required for the acetylation reactions. Coenzyme A accepts acetyl groups from one metabolite and donates them to another in the presence of specific enzymes.

The most important CoA compound is acetyl. CoA (activated acetate), which takes part in many acyl transferring reactions. In this, the acyl residue is bound to free—SH groups to form very reactive thioester.

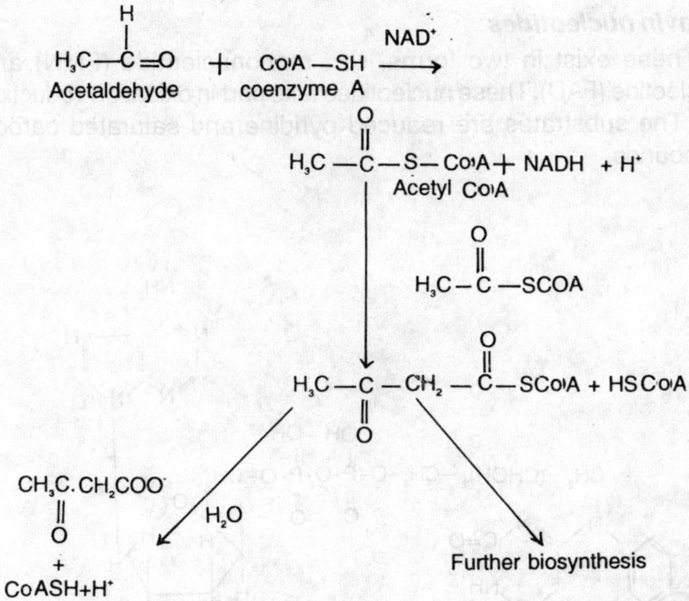

In the formation of aceto acetyl CoA from acetate thioester there occurs nucleophilic displacement at the carbon and one more acetate group gets added at the place of coenzyme A.

Besides, acylation, coenzyme A can also undergo phosphorylation.

$$\text{CoASH} + \text{X—O—}\overset{\overset{\Large O}{\|}}{\underset{\underset{\Large OH}{|}}{\text{P}}}\text{—OH} \rightarrow \text{CoA—S—}\overset{\overset{\Large O}{\|}}{\underset{\underset{\Large OH}{|}}{\text{P}}}\text{—OX} + H_2O$$

In the reactions catalysed by phosphotransacetylase, thiokinase and phosphokinase, phosphorylation of CoA occurs. Thiokinases have been found to catalyse the following reaction by involving CoA.

acetyl CoA+adenylic acid+$2H_3PO_3$→ATP+acetate+CoA—SH

Coenzyme A also takes part in the elimination reaction. For example in the formation of unsaturated fatty acids from saturated fatty acid coenzyme is involved. This needs another coenzyme FAD.

$$\text{FAD} + \text{—C—C—}\overset{\overset{\Large O}{\|}}{\text{C}}\text{—SCoA} \rightarrow \text{—}\overset{|}{\text{C}} = \overset{|}{\text{C}}\text{—}\overset{\overset{\Large O}{\|}}{\underset{\underset{\Large SCoA}{|}}{\text{C}}} + \text{FADH}_2$$

7. Flavin nucleotides

These exist in two forms, i.e., mononucleotide (FMN) and dinucleotide (FAD). These nucleotides take part in oxidation-reduction also. The substrates are reduced pyridine and saturated carbon compounds.

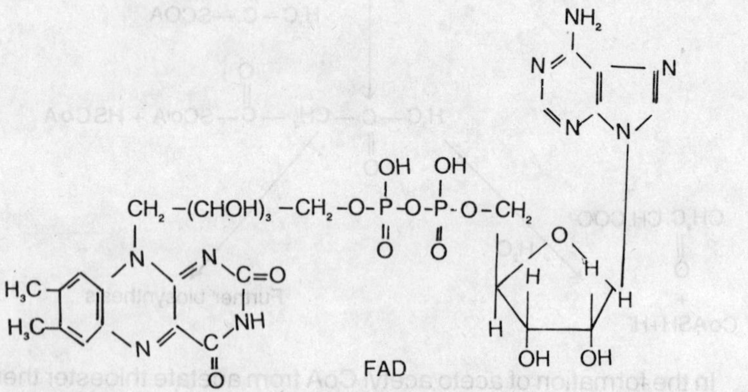

FAD

Role of FMN and FAD. Similar to NAD^+ and $NADP^+$, FMN and FAD are dehydrogenases. However, there is a distinction between NAD^+ catalysis and FMN and FAD catalysis. NAD^+ (or $NADP^+$) carries hydrogen from substrate AH_2 to substrate B with no immediate need for oxygen; FMN and FAD, on the other hand, carry hydrogen from AH_2 to oxygen directly as shown schematically by means of curved arrows:

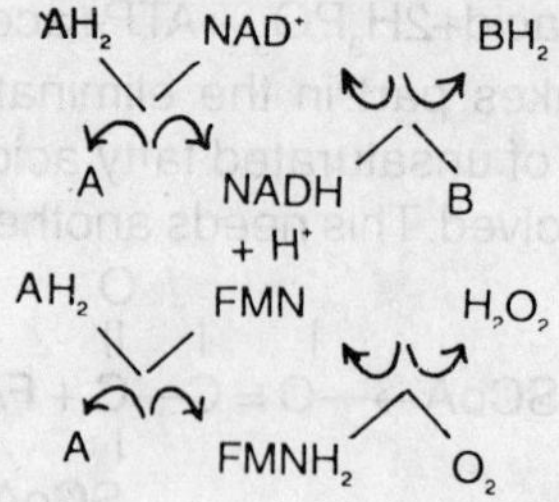

Curved arrows are used to represent the flow of hydrogen.

In FMN and FAD, the flavin is reduced. The reduction is accompanied by a loss of yellow colour. This colour is due to the system of conjugated double bonds seen in the oxidised form and lost upon reduction.

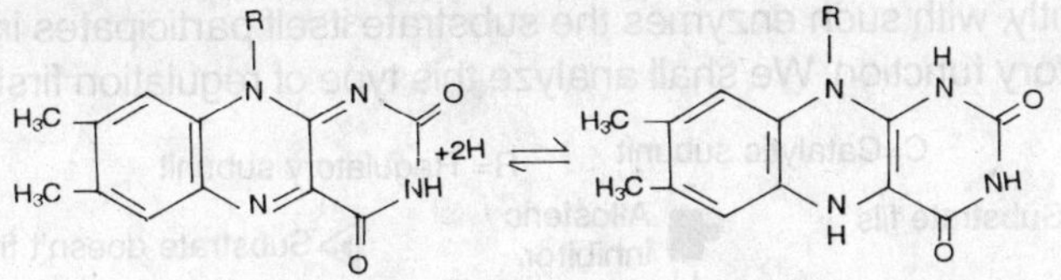

Kinetic Properties of Enzyme Systems Related to their Quaternary Structure

Allosteric Enzymes

Although most enzyme activities have apparent temperature optima when measured in vitro, the isothermal conditions characteristic of higher forms of life would preclude any major contribution of temperature-dependent rate changes in the control of enzyme action. Slight changes in $[H^+]$ in the environment of an enzyme, on the other hand, can have significant effect on activity. Often, the $[H^+]$-activity profile of an enzyme has the following shape:

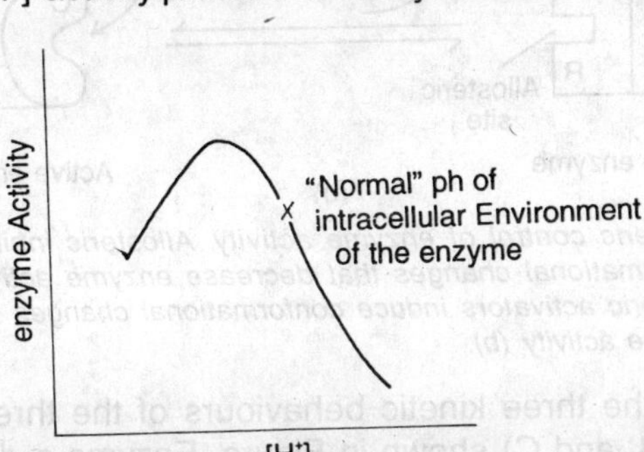

If the intracellular environment of the enzyme were on either the up-slope or down-slope of the $[H^+]$-activity curve, it is reasonable that the activity could be sensitive to changes in $[H^+]$. Because they are macromolecular electrolytes, enzymes are also affected by ions such as K^+ and Mg^{2+}. Such influences can contribute to cellular control of an enzyme's activity and, therefore, are relevant to metabolic regulation.

A more specific mode of regulation of catalytic activity is achieved with enzymes that catalyze the first and the irreversible step in a multireaction sequence. These enzymes are oligomeric and respond extremely rapidly to either inhibitory or stimulatory modulators. Frequently, with such enzymes the substrate itself participates in the modulatory function. We shall analyze this type of regulation first.

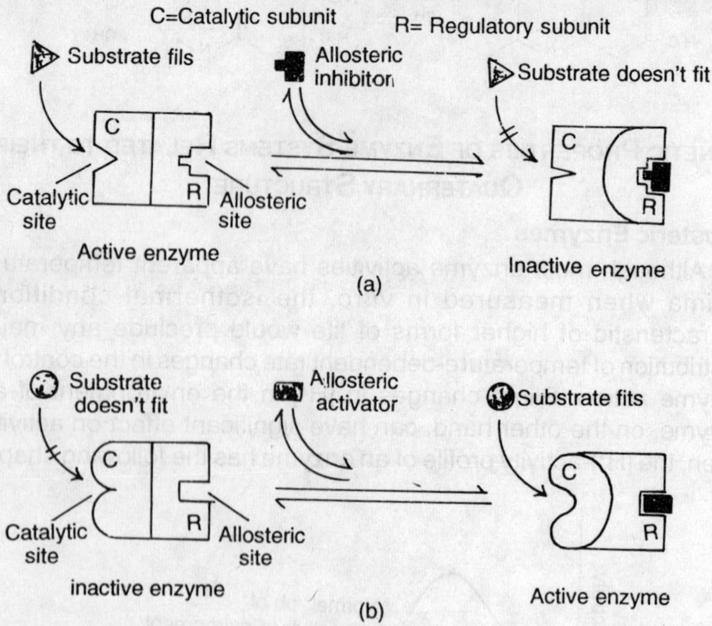

Fig. 10.24. Allosteric control of enzyme activity. Allosteric inhibitors induce conformational changes that decrease enzyme activity (a) while allosteric activators induce conformational changes that increase enzyme activity (b).

Consider the three kinetic behaviours of the three different enzymes (A, B, and C) shown in Figure. Enzyme a displays the Michaelis-Menten relationship among [S], v_0 and V_{max}. B does not give the rectangular hyperbolic plot of v_0 versus [S] but instead shows a sigmoid curve relationship between these two variables. In the case of C, the plot of v_0 versus [S] is a flattened hyperbola and the attainment of V_{max} requires that [S] be many times greater than in case A. It will be assumed that the value of V_{max} is the same for all three enzymes. Comparing enzymes A and B, it is seen that in the steep portion of the sigmoid curve, the change in v_0 as a function of increasing [S] is many times greater than for the corresponding part of the hyperbolic curve. One conclusion that can be drawn is that when enzyme B begins to bind its substrate, combination with additional substrate is enhanced. The consequence of this facilitation of substrate binding is a marked acceleration in the v_0 of the enzyme reaction as evidenced by the sharp upturn in the sigmoid plot. This effect has been termed *positive cooperativity.* In the plot of v_0 versus [S] for C, it is apparent that overall, the reaction rate with this enzyme is not as sensitive to small increases in [S] as in the case of A or B. V_{max} is approached much more slowly and at a much higher [S] than with A and B. These data are consistent with a *negative cooperativity* effect in which the binding of one substrate molecule makes the union and additional substrate more difficult.

As pointed out above, the double-reciprocal plots ($1/v_0$ versus 1[S]) that are useful for analyzing Michaelis-Menten kinetics are not applicable to enzyme systems showing either positive or negative cooperative behaviour. For example, the transposition of the curves to the reciprocal form would give the following plots:

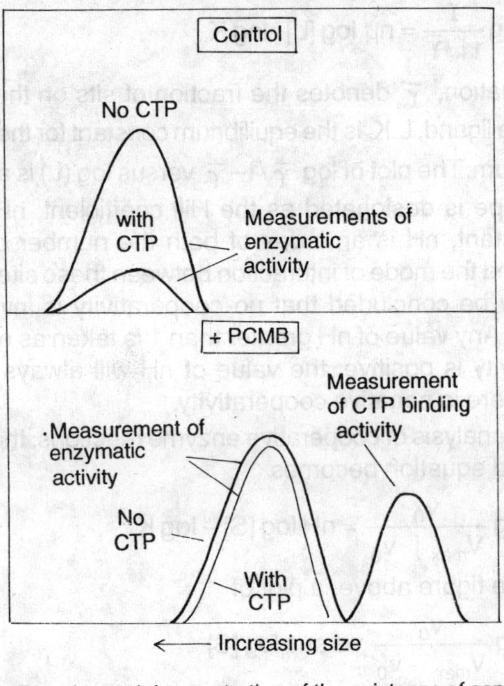

Fig. 10.25. Experimental demonstration of the existence of separate catalytic and regulatory subunits in the allosteric enzyme aspartate transcarbamoylase.

The most direct treatment of kinetic data such as those illustrated by curves (B) and (C) makes use of a logarithmic relationship, as shown below :

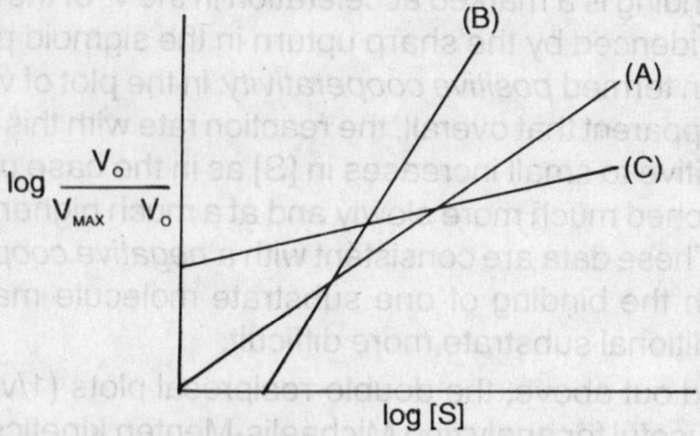

These plots are modeled after the *hill plot.* Hill equation, which has been useful in describing the relationship between ligand binding by an oligomeric protein, can be formulated as :

$$\log \frac{\overline{\Upsilon}}{1\,i\,\overline{\Upsilon}} = n_H \log [L]^{-} \log K$$

In this equation, $\overline{\Upsilon}$ denotes the fraction of sits on the protein occupied by the ligand, L.K. is the equilibrium constant for the protein-ligand equilibrium. The plot of log $\overline{\Upsilon}/1-\overline{\Upsilon}$ versus log (L) is a straight line whose slope is designated as the Hill coefficient, nH. As an empirical constant, nH is an index of both the number of ligand binding sites and the mode of interaction between these sites. When nH = 1, it may be concluded that no cooperativity is involved in ligand binding. Any value of nH greater than 1 is taken as evidence that cooperativity is positive; the value of nH will always be less than 1 when there is negative cooperativity.

For kinetic analysis of cooperative enzyme reactions, the analog of the Hill ligand equation becomes

$$\log \frac{v_0}{V_{max}\ j\ v_0} = nH \log [S] - \log K$$

As shown in the figure above, a plot of

$$\log \frac{v_0}{V_{max} - v_0} \text{ versus } \log [S]$$

will give a straight line whose slope is the value of nH. In applying the Hill equation to enzyme kinetic data, two assumptions are made

: V_{max} is attained when all of the substrate binding sites are occupied and the rate-limiting step in the entire reaction sequence is the binding of S by the enzyme.

The kinetic data presented in the figure illustrate how the activities of allosteric enzymes can be modulated or regulated solely by their substrates and that the degree of modulation is determined by the number of the multiple-substrate binding sites occupied. Allosteric enzymes can also be modulated by molecules other than their substrates and such regulation may be expressed either as stimulation or as an inhibition of activity. Their ability to respond to modulator molecules other than their substrates is the central reason for calling these enzymes *allosteric*. The term "allosteric" means "another place or structure" and aptly describes these regulatory enzymes which have been shown to have modulator-binding sites, in addition to the catalytic substrate-binding sites. Like the substrate-binding site(s), the allosteric site(s) can exhibit a high degree of specificity in binding a modulator molecule.

Modulators be positive (stimulatory) or negative (inhibitory) and the response of the affected enzymes will be reflected in the plots of v_0 versus [S] as shown in the figure. Of two types of allosteric

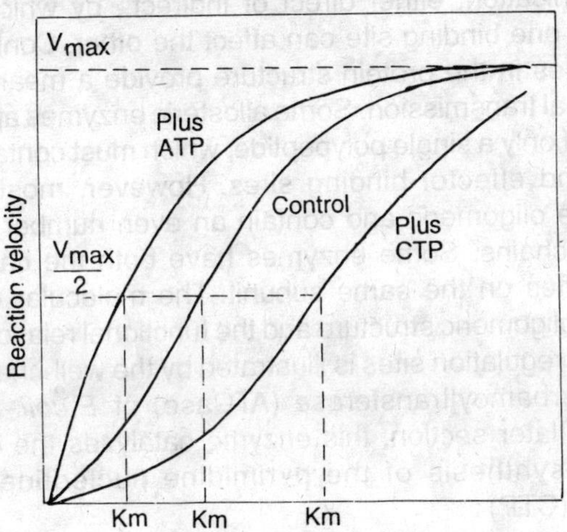

Fig. 10.26. Allosteric regulation of aspartate transcarbamoylase. The allosteric inhibitor CTP increases the K_m of the reaction while the allosteric activator decreases the K_m. The sigmoidal shape of the curve obtained for the control enzyme indicates the existence of cooperativity.

enzymes described here, (a) is one in which S binding is enhanced by positive cooperativity, and (b) is an enzyme in which negative cooperativity hinders S binding. In contrast to enzyme reactions showing Michaelis-Menten kinetic relationship between v_0 and [S], the K_M values for alloseric enzymes must be labeled as "apparent K_M" values and are equated to the [S] values giving half-maximal velocity. The positive and negative modulator effects on the apparent K_M and V_{max} values of the two enzymes described in the figure, are summerized by the following generalizations.

Type of enzyme	Effect of Stimulator, or (+) V_{max}	Modulator on : Apparent K_M	Effect of Inhibitory or (–) V_{max}	Modulator on : Apparent K_M
S-binding(+) cooperative	Unchanged	Decreased	Unchanged	Increased
S-binding (–) cooperative	Increased	Unchanged	Decreased	Unchanged

With either negative or positive modulator or enzyme activity, the catalytic and regulator sites must have some facility for intercommunication, either direct or indirect,, by which changes occurring in one binding site can affect the other. Conformational steric changes in the protein structure provide a means for such intersite signal transmission. Some allosteric enzymes appear to be comprised of only a single polypeptide, which must contain both the substrate-and effector-binding sites. However, most allosteric enzymes are oligomeric and contain an even number of subunit polypeptide chains. Some enzymes have both the catalytic and regulatory sites on the same subunit. The molecular complexity typical of an oligomeric structure and the functional relation between catalytic and regulation sites is illustrated by the well-characterized aspartate carbamoyltransferase (ATCase) of *E.coii.* As will be detailed in a later section, this enzyme catalyzes the committed step in the synthesis of the pyrimidine nucleotide, cytidine triphosphate (CTP) :

$$\text{L-Asparate + carbomoyl phosphate} \xrightarrow{\text{ATCase}} \text{carbamoyl}$$

$$\text{aspartate} \xrightarrow{\text{seven step}} \!\!\to\!\!\to\!\!\to\!\!\to\!\!\to\!\!\to \text{CTP} \to\!\!\to\!\!\to \text{nucleic acids}$$

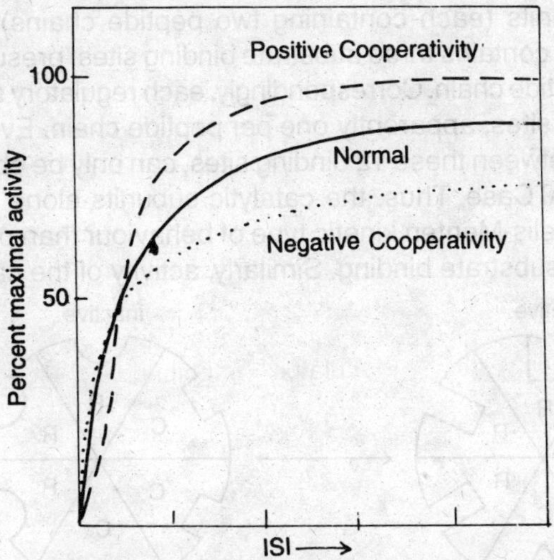

Fig. 10.27. Effects of negative and positive cooperativity on enzyme kinetics. With positive cooperativity, enzyme activity is more sensitive to changes in substrate concentration and achieves maximal velocity at lower substrate concentrations than in the absence of cooperativity. Negative cooperativity achieves the opposite effect.

The allosteric character of ATCase is illustrated by the kinetic data. As with any enzymatic reaction involving two substrates, the initial rates shown here be measured under conditions in which the concentration of only one of the substrates is made the variable. In such analyses, the ATCase is saturated with carbamoyl phosphate and v_0 thus becomes dependent on the aspartate concentration.

As ATCase is an enzyme that exhibits a positive cooperativity in the binding of substrate. This enzyme also presents a model system for the demonstration of both positive and negative modulation. In the presence of ATP, a positive modulator of ATCase, the apparent K_M is decreased and the initial rates are higher than in the absence of ATP. The converse is found when CTP is added to the system. Acting as a typical negative modulator, the CTP can bind to the enzymes and causes a decrease in activity and an increase in the apparent K_M.

The kinetic behaviour of ATCase has been found to be sensitive to the quaternary structure of the enzyme. The *E.coli* enzyme for example, can be dissociated into two identical catalytic subunits (each containing three polypeptide chains) and three identical regulatory subunits (each containing two peptide chains). Each catalytic subunit contains three substrate binding sites, presumably one on each peptide chain. Correspondingly, each regulatory subunit has two binding sites, apparently one per peptide chain. Evidence for interaction between these 12 binding sites, can only be obtained with the intact ATCase. Thus, the catalytic subunits alone exhibit more of a Michaelis-Menten kinetic type of behaviour than positive cooperativity in substrate binding. Similarly, activity of the catalytic

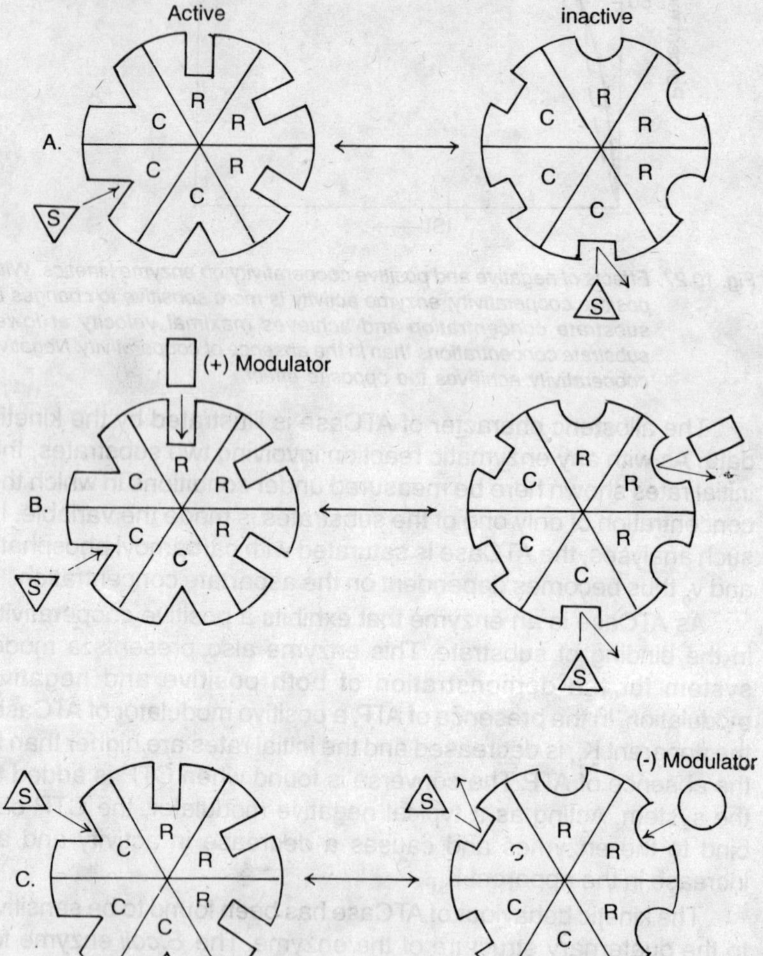

Fig. 10.28. Conformational states of allosteric enzymes.

subunits is not affected by the addition of CTP. Modulatory control by CTP is only observed when the catalytic and regulatory subunits are physically associated as permitted by the oligomeric form of the enzyme.

All models proposed for relating kinetic behaviour and macromolecular structure of oligomeric allosteric enzymes assume that both the catalytic and regulatory subunits can have two conformational states in equilibrium mixture : one, which is inactive, with a low affinity for S, and the other active, with a high affinity for S. The scheme given in figure illustrates the interconversion of these forms in positive cooperative binding of S and allosteric regulation by positive and negative modulators. As shown in figure when substrate combines with the catalytic site, the enzyme is "locked" into the active form, causing a shift in equilibrium of the two states as well as facilitating union of S with the remaining site on active forms. This model also describes the role of the regulatory subunits in the modulation of activity of the oligomeric enzyme. The binding site on the regulatory subunit of the activity form of the enzyme has the proper configuration for binding a positive modulator. The direct consequence of this binding is to facilitate locking of the enzyme in the active form. In contrast to the accelerating effect of a positive modulator, binding of an inhibitory modulator by the regulatory subunits is favoured by the enzyme's inactive form and, therefore, will stabilize this conformation. The initial rate of the enzyme reaction is inhibited accordingly.

The contribution of enzymes to metabolic regulation will be treated in a broader context is a later, section. In the present discussion devoted primarily to the molecular basis of such regulatory mechanisms, the principal focus has been on the peculiar changes in conformation of some enzymes and how such changes have the potential for fine turning of metabolic controls. The inextricable relationship between structure and function that is so beautifully illustrated by the allosteric enzymes can also be expressed in other ways. Two additional examples of nature's versatility in enzymatic regulation are (1) the enzymes whose activity depends on covalent modification and (2) the isozymes.

Enzymes that can be Modified Covalently

Covalent modification may take several forms. In the case of the less active form of phosphorylase (phosphorylase *b*), the enzymes responsible for catalyzing the conversion of glycogen to

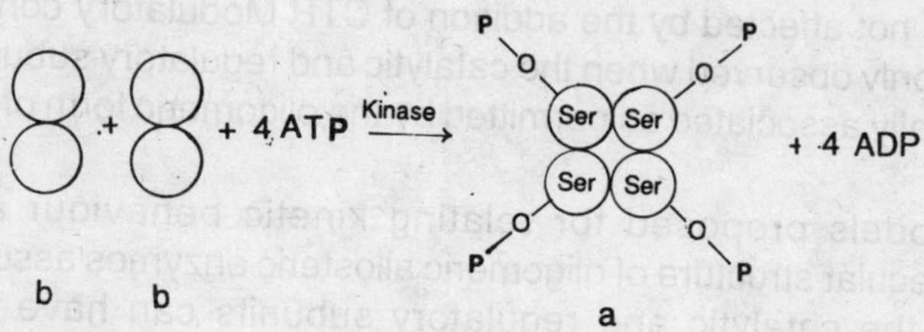

glucose-1-phosphate, serine residues are phosphorylated enzymatically to convert the enzyme *b* to its active oligomer, enzyme *a* : Another example of covalent modification is the formation of hydrolytic enzymes from inactive precursors. These enzymes are essential for digestion of proteins in the gastrointestinal tract and are synthesized and secreted in catalytically inactive forms designated as zymogens. The hydrolytic rupture of specific peptide bonds in the zymogens renders them enzymatically active:

$$\text{Zymogen} \xrightarrow[\text{enzymatic}]{\text{HOH}} \text{active enzyme + smaller peptide(s)}$$

The details of this activation process and its physiologic implications for digestion are treated later.

Isozymes

First recognized by the powerful analytical tool of electrophoresis, the isozymes provide another example of low biologic processes can endowed with sensitive controls because of the functional variations which become possible when polypeptide chains or subunits of an oligomeric enzyme are harmoniously associated. Lactase dehydrogenase, the terminal catalyst in glycolytic conversion of glucose to lactate, is such as enzyme.

Characterization of the lactate dehydrogenase of rat tissues has demonstrated that the enzyme can occur in as many as five isozymic forms. Each of these enzymes contains four polypeptide chains. These polypeptide subunits have been shown to be of only two kinds and are commonly labelled as the H (for heart) and the M (for muscle) chains. The single chains are inactive. The five tetrameric isozymes of lactate dehydrogenase are the five different combinations of these subunits :

HHHH HHHM HHMM HMMM MMMM

The K_M and V_{max} values and the degree of product inhibition of the homologous tetramers differ significantiy. The advantages of these differences in kinetic properties of the lactate dehydrogenase isozymes are described in a later section which treats regulation of metabolic activity in the heart and skeletal muscle.

Table 10.4 Characteristics of lactate dehydrogenase isozymes

Characteristics	M_4	H_4
K_M for pyruvate	Low	High
V_{max} for pyruvate → lactate	High	Low
Inhibition by pyruvate of lactate → pyruvate	Weak	Strong

11

Nucleic Acid (Deoxyribonucleic Acid)

The fundamental investigations which led to the discovery of the nucleic acids were made by *Fredrich Micscher* (1844-95) who may be regarded as the founder of our knowledge of the chemistry of the cell nucleus. In early work carried out in 1868, in the laboratory of Hoppe Seyler in Tubingen, he isolated the nuclei from pus cells obtained from discarded surgical bandages and showed the presence in them of an unusual phosphorus containing compound that he called '*nuclein*' and which we now know to have been nucleoprotein. It was subsequently shown that nucleic acids were normal constitutents of all cells and tissues and Miescher's investigations of the nucleic acid were continued by *Altmann,* who is 1889 described as method for the preparation of protein free nucleic acids from animal tissues and from yeast. The term *nucleic acid* was introduced by *Altmann. Albrecht Kossel, P.A. Levine* and *Walter Jones* described the chemical nature of nucleic acid. According to them the nucleic acid si composed of phosphoric acid, a sugar and nitrogenous bases. *Franklin W. Stahl* presented first evidence that nucleic acid forms the genetic material. *Chargaff* (1961) described the occurrence of nitrogenous bases in equal proportion. *Dotty* (1961) has emphasized much on the physical properties of DNA. In 1962 *Willkins, Wlatson* and *Crick* proposed a model to explain the double helicular structure of DNA. They also shared the Nobel prize for the same.

Deoxyribonucleic Acid or DNA

DNA is found in the cells of all living organisms except plant viruses, where RNA forms the genetic material and DNA is absent. In bacteriophages and viruses there is a single molecule of DNA, which remains coiled and is enclosed in the protein coat. In bacteria, mitochondria and plastids of eukaryotic cells DNA is circular and lies naked in the cytoplasm. In the nuclei of eukaryotic cells DNA occurs in the form of long spirally coiled and unbranched threads. The number of DNA molecules is equivalent to the number of chromosomes per cell. In them DNA is found in combination with proteins forming nucleoproteins—the *chromatin* material and is enclosed in the nucleus.

DNA is a polymer, the monomeres of which are the deoxyribonucleoside monophosphates. Most DNA have high molecular weight, varying from 10^6 to 10^9 or more. Chemical analysis of DNA molecules from different kinds of orgnaisms suggest that in one kind of DNA the number of adenine (A) nucleotides was always the same as the number of thymine (T) nucleotides. Similarly the number of guanine (G) nucleotide was always the same as the number of cytosine (C) nucleotides. This evidence suggest that nucleotides occur in pairs in the DNA molecule. Further, quantity of A present in one organisms's DNA (and it varies from species to species) is equal precisely to the amount of T. Similarly the amount of G present is always equal to the amount of C.

Physical Structure of DNA

According to the model proposed by *Watson* and *Crick,* the DNA contains two polynucleotide chains running antiparallel and twisted around each other in the form of a regular double helix. The diameter of the helix is about 20Å and each chain makes a complete turn every 34Å. Each chain is polynucleotide in which the sugar of each nucleotide is linked by a phosphate group to the sugar of the adjacent nucleotide. There are 10 nucleotides on each chain in every turn of the helix. The two helices are coiled in such a way that they cannot be separated without unwinding. The two chains of helix are of opposite polarity. If one chain runs 3'-5' direction, then the other will run as 5'-5' direction.

The molecular weight for purified DNA is found in the range 1000,000 to 120 x 10^6. The amount of DNA in a mammalian cell corresponds to about 25,000 molecules of molecular weight 120 x 10^6, a total of about 10,000 million nucleotides.

Monomeric Components of DNA

Before any account is given of the structure of DNA proper it is necessary to describe the structure of its component parts. The DNA molecule is a polymer consisting of several thousand pairs of nucleotide monomeres. Each nucleotide consists of the pentose sugar, *deoxyribose, a phosphate* group and a *nitrogenous base,* which may be either a *purine* or a *pyrimidine.*

1. Deoxyribose sugar

Levine (1909) identified ribose sugar as a main constituent of nucleic acid structure. *Mori* (1929) isolated sugar from the guanine nucleoside and showed that it is deoxyribose sugar. It is pentose type which in DNA molecule shows the absence of one oxygen molecule from carbon-2 position of ribose sugar. Both deoxyribose and ribose (pentose sugars of nucleic acids) have a pentagonal ring with five carbons, among which two (i.e., 3' and 5') are attached to phosphoric acid and three (1') to the base. This sugar is called *deoxyribose* and it simply acts as a support column to which bases are attached. When the pentose sugars occur in nucleic acids nucleotides the carbon atoms an numbered as 1', 2', 3' etc. to avoid confusion with the numbering of the ring atoms of the bases.

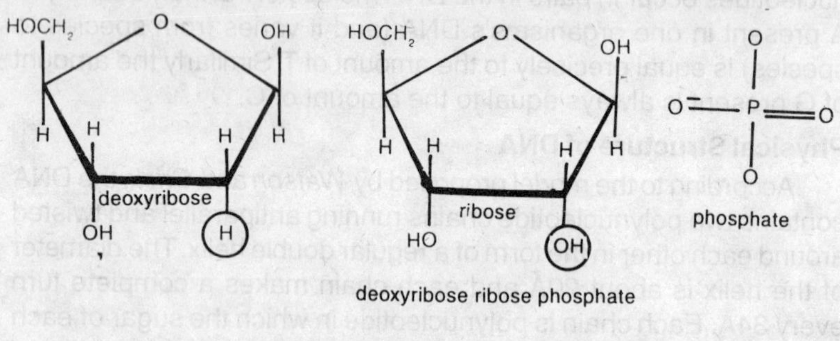

Fig. 11.1. Pentose sugars and phosphate.

2. Phosphate

In the DNA strand the phosphate groups alternate with deoxyribose. Each phosphate group is joined to carbon atom 3' of one deoxyribose and to carbon atom 5' of another. Thus each strand has a 3' end and a 5' end. The two strands are oriented in opposite directions. The 3' end of one strand corresponds to the 5' end of the other. Consequently the oxygen atoms of deoxyribose point in opposite directions in the two strands. *Phosphodiester bonds* are formed between the sugars of two different nucleotides and a phosphate group.

3. Nitrogenous bases

The nitrogenous bases of nucleic acid are of two types : *Purine* and *Pyrimidine.* Purine bases comprise mainly adenine (A) and guanine (G) which are common to both DNA and RNA while pyrimidine bases comprise cytosine (C) and thymine (T) in DNA and in the case of RNA, thymine (T) is replaced by uracil (U).

Table. 11.1. Four Nitrogen Bases, Nucleosides and Nucleotides of DNA Molecule

Nitrogen base	*Base + deoxyribose =deoxyribonucleoside*	*Deoxyribonucleoside + Phosphoric acid = Deoxyribonucleotide*	*Abbreviation for nucleotide*
1. Adenine (A)	Deoxyadenosine	Deoxyadenylic acid (Deoxyadenosine monophosphate)	3'-dAMP
2. Guanine (G)	Deoxyguanosine	Deoxyguanylic acid (Deoxyguanosine monophosphate)	3'dGMP
3. Cytosine (C)	Deoxycytidine	Deoxycytidylic acid (Deoxycytidine monophosphate)	5'-dCMP
4. Thymine (T)	Deoxyhymidine	Deoxy Thymidylic acid (Deoxythymidine monophosphate)	5'-dTMP

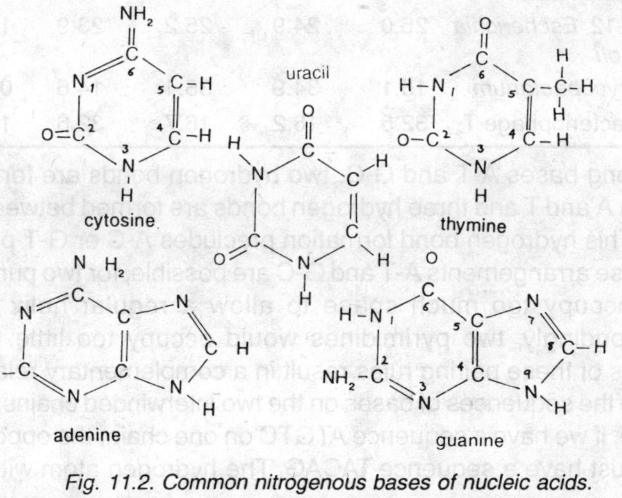

Fig. 11.2. Common nitrogenous bases of nucleic acids.

These bases are arranged linearly as the back bones of chains which (back bones) are composed of alternating sugar and phosphate. The bases have specific way of combining, if C is found in one chain G will be opposite to it and in other if A is on one T must be on the other. Or in other words adenine (A) is always paired with thymine (T) and guanine (G) with cytosine (C).

Table 11.2. Showing Amount of Various bases in Different Tissues

S.No.	*Source*	*Adenine*	*Guanine*	*Cytosine*	*Thymine*	$\frac{A+T}{G+C}$
1	2	3	4	5	6	7
1.	Human Liver	30.3	19.5	19.9	30.3	1.53
2.	Human sperm	30.7	19.3	18.8	31.2	1.62
3.	Hen red cells	28.8	20.5	21.5	29.2	1.38
4.	Rat bone marrow	28.6	21.4	21.5	28.4	1.33
5.	Herring sperm	27.8	22.2	22.6	27.5	1.23
6.	*Paracentrotus lividus* (sea urchin) sperm	32.8	17.7	18.4	32.1	1.85
7.	Salmon	29.7	20.8	20.4	29.1	1.43
8.	Wheat germ	26.5	23.5	23.0	27.0	1.19
9.	Yeast	31.3	18.7	17.1	32.9	1.79
10.	*Diplococcus pneumonial*	29.8	20.5	18.0	31.6	1.59
11.	K-12 *Escherichia Coli*	26.0	24.9	25.2	23.9	1.00
12.	*Trycobactarium*	15.1	34.9	35.4	14.6	0.42
13.	Bacteriophage T_2	32.5	18.2	16.7	32.6	1.86

Among bases A-T and G-C, two hydrogen bonds are formed between A and T and three hydrogen bonds are formed between C and G. This hydrogen bond formation precludes A-C or G-T pairs. Only these arrangements A-T and G-C are possible, for two purines would occupy too much space to allow a regular helix and correspondingly, two pyrimidines would occupy too little. The stickness of these pairing rules result in a complementary relation between the sequences of bases on the two interwinded chains. For example, if we have a sequence ATGTC on one chain, the opposite chain must have a sequence TACAG. The hydrogen atom with its positive charge is shared between an oxygen atom and a nitrogen atom, both with slight negative charges. Although hydrogen bonds are weak, the fact that there are so many gives stability to the DNA molecule. The weak hydrogen bonding enables the two strands of the DNA to separate during replication.

The bases are joined to the pentoses by N-C *glycosidic bonds.* For the purine, the glycosidic bond is between the C_1 position of the pentose sugar and the N_9 position of the bases. For the pyrimidines the linkage joints the C_1 and N_3 positions.

Nucleosides

When a purine or a pyrimdine base is linked to ribose (in case of RNA) or deoxyribose (in case of DNA) the resulting compound is known as a *nucleoside.* Thus adenine condenses with ribose to form the nucleoside adenosine, guanine forms guanosine, cytosine forms

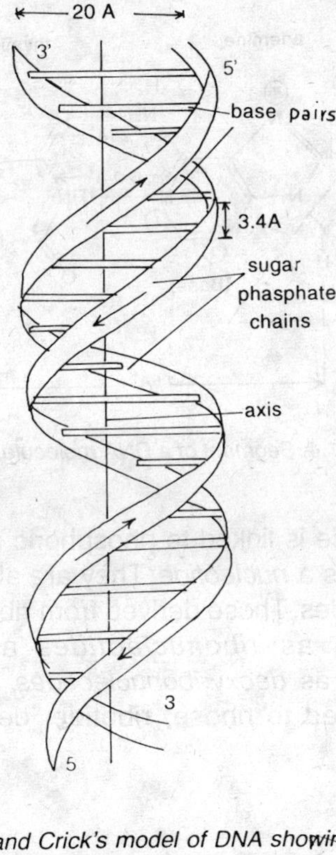

Fig. 11.3. Watson and Crick's model of DNA showing double helical structure.

cytidine and uracil form uridine. These ribonucleosides can be formed on partial hydrolysis of RNA. The ribonucleoside from hypoxathnine is named inosine. The nucleosides derived from 2-deoxyribose are known as deoxyribonucleosides-deoxyadenosiné, deoxyguanosine, deoxycytidine, deoxythymidine and so on.

In addition to the nucleosides listed above several others are found in very small amounts in certain classes of nucleic acids.

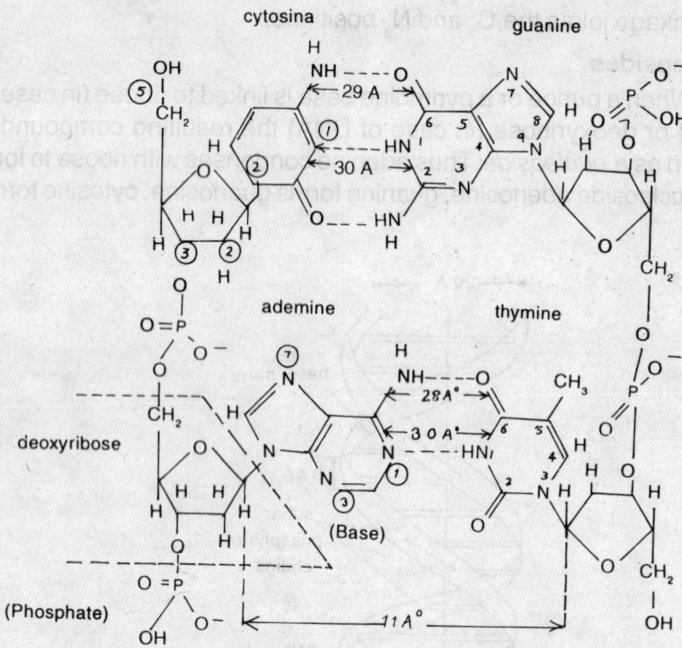

Fig. 11.4. Segment of a DNA molecule.

Nucleotides

When a nucleoside is linked to phosphoric acid the resulting compound is known as a *nucleotide.* They are all phosphoric acid esters of the nucleosides. Those derived from ribonucleosides are usually referred to as *ribonucleotides* and those from deoxyribonucleosides as *deoxyribonucleotides.* These terms are sometimes abbreviated to ribose, ribotide, deoxyriboside and deoxyribotide

Table 11.3. The Relative (G +C) Content of DNA's from Various Sources

Source of DNA	*Per cent (G + C)*
Dictyostelium (slime mould)	22
M. pyogenes	34
Bacillus cereus	37
B. megaterium	38
Haemophilus influenzae	39
Saccharomyces cerevisiae	39
Calf thymus	40
Rat liver	40
Bull sperm	41
Diplococcus pneumoniae	42
Chicken liver	43
Mouse spleen	44
Salmon sperm	44
B. subtilis	44
T_1 *bacteriophage*	46
Esch. coli	51
T_3 *bacteriophage*	53
Neurospora crassa	54
Pseudomonas aeruginosa	68
Micrococcus lysodeikticus	72
Herpes simplex virus	72
Mycobacterium phlei	73

The Primary Structure of DNA

As stated already that nucleic acids are polymers made up of hundreds, thousands of millions of nucleotides coupled together by *phosphodiester linkages.* In the case of DNA where C-4' in the sugar is occupied in ring formation and C-2' carries no hydroxyl group, only the hydroxyl groups at positions 3' and 5' are available for internucleotide linkages. The absence of a hydroxyl group at C-2' makes cyclic phosphate formation impossible and DNA is therefore not hydrolysed by alkali in the same way as in RNA. The primary structure of the polynucleotide chain in DNA is shown in the following fig.

The representation of polynucleotide chains by complete formulae is clumsy and it has become customary to use the schematic system illustrated in the above figure, where the chain is abbreviated as on

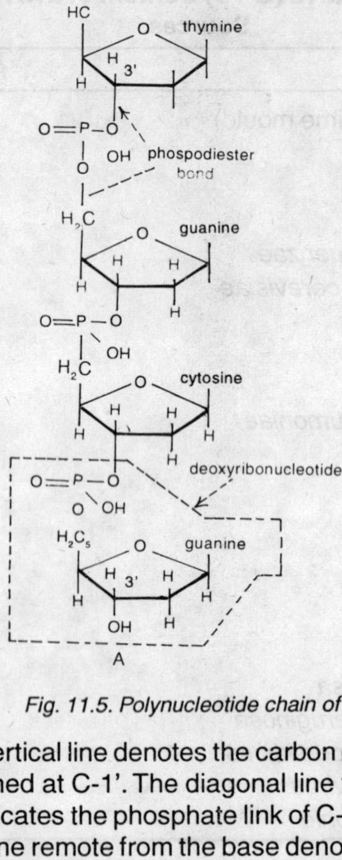

Fig. 11.5. Polynucleotide chain of DNA.

the right. The vertical line denotes the carbon chain of the sugar with the base attached at C-1'. The diagonal line from the middle of the vertical line indicates the phosphate link of C-3' while that at the end of the vertical line remote from the base denotes the phosphate link at C-5'. This system may be used for either RNA or DNA.

Molecular Weight of DNA

The molecular weight of DNA molecules are very difficult to determine accurately by the methods of classic chemistry since they range from 10^6 to more than 10^{10}. Conventional analytical equilibrium ultracentrifugation is unsatisfactory since the available instruments are not stable at speeds low enough to balance centrifugal forces against diffusion forces. The best absolute method of DNA molecular weight determination are light scattering on low angle instruments equilibrium analytical ultracentrifugation in Caesium chioride gradient and viscoelastic relaxation. Most other methods have an empirical bases and rely on the few light scattering experiments which have been performed for their calibration. These more empirical methods include the measurement of intrinsic viscosity, sedimentation rate, electron microscopy, electrophoresis and radio autography. Generally a combination of two or more of these methods can be used. The advent of the laser has led to the development of a new technique for determining diffusion coefficients of large molecules from the Doppler shift in the wavelength of the scattered laser night. This technique together with conventional analytic velocity centrifugation to determine sedimentation coefficients should lead to reliable absolute molecular weight determinations. The determination of the molecular weight of DNA has also been complicated by the difficulties experienced in the preparation of whole DNA molecules since high molecular weight DNA is very susceptible to hydrodynamic shearing forces and to the action of contaminating nucleases. Thus all molecular weight determinations on an unknown DNA must be carried out in nuclease free conditions, without pipetting or any other manipulation which puts shear stress on the DNA, any methods which might themselves shear the DNA, such as the viscosity determinations, must be used with caution. The size of large DNA molecules from individual chromosomes of eukaryotic cells has recently been determined by viscoelastic relaxation, and if appear that in *Drosophila* at least the DNA is contained in one complete molecule per chromosomes.

Table 11.4. DNA Molecular Weights

Source	*Mol. wt*	*Length*	*Number of nucleotide pairs*	*Conformation*
E.coli chromosome	2×10^9	1 mm	3×10^6	Cyclic, duplex
H. influenzae chromosome	8×10^{-3}	300 μm	12×10^5	—
Mycoplasma PPLO strain H-39	4×10^8	150 μm	6×10^5	—
Bacteriophage T2 or T4	1.3×10^3	50 μm	2×10^5	Linear, duplex
Bacteriophage λ	33×10^6	13 μm	0.5×10^5	Linear, duplex
Bacteriophage φX174	1.6×10^6	0.6* μ	—	Cyclic, single-stranded
Polyoma virus	3×10^6	1.1 μm	4.6×10^3	Cyclic, duplex
Drosophila melanogaster chromosome	43×10^9	20 mm	65×10^6	—

In more recent years if has become customary to describe the size of DNA molecules in terms of kilobase pairs (kbp), rather than daltons or molecular weight. A molecular weight of one million corresponds approximately to 1.5 kilobase pairs.

Secondary Structure of DNA

Using early information obtained by X-ray diffraction and chemical observations, *Watson* and *Crick* in 1953 put forward the view that the DNA molecule is *double stranded* and in the form of a right handed helix with the two polynucleotide chains would round the same axis and held together by hydrogen bonds between the bases. By using scale models they were able to show that the bases couuld fit in if they were arranged in pairs of one pyrimidine and the purine. The equivalence of adenine and thymine, and guanine and cytosine in most naturally occuring DNA molecules. First absorved by *Chargaff* suggested that the most likely hydrogen bonding configuration of base pairs of these two types. These base pairing arrangement have been confirmed as the only ones possible in the B-form of DNA. Although other base pairing arrangements have since been suggested they have only been shown to occur in RNA where they are involved in secondary and tertiary structure stabilization.

The most important consequence of the base pairing configuration formed in DNA is that the order in which the bases occur in one chain automatically determines the order in which they occur in the other, *complementary chain.* Apart from this essential condition, there are no restrictions on the sequence of the bases along the chains.

The shorthand method of describing two DNA strands has been to have the strand of 5'→3' polarity on the top line of the sequence with the complementary strand of opposite polarity lying below. For example:

5' - AGGTC -3'

3' - TCCAG -5'

The pairs of bases are flat and may be stacked on above the other like a pile of plates so that the molecule is readily represented as a spiral staircase with the base pairs forming the treads. The two polynucleotide chains are of opposite polarity in the sense that the internucleotide linkage in one strand is 3'→5' while in the other it is 5'→3'. The two helices are right handed (i.e., (a) turn in a clockwise direction as viewed from the near end or (b) turn in the direction of the fingers of the right hand when the thumb indicates the line of sight) and cannot be separated without unwinding. The pitch of the duplex is 3.4 nm and since there are 10 base pairs on each turn of the helix, there is a distance of 0.34 nm between each base pair. The diameter of the helix is 2 nm.

The base sugar linkage is normally *anti* with the C8 of the purine or the C6 of the pyrimidine ring over the sugar. The sugar puckev is variable but a base pair usually has the sugar attached to the pyrimidine in the C3' *endo* configuration. There is lot of water associated with DNA. Upto 72 molecules of water per 12 base pairs are found largely in the *minor groove* where they form a spine.

The stacked bases form a hydrophobic core but the amino and keto groups point upto the grooves and allow interaction with the solvent. The sugar backbone points out to give hydrophilic with one negative charge per phosphate.

While the basic model put forward of *Watson* and *Crick* remains close to the accepted structure of the DNA molecule in solution, the more refined X-ray diffraction studies of *Wilkins* and his colleagues demonstrated that DNA fibres can have three possible structures with the B structures corresponding closely to the original Watson-Crick model. The A and C structures are also right handed double helices but differ in the pitch and in the number of bases per turn. In both A and C conformations the bases are not flat but are tilted. Their biological significance is not clear but the A conformation is believed to be very close to the structure adopted by double stranded RNA and by DNA-RNA hybrids. Because of the presence of extra 2' hydroxyl group, RNA appear to be unstable to adopt the B conformation. Thus when it is engaged as a template for making RNA, the DNA molecule must adopt the A conformation. Various organic solvents and proteins have been suggested to force the DNA from B to the A form, and the transition from the B conformation to the C conformation appear to occur in concentrated salt solution and ethylene glycol.

Various forms of DNA

A

B

C

D

Z

The Tertiary Structure of DNA

The DNA from many sources, particularly cellular chromosomal DNA, cannot be isolated in the form of intact molecules because of its large size and consequent sensitivity to shear and enzymatic degradation. However, the DNA from many viruses from mitochondria and chloroplast and from a few other sources (e.g. bacteral plasmid DNA) can be isolated in an ungraded form. The examination of such ungraded DNA has shown that the double helical secondary structure of DNA can in some instances be further contained giving rise to tertiary structure such as super coiled and open cyclic molecules. Linear DNA occurs naturally and is also a product of the degradation or cyclic DNA. Single stranded DNA, both in linear and cyclic forms, is found in some viruses, and can be formed artifically by denaturing double stranded DNA.

Types of DNA's

Eukaryotic DNA

It is the filamentous types of DNA found in all eukaryotic cells of animals and plants. In these cells, DNA synthesis takes place in a restricted time period during the cell cycle. In mitosis the cells divide and form two daughter cells. This is then followed by a gap period designated G1 during which no DNA synthesis occurs. The length of the various gap periods and indeed the length of the entire cell cyclic depends very much upon the cell type. The synthesis of DNA occurs in S phase and then follows a second shorter gap period known as G_2. The cells then go into mitosis again.

Prokaryotic DNA

Also called the *circular* a *superhelicular DNA.* The DNA molecules of prokaryotes and most viruses are circular. A circular molecule may be a covelently closed circle, which consists of two unbroken complementary, single strands, or it may be a *nicked circle,* which has one or more interruption (nicks) in one or both strands. With few exceptions, convalently closed circles are twisted. Such a circle is said to be a *superhelix* or a *supercoil.*

Extranuclear DNA

Also known as non-chromosomal or cytoplasmic DNA. In cytoplasm this DNA is associated with some of cell organelles as given belows:

(a) Mitochondria DNA

Mitochondrial DNA is usually found in cyclic double stranded, supercoiled molecules, the exceptions being the linear mitochondrial DNA1 molecules from *Tetrahymena* and *Parameciui.* Marmialian mitochondrial DNA molecules are not packaged into nucleosomes. They are about 15 kbp long and can therefore code for 15 to 20 proteins. However, some yeast ones are considerably larger and *Saccharomyces* mitochondrial DNA is 75 kbp long. Plant mitochondrial DNA is much longer still. Yeast mitochondrial DNA molecules have been widely studied since they are readily amenable to genetic analysis. Mitochondrial DNA can only codes for a small proportion of mitochondrial proteins.

Mitochondrial DNA from several mammals, *Xenopus* and *Drosophila,* has been sequenced and the sequence analysed. There are genes from two ribosomal RNAs, 22 tRNAs and for 13 protein most or all of which are involved in electron transport. The only region of mammalian mitochondrial DNA which is non-coding is the *D-loop region* involved in the initiation of DNA replication. This is also the region at which transcription of both strands is initiated. Transcription continues uninterrupted around the cyclic molecule and the transcripts are the processed to give individual messenger, ribosomal and tRNAs. The tRNA forms the punctuation between the various protein coding regions and provides the signals for the processing enzymes.

Plant mitochondrial DNA is apparently much more complex than animal mitochondrial DNA. It consists of per mutation of basic structure related to each other by recombination.

(b) Chloroplast DNA

Chloroplast DNA is in general much larger than mitochondrial DNA, being in the molecular weight range 100 million (150 kbp). In contrast to mitochondria which have from one to ten molecules of DNA per organelie, chloroplasts tend to have a very large number of copies of the DNA molecules in each organelle, in some cases greater than one hundred. Like mitochondrial DNA, the DNA in chloroplast carries the coding information for essential membrane components, tRNA and rRNA. All known chloroplast DNA molecules are cyclic and supercoiled.

(c) Kinetoplast DNA

The kinetoplast is part of highly specialized mitochondrion found in certain groups of flagellated protozoa such as trypanosome. Its

DNA (kDNA) consists of an interlinked series of many thousands of cyclic DNA molecules which vary in size from 0.6 kbp to 2.4 kbp depending from the type of trypanosomes from which they are obtained. These components, whch are known as *minicircles,* are further interlinked with a much smaller number of larger circular DNA molecules of above 30 kbp in size known as *maxicircles* in the majority of system analyzed. While maxicircles appear to perform the connectional functions of mitochondrial DNA in trypanosomes, minicircles are microheterogeneous in sequence and size and there is no evidence to suggest that they are ever transcribed so that it is possible that they may fulfil some structural rather than coding role.

Unique DNA

In human cells it is generally classified as that DNA which has a $Cot_{1/2}$ value of about 1000 moles of nucleotide seconds $litre^{-1}$. It comprise about half of the total haploid DNA content and is thought to consist of the sequences coding for most enzyme functions for which there is only one or a small number of genes per haploid genome. The genes coding for the various chains of globin or the enzyme glucose 6-phosphates fall into this category.

Repetitive DNA

In this type of DNA certain bases or nucleotides are repeated many times. Among the repetitive DNA is a fraction which reanneals with a Cot value of between 100 and 1000 moles of nucleotides seconds $litre^{-1}$. The sequences represented in this group are generally thought to be those coding for proteins which form major structural components of the cell such as the histones. The genes for rRNA and tRNA also fall into this category.

Satellite DNA

At the other extreme DNA with a $Cot_{1/2}$ value as low as 10^{-3} moles of nucleotide seconds $litre^{-1}$ consists largely of *satellite DNA.* This represents highly repeated sequences of which there may be a million or more copies per haploid genome, which are usually quite short and are arranged in tandem arrays. The origin of the name satellite relates to the method of its isolation on caesium chloride buoyant density gradients of sheared DNA where it will sometimes form a satellite band separate from main DNA band, due to its differing content of adenine and thymine residues. The simplest known DNA is poly [d (A-T)], which occurs in certain crabs. Other satellites can have any number upto several hundred base pair which are repeated in tandem fashion alsong the genome.

The distribution of satellite DNA among chromosomes varies. Some chromosomes have virtually no satellite sequences while others are largely composed of satellite sequences. In general, satellite DNA appears to be concentrated near the centromere of the chromosomes in the heterochromatin fraction. DNA sequence analysis has shown that the basic repeat unit of satellite DNA is itself made up of sub repeats. For example, the major momse satellite has a repeating structure of 234 base pairs made up of four related 58 and 60 bp segments each in turn made up of 28 and 30 bp sequences. The satellite between and within the related species are themselves related in an evolutionary sense by cyclic rounds of multiplication and divergence of an initial short sequence.

Fold-back DNA

Also known as *palindromic DNA.* It is a special class of DNA sequences comprising 3-6% eukaryotic DNA. The size range is from 300 to 1200 base pairs and the molecules have a $Cot_{1/2}$ value of less than 10^{-5} moles of nucleotide second $litre^{-1}$. This palindromic DNA is represented in all frequency classes and is widely distributed throughout the metaphase chromosomes. Some palindromic DNA arises when two copies of the *Alu* sequences are present close to one another in opposite orientations and it was a result of the ability of such DNA to renative instantaneously.

Z-DNA or Left Handed DNA

On the basis of the arrangement of pentose sugar forming chains of DNA two forms of DNA have been reported. Usually the sugar chain has the pentose sugar with ascending sequence of carbon atoms. This is the right handed DNA or B-DNA. Recently in 1979, another form of DNA have pentose sugar with ascending arrangement of carbon atom has been reported and named Z-DNA or left handed DNA.

Tautomeric form of DNA

An important chemical feature of DNA is the position of the hydrogen atoms in the purine and pyrimidine bases. Before 1953, many chemists through that some of these hydrogen atoms randomly moved from one ring nitrogen or oxygen atoms to another and os could not be assigned a fixed location. Now we realize that althoguh such movements, called *tautomeric shifts* do occur, these hydrogens have preferred atomic locations. The nitrogen atoms attached to the purine and pyrimidine rings are usually in the *amino* (NH_2) form and only rarely assume the *imino* (NH) configuration. Likewise, the oxygen atoms attached to the C6 atoms of guanine and thymine normally have the *keto* (C-O) form and only rarely take up the *enol* (COH) configuration.

It is essential to the biological functioning of DNA that the hydrogen atoms have relatively stable locations. If the hydrogen atoms had no fixed positions, adenine could often pair with cytosine and guanine with thymine, and sequence of the bases on the two interwined chains would not necessarily be complementary. If that were the case, the DNA could not function as a genetic molecule, for it is the complementary relationship between the opposing chain sequences that gives DNA its capacity for self-replication. Replication

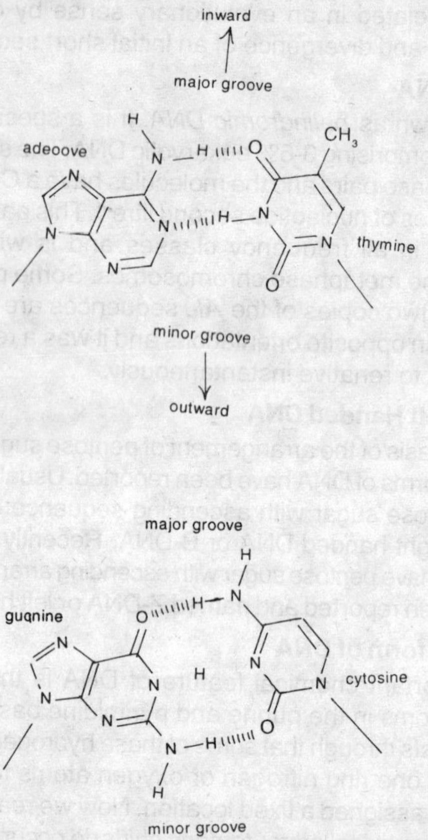

Fig. 11.6. The relative directions in which the base-pair nitrogens and oxygens project in each groove, the nitrogens and oxygens of the floor of the major groove project toward the center of the DNA molecule, while those of the minor groove project toward the outer edge of the DNA molecule.

of the double helix involves strand separation followed by formation of complementary DNA chains using the now free single strands as

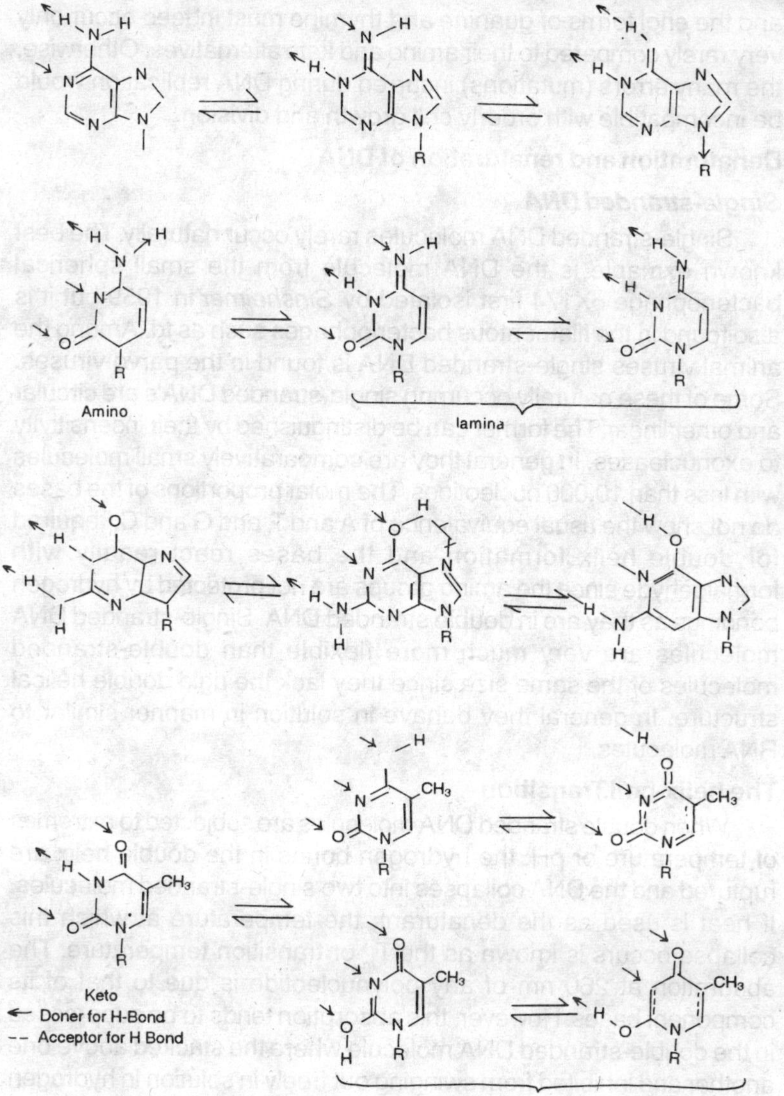

Fig. 11.7. Amino=imino and keto=enol tautomerism. (a) The A and C nucleotide bases are usually in the amino form but rarely form the imino configuration. (b) The bases G and T are usually in the keto form but are rarely found in the enol configuration. Atoms shown in colour may participate in hydrogen-bond formation.

templates to attract the appropriate partner bases dictated by AT and GC base-pairing rules. The imino forms adenine and cytosine and the enol forms of guanine and thymine must indeed occur only very rarely compared to their amino and keto alternatives. Otherwise, the many errors (mutations) incurred during DNA replication would be incompatible with orderly cell growth and division.

Denaturation and renaturation of DNA

Single-stranded DNA

Single-stranded DNA molecules rarely occur naturally. The best known exmaple is the DNA molecule from the small spherical bacteriophage ϕX174 first isolated by *Sinsheimer* in 1959 but it is also found in the filamentous bacteriophages such as fd. Among the animal viruses single-stranded DNA is found in the parvo viruses. Some of these naturally occurring single-stranded DNA's are circular and other linear. The former can be distinguished by their insensitivity to exonucleases. In general they are comparatively small molecules with less than 10,000 nucleotides. The molar proportions of the bases do not show the usual equivalence of A and T, and G and C, required for double helix formation and the bases react readily with formaldehyde since the amino groups are not protected by hydrogen bondings as they are in double stranded DNA. Single-stranded DNA molecules are very much more flexible than double-stranded molecules of the same size since they lack the rigid double helical structure. In general they behave in solution in manner similar to RNA molecules.

The helix-coil Transition

When double stranded DNA molecules are subjected to extremes of temperature or pH, the hydrogen bonds in the double helix are ruptured and the DNA collapses into two single-stranded molecules. If heat is used as the denaturant, the temperature at which this collapse occurs is known as the T_m or transition temperature. The absorption at 260 nm of any polynucleotide is due to that of its component bases. However, this absorption tends to be suppressed in the double-stranded DNA molecule where the stacked above one another and inhibited from swinging out freely in solution in hydrogen bonds, and consequently it is very much lower than that of equimolar amounts of the component bases or nucleotides free in solution. This inhibition of absorption is to some extent relieved when the DNA molecule goes through the helix-coil transition and the bases

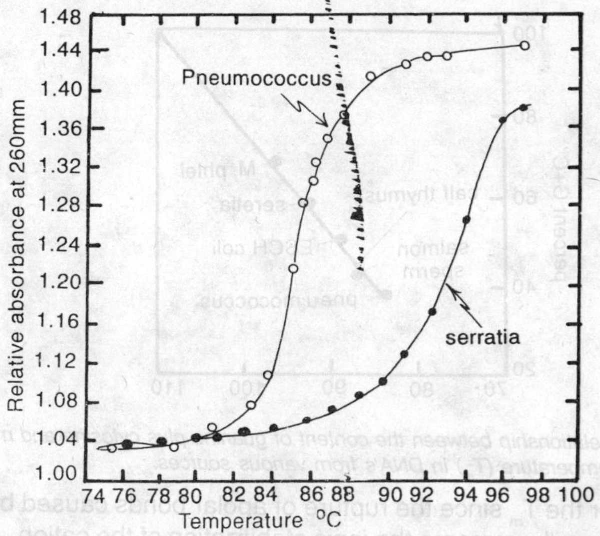

Fig. 11.8. Increase in absorbance at 260 nm with rising temperature for DNA from two different sources.

are no longer so rigidly stacked although they still interact to some extent. As a consequence of this absorption of DNA solutions rises by about 20-30 per cent as the DNA undergoes the melting process, and the transition process is usually measured by ultraviolet spectroscopy. The increase in absorbance on melting is known as the *hyperchromic effect.*

Such a transition can be characterized by several factors:

1. *The nature of the DNA.* Homogenous DNA, such a viral DNA, melts over a short temperature range but heterogenous DNA melts over a longer range.
2. *The (G + C) content of the DNA.* The melting temperature of any DNA can be related to its (G + C) content since this base pair confer extra stability on the molecule. In DNA molecules such as those from bacteriophage l, in which some regions are richer in (G + C) than others, the transitions of both regions can be observed.
3. *The nature of the solvent.* In low concentrations of counterion the transition temperature is low and its width broad. At high concentrations of counterion the T_m is raised and with width of the transition becomes sharp. However, if 'denaturing' salts such as sodium perchlorate are present, addition of more salt will

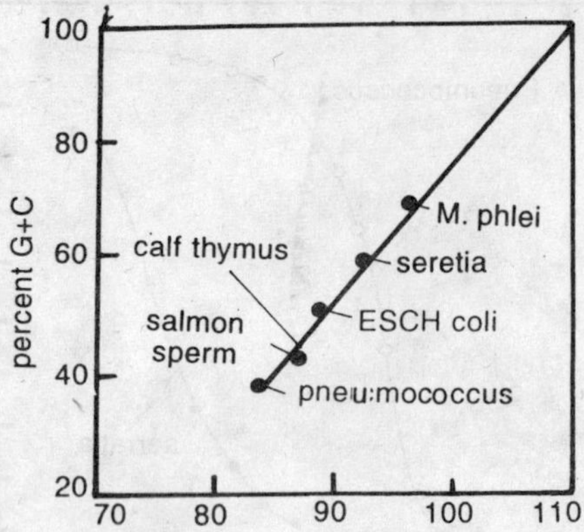

Fig. 11.9. Relationship between the content of guanine plus cytosine and melting temperature (T_m) in DNA's from various sources.

lower the T_m since the rupture of apolar bonds caused by the anion will overcome the ionic stabilization of the cation.

The helix-coil transition is also associated with a change in density of the DNA molecule, the single-stranded DNA being more dense than the equivalent double stranded from with the same (G + C) content.

The Renaturation of DNA

When two DNA strands are returned from the extreme conditions which caused them to melt to their original state they may reassociate to form a double helix again. However, whether or not they do so depends on a variety of factors. In principle the process should follow simple second-order kinetics but, because of the high probability of two sequences which were not previously matched forming an imperfect union which is then difficult to dissociate, the degree of renaturation can vary from less than 1 per cent according to the nature of the sample and the conditions of renaturation.

1. The nature of the sample

Simple sequence DNA molecules such as $d(G)_a\ d(C)_n$ have no difficulty finding the appropriate sequence with which the reanneal and do so without difficulty. However, if a eukaryotic DNA sample from a large genome is reannealed, clearly some of the sequences will have to encounter many other non-complementary sequences in solution before they find the correct partner. Thus if most of the DNA from eukaryotic cells is quickly cooled to a low temperature, so that the diffusion of the DNA in solution is inhibited, few of the single strands will reassociate.

2. The temperature of the reassociation process

At very low temperature (4°C) not only a diffusion limited but, for a DNA molecule which has become mismatched with a strand having only a few complementary bases, the opportunities to break away and continue its search for th correct complementary strands are reduced. Consequently DNA which is heated beyond the T_m and then quickly cooled to a low temperature will be denatured whereas solutions maintained at high temperatures below the T_m may renature.

3. The size of the DNA fragments

Large string-like fragments of single-stranded DNA encounter diffusion problems and frequently cannot reanneal correctly since this effectively reduces their opportunities for finding complementary strands. For this reason DNA is often sheared for reannealing experiments.

4. The ionic strength of the solution

Two highly charged DNA molecules are likely to repel one another, and the presence of salt is necessary to mask this repulsion in renaturation.

5. The concentration of the DNA

Clearly, at higher concentrations the probability of two complementary strands encountering each other is raised.

6. The time allowed for reannealing experiments

If renaturation is allowed under ideal conditions, two DNA samples of identical concentration should take different times to reanneal according to the genome size. For this reason the term *Cot value* has been defined for the study of reannealing of DNA and also for the study of the formations of DNA-RNA hybrids. *Co* represents the DNA concentration and *t* represents time in seconds. This is again monitored by ultraviolet spectroscopy or, more frequently, by hydroxyapatite chromatography since this can be used to separate double- and single-stranded DNA molecules and can be used at higher concentrations and hence for shorter times.

Single-Stranded DNA or Viral Chromosomes

At first it was thought that all DNA molecules are double stranded except during replication, when a small region around the replicating fork is temporarily in a non-hydrogen-bonded, single-stranded form. It therefore came as quite a surprise when experiments revealed that the DNA of several groups of small bacterial viruses exists as single-stranded molecules in which the amount of A is not equal to the amount of T and the amount of G does not equal the amount of C. Among these single-stranded phages are the spherically shaped ϕX174 and S13 virsuses and the rod-shaped f1 and M13 viruses. The chromosomes of the parvoviruses, minute virsuses that infect many vertebrate organisms, are also single stranded DNA molecules.

When these single-stranded viral chromosomes enter their host cells, they serve as templates for the formation of complementary strands. The resulting double helices in turn serve as templates for new single strands that then become incorporated into new virus particles. Thus, the fundamental mechanism for ordering nucleotides during the synthesis of single-stranded DNA is basically the same as that used for double-hellical DNA. Nucleotide selection always occurs by attraction of the complementary base. Single-stranded DNA replication usually differs from double-helical replication in that

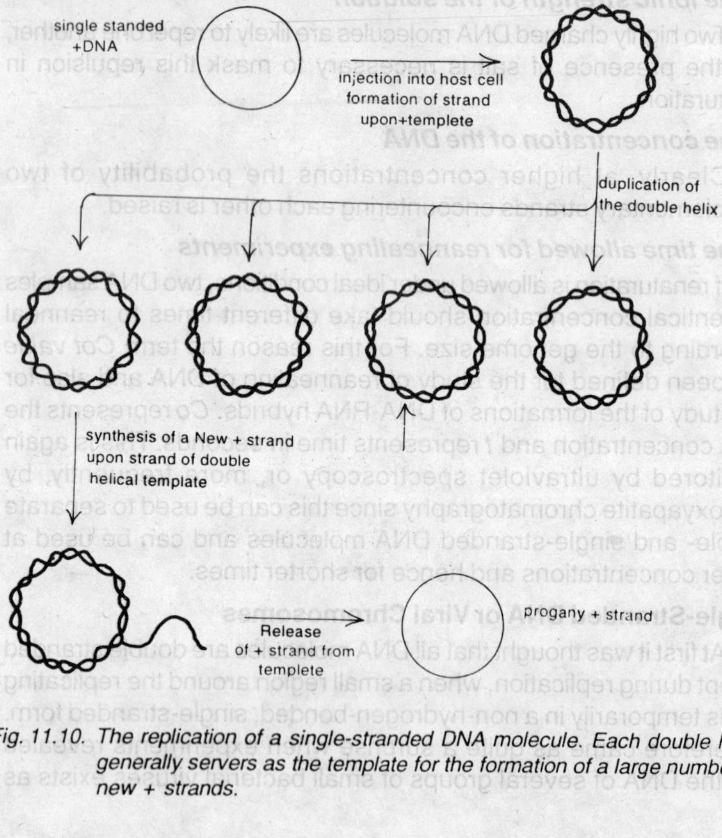

Fig. 11.10. The replication of a single-stranded DNA molecule. Each double helix generally servers as the template for the formation of a large number of new + strands.

it uses only one of the two complementary strands (the—strand) as a template for the progeny (+) strand.

Although single-stranded DNA can serve as the chromosome of a small virus, it would not be very effective in storing genetic information within the chromo-somes of cells. In bacteria and in many eukaryotic cells (e.g., the haploid phase of yeast cells), there usually is only one copy of a given gene. If interphase chromosomes contained single-stranded DNA, with the double helix existing only briefly during mitosis, then the daughter cells would contain two completely different sets of genetic informations. This follows from the fact that the two complementary chains do not have identical base sequences and hence would code for entirely different amino acid sequences. The double-strand ensures that each daughter cell will contain the same genetic information. The double-stranded form also permits effective DNA repair mechanisms to exist. For example, when one strand is damaged by exposure to X-rays, the remaining strand can provide the genetic information to rebuild the damaged section. It is probably no accident that only very small viral chromosomes are single-stranded. If they were to become, say, the size of T2, they would present too large a target for chain-damaging events.

Structure of Single-stranded DNA

Single-stranded DNA molecules have a strong tendency of fold back on themselves to from irregular double-helical hairpin loops whenever their sequences permit signficant numbers of nucleotides to base-pair. Imperfect as these loop are, they nonetheless are energetically favoured under physiological salt conditions, since they allow much more effective stacking of the flat hydrophobic surfaces of the bases then is possible in any fully extended single-stranded structure. Often, as many as half the bases of single-stranded DNA are so hydrogen-bonded, and their molecules are highly compacted. If, however, there are only minimal numbers of, neutralizing cations (e.g., less than or equal to 0.01 M Na^+) then the electrostatic repulsion between the partially unneutralized phosphates will kept the single-stranded chains highly extended, and there will be fewer hairpin loops.

Single-stranded DNA is always denser than double-helical DNA. Much more complete hydrogen bonding to water is possible for the atoms of the highly regular double helix than for the atoms of the inherently irregularly shaped single-stranded DNA. In the absence of base-pairing hydrogen bonds, Van der Waal's interactions take over and bring the atoms of DNA closer together than would be the case if they formed regular hydrogen bonds with each other. An analogy can be made with the behaviour of water, which becomes less dense as it forms the regular, hydrogen-bonded lattice of crystalline ice.

Rigorous Crystallography

X-ray diffraction patterns from parallel-oriented fibers of DNA provided the first clues that the polynucleotide chains of DNA have helical conformations. They also told us that DNA was a multichained molecule composed of two or three polynucleotide chains. However, the double-helical nature of DNA did not directly emerge from X-ray diffraction analysis alone. Equally important was the use of model building in which the question was asked. Given the chemical features of a single DNA chain, into what three-dimensional helical conformations could it fold? The simplest models were those in which the sugar-phosphate backbones were on the outside with the bases stacking in the center. The double helix emerged when it was realised that thymine and guanine had keto, not enol, configurations and that by base-pairing adenine to thymine and guanine to cytosine, a stereochemically pleasing, regular helical molecule resulted (i.e., the AT base pairs had the same "shape" as the GC base pairs).

They were originally discovered through their antimicrobial and anticancer attributes. Each time these intercalating agents become inserted into a DNA molecule, they necessarily *increase* the spacing of successive base pairs along the helical axes to roughly 7Å, almost the distance between phosphate atoms in a fully extended a polynucleotide chain. In this situation, very little rotation is possible around the helical axis, and so intercalating agents not only extend double helices, but extensively unwind them. For example, when a molecule of either ethidium bromide (an inhibitor of DNA synthesis) or actinomycin D (a powerful inhibitor of RNA synthesis) intercalates, the rotation angle between the two adjacent base pairs is reduced from 36^0 to 10^0. The fact that intercalation occurs of readily indicates that it must be energetically favoured, with the van der Waals bonds holding the inserted molecules to the base pairs beings ronger than those found between conventionally stacked base pairs. Intercalation is additional evidence for the *metastability* of the double-helical structure—its ability to temporarily assume many inherently unstable configurations that normally quickly revert back to the standard B conformation. One such metastable variant must be short stretches of extended chains with gaps separating adjacent base pairs, into which an intercalating agent may bind.

The Chromosomes of Viruses, E.coli and Yeast are Single DNA Molecules

The first estimates of the average molecular weights of DNA centered at about a million, the size needed to encompass in average-size bacterial gene coding for some 300 to 400 amino acid. Therefore, it seemed natural to equate single DNA molecules with single genes. However, these early reports were inaccurate owing to DNA breakage during its isolation and study. Now it is clear that virtually all undergraded DNA molecules contain the information of a least several genes. The most certain molecular weight values come from DNA-containing viruses. Regardless of whether the DNA content is relatively small or large, each virus particle contains a single DNA molecule. For example, all the DNA of the small monkey (simian) virus SV40 is present within a single molecule of molecular weight ~ 3×10^6 daltons (5×10^3 base pairs, or 5 kbp), while the DNA molecule of the large bacterial virus T2 has a molecular weight of $1.2. \times 10^8$ dalton (2×10^3 kbp). In these cases, the entire viral chromosome, rather than separate genes, corresponds to a single DNA molecules. Likewise, the chromosome of an *E. coli* cells is a single DNA molecule whose molecular weight is about 2.5×10^9 daltons (about 4×10^3 kbp) and whose extended length is, roughly 1 mm.

DNA molecules of the yeast *Saccharomyces cerevistiae* range is size between the T2 and *E.coli* DNAs, with their average size being that expected if each of the yeast's 17 chroimosomes contain one DNA molecule. Thus, the centromeres or yeast chromosomes (if not those of all eukaryotic chromosomes) do not represent discontinuities between separate DNA molecules, but instead are specialized regions of DNA evolved to interact with the spindle bodies upon which chromosomal segregation occurs during cell division. As of yet, there is no direct proof that the very much larger chromosomes of higher plants and animals (often 50 times larger than the *E. coli* chromosomes) also contain only one DNA molecule. However, we now expect that the one chromosome—one DNA molecule rule will hold for the chromosomes of all organisms. In any case, some DNA molecules are larger than any other biological molecules by several powers of ten.

Circular Versus Linear DNA Molecules

Our evidence, which came largely from autoradiography and electron microscopy, initially suggested that all DNA molecules are linear and have two free ends. But when it became possible to take a better look at undergraded DNA, many DNA molecules were found to be circular. For instance, the small monkey DNA virus SV40 has a 5000-base-pair circular double-helical chromosome; the similarly short chromosomes of single-stranded phases are likewise circular, as are almost all autonomously replicating plasmid DNAs. Most, if not all, bacterial chromosomes are also circular. We suspect that the DNA found in the rare circular chromosomes of higher cells are likewise circular molecules. Circular shapes were initially puzzling, since they seemed to present obstacles to the untwisting of double helices during DNA replication. Only when specific enzymes that first snip and then join DNA chains were discovered did the untwisting dilemma disappear.

Equal important is the realization that linear DNA molecules do not have the uncomplicated structures first envisioned for them. As we see inthe next chapter, replicating the ends of DNA molecules is not a straight forward process. Thus, all linear DNA molecules have evolved special tricks to replicate their ends, as well as to prevent their free ends from either being nibbled back by cellular enzymes or being enzymatically ligated together to form even larger chromosomes. For example, the ends of the linear adenovirus chromosomes have specialized proteins sovalently attached to the 5' ends of strands; and poxvirus chromosomes have closed hairpin loops at their ends, with their basic structures being that of largely self-complementary circular single strands. Still other linear viral chromosomes (e.g., T2 and T7) have the same sequence at both ends, allowing the ends of the chromosome to recombine together during DNA replication to form very long, end-to-end aggregates (concatamers).

Moreover, some DNA molecules that are linear when isolated from a virus particle (e.g., phase λ) are found as circles inside the host cell. This tells us that the linear and circular forms of such DNA molecules are interconvertible. Starting with a closed circle, a specific enzyme introduces two breaks, one in the + strand and one in the – strand. As thse breaks are very close to each other, the intervening hydrogen bonds occasionally are broken by thermal agitation, and then the circle unfolds. Most importantly, the resulting linear form contains single-stranded ends that have complementary nucleotide sequences (:*stickly ends*"). Thus, at a later time the linear form can base-pair and resume a circular configuration. If the mission phosphodiester bonds are then reformed, the covalent circle is regenerated.

There is solid evidence that phase λ DNA interconverts between its circular and linear forms using precisely this mechanism. Later, we shall see that the duplication of such "sticky" DNAs involves circular replicative intermediates. Thus, the linear form is most likely an adaptation for injection of the viral chromosome through the narrow phase tail.

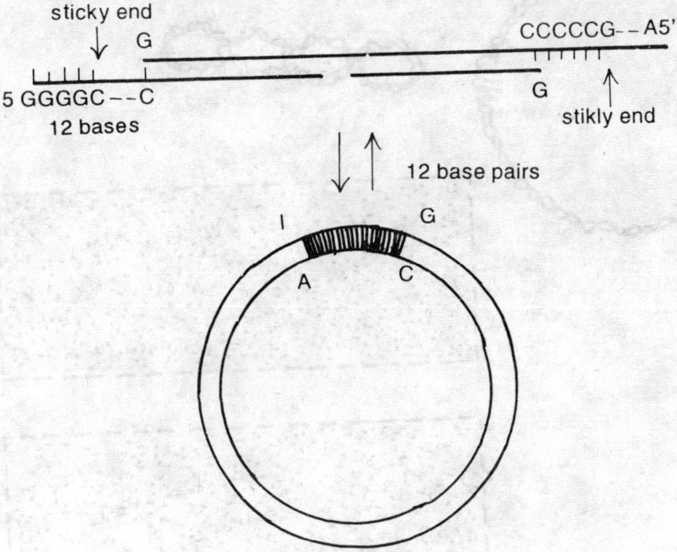

Fig. 11.11. Interconversion of the linear and circular forms of λ DNA.

Supercoiling of Circular DNA

As DNA is a very flexible structure, its exact molecular parameters are a function of both the surrouding ionic environment and the nature of the DNA-binding proteins with which it is complexed. Because their ends are free, linear DNA molecules can freely rotate to accommodate changes in the number of times the two chains of the double helix twist about each other. But once the two ends become covalently linked to form circular DNA molecules, the absolute number of times the chains twist about each other (the *linkage*

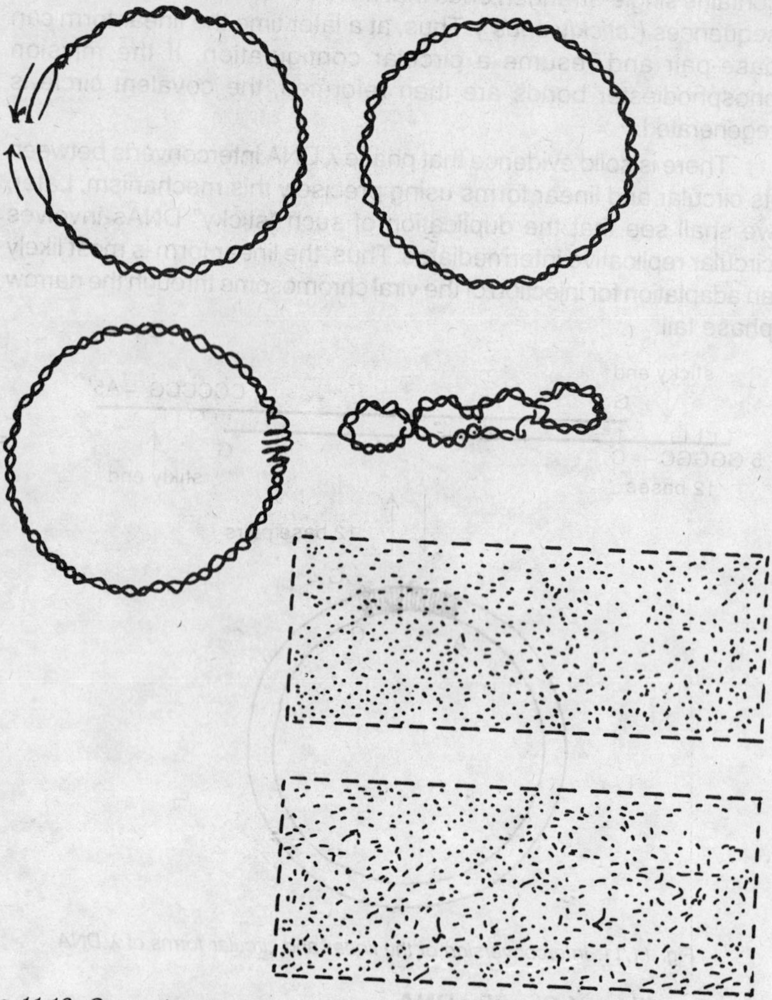

Fig. 11.12. Generation of supercoiled DNA by cutting and ligating. In a circular DNA molecule, the number of times the DNA strands twist about each other is fixed. If both strands are cut (1) and rotated in the opposite direction from that of the twist of the helix (2), and if the cut ends are then rejoined (3) then the DNA molecule negatively supercoiled. (4) The upper electrons micrograph is relaxed (non-supercoiled) DNA molecule of bacteriophage PM2. The lower electron micrograph shows the phages in its supertwisted form.

number) cannot change. For the most part, changes in the average number of base pairs per turn of the double helix will necessarily be accomodated by the formation of the appropriate number of supercoils in the opposite direction. *Supercoiling* refers to the further twisting of double-helical DNA molecules. Untwisting of the double helix usually leads to supercoiling in the negative (left-handed) direction while ovetwisting leads to positive (right-handed) supercoiling.

The supercoiled state is inherently less stable than uncoiled DNA, and the cutting, or *nicking*, of a single strand instantly converts a supercoiled molecules into its simple (*relaxed*) circular state. Because circular DNA molecules become increasingly compacted as they become more supercoiled, relaxed DNA is easily distinguished from supercoiled DNA by its slower sedimentation in a centrifugal field and by the longer time it takes to move through an agarose gel. At the present time gel electrophoresis affords the most direct way to measure supercoiling, since it can separate molecules that differ by only single turns.

Protein Containing Supercoils

Just becuase isolated DNA molecules would energetically prefer to be in the relaxed state does not mean that the "natural state" of DNA is relaxed. In fact, virtually all DNA within both prokaryotic and eukaryotic cells exists in the negative supercoiled state. As such it seldom, if ever, occurs as free DNA, but is complexed with specific DNA-binding proteins to form compacted molecules called *chromatin*. How this compaction occurs in best understood for eukaryotic chromatin, whose most prominent DNA-binding components are the histones. *Histones* are relatively small, positively charged, arginine- and lysine rich proteins that aggregate together to form discrete ellipsoid-shaped packets (histone cores) around which the DNA supercoils. The resulting *nucleosomes* give to chromatin its beaded appearance.

Since the same collection of histones binds to nearly all sections of DNA, histones are thought to play essentially structural roles, as opposed to enzymatic or regulatory roles. Only about half the mass of chromatin proteins is histone. The remaining proteins are not nearly as well characterized with their exact functions yet to be determined.

Viral DNA chromosomes are also complexed with proteins, both when multiplying within cells and when packaged into virus particles. The SV 40 viral chromosome is at all time compacted into 24 histone-containing nucleosomes identical it structure to those existing in the chromosomes of their host cells. When freed from its protein components, SV 40 DNA is found as a negative supercoil. In contract, no histones are complexed with the DNA found in adenovirus particles. These vertebrate viruses encode a unique DNA binding protein to help package their DNA within their protein capsules. When, however, an adenovirus DNA functions within a cell, its own DNA-binding protein is released and replaced with the same histone components that bind to all other cellular DNA.

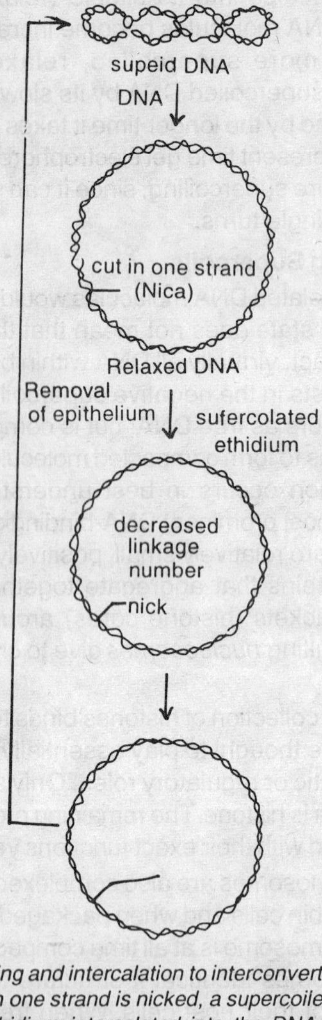

Fig. 11.13. Use of nicking and intercalation to interconvert supercoiled and relaxed DNA. When one strand is nicked, a supercoiled circular DNA molecule relaxes. Ethidium intercalated into the DNA molecule incerases the spacing of base pairs, unwind the double helix by 26⁰, and thus decrease the linkage number. DNA ligase can rejoin the free ends. When ethidium is removed, the DNA molecule supercoils negatively.

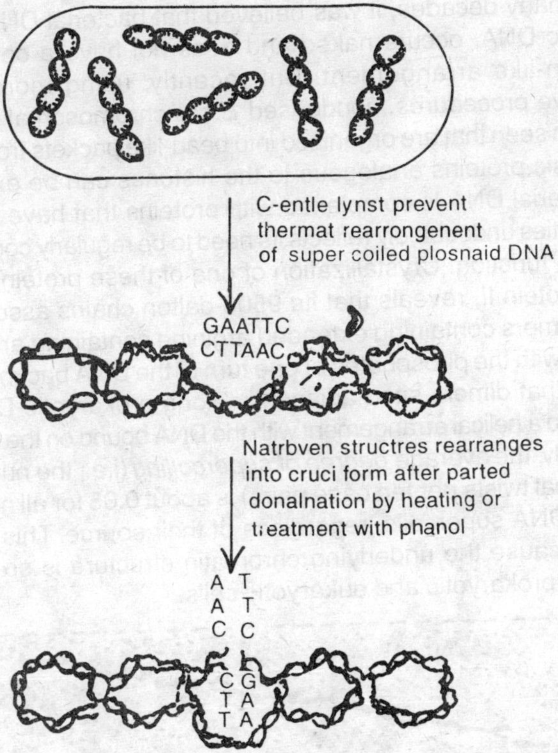

Fig. 11.14. Formation of stable cruciform structures from inverted palindromic sequences. It is now questionable whether these do indeed form within cells.

DNA Supercoils Twice Around Each Nucleosome

In forming the nucleosomes of eukaryotic cells, 200-base-pair-long segments of DNA negatively (left-handedly) supercoil twice around the histone cores. The energy lost by supercoiling may be partially compensated by the energy gained from the ionic and hydrogen bonds formed between the histones and the DNA. The number of supercoils in solution need not be the same as found in chromatin. In chromatin, the double helix is slightly more twisted than in solution, with 10 base pairs per helical turn in chromatin versus 10.5 base pairs per helical turn in naked DNA in solution. As a result, there are fewer super helices found in solution than found in chromatin.

Histone-like DNA-binding Proteins in Prokaryotes

For many decades, it was believed that bacterial DNA, unlike eukaryotic DNA, occurs naked and does not have a compacted chromatin-like arrangement. But recently, using more gentle preparative procedures, condensed *E.coli* chromosomal sections have been seen that are organized into bead-like packets from which small, basic proteins analogous to the histones can be extracted. That bacterial DNA is complexed with proteins that have histone-like properties undoubtedly reflects its need to be regularly compacted in order to function. Crystallization of one of these proteins, DNA-binding protein II, reveals that its 9500-dalton chains asscciate in pairs as dimers containing extended arginine containing arms able to interact with the phosphates of one turn of the DNA backbone. So it is likely that dimers sited adjacently along prokaryotic DNA are oriented into a helical arrangement with the DNA bound on the outside. Interestingly, the average *degree of supercoiling* (i.e., the number of super-helical twists per ten base pairs) is about 0.05 for all naturally occurring DNA supercoils, regardless of their source. This may in part be because the underlying chromatin structure is so similar within both prokaryotic and eukaryotic cells.

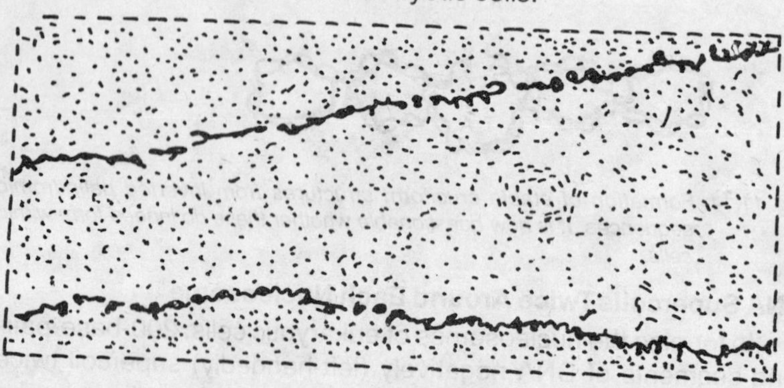

Fig. 11.15. Nucleosomes give chromatin strands a beadlike appearance. Magnification 100,000.

Topo-isomerases Change the Linkage Numbers of Supercoiled DNAs

That supercoiling not only occurs but is biologically very important is affirmed by the existence of the enzyme topoisomerase II, which specifically generates the negative supercoils that characterize so much of chromosomal DNA. *Topoisomerases* are a group of enzymes that convert (isomerize) one topological version of DNA into another. They do so by changing the linkage number, which is the number of times two DNA chains twist around each other. Prokaryotic topoisomerase II (sometimes called *gyrase*) uses the energy of ATP to generate negative supercoils by untwisting DNA in the left handed direction. Its eukaryotic counterpart, however, may be capable of using ATP to generate negative supercoils only in the presence of other, yet to be described proteins. The supercoiling action of topoisomerase II is counterbalanced by a second enzyme, topoisomerase I, which converts supercoiled DNA to the unstrained, energetically more favourable relaxed state. The relative amounts of topoisomerase I and topoisomerase II in cells are finely tuned, tending to create just the right amount of negative supercoiling. Thus, mutations that lower the number of topoisomerase I molecules are viable only if the number of topoisomerase II molecules also decreases.

Both topoisomerase I and II work by catalyzing the breakage and rejoining of DNA phosphodiester bonds. Unlike virtually all other enzymes, their action does not lead to new patterns of covalent bonds. Rather, their role is to create temporary gaps in polynucleotide chains. When so acting topoisomerases do not create free ends but instead become themselves covalently attached to one of the two broken ends (depending on the specific enzymes, either the 3' and 5' end). Through such catalysis, a tyrosine group on the topoisomerase becomes linked to the terminal pohosphate group of the cut polynucleotide chain. The free energy present in the original phosphodiester bond is thus preserved so that it can be used to rejoin the broken chain.

When a DNA chain is broken by a topoisomerase, its broken ends do not fall apart but are held together by the topoisomerase. At no time do the broken ends have the capacity to rotate freely. If that happened, topoisomerase would be limited to only completely relaxing double helices. Instead, individual topoisomerases function to make discreate changes of double helices. Instead, individual topoisomerases function to make discrete changes of either one positive turn (topoisomerease I) or two negative turns (topoisomerase II) in the linkage number. Such discrete steps result from DNA chains passing through either transient single-stranded breaks (topoisomerase I) or transient double-stranded breaks (topoisomerase II).

Table 11.5. DNA Topoisomerases

	Prokaryotic (E.coli)		*Eukaryotic (Hela)*	
	Type I	*Type II*	*Type I*	*Type II*
Approximate MW	100,000	400,000	100,000	309,000
Subunits	Monomer	Tetramer (A_2B_2)	?	Dimer (subunit MW=172,000)
Gene	top A	*gyr A* Polypeptide MW 105,000; activity inhibited by nalidixic acid and oxolinic acid *gyr B* Polypeptide MW 95,000; activity inhibited by cumermycin and novobiocin	?	?
Covalent intermediate with DNA	At 5' P	Subunit A Tryosine covalent bond to 5' P	At 3' P	At 5' P
Relaxation	Mg^{2+} required; ATP-independent	ATP-independent	No Mg^{2+} required; ATP-independent	ATP-dependent
Negative supercoiling	None	ATP-dependent	None	None

Topoisomerases are constructed in such a way that they can undergo reversible conformational changes that create cavities through which DNA chains can pass. The ATP requirement for bacterial topoisomerase II (gyrase)-induced supercoiling most likely reflects the use of ATP in mediating the necessary conformational changes that lead to chain passage. Topoisomerase I, however, relaxes supercoiled DNA without requiring a cyclical input of energy. Complete understanding of how topoisomerase I and II work can occur only when their precise structures and determined through X-ray crystallographic procedures. Even now, however, the prediction can be made that double helices wrap themselves around topoisomerase II in positive supercoils prior to the start of the cutting-rejoining cycle. Without this restraint on the orientation of the DNA, topoisomerase II would break as well as make negative supercoils, and the net result of its action would be eventual relaxation of the double helical substrate to the uncoiled state.

Looped and Supercoiled Structure

The fact that linear DNA molecules have free ends would seem to preclude their possession of supercoiled segments. Yet, examination of gently isolated linear eukaryotic chromosomes shows that their chromatin is organized into a large number of successive looped domains organized along a scaffold containing two major DNA-binding proteins. Somehow, attachment ot the scaffold prevents the rotation of one domain from being transmitted to adjacent sections, since the DNA within each of these loops independently supercoils and may be under different torsional strain. Now there is much evidence both from *Drosophila's* giant salivary chromosomes and from the looped lamph rush chromosomes of the salamander that the individual loops represent functional units of chromatin, with given loops usually being transcribed into one or more very long RNA molecules. How the protein scaffold might maintain independently supercoiled loops was a complete mystery until recently. Now it is less so, following the discovery that a major component of the scaffold is none other than topoisomerase II.

Not only are the long, linear chromosomes of higher organisms divided into looped, functional domains, but the circular *E. coli* DNA chromosome also is somehow divided into some 50 looped domains (each about 20 µm long) whose supercoiling is independently controlled. No scaffold has yet been isolated from any bacterial chromosome. Indeed, the best way now to search for it may be to look for topoisomerase II activity in very gently ruptured cells.

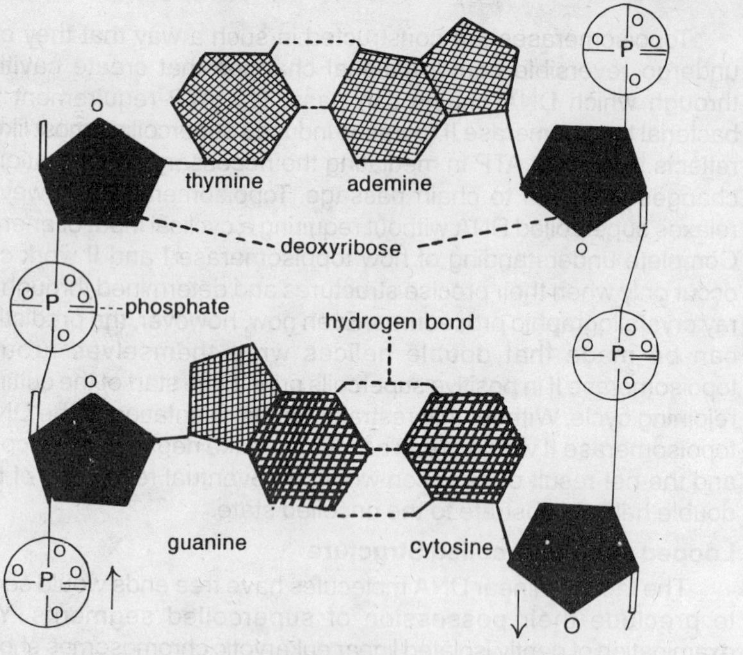

Fig. 11.16. Showing antiparallel nature of backbones of DNA.

Biological Replication of DNA

Study of replication of DNA is one of the interesting aspect of molecular control of heredity. Various study of the duplication of genetic material was presented from time to time in different organisms. But most of them are discarded because no concrete ideas have been proposed till *Watson* and *Crick* proposed a detailed scheme for the molecular sructure of DNA. Later on *Korenberg, Stahl* and *Taylor* have confirmed the Watson-Crick hypothesis. They also made a suggestion for its mode of replication which follows in a natural and attractive fashion from the proposed structure. During the replication, the DNA duplicates exactly into two chains. In the beginning both the strands of double stranded DNA, apart from each other by the disolution of hydrogen bonds and thus uncoiling of the chains occur. And then new chains, "the complementary chains" are built on them so that the newly formed daughter chains are the mirror images of each other.

Mechanism of DNA Replication

The entire process of DNA replication involves following steps in *E.coli:*

1. *Recognition of the initiation point.* Replication of DNA begins at a specific point, called initiation point. Specific initiator proteins are required to recognise the initiation point on DNA.

   The initiator protein along with DNA directed RNA polymerase initiates the synthesis of RNA primer for the formation of new DNA chain.
2. *Unwinding of DNA.* The unwinding proteins bind to the nicked strand of the duplex and open up a loop or buble, separating the two strands of DNA duplex.
3. *RNA priming.* The DNA directed RNA polymerase now synthesizes the primer strands of RNA (*RNA primer*). The priming RNA strands are complementary to the two strands of DNA and are formed to 50 to 100 nucleotides.
4. *Formation of DNA on RNA primers.* The new strands of DNA are formed in the 5'-3' direction from the 3'-5' template DNA by the addition of deoxyribonucleotides to the 3' end of RNA primer. The addition is affected by DNA polymerase III in the presence of ATP. The unwinding proteins separate the duplex strands ahead of the replication fork.

   The leading strand of DNA synthesized in 3'-5' direction. The following strand of DNA is formed in its opposite direction. The

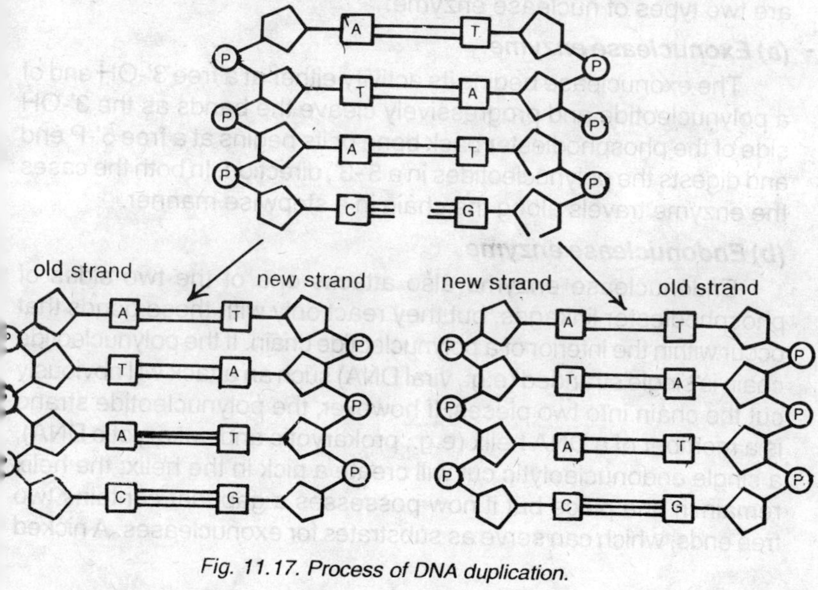

Fig. 11.17. Process of DNA duplication.

DNA strands are synthesized in short segments, consisting of 1,000 to 2,000 nucleotides. These segments are called *Okazaki fragments.*

5. *Excision of RNA primers.* Once a small segment of an Okazaki fragment has been formed, the nucleotides of RNA primer are removed from the 5' end one by one by the action of 5'-3' exonuclease activity of DNA polymerase-I.
6. *Joining of Okazaki fragments.* The gaps left between Okazaki fragments are filled with complementary deoxyribonucleotide residues by DNA polymerase-I. Finally the adjacent 5' and 3' ends are joined by *DNA-ligase.*

In eukaryotic cells, the mechanism of DNA replication is expected to be more complex than in prokaryotes. The replication of eukaryotic DNA begins at multiple points of origin. Their RNA primer is formed of about 10 nucleotides and the Okazaki fragments are much shorter (about 100 to 150 nucleotides).

Enzymes for DNA Synthesis

Different prokaryotic and eukaryotic cells have been formed to contain three kinds of enzymes that act on DNA namely nucleases, polymerases and ligases.

1. Nuclease Enzymes

These enzymes attack on the DNA molecules directly. There are two types of nuclease enzyme:

(a) Exonuclease enzyme

The exonuclease begins its activity either at a free 3'-OH end of a polynucleotide and progressively cleave the bands as the 3'-OH side of the phosphodiester back bone or its begins at a free 5'-P end and digests the polynucleotides in a 5'-3', direction. In both the cases the enzyme travels along the chain in a stepwise manner.

(b) Endonuclease enzyme

Endonuclease enzyme also attacks one of the two sides of phosphodiester linkages, but they react only with those bonds that occur within the interior of a polynucleotide chain. If the polynucleotide chain is single stranded (e.g., viral DNA) such an attack will obviously cut the chain into two pieces. If however, the polynucleotide strand is a member of a DNA helix (e.g., prokaryotic and eukaryotic DNA), a single endonucleolytic cut will create a nick in the helix; the helix remain in one piece but it now possesses a gap that contains two free ends, which can serve as substrates for exonucleases. A nicked double helix suffers a localized disrupted of its secondary structure. The molecules becomes free to bend or rotate around its intact strand and the two broken ends are free to dangle. The increased molecular motion in the region of the nick will very likely interrupt hydrogen bonding between bases in the vicinity of the nick, thus effecting a limited "unraveling" of the helix.

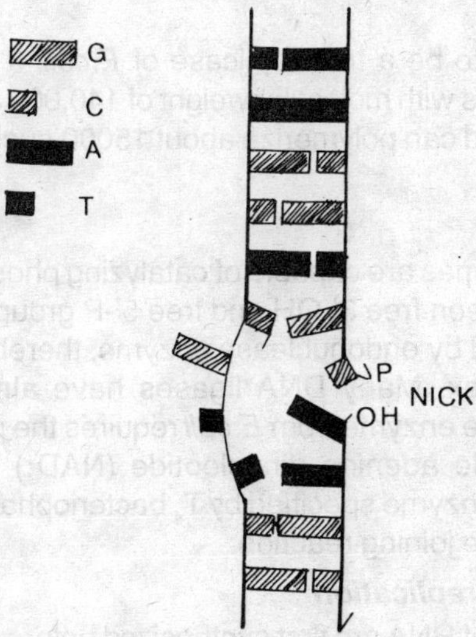

Fig. 11.18. A nicked duplex DNA molecule such as is formed by an endonuclease enzyme.

2. DNA Polymerase

For many years, *Korenberg polymerase* was believed to be the enzyme responsible for the replication of the *E. coli* chromosomes. But recently *Delucia* and *Cairus* isolated a true replicase from *E. coli* and named as *DNA Polymerase.* The polymerase catalyses the formation of a polymer. The synthesis of a new complementary chain becomes possible in the presence of polymerase enzyme. There are three categories of DNA polymerase enzymes.

DNA polymerase I

Discovered by *Korenberg* in 1955 and has been studied in *E. coli.* It is believed to take part in the repair of DNA. DNA polymerase I is a single polypeptide chain with a molecular weight of about 109,000. There may be as many as 400 molecule of DNA polymerase I in *E coli.*

DNA polymerase II

It is also made up of a single polypeptide chain having a molecular weight of about 90,000. The number of DNA polymerase II molecule is about 40 in *E.coli.* The enzyme can elongate Okazaki segments even in the absence of polymerase I.

DNA polymerase III

It is considered to be a true replicase of *E.coli.* It consists of two polypeptide chains with molecular weight of 140,000 and 40,000. It is most effective and can polymerize about 15000 nucleotides per minute in *E. coli.*

3. DNA Ligases

DNA ligase enzymes are capable of catalyzing phosphodiester bond formation between free 3'-OH and free 5'-P groups of nick of DNA which is created by endonuclease enzyme, thereby restoring an intact DNA duplex. Many DNA ligases have already been discovered. The ligase enzyme from *E.coli* requires the presence of oxidized nicotinamide adenine dinucleotide (NAD^+) a cofactor, whereas, the ligase enzyme specified by T_4 bacteriophage requires ATP to bring about the joining reaction.

RNA primer in DNA replication

Short segments of RNA are first synthesized before actual DNA replication starts. These RNA segments are later excised and filled with DNA with the held of enzyme DNA polymerase I. It was found that if RNA synthesis is stopped by a drug rifampicin, DNA replication stops thus providing the role of RNA primers in DNA replication. There are several other evidences which established this point.

Types of Replication

Three different types of replication of DNA have been discovered:

1. Semiconservative
2. Conservative
3. Dispersive

1. Semiconservative

Here the replication of DNA involves the unwinding of the double helix, but no rupture of the separated polynuclctide strands. Thus the primary DNA is causevned during the replication process. According to this mechanism, the unwinding of the original strands would be accompanied or followed by the guided synthesis of the complementary strands of each of the two separated polyucleotide

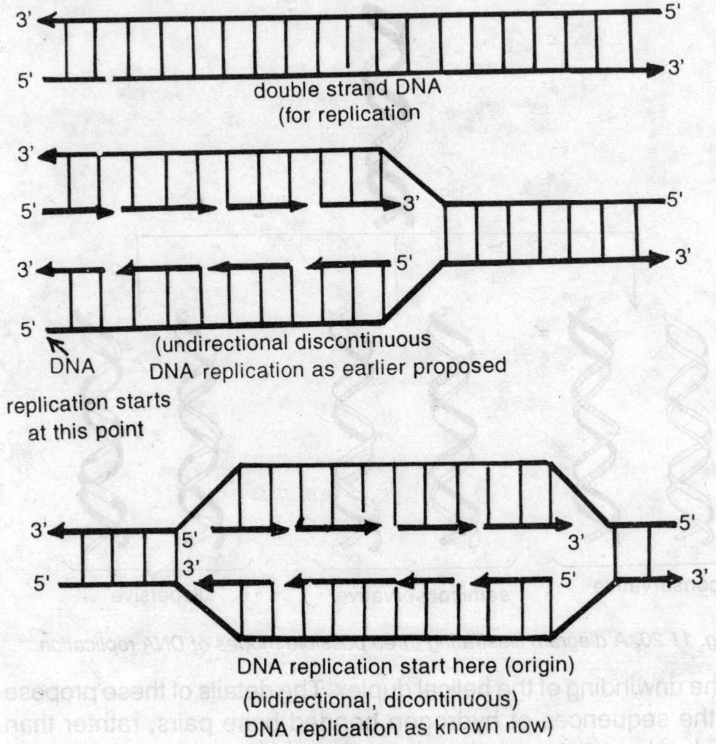

Fig. 11.19. Discontinuous DNA replication.

strand. The linearly assembly, catalyzed by enzymes, of the nucleotides of the complementary sequence would be guided by their paining with the Watson-Crick partners on the original strands. In this manner, the proper nucleotide sequence would be automatically assumed. The DNA helix would thereby regenerated each of the strand of the original helical duplex uniting with its freshly synthesized complement to form a total of two new double stranded structures. The semi-conservative model thus predicts that entire polynucleotide strands of the parental DNA will be passed on to the progeny in intact form. Each molecule of progeny DNA will receive either one half or none of its nucleotides from the parental DNA.

2. Conservative

According to this mechanism both the primary and secondary structure of the parental DNA are conserved. Thus the replication process would involve neither the rupture of the individual strands

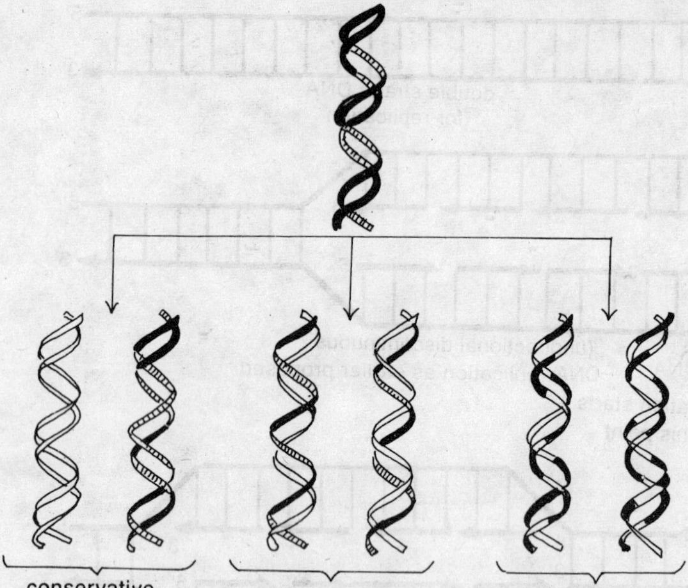

Fig. 11.20. A diagram illustrating three possible modes of DNA replication.

nor the unwinding of the helical duplex. The details of these propose that the sequences of hydrogen bonded base pairs, rathter than individual bases, serves the genetic determinant. According to this picture, particular base pairs might preferentially bind specific nucleotides and determine their sequential position in the daughter polynucleotide.

The conservative model allows for no redistribution of parental DNA among the progeny. The original DNA is transferred as an intact unit or not at all.

3. Dispersive

Into this class fall all mechanisms which postulate a general break up of the parental DNA and its more or less uniform distribution among the progeny. It is difficult to propose a model of this type which can possible allow the preservation of nucleotide sequence. Moreover, the existing experimental evidence is uniformly inconsistent with this kind of scheme.

12

BIOLOGICAL REPLICATION OF DNA

The Replication of DNA

When the double helix was first discovered, the feature that most excited the world of genetics was the complementary relationship between the sequences of bases on its interwined polynucleotide chains. It seemed unimaginable that such a complementary structure would not be utlized as the basis for DNA replication. In fact, it was the self-complementary nature of DNA that finally led most geneticists and biolgoists to accept Avry's conclusion that DNA, not some form of protein, was the carrier of genetic information.

Earlier in our discussion of how templates act, we emphasized that two identical surfaces will not attract each other and that it is much easier to visualize the attraction of groups with opposite shape or charge. Thus, without any detailed structural knowledge, we might guess that a molecule as complicated as the gene could not be directly copied. Instead, replication would involve the formation of a molecule complementary in shape, and this, in turn, would serve as a template to make a replica of the original molecule. So some geneticists, in the days before detailed knowledge of protein or nucleic acid structure, wondered whether DNA served as a template for a specific protein that in turn served as a template for a corresponding DNA molecule.

But as soon as the self-complementary nature of DNA became known, the idea that protein templates might play a role in DNA replication was discarded, for it was immensely simpler to postulate that each of the two strands of very parental DNA molecule served as a template for the formation of a daughter strand complement. Although from the start this hypothesis seemed too good not to be true, experimental support nonetheless had to be generated. Happily within five years of the discovery of the double helix, there emerged both descisive evidence for the separation of the complementary strands during DNA replication and firm enzymological proof that DNA alone can function as the template for the synthesis of new DNA strands.

With these results, the problem of how genes replicate was in one sense solved. But in another and equally valid sense, the study of DNA replication had only begun. As this chapter will show, the replication of even the simplest DNA molecule is a complicated, multistep process, involving many more enzymes than we initially anticipated following the discovery of the first DNA polymerizing enzyme. Fortunately, it now seem likely that all the key enzymatic steps involved in the replication of both linear and circular DNA have been described and that the general picture of DNA replication presented here will not be seriously modified.

Strand Separation Requires Untwisting of the Double Helix

In even the smallest DNA molecules, the two chains of the double helix are twisted around each other many times and must be untwisted an equal number of times for DNA replication to occur. This feature initially troubled many scientists who were otherwise enamored of the implications of DNA's double-helical structure. But untwisting linear DNA molecules need not lead to insuperable problems if the unreplicated sections of DNA can rotate around their axis, much like the rotating cables of speedometers. Since the DNA molecule is very thin, if should encounter virtually no frictional resistance, and thus the energy input needed for rotation should be negligible.

No conceptual difficulty arises from the need to break the hydrogen bonds joining the base pairs in the parental templates. Although these bonds are highly specific, they are also relatively weak, and no enzymes are required to make or break them. Hydrogen bonds thus provide highly suitable means of specifically holding together a template and its complementary replica.

Base Pairing should Permit very Accurate Replication

It has already been stated that the chemistry of the various amino acid side groups argued against their employment as templates. Just the opposite, however, is true of the purine and

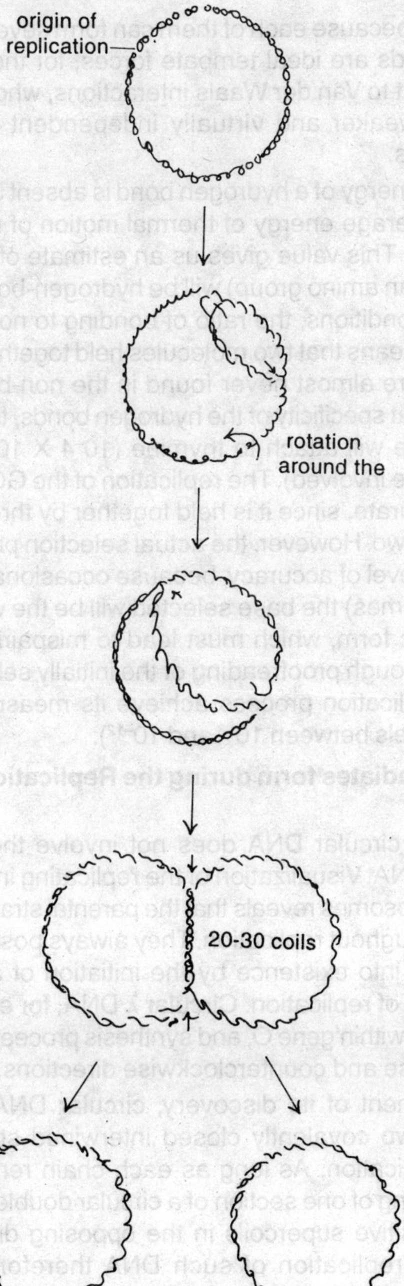

Fig. 12.1. Simplified version of the bidirectional replication of circular DNA molecules..

pyrimidine bases, because each of them can form several hydrogen bonds. These bonds are ideal tempate forces, for they are highly specific, in contrast to Van der Waals interactions, whose attractive forces are both weaker and virtually independent of particular chemical groupings.

The average energy of a hydrogen bond is absent 3 kcal/mol, or eight times the average energy of thermal motion of molecules at room temperature. This value gives us an estimate of how often a given group (e.g., an amino group) will be hydrogen-bonded. Under ordinary cellular conditions, the ratio of bonding to non-bonding is about 10^4:1. This means that two molecules held together by several hydrogen bonds are almost never found in the non-bonded form. Because of the great specificity of the hydrogen bonds, the frequency with which adenine will attach to thymine ($10^{-}4$ X 10^{-4} since two hydrogen bonds are involved). The replication of the GC pair should be even more accurate, since it is held together by three hydrogen bonds rather than two. However, the actual selection process never reaches this high level of accuracy, because occasionally (perhaps about once in 10^4 times) the base selected will be the wrong (imino or enol) tautomeric form, which must lead to mispairing. Now we realize that only through proofreading of the initially selected bases does the DNA replication process achieve its measured level of accuracy (error levels between 10^{-8} and 10^{-12}).

Θ-Shaped Intermediates form during the Replication of Circular DNA

Duplication of circular DNA does not involve the temporary creation of linear DNA. Visualization of the replicating intermediates of many ring chromosomes reveals that the parental strands maintain a circular form throughout replication. They always possess a 0-like shape that comes into existence by the initiation of a replicating bubble at the origin of replication. Circular λ DNA, for example, has its replication origin within gene O, and synthesis proceeds outwardly in both the clockwise and counterclockwise directions.

From the moment of its discovery, circular DNA posed the question of how two covalently closed interwined strands could unravel during replication. As long as each chain remains intact, even minor untwisting of one section of a circular double helix forces the creation of positive supercoils in the opposing direction. The semiconservative replication of such DNA therefore demands temporary cuts in one or both of its polynucleotide backbones. The problem thus becomes how and when the breaks are made. About this there could only be speculation until the discovery of the topoisomerases, DNA nicking/closing enzymes that can either further supercoil or relax DNA. The transient single-stranded cut made by topoisomerase I, for example, can create a molecular swivel around which supercoiled parental DNA strands can twist.

Now it seems clear that both parental strands remain intact throughout most of the replication cycle. The unreplicated portion of the 0-shaped intermediate becomes periodically supercoiled in either of two possible ways: Negative superhelices (untwists) maybe made by topoisomerase II (gyrase) in anticipation of subsequent chain elongation, and positive superhelices may be made by elongation of replicating bubbles in the absence of chain cuts. The later, positive supertwisting is clearly limited in scope and should halt replication bubble growth owing to the steric impossibility of creating further positive supercoils. Such halted bubbles could resume growth if a topoisomerase type I molecule made the single-stranded cuts that convert the highly supercoiled unreplicated section into the relaxed configuration. Alternatively, the essential cutting agent needed for further movement may be topoisomerase II (gyrase), acting to untwist the positively supertwisted DNA.

Overall Chain Growth Occurs in both 5'→3' and3'→5' Directions

The opposing chain directions (5'-3' and 3'-5') of the two strands of a double helix mean that the two daughter strands being synthesized at each replicating form must also run in opposite directions. Therefore, the overall direction of chain growth must be 5'→3' for one daughter strand and 3'→5' for the other daughter strand. However, all known forms of DNA polymerase, the only type of enzyme that can add nucleotide precursors of DNA, extend chains only in the 5'-3' direction, since the chemical reaction catalyzed by these enzymes allows a nucleotide triphosphate to react only with the free 3'-OH end of a growing polynucleotide strand. The question thus arose whether other replication enzymes had been missed, specifically, those that would add nucleotides onto free 5' ends. Despite a very extensive search, no enzyme with such specificity has ever been found.

Small Fragments are Precursors of many Long Chains: The Leading Versus the Lagging Strand

Resolution of the apparently paradoxical 3'→5' direction of synthesis came with the discovery that deoxyribonucleotide substrates are not always added directly to the very long daughter polynucleotide. Instead, they are sometimes found as parts of much shorter DNA chains 9100 to 1000 bases long) that later link up to the main daughter strands through the action of the polydeoxyribonucleotide-joining enzyme, DNA ligase. This finding opened up the possibility that a chain whose final direction of growth is 3'→5' might, in fact, be formed by the joining of smaller chains (often called *Okazaki f agments,* after their discoverer), each of which grows in the coventional 5'→3' manner. In this way, the same set of enzymes could be responsible for making all progeny strands. Proof that this indeed in the case came from direct studies of the ends of growing Okazaki fragments to which precursor nucleotides had recently been added. All such experiments show growth exclusively at the 3' end.

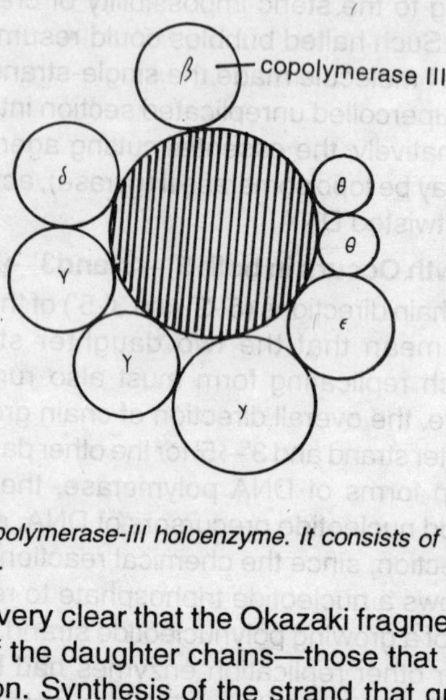

Fig. 12.2. DNA polymerase-III holoenzyme. It consists of subunits α, β, γ, ε, θ, δ and ∈.

Now it is very clear that the Okazaki fragments are precursors only to half of the daughter chains—those that grow overall in the 3'→5' direction. Synthesis of the strand that grows overall in the 5'→3' direction for the most part precedes that of the strand appearing to grow 3'-5'. This difference in rates of growth between the leading (5'→3') and lagging (3'→5') strands produces a small section of single-stranded parental DNA at the replication fork. It becomes converted to double-stranded DNA upon the initiation of an Okazaki fragment whose synthesis continues until it reaches the terminus of the previously synthesize Okazaki fragment, to which it then becomes joined. Okazaki fragments only commence synthesis at precise points along single stranded DNA templates and it is the distance between these signals that determine the average length of Okazaki fragments formed during replication of any particular DNA.

DNA Replication in Prokaryotes

Nucleotide Synthesis

To be efficient, virtually all cells use two different kinds of pathway for the synthesis of nucleotides. One is a *de novo pathway,* in which a sugar (ribose 5'-phosphate), certain amino acids, carbon dioxide, and NH_3 are combined in a series of reactions to form the nucleotides directly. In this case neither the free purines or pyrimidines nor the nucleosides are intermediates in the pathway.

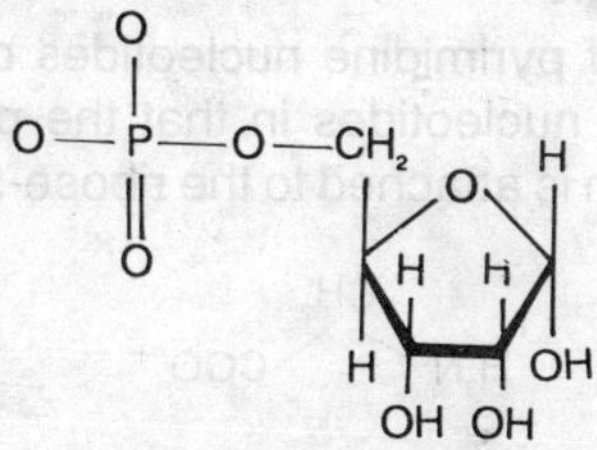

Fig. 12.3. The structure of ribose 5'-phosphate.

The second kind of pathway is called a *salvage pathway,* in which the purines, pyrimidines, and nucleosides released by the break-down of nucleic acids can, by a variety of routes, be converted back to the nucleotides needed for nucleic acid synthesis. Both pathways are important to the cell, and how active each pathway is depends upon the resources available to the cell.

Purine Biosynthesis

The purine biosynthetic pathway follows essentially the same sequence of reactions in a wide range of prokaryotic and eukaryotic organisms, including *E. coli*, yeast, plants and man. The purine ring structure is assembled on the sugar, ribose 5'-phospahte. A number of components are used to produce the purine ring, including some amino acids. The primary product of the purine biosynthesis pathway is the ribonuclotide inosine 5'-monophosphate (IMP) and it is from this compound that the adenine and guanine ribonucleotides are derived. As will be described later, these later two compounds are phosphorylated further to convert them to the immediate RNA and DNA precursors.

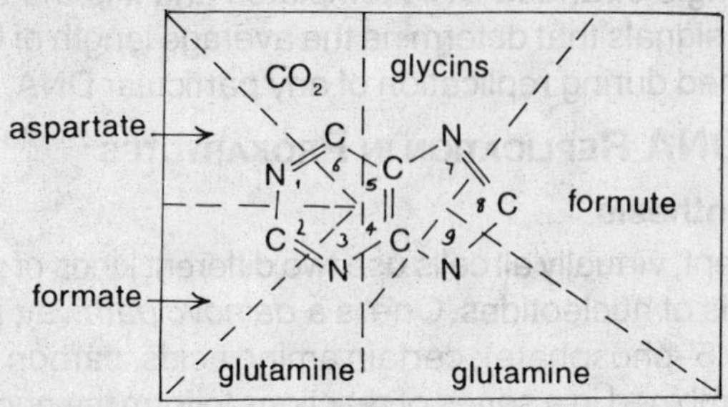

Fig. 12.4. Origin of the carbon and nitrogen atoms of the purine ring.

Pyrimidine Biosynthesis

The biosynthesis of pyrimidine nucleotides differs in general from that of the purine nucleotides in that the pyrimidine ring is assembled first and then is attached to the ribose-5'-phosphate. As

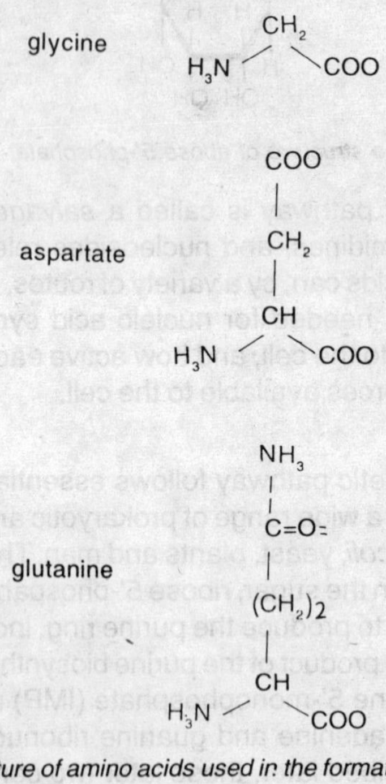

Fig. 12.5. Structure of amino acids used in the formation of purine (and pyrimidine) rings.

was the case with the purine ring, a number of compounds are used in the construction of the pyrimidine ring. The pyrimidine biosynthesis pathway is also shown diagrammatically. The key pyrimidine intermediate in this pathway is orotate, which becomes attached to an activated form of ribose 5'-phosphate, phosphoribosyl pyrophosphate (PRPP), to form a nucleotide, orotidine 5'-monophosphate (OMP). Removal of a carbon dioxide molecule generates uridine 5'-monophosphate (UMP). The cytosine nucleotide then is produced from UMP by the addition of an amino (NH_2) group.

The nucleoside monophosphates generated by the de novo purine and pyrimidine biosynthesis pathways do not participate directly in nucleic acid biosynthesis. Instead, they are first converted to the triphosphate derivatives via diphosphate intermediates. The phosphate donor for all of these reactions is the ribonucleotide, adenosine 5'-triphosphate (ATP), and the reactions are catalyzed by enzymes called *kinases.* (These enzymes are sites for regulation of the production of DNA and RNA precursors). The conversion of

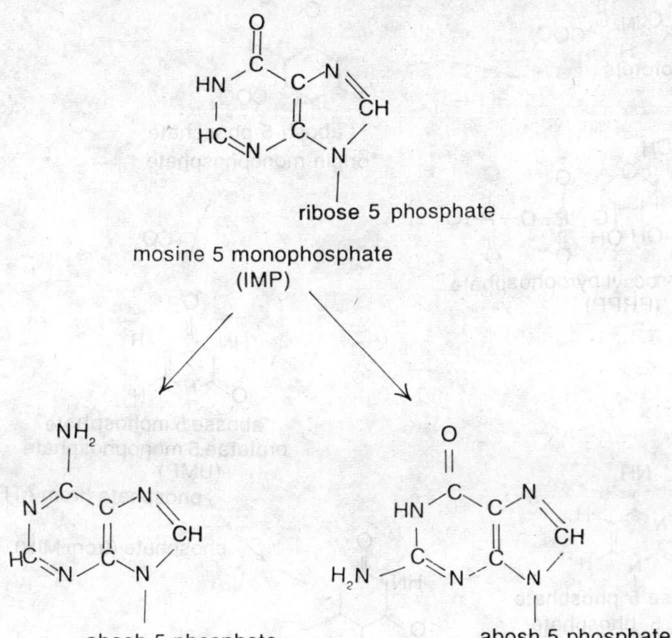

Fig. 12.6. Purine biosynthesis : production of the adenine and guanine nucleotides from inosine 5'-monophosphate.

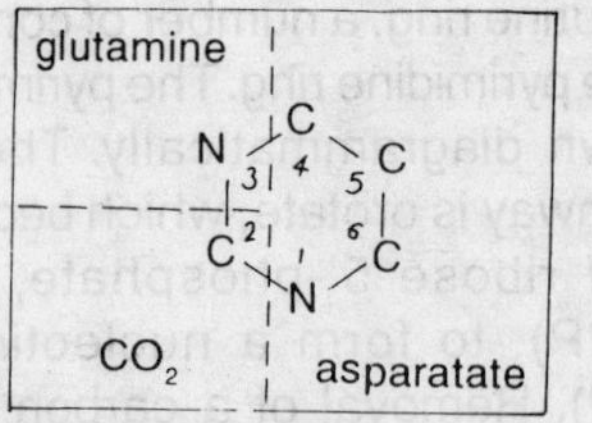

Fig. 12.7. Origin of the carbon and nitrogen atoms of the pyrimidine ring.

nucleoside monophosphates to diphosphates involves kinases that are specific for each base but non-specific with regard to the ribose or deoxyribose sugar. Thus, they are involved in the synthesis of both RNA and DNA precursors. On the other hand, the synthesis of nucleoside triphosphates from the diphosphates is catalyzed by a kinase that is non-specific for both the base and the sugar.

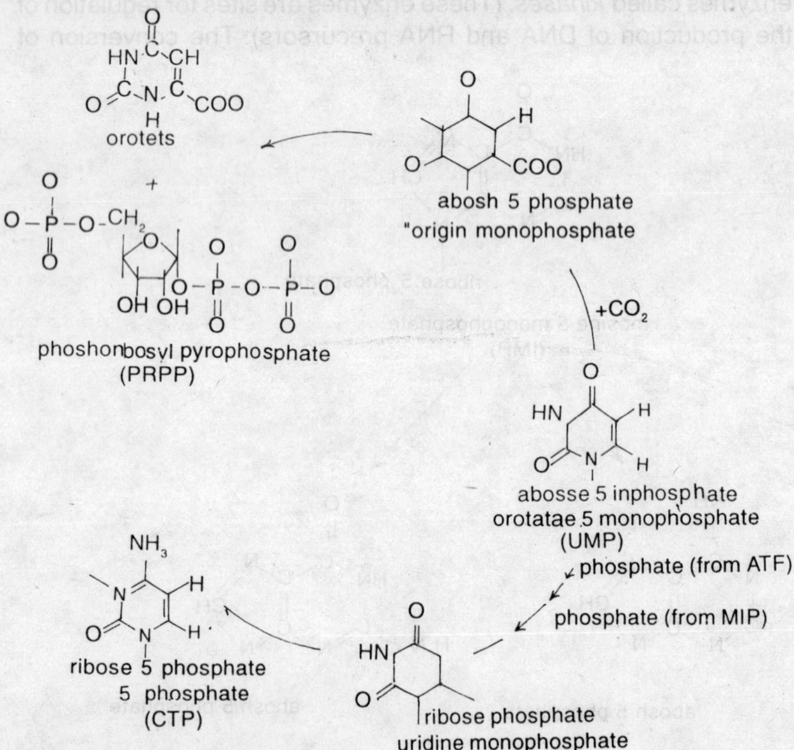

Fig. 12.8. General outiine the pyrimidine biosynthesis pathway.

By the kinase-catalyzed reactions, the ribonucleoside 5'-triphosphates are produced from the ribonucleoside monophosphate end products of the purine and pyrimidine biosynthesis pathways. These triphosphates are the immediate precursors of RNA synthesis, a process which will be described in a later chapter. The precursors for DNA synthesis are also derived from the ribonucleoside 5'-monophosphates. First, the diphosphates are produced and then the ribose sugar is reduced to the deoxyribose sugar in a reaction catalyzed by the enzyme ribonucleoside phosphate reductase. With the exception of dUDP, the resulting deoxyribo-nucleoside 5'-diphosphate (dADP, dGDP, dCDP) are then phosphor-lyated to produce 5'-triphosphates. As was discussed earlier, the base uracil is specific for RNA and this pyrimidine is replaced in DNA by thymine, which is actually 5'-methyluracil. To produce the thymine deoxyribonucleotide, dUDP is dephosphorylated to dUMP and the 5-position carbon of the pyrimidine ring is then methylated in a reaction catalyzed by thymidylate synthetase. The resulting deoxythymidine 5'-monophosphate is phosphorylated to produce the triphosphate. In this way, the for DNA precursors, dATP, dGTP, dTTP and dCTP, are produced in reactions catalyzed by the kinases discussed previously.

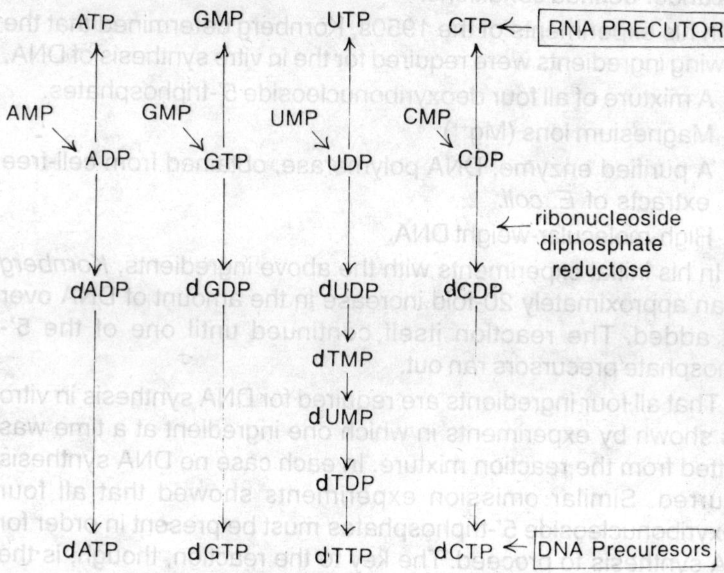

Fig. 12.9. Formation of ribonucleotide precursors for RNA and deoxyribonucleotide precursors for DNA.

Since the synthesis of RNA and DNA precursors are intimately related, one might expect that the regulatory elements of the system would be complex and indeed that is the case. For example, since cells must distribute their resources appropriately between RNA and DNA synthesis, one of the pivotal enzymes is ribonucleoside diphosphate reductase. This enzyme receives complex inhibitory and stimulatory signals by a variety of deoxyribonucleotides so as to provide a balanced supply of precursors for DNA synthesis. The synthesis of the enzyme can also be repressed by the deoxyribonucleotides if their poor sizes become large.

DNA Synthesis in Vitro

It was shown that the basic building blocks for DNA synthesis are the deoxyribonucleoside 5'-triphosphate, and the precursors are polymerized into DNA. In DNA synthesis, the polynucleotide chain grows from one end by the stepwise addition of deoxyribonucleoside triphosphates. In this, the direction of synthesis is determined by the fact that the 5'-triphosphate precursors must bond to the 3'-OH of the preceding deoxyribose. Before discussing recent information about DNA synthesis, we shall review the historical work of A. Kornberg and his colleagues that resulted in the in vitro synthesis of DNA under defined conditions.

In his experiments of the 1950s, Kornberg determined that the following ingredients were required for the in vitro synthesis of DNA.

1. A mixture of all four deoxyribonucleoside 5'-triphosphates.
2. Magnesium ions (Mg^{++})
3. A purified enzyme, DNA polymerase, obtained from cell-free extracts of *E. coli.*
4. High-molecular-weight DNA.

In his initial experiments with the above ingredients, *Kornberg* got an approximately 20-fold increase in the amount of DNA over that added. The reaction itself continued until one of the 5'-triphosphate precursors ran out.

That all four ingredients are required for DNA synthesis in vitro was shown by experiments in which one ingredient at a time was omitted from the reaction mixture. In each case no DNA synthesis occurred. Similar omission experiments showed that all four deoxyribonucleoside 5'-triphosphates must be present in order for DNA synthesis to proceed. The key to the reaction, though, is the DNA polymerase, which promotes synthesis by effecting the 3'-5'-phosphodiester linkages.

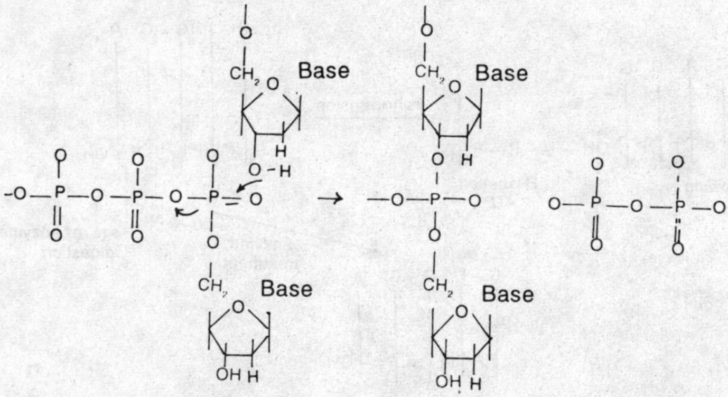

Fig. 12.10. Mechanism of DNA Polymerization.

Role of DNA in the Reaction

Two possibilities were proposed originally for the role of high-molecular weight DNA in the *in vitro* synthesis of DNA:

1. The DNA acts as a primer, a growing point for the terminal random addition of nucleotides.
2. The DNA acts as a template; that is, the base sequence of the DNA chain specifies the complementary base sequence of the new DNA chain with which it is bound by hydrogen bonds in the double helix structure. The requirement for all four DNA precursors to be present for any DNA synthesis to occur suggests that DNA has a template role.

All the available evidence indicates that the second possibility is correct: DNA acts as a template. The initial evidence came from studies that showed that the (A+T)/(G+C) ratio of in vivo synthesized DNA was the same as the DNA used to start the reaction. However, an objection to this is that the results could be a coincidence since base ratios do not indicate the arrangement of bases along the DNA. A more exacting comparison of newly synthesized and template DNA can be made by nearest neighbour analysis, which shows the relative frequency with which each of the four bases is located adjacent to each of the four bases in the DNA chain (i.e., A next to A, G, C, and T; C next to A,G,C and T; G next to A,G,C, and T; and T next to A,G,C, and T). Nearest neighbour analysis does nto give some information about the sequence of bases along the DNA, but it does give some information about the general arrangement of the bases in the DNA.

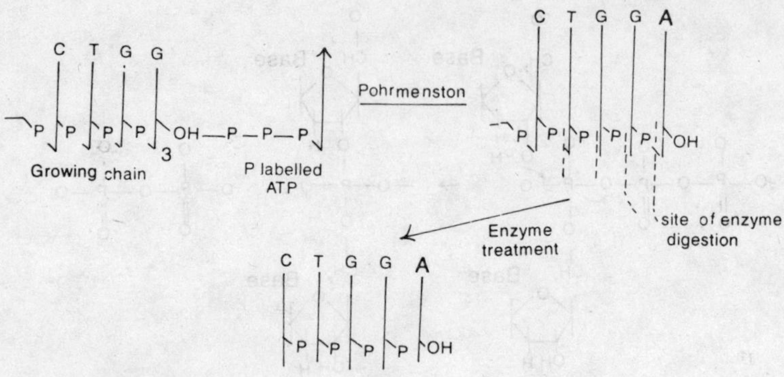

Fig. 12.11. The procedure of nearest neighbour analysis.

In nearest neighbour analysis DNA is isolated from the organism of interest and then is used as the template in Kornberg's in vitro synthesizing system. In this case, only one of the four deoxyriboncleoside 5'-triphosphates (dATP in the figure) is labeled with ^{32}P, a radioactive isotope of phosphorus. The result DNA is treated with a mixture of micrococcal deoxyribonuclease and spleen diesterase. These enzymes break the backbone of DNA between the 5'-carbon and the phosphoryl group, producing deoxyribonucleoside 3'-monophosphates, which can be analyzed by paper electrophoresis to determine the amount of radioactivity each contains.

In the example shown, the ^{32}P is now found in the form of DG 3' MP (deoxyguanosine 3'-monophosphate), which was the nearest neighbour to the 5' side of the adenine nucleotide in the DNA. Extrapolating this discussion to the entire DNA, the method provides data concerning the relative amount of radioactive label transferred from ^{32}-labeled dATP to the four possible neighbours. This gives information about the relative frequency with which A is next to A,G,C and T in the DNA. These four nearest neighbours are usually written as 5'-ApA-3', GpA, CpA and TpA. The experiment is then repeated with each of the other three DNA precursors such that in the end a matrix of 16 nearest neighbour values is produced. The 16 frequencies found are generally unique to the DNA is question. An example of the type of data that is obtained from nearest neighbour analysis is shown in Table.

Table 12.1. Nearest Neighbour Frequencies of Mycobacterium phlei DNA

Labelled triphosphate	*Deoxyribonucleoside 3'-monophosphate isolated*			
	Tp	*Ap*	*Cp*	*Gp*
dATP	TpA 0.012	ApA 0.024	CpA 0.063	GpA 0.065
dTTP	TpT 0.026	ApT 0.031	CpT 0.045	GpT 0.060
dGTP	TpG 0.063	ApG 0.045	CpG 0.139	GpG 0.090
dCTP	TpC 0.061	ApC 0.064	CpC 0.090	GpC 0.122
Sum	0.162	0.164	0.337	0.337

The data presented illustrate several points:

1. The sixteen possible nearest-neighbour frequencies occur with a large number of frequencies.
2. The sums of the vertical columns show that the amount of A is equal to the amount of T and that G equals C, thus indicating the DNA was probably replicated correctly.
3. The two DNA strands are of opposite polarity. This is borne out by the frequency equivalence of the pairs of sequences: CpT and ApG, GpT and ApC, GpA and TpC, and CpA and TpG, as predicted by antiparallel DNA strands. Were the two strands of the same polarity, different matching sequences would have been predicted, for example, TpA and ApT, GpA and CpT, CpA and GpT etc.

To show that newly synthesized DNA is made using a DNA template requires two rounds of nearest neighbour analysis. In the first round the originally isolated DNA is used in the reaction and a set of 16 nearest neighbour frequencies are obtained as described. In the second round the enzymatically synthesized DNA is used in the reaction and a second set of frequencies is produced. The results show good agreement between the two sets of frequencies, thus indicating that DNA plays a template role in DNA replications in vitro.

Since the early, comparatively crude in vitro DNA synthesis experiments, a large number of refinements have been made. For example, the reaction mixtures now duplicate the conditions that are present in growing cells such that it is possible to synthesize DNA in vitro that is identical in all respects with DNA produced in vivo. The reaction mixtures required for efficient DNA synthesis vary, depending on the DNA used as the template. In general, there is a basic set of enzymes and proteins required for the synthesis of all DNA sources. Beyond that, each DNA requires a number of specific proteins for new synthesis to occur, and the actual set of proteins needed depends on the DNA in question.

EVIDENCES IN SUPPORT OF SEMICONSERVATIVE REPLICATION OF DNA

1. Meselson and Stahl's Experiment

(i) *Meselson and Sthal* (1958) cultured a species of bacteria (*Escherichia coli*) in a culture medium containing N^{15} isotopes of nitrogen. After these had replicated for a few generations in that medium both the strands of their DNA contained N^{15} as constituent

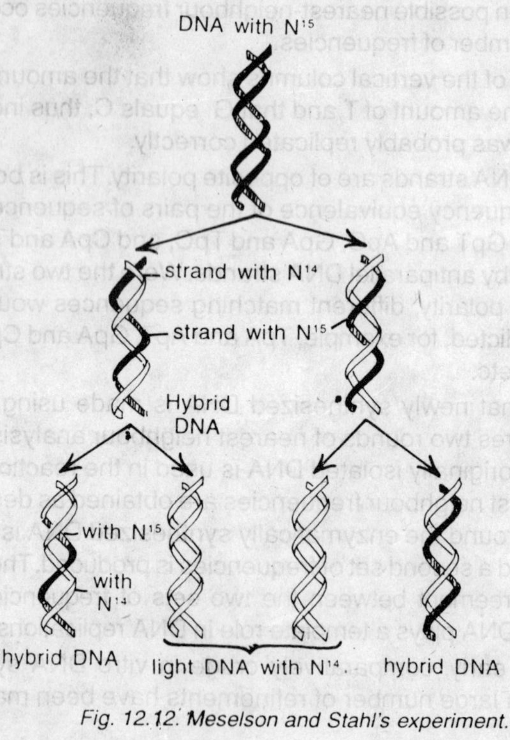

Fig. 12.12. Meselson and Stahl's experiment.

of purines and pyrimidines. When these bacteria with N^{15} were transferred in cultural medium containing N^{14}, it was found that DNA separated from fresh generation of bacteria possesses one strand heavier than the other. The heavier strand represents the parental strand and lighter one is the new one synthesized from the culture indicating semi conservative method of DNA replication.

(ii) Density determination experiments were also conducted. The DNA of parental generation (labelled with N^{15}) was found to be heaviest. In first generation it was lighter than the first one. In second generation two bands were observed indicating the presence of DNA with two different densities, as is expected according to semiconservative method of DNA replication. These two bands were of equal intensity in the second generation. In the subsequent generations the intensity of light density bands gradually increased and the intensity of hybrid density band gradually decreased.

2. Cairn's Autoradiography Experiment

J. Cairns demonstrated the semiconservative type of replication by using a technique of autoradiography as given in the following lines:

In autoradiography technique, the material is first supplied with a suitable radioactive material like tritiated thymidien (H^3-TdR; H^3 is heavy isotope of hydrogen and it replaces normal hydrogen in thymidine to give rise to tritiated thymidine). Tritiated thymidine is used since this will selectively label only DNA and will not label RNA, since thymine base is absent in RNA. The tritiated thymidine gets incorporated into DNA and replaces ordinary thymidine. The material is then sectioned or else the cells may be broken down to release the intact bacterial chromosome on slides. These slides are then covered by photographic emulsion and stored in the dark. During this storage the particles emitted by tritiated thymidine will expose the film, which can be developed. This photograph will then show the regions of the presence of tritium and thus indirectly the presence of labelled DNA.

Using this above technique, the replication of DNA could be easily followed by *J. Cairns* and the results were reported in 1963. During incorporation of tritiated thymidine, autoradiographs could be prepared at regular known intervals. Autoradio-graphs from this replicating material prepared at regular known intervals demonstrated semi-conservative mode of replication. It would be expected that in regions, where one cycle of replication of DNA has been completed,

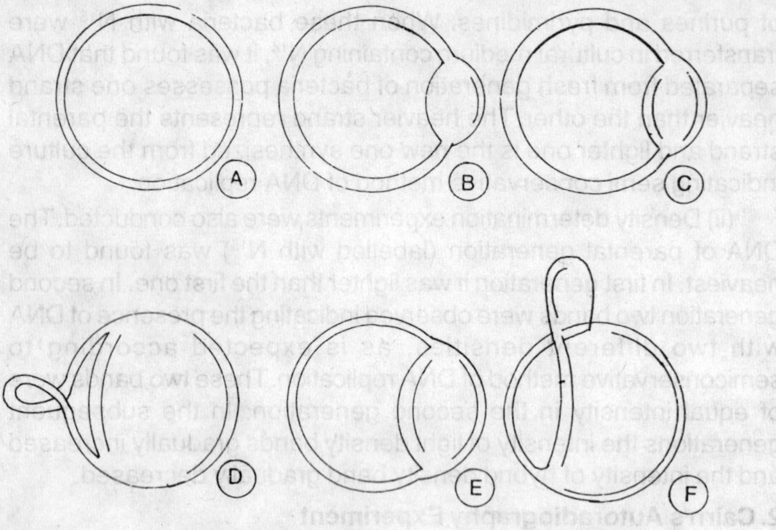

Fig. 12.13. Carn's model for circular DNA replication.

the density of dots will be higher than in the region where the replication has not taken place. The lighter density of dots is considered to indicate that only one of the two strands is labelled, while a heavier density of dots will indicate that both strands are labelled. Such a situation was actually observed. The rate at which the replication proceeds could also be worked out by measuring the length of DNA undergoing replication in a known interval of time. The generation time was worked out by *Cairns* as 30 minutes in *E.coli.* The length of the chromosome was worked out to be about 1 mm. The rate of replication as obvious would be approximately 30μ to 40μ per minute (1 mm = 1000μ).

That after replication, one of the two strands in the daughter DNA molecules is derived from the parent molecule and the other is synthesized afresh. In θ shaped figure, which is obtained the second cycle of replication in presence of label, the two areas in the split region would never be equally labelled. For instance, one arc would be twice as heavily labelled as the other arc. This is what actually observed by *Cairns*. Thus the observation support the semiconservative type of replication of DNA.

3. Taylor's Experiment on Vicia faba Root Tips

J.H. Taylor (1957) and his coworkers also demonstrated the semiconservtive method of DNA duplication in the root tip cells of

Vicia faba by autoradiography. The roots were grown in a medium containing radioactive thymidine, so that the radioactivity is incorporated in the DNA of these cells. The outline of this labelled chromosome on a photographic film appears in the form of scattered black dots of silver grains. When these root tips with labelled chromosomes were transferred to the unlabelled medium containing colchicine and studied for radioactivity, the following observations were made:

1. In the chromosomes of first generation the radioactivity was found to be uniformly distributed in both the chromatids, because in them the original strand of DNA double helix was labelled with radioactivity and the new one was non-labelled.
2. In the chromosomes of second generation only one of the two chromatids in each chromosome was radioactive.

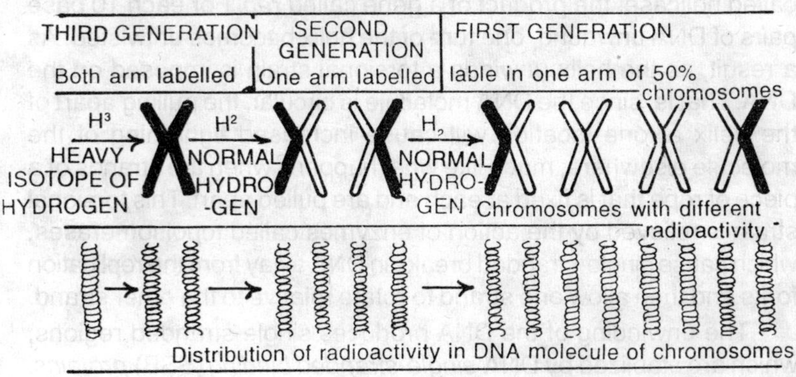

Fig. 12.14. Taylor's experiment on Vicia faba.

This experiment demonstrates the semiconservative method of chromosome replication. But as it is known that each chromatid is composed of only one double helical molecule of DNA, it is the replication of DNA.

Enzymes and Proteins Involved in DNA Replication

The complicated process of DNA replication in *E.coli* requires many different proteins and enzymes, some of which also function in other cellular process such as the repair of DNA damage and genetic recombination. The key enzyme in the synthesis of new DNA is DNA polymerase.

In *E. coli* there are three DNA polymerases. Each catalyzes the DNA template-directed condensation of deoxyribonucleoside 5'-triphosphates. There are significant differences in their activity in DNA synthesis and in the nuclease (DNA or RNA break-down) activities associated with them.

All three polymerases catalyze new DNA synthesis in the 5'-3' direction. The rate of DNA synthesis catalyzed by the three enzymes varies circular prokaryotic DNA molecules replicate bidirectionally (in both directions away from the origin of replication). In these cases there are two replication forks that are mirror images of one another picture two Ys joined head-to-head by the two pairs of arms. The area of a double-stranded DNA molecule that has denatured for replication is called a *replication bubble.*

In *E. coli* the unwinding of the DNA is catalyzed by an enzyme called helicase, the product of a gene called *rep.* For each 10 base pairs of DNA unwound, one turn of the helix becomes untwisted. As a result, as the helix unwinds a torsional strain is imposed on the DNA. That is, since the DNA molecule is circular, the pulling apart of the helix at one location will cause increased tightening of the molecule elsewhere, much like what happens when the strands of a piece of rope that is fixed at each end are pulled apart. This torsional strain is relieved by the action of enzymes called topoisomerases, which cause single-stranded breaks in DNA away from the replication forks and then allow one strand to rotate relative to the other strand.

The unwinding of the DNA produces single-stranded regions, which are stablized by DNA *single-stranded binding* (SSB) *proteins.* Each SSB protein is a tetramer of a 74,000 dalton subunit, and this tetramer binds to eight bases of single stranded DNA. Over 250 tetrameres bind to each replication fork. Once the strands have started to unwind, the internal bases become available for the formation of bonds with bases in new chain.

Initiation of DNA replication next takes place. In *E. coli* primases bind to the single-stranded DNA of both arms of the replication fork and synthesize short primers. The primers are lengthened by the action of DNA polymerase III, which synthesizes the complementary DNA chains to the template strands. Note that, because of the opposite polarities of the two DNA template strands and the requirement for a 5'- to -3' direction of new DNA synthesis, the primers are located at different relative positions on the two template strands.

The DNA helix continues to unwind. On the template on the left the new strand continues to be synthesized continuously (in the same direction as the direction of undwinding). However, because DNA polymerases can only synthesize DNA in the 5'-to-3' direction, the new synthesis on the template on the right has gone as far as it can. Thus, a new initiation of DNA synthesis occurs on the right-hand template on the newly exposed single-stranded template close to the site of helicase activity. As before, a primer is made, which is lengthened by DNA polymerase III action. The overall result is that the new DNA strand being made on the left-hand tempate is synthesized continuously while the new DNA being made on the right-hand template is synthesized discontinuously so that Okazaki fragments are produced.

Eventually the Okazaki fragments are joined together into a continuous DNA strand. This process requires the activities of the enzymes DNA polymerase I and DNA ligase. If we consider two adjacent Okazaki fragments, the 3' end of the newer DNA fragment is adjacent to but not joined to the primer 5' end of the previously synthesized fragment. The DNA polymerase III that has just completed the synthesis of the newer fragment now dissociates from the DNA and DNA polymerase I takes its place. This enzyme continues the 5'- to -3' synthesis of the newer DNA fragment, at the same time removing the primer of the older fragment through the action of the 5'-to-3' exonuclease function of DNA polymerase I. When DNA polymerase I is finished, there is a gap between the two new DNA fragments, and this gap is sealed in a sealed in a reaction catalyzed by DNA ligase, thereby producing a longer DNA strand. The whole process is completed until all the DNA is replicated.

13

RIBONUCLEIC ACID (RNA)

Ribonucleic acid usually occurs in nature as long unbranched polymeric molecule formed of single chain. RNA is generally present in the cytoplasm as well as in nucleolus. *Casperson* 1950 and *Brachet* in 1955 suggested that there is an association between RNA and protein synthesis. A large number of studies have been undertaken to characterise physically the nature and structure of RNA and their biochemical properties and by now we have acquired a lot of knowledge about the various types of RNA and their molecular structures but still there are various gaps to be filled to make a complete picture.

Table 13.1. Genetic RNA in Different Viruses

Types of Viruses	*Name of Viruses*	*Type of RNA*
1. Plant viruses	1. TMV	1. Single stranded
	2. Wounded tumour	2. Double stranded
2. Animal viruses	3. Influenza virus	3. Single stranded
	4. Poliomyelitis	4. Single stranded
	5. Reo virus	5. Double stranded
3. Bacteriophages	6. MS_2	6. Single stranded
	7. F_2	7. Single stranded
	8. r_{17}	8. Single stranded

Some plant viruses (e.g., TMV, turnip yellow mosaic viruses, wound tumour viruses, etc.), animal viruses (e.g., influenza viruses, foot and mouth viruses; rous sarcoma viruses, poliomyelitis viruses reoviruses, etc.) and bacteriophages (e.g., MS2, etc.) contain ribonucleoic acid (RNA) as their genetic material. That is why in these cases it is called the *Genetic RNA.*

Molecular Weight of RNA

The molecular weight of RNA may be determined by light scattering measurements, by sedimation velocity measurements, by viscosity measurements, by equilibrium ultracentrifugation or by polyacrylamide gel electrophoresis in formawide.

Bacteria, *Actinomycetes,* blue green algae and higher plant chloroplasts all have rRNA's of molecular weight 1.1 X 10^6 (23S) and 0.55 X 10^6 (16S) whereas the corresponding values for higher plants, ferns, algae, fungi and some protozoa are 1.3 X 10^6 (26S) and 0.7 X 10^6 (18S). The 18S components is common to all animals, but the large 28S component has evolved from 1.4 X 10^6 daltons in sea urchins to 1.75 X 10^6 daltons is mammals. In mitochondria (mammalian), the rRNA's are somewhat smaller's the 21S species is 0.56 X 10^6 daltons and the 12S species 0.36 X 10^6 daltons. Whlst the ribosomal RNAs form most organisms possess broadly similar base compositions with G plus C content between 50 and 60 percent, there are some exceptions to this, e.g. *Drosophila* (40%) and *Tetrahymena* (43%).

Primary Structure of RNA

RNA usually occur in nature as long unbranched polymeric molecule made up of single chain or helix of polynucleotide strand: Ribnucleic acid is composed of pentose sugars, phosphoric acid and nitrogenous bases. The pentose sugar of RNA is ribose sugar. The sugar forms the back bone along with the phosphoric acid just in a manner studied in DNA.

Phosphoric acid is also one of the constituent of the RNA which along with the pentose sugar forms the back bone, on which the nitrogenous bases are linearly arranged, as shown in the Fig. On the bases of phosphate or phosphoric acid, the chemical constituents of RNA can be written in two ways:

Nitrogenous Base + Ribose Sugar + Phosphate = Nucleotides

Nitrogenous Base + Ribose Sugar = Nucleoside

The nitrogenous bases of ribonucleic acid are of two types:

(i) Purine

(ii) Pyrimidine.

Purine bases comprise mainly adenine (A) and guanine (G) while Pyrimidine bases comprise cytosine (C) and Uracil (U).

These bases are joined to the pentoses by N-C *glycosidic bonds.* For the purine bases the glyclosidic bond is between the C_1 position of the pentose sugar and the N_9 position of the base. For the pyrimidines, the linkage joins the C_1 and N_3 positions in figure given below:

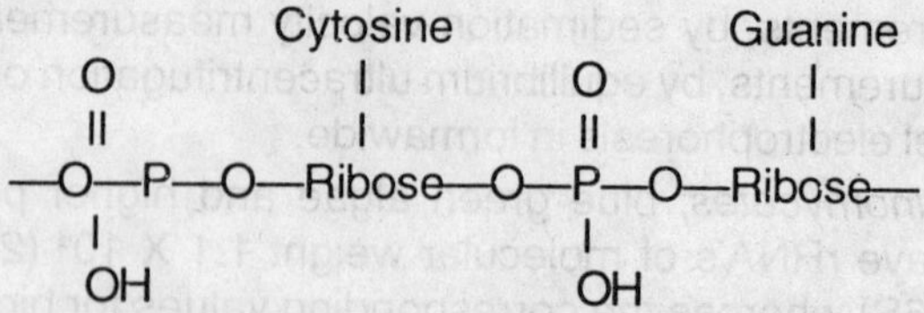

The bases are arranged on the back bone of the chain which is composed of alternating sugar and phosphate. The bases are attached to the ribose sugar in the same way, the bases in DNA are attached. The arrangement of the bases on the back bone is shown in the following diagram:

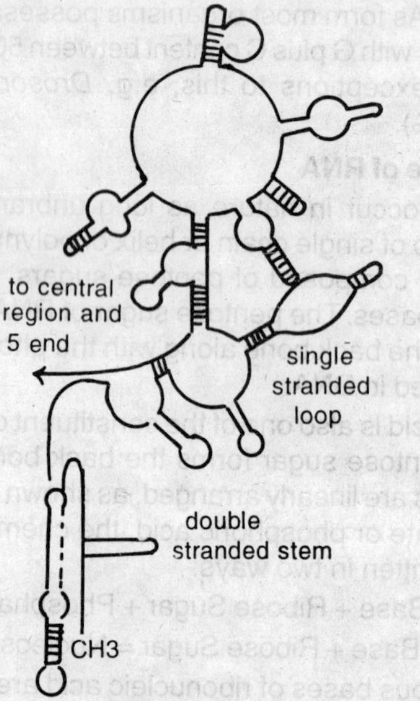

Fig. 13.1. Showing the Secondary structure of RNA.

Secondary Structure of Ribonucleic Acid

Generalization as to the secondary structure of RNA is probably less justified than in the case of DNA, as the available information is much less extensive. The available information does permit the following conclusions:

1. There is no evidence for the existence of more than one polyribonucleotide strand in the natural RNAs.
2. Hence the helical regions of natural RNA arise from the hair pin like bending of the single strand back upon itself.
3. The helical regions are stabilized by A—U and C—G interactions similar to those which stabilize the confirmation of DNA. However, the irregular base composition of RNA does not allow the incorporation of the entire molecule into double perfect helix of this type, and the helical zone co-exist with the random segments.
4. Two alternative models for the structure of RNA may be postulated:

(a) The non-helical fraction of RNA may take the form of extended amorphous zones separating segments of nearly perfect helical organization. This model pre-suppose that extensive complementary sequences exist within the single strand. These tend to seek each other out to form perfect or nearly perfect, helical regions.

(b) Alternatively, nucleotides which do not participate in the helical organization may occur largely as numerous small loops external to the helically paired structure.

Types of Ribonucleic Acid

Four types of ribonucleic acids have been studied and they are as follows:

1. mRNA
2. tRNA
3. rRNA
4. vRNA

1. Messenger RNA (mRNA)

Studies by *Huxley* and later by *Volkin* and *Astrachan* with T_2 bacteriophage infected (bacterial virus, *E.coli*) in 1956 led to the first existence for the unique kind of the RNA, later called messenger RNA. *Jacob* and *Monod* (1961) who first proposed this name, believed messenger to be necessary to convey the message of the genetic code from the DNA of the nucleus to the cytoplasm. This messenger is known as mRNA and it acts as a template for the translation of the DNA code into a specific protein. Thus it helps in determining the structure of protein.

mRNA molecules exist in a large variety of lengths, depending on the length of the polypeptide chain for which they code. Its molecular weight and sedimention value vary considerably. Generally the life span of mRNA is relatively short, although in some eukaryotic cells there is evidence of some degree of stability.

Further mRNA is always complementary to a portion of strand of DNA, with the exception that uracil replaces thymine. It should therefore hybridize with the DNA from which it was transcribed.

It is now an established fact that when the code has been transcribed from DNA on to mRNA, the later leaves the nucleus passes through the nucleus membrane into the cytoplasm. Here it moves to the ribosomes, the site of protein synthesis. mRNA molecules attaches reversibly to the surface of ribosome always binding to the smaller sub-units. The magnesium cation, Mg^{++} is involved in the formation of a complex between mRNA and ribosomes.

Types of mRNA

On the basis of size and the coding two types of mRNA has been described:

(a) *Monocistronic mRNA.* Mostly the mRNA carries the codons of single cistron (i.e., codes for one complete protein molecule) of the DNA. Such mRNA molecule is called monocistronic mRNA. For example, for the synthesis of a polypeptide chain of 300 to 500 amino acid residues, a monocistronic mRNA of *E.coli* contains 900 to 1500 nucleotides in its molecule.

(b) *Polygenic or polycistronic mRNA.* Sometimes a mRNA molecule carries the codes from several adjacent DNA cistrons and become much longer in size. This type of mRNA is called polygenic or polycistronic mRNA. For example, for the metabolism of the histidine protein the cell synthesizes about 10 specific enzymatic proteins and a mRNA in this case may carry codons for all the 10 enzymes.

Informosomes

In eukaryons mRNA is associated with protein forming ribonucleo-protein complexes. Some of these complexes remain free in the cytoplasm without being attached to the polyribosomes. These are named as informosomes by *Spirin.* These are very stable and can remain in the cell cytoplasm for several days. The stability is attributed by the presence of protein sheath around mRNA molecule. In informosomes proteins and mRNA are found in the ratio of 4:1. These occur in the form of particles of 80S or less.

Since informosomes are very stable, these are used by the cell is protein synthesis only when the is delay in translation. In embryonic cells, genetic apapratus starts functioning only late during organogenesis. Therefore, in early stages of cleavage when there is no rRNA, informosomes govern the process of protein synthesis.

Stability

The cell does not contain quantities of mRNA. This is because mRNA, unlike other RNAs is constantly undergoing breakdown. It is broken down to its constituent ribonucleotides by ribonucleases. In bacteria, mRNA may be so short-lived that while one end is translating protein the other end may be undergoing breakdown. In *E.coli* the average half-life of some mRNA is about two minutes.

Eukaryotic cells, on the other hand, contain metabolically stable mRNA. Mammalian reticulocytes are immature RBC which have lost their nuclei but have retained ribosomes. These cells can synthesize haemoglobin for many hours and even days by utilizing the mRNA transcribed by the nuclei when they were present.

2. Transfer Ribonucleic Acid

In the literature, t-RNA molecule have been variously termed tRNA, sRNA for soluble RNA, or adaptor RNA. It forms about 10-15% of the total RNA in most of the existence. Each tRNA contains approximate 80 nucleotides. The soluble or t-RNA forms large varieties of different cells, is homogenous in its molecular size having a molecular weight of about 25000.

The nucleotide sequence of tRNA was first worked out by *Holley et.al.* (1965) for yeast alanine tRNA. Since then the sequence of about 75 different tRNAs have been established. According to the Holley's leaf model the tRNA is a polynucleotide chain which is folded upon itself to form 5 arms. As a result of the folding the 3' and 5' ends of the chain come near each other. An arm consists of a stem and a loop. Bases are arranged in pairs in arms. There is no base pairing in the loops.

The 3'end of the polynucleotide chain ends in CCA base sequence. This represents site for the attachment of activated amino acid. The end of the chain terminates with guanine base. The bent in the chain of each tRNA molecule contains a definite sequence of

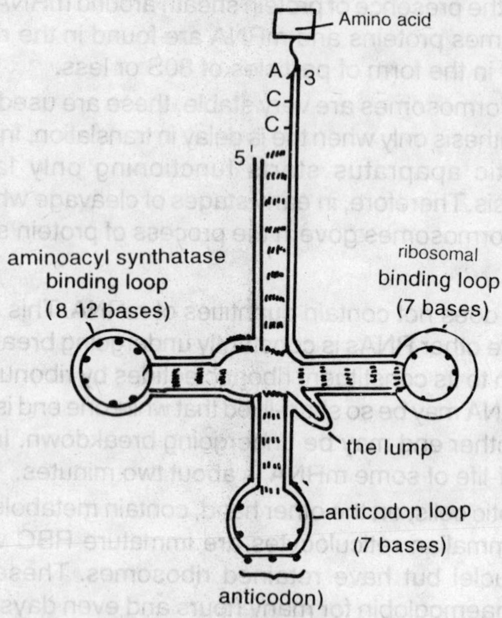

Fig. 13.2. Clover leaf model of tRNA.

three nitrogenous bases which constitute the anticodon. It recognizes the codon on mRNA.

Four different regions or special sites can be recognized in the molecule of tRNA. These are:

(i) *Amino acid attachment site.* It occurs at the 3'end of the tRNA chain and has -OH group of ribose molecule free to pair. This -OH group combines with the specific amino acid in the presence of ATP forming amino-acyl tRNA. It is common to all the tRNA molecules.

(ii) *Recognition site.* It contains a specific base sequence which dictates the attachment of correct amino acid to the tRNA molecule. It matches with the amino acid activating enzyme through which attachment of amino acid to tRNA takes place.

(iii) *Anticodon or codon recognition site.* This site has three unpaired based (triplet of base) whose sequence is complementary with a codon (triplet) in mRNA. Therefore, it determines the pairing of tRNA with the specific codon (triplet) of mRNA. It is therefore, the most specific region of tRNA molecule.

(iv) *Ribosome recognition site.* This helps in the attachment of tRNA to the ribosome. This site is common to all the molecules of tRNA.

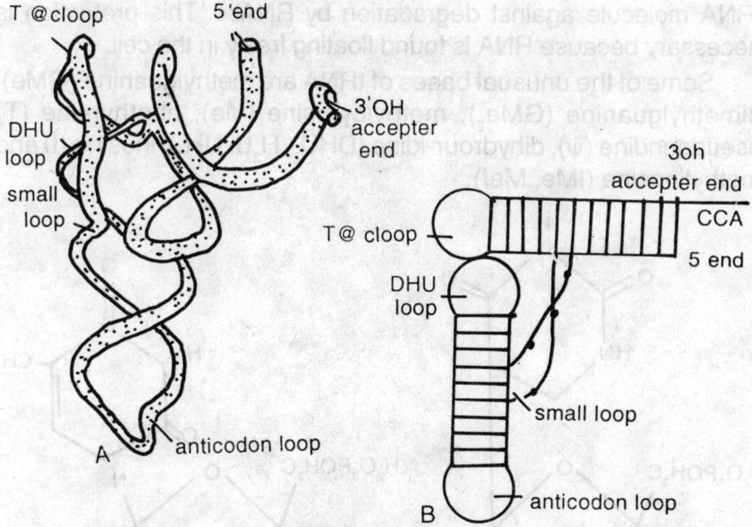

Fig. 13.3. Three dimensional structure of yeast phenylalanine tRNA.

Three dimensional structure of tRNA. In order to understand the structure-function relationship of tRNA, its three dimensional structure (TDS) should be known. With the availability of tRNA in crystalline form in 1968, TDS of tRNA could be studied through X-ray crystallography as was done for DNA double helix. About a dozen models for TDS were proposed, but the latest and most acceptable model was proposed in 1973 by *S.H. Kim.* He suggested that TDS of tRNA takes the shape of letter L with a thickness 20Å. This can be easily derived from two dimensional clover leaf model. The extra arm which varies in different tRNA molecules can be extended without distoring TDS conformation. The CCA stem projects out and can take different orientations.

Unusual bases

In addition to the usual basses, A, U, G and C, tRNA contain a number of unusual bases, and in this respect differs from mRNA and rRNA. The unusual bases of tRNA account for 5-20% of the total RNA of cell. Most of the unusual bases are formed by methylation (addition of $-CH_3$ or methyl group to the usual bases) e.g., cytosine and guanine on methylation yield methylcytosine and methylguanine, respectively. Precursor tRNA molecules transcribed on the DNA template contain the usual bases. These are then modified to unusual bases. The unusual bases are important because they protect the tRNA molecule against degradation by RNAse. This protection is necessary because RNA is found floating freely in the cell.

Some of the unusual bases of tRNA are methylguanine (GMe), dimethylguanine (GMe_2), methylcytosine (Me), ribothymine (T) pseudouridine (ψ), dihydrour-idine (DHU_2, H_2U, UH_2), inosine (I) and methylinosine (IMe, MeI).

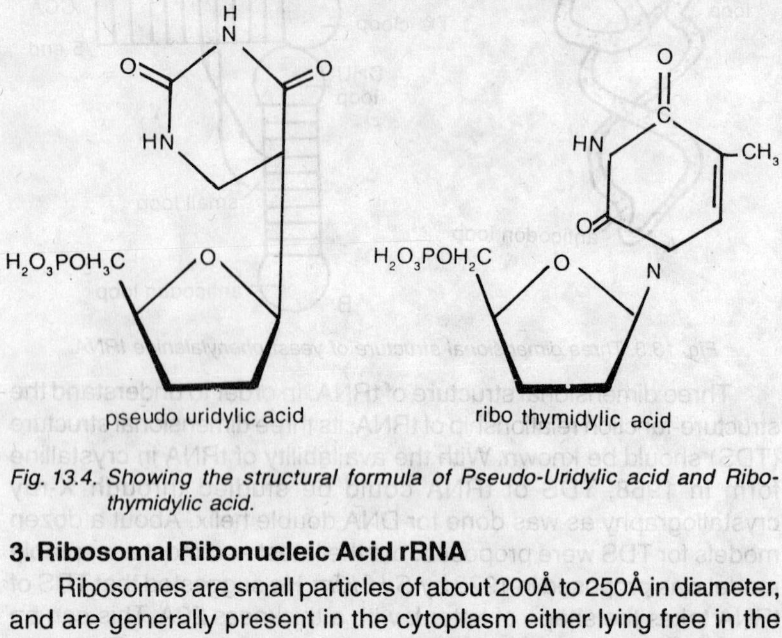

Fig. 13.4. Showing the structural formula of Pseudo-Uridylic acid and Ribo-thymidylic acid.

3. Ribosomal Ribonucleic Acid rRNA

Ribosomes are small particles of about 200Å to 250Å in diameter, and are generally present in the cytoplasm either lying free in the cytoplasm or attached to the outer membrane of endoplasmic reticulum or nuclear membrane. These particles are ribosomes containing about 60% ribonucleic acid or RNA which is ribosomal RNA ro rRNA and 40% protein. It is generally agreed that cytoplasmic ribosomes or their subunits are made in the nucleus and are subsequently transferred to the cytoplasm.

However, the mode of their passages across the nuclear membrane is still a matter of controversy.

Even from the morphological evidence, it would appear that the reticular or microvesicular spheres of the nucleus could be engaged in either the synthesis of rRNA, or its assembly, or both.

Table 13.2. Comparison of the Different Types of RNA

	Ribosomal RNA (rRNA)	*Messenger RNA (mRNA)*	*Transfer RNA (tRNA)*
1. Percentage of total RNA of cell	80%	3 to 5%	10 to 20%
2. Sedimentation Coefficient	28S, 18S, 5.8S, 5S 23S, 16S, 5S	8S	3.8S
3. Number of nucleotides	5S RNA: 120 nucleotides 16- 18S RNA: 1,600 to 2,500 nucleotides 23-28S RNA: 3,200 to 5,500 nucleotides	*E.coli*: 900 to 1,500 nucleotides	73 to 93 nucleotides.
4. Molecular weight	23S RNA: 1.1 x 10^6 30S RNA: 0.55 x 10^6	500,000	25,000-30,000
5. Unusual bases	Small amount of methylated bases.	Small amount of unusual bases. 1 per 30-40	High content unusual bases. (*E.coli*: 1 per 100-150 nucleotides).
6. Site of synthesis	Derived from nucleolar DNA	Synthesized in nucleus on DNA template.	Synthesized in nucleus on DNA template.
7. Beginning of synthesis	Synthesis begins at gastrulation, and increases as development proceeds	Some mRNA is found in the ovum. New mRNA is synthesized during early cleavage.	tRNA synthesis occurs at the end of cleavage stage
8. Base relationship of DNA	No obvious base relationship to rDNA. rRNA is from only small section of DNA	mRNA shows base relationship to DNA. It is formed from all sections of DNA	Same as in rRNA.
9. Function	Unpaired bases may bind mRNA and tRNA to ribosomes.	Conveys genetic information from DNA of chromosomes where it takes part in protein synthesis.	Adaptor for attaching amino acids to mRNA template.

The finding that hybridization of rRNA takes place with the chromatin associated with the nucleolar region suggests that rRNA is synthesized on the part of the chromosome traversing the nucleolus (the nucleolar organizer) or, in some cases, on DNA detached from the chromosome and associated with the nucleolus.

Ribosomal RNA in size and base contents from tRNA and from other RNA classes of most cells. rRNA from most organism is relatively rich in guanine and cytosine, and it lacks many of the rare base found in tRNA, although some methylated bases are present.

The function of rRNA is protein synthesis is not yet completely clear. The unpaired bases in the molecule may bind mRNA and tRNA to ribosomes possibly by Mg^{++} linkages between phosphate groups on the two molecules.

Ribosomal RNA also gives some evidence for developing the double helical structure.

4. Virual RNA (vRNA)

Viruses may contain either DNA or RNA. However, each particular virus generally contains nucleic acid of only one type. Some examples of RNA viruses are influenza, turnip yellow mosaic, tomato bushy stunt, tobacco necrosis, and the particularly well studied tobacco mosaic virus (TMV).

Table 13.3. Comparison between DNA and RNA

Deoxyribonucleic acid	*Ribonucleic acid*
1. DNA is the usual genetic material	1. RNA is the genetic material of some viruses.
2. DNA is usually double-stranded. (In certain viruses DNA is single stranded, $\phi\chi$ 174)	2. Most cellular RNA is single stranded. (Some viruses e.g., reovirus, have double stranded RNA).
3. The pentose sugar is double stranded	3. The pentose sugar is ribose.
4. The common organic bases are adenine, guanine, cytosine and thymine.	4. The common organic bases are adenine, guanine, cytosine and uracil.
5. Base pairing: adenine pairs with thymine and guanine with cytosine.	5. Adenine pairs with uracil and guanine with cytosine.
6. Pairing of bases is throughout the length of the molecule.	6. Pairing of bases is only in the helical region.

7. There are fewer uncommon bases.	7. There are more uncommon bases.
8. DNA is only of one type.	8. There are three types of RNA, messenger RNA, ribosomal RNA and transfer RNA.
9. Most of the DNA is found in the chromosomes. Some DNA is also found in the cytoplasm e.g. in mitochondria and chloroplasts.	9. Messenger RNA is formed on the chromosomes, and is found in the nucleolus and cytoplasm. rRNA and tRNA are also formed on the chromosomes, and are found in cytoplasm.
10. Denaturation (melting) is partially reversible only under certain conditions of slow cooling (renaturation).	10. Complete and practically instantaneous reversibility of the process of melting.
11. Sharp, narrow temperature interval of transition in melting.	11. Board temperature interval of transition in melting.
12. DNA on replication forms DNA, and on transcription forms RNA.	12. Usually RNA does not replicate or transcribe. (In certain viruses RNA can synthesize on RNA chain).
13. Genetic messages are usually encoded in DNA.	13. The usual function of RNA is translating messages encoded in DNA into proteins.
14. DNA consists of a large number of nucleotides, upto 4.3 million.	14. RNA consists of fewer nucleotides, upto, 12,000.

14

THE BIOSYNTHESIS OF RNA : TRANSCRIPTION

In the process of DNA transcription the positioning of nucleotide units in the RNA molecules that are being made is under the control of the DNA which acts as template.

Each RNA chain is initiated at a specific site on the DNA template and subject to termination at another unique type of site on the template. In other words there are defined units of transcription. Specific signals in the DNA template are recognized by the transcription apparatus. Initiation is governed by *promoter* regions in the DNA, and a region governing termination is designated a *terminator.*

Transcription mediated by DNA-dependent RNA polymerases. The properties of the enzyme from *E.coli* have however been the most widely explored.

The products of DNA-dependent RNA polymerases have ribonucleotide sequences complementary to *one* of the strands of the DNA which was used as 'template'. A guanine residue in the DNA template strand dictates the insertion of a cytosine nucleotide in the RNA strand under construction, whilst a cytosine in the DNA causes a guanine nucleotide to appear in the new RNA strand. Similarly, a thymine in the DNA results in an adenine nucleotide in the RNA, and an adenine in the DNA results in a uracil nucleotide in the RNA.

The DNA-dependent RNA polymerase has now been extensively purified form *coli.* The holoenzyme is a complex zinc-containing protein of molecular weight 480 000-500 000 which can be dissociated, for instance in 6M-urea, into a number of polypeptide chain subunits as follows:

Two α-chains each of mol. wt. 39 000. One β-chain of mol. wt. 155 000. One β'-chain of mol. wt. 165 000. One molecule of σ(sigma)-factor, mol. wt. 95 000.

The holoenzyme may therefore be represented as $\alpha_2\beta\beta'\sigma$. The hole enzyme without α-factor is termed 'core enzyme', or $\alpha_2\beta\beta'$. If the holoenzyme is separated into the core enzyme and the α-factor, the core enzyme will catalyse the synthesis of RNA chains from random sites on a DNA template *in vitro.*

Addition of σ-factor however restores the synthetic activity of the core enzyme by activating selective initiation of RNA synthesis. It complexes with the core enzyme to yield holoenzyme which interacts specifically with promotor regions on the DNA template.

Binding and Initiation

The sequence of molecular events for accurate initiation of RNA synthesis can be envisaged as follows. After a series of unproductive random interactions with the DNA template, the holoenzyme 'recognizes' a specific structure, or sequence, within a promoter region. This enables the holoenzyme to bind to the DNA in this region at least an order of magnitude more tightly than in the non-specific interaction mentioned above.

The next step in the Initiation process is a change in the conformational state of the DNA in this promoter. Once this transition has occurred, the holoenzyme is poised to commence the manufacture of an RNA chain, but actually using only *one* of the strands as template.

Although it is the β'-subunit that is implicated in the binding of the polymerase of the DNA and perhaps has the required specificity, σ appears to bind to the β-subunit. It may be best to consider the σ-factor as an allosteric effector of a multi-submit enzyme. The β-subunit itself is probably most concerned with at least the initial catalytic reactions. Once the initial internucleotide link between either GTP or ATP and the next nucleotide specified by the template has been made, the σ-factor is released, thus possibly reducing the affinity of the polymerase from the promoter site and allowing the remaining core to move along the DNA template, spinning off a strand of RNA as it proceeds. The process can be visualized in the electron microscope.

Elongation of RNA Chains and Direction of Transcription

Basically the new chain grows by the subsequent addition of ribonucleotides to the free 3'- hydroxyl group. For instance, the RNA has a triphosphate group at position 3' on the other, end. Alkaline hydrolysis of this particular RNA will yield a molecule of the nucleoside

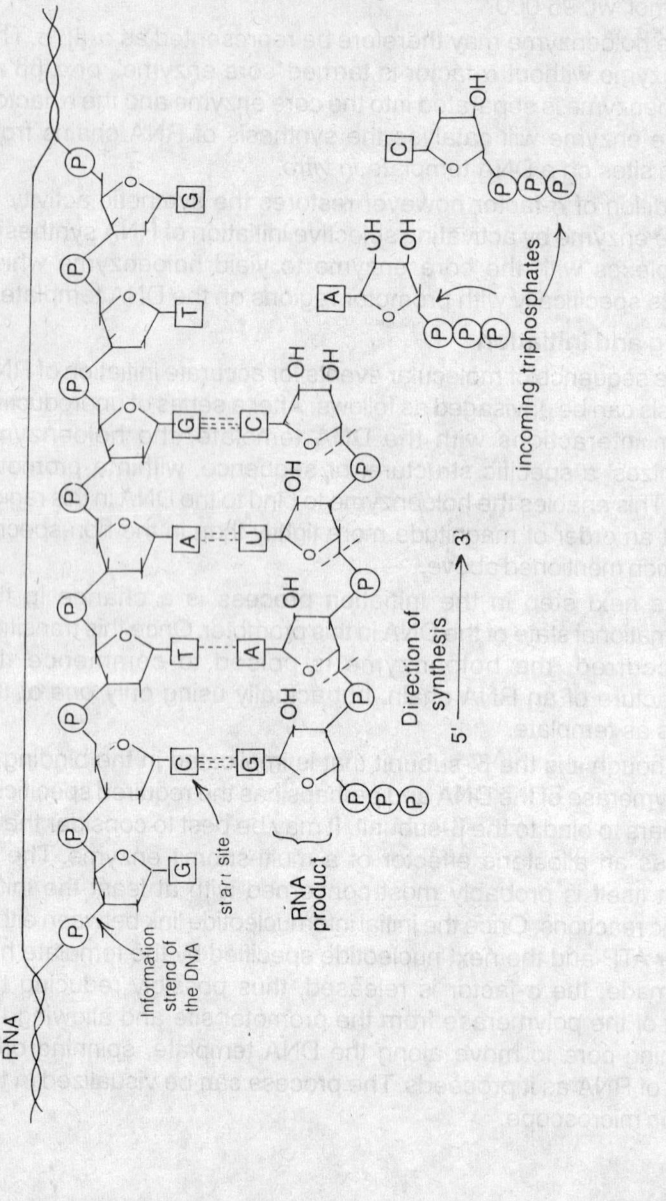

Fig. 14.1. A diagrammatic representation of the biosynthesis of RNA on one strand of DNA acting as template. The broken lines indicate sites of hydrolysis with alkali.

cytosine from the 3'-hydroxyl end of the molecule, uridine and adenosine monophosphates and a molecule of guanosine tetraphosphate from the 5'-end. Synthesis proceeds from the 5'-end to the 3'-end of the RNA molecule.

Still unclear are the precise characteristics of DNA structure at the promoter where the DNA is actually used as template rather than just for the tight binding of the RNA polymerase. In any even the nucleotide residue at the 5'-end of the nascent chain always contains a purine base; thus any DNA sequence involved in the actual commencement of template-directed RNA synthesis must contain a thymine or cytosine nucleotide.

As the RNA chain is formed it peels of the DNA template and, in *coli* at least, immediately becomes associated with the ribosomes so that polysomes may be formed before transcription is complete.

Chain Termination

The selective termination of RNA chains in *vitro* is less well understood. It appears in the first place that transcription can be terminated directly by virtue of specific sequences on the DNA template acting directly on the core enzyme. In principle the process of termination involves (a) the cessation of RNA chain elongation, (b) the release of the newly formed RNA, and (c) the release of the RNA polymerase from the DNA. All three of these reactions can occur at 'direct terminator' sites *in vitro.* Little is known regarding the structure of such sites. A second type of termination can be induced *in vitro* by rho (ρ) factor. This factor from *coli* is an oligomeric protein (monomer molecular weight 50 000) which depresses the amount of RNA formed in the *in vitro* reactions by causing RNA chain termination. The resulting RNA chains can be released from the DNA but the RNA polymerase remains bound to the DNA.

As already mentioned, relatively little is known about the structure of terminators; however, by *in vitro* chemical and enzymic methods, some information has been gained about the nucleotide sequences of a possible *terminator* region of the *coli* tyrosine suppressor tRNA gene.

The Eukaryotic RNA Polymerase

In eukaryotic cells the three classes of RNA polyerases have been isolated after solubization from chromatin, chromatography of the enzyme activity from variety of animal tissues on DEAE-Cellulose. These are RNA polymerase I, II and III. RNA polymerase III has been further resolved into IIIA and IIIB.

Class I RNA polymerases has been established to be nucleolar origin, whereas classes II and III are of nucleoplasmic origin. Additionally these enzymes operate optimally under somewhat different conditions. The nucleolar enzymes work best at low ionic strength and utilize Mn^{++} and Mg^{++} equally well. On the other hand, higher Mn^{++} concentrations and ionic strengths are required for maximum activity of the nucleoplasmic enzymes.

Since factors such σ and ρ appear to play an important role in the control of transcription, a search for such factors was made in mammalian systems. A protein factor (mol. wt. 70 000) which can specifically bind to, and stimulate, the activity of nucleoplasmic class II enzymes from rat liver and calf thymus has been tentatively indentified in the cytoplasm of rat liver. However, the question of whether or not the factor can promote initiation of RNA synthesis at specific sites on the DNA template cannot yet be answered. With regard to termination in eukaryotes no information is presently available.

The structural analyses have shown that the molecular weights of the large subunits detected the class III enzymes (138 000 and 155 000) differ from those of the class II enzymes (140 000 and either 170 000 or 205 000, or 240 000) and from those of the class I enzymes (117 000 and 195 000). Some low molecular weight subunits are also unique to each enzyme class. These data clearly distinguish class I, II and III enzymes on a structural basis. In addition, polypeptides of molecular weight 29 000 and 19 000 were found in all classes, a polypeptide of molecular weight 52000 was found only in class I and III enzymes, and a polypeptide of molecular weight 41 000 was found only in class II and III enzymes. Additionally it appears that the three classes of polymerase are assembled primarily from distinct gene products, and that they are not interconvertible by simple structural alterations.

Despite these advances, the role of these various polymerases in cellular DNA synthesis remains to be clarified. It is quite possible that the nucleolar enzymes (I) are involved in the synthesis of the ribosomal precursor RNA whereas the nucleoplasmic enzymes (II and III) make messenger RNA precursors, tRNA, and 5S RNA. Indeed, product analysis indicates enzyme III to be specifically responsible for the biosynthesis of 5S RNA and the precursor to tRNA in mammalian cells, thus leaving the possibility that enzyme II is involved in mRNA precursor production.

The Biosynthesis of rRNA

That the composition and nucleotide sequence of cytoplasmic RNA is under the control of nuclear DNA was supported by molecular hybridization experiments in which it was shown that there were sequences of DNA precisely complementary to the nucleotide sequences of ribosomal RNA, tRNA, and messenger RNA.

Whilst there is RNA in the nucleus of eukaryoteis, most of this is located in the nuucleolus. In fact the nucleolus is very active in the manufacture of RNA. Moreover, molecular hybridization experiments demonstrate that the parts of the DNA genome which contains the deoxyribonucleotide sequences complementary to the sequences that make up ribosomal RNA are located in the nucleolus, or nucleolar organizer region of the genome, of certain eukaryotes.

Siherrer and *Darnell* (1962) have demonstrated that these exists a short-lived species of RNA in the cultures of human cells which were exposed for a short time to ^{3}H-uridine. This RNA was distinguishable from tRNA and rRNA. It has a sedimentation constant of 45S and having a molecular weight of about 41 000 000 daltons. The site of its synthesis found to be nucleolus. Shortly after its synthesis, or transcription from the nucleolar DNA, this long single-stranded molecule undergoes a series of molecular 'tailoring' events in the nucleolus whereby its molecular dimensions are progressively reduced to give rise eventually to the 28S and 18S ribosomal RNA species found in cytoplasmic ribosomes.

Whilst the 45S species is found exclusively in the nucleolus, after longer labelling times radioactivity begins to appear in other species of RNA, firstly in 41S and 32S RNA (which are also confined to the nucleolus) and 20S RNA, and then in 18S RNA which passes rapidly to the cytoplasm'. Finally, radioactivity appears in 28S RNA initially in the nucleolus, then in the nucleoplasm, and shortly afterwards in the cytoplasm. A precursor-produced relationship between these RNA's has been proposed based on a variety of data including recent electron microscopic examination of the various RNA species.

The '5.8S RNA' fragment appears to be generated simultaneously with the 32S→28S RNA transition. This has been interpreted by some as indicating the possibility that, in the conversion, a loop of 'spacer' RNA located on the precursor chain between the 28S and 5.8S species is degraded.

With regard to the sequence modifications to rRNA (i.e. the methylation and pseudouridine formation), the majority of these occur rapidly in the nucleolus on the 45S precursor. Interestingly, the methylation occur only in the portions of the precursor destined to become either the 28S RNA or 18S RNA components, and are mostly in the form of 2'-O-methylribose.

5S rRNA originates independently in the non-nucleolar part of the nucleus and later becomes permanently associated with the large ribosomal subunits.

Although the ribosomal proteins appear to be synthesized in the cytoplasm, it is reasonably certain that they become associated with the ribosomal precursor RNA's in the nucleolus where the actual assembly of the ribosome takes place. It is believed that the cleavage of the precursors takes place within 'nascent ribosomal particles'.

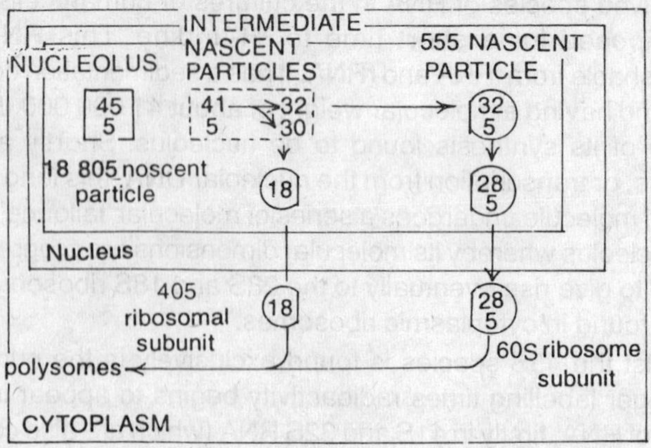

Fig. 14.2. An outline of eukaryotic ribosome formation illustrating the processing of RNA taking place within 'nascent ribosomal particles'. The number within the particles refer to the sedimentation coefficients of their constituent RNA's. Note the occurrence of 5S RNA even within '80S nascent particles'.

Biosynthesis of tRNA

Next to rRNA, tRNA is most plantiful RNA by weight; it makes up about 15% of the total RNA of the cell, compared to 80% for rRNA. The total tRNAs average only 75-80% nucleotides in length, however, so there are actually many more tRNA than rRNAs. Many tRNA genes exist in multiple copies in chromosomal DNA. Although they are not so highly repeated as 5S rRNA genes are. All of the primary transcription products of tRNA genes studied so far are, like pre-rRNA, larger than the final products. Thus, precursor tRNA molecules require shortening as they mature. They also require many other chemical modifications, such as the addition of methyl and isopentenyl groups and the conversion of uridines into pseudouridines. In fact, about 1 base in every 10 of a tRNA molecule is modified in some way.

Some tRNA genes in eukaryotic cells to not exist as a consecutive stretch of DNA. For example, to form the yeast phenylalanine and tyrosine tRNAs, a segment must be removed from the middle of the primary transcripts, and the two ends must be spliced together.

The transcription of tDNA (the genes that encode tRNA) is calalyzed by RNA polymerase III. As has been done with 5S rDNA, the tDNA sequence from a number of different eukaryotic organisms have been cloned as recombinant DNA molecules in *E. coli.* The processing steps include the addition of CCA to the 3' ends of the yeast tRNAs, and the addition of various other chemical structures such as methyl and isopentenyl groups. Thus the recognition of the

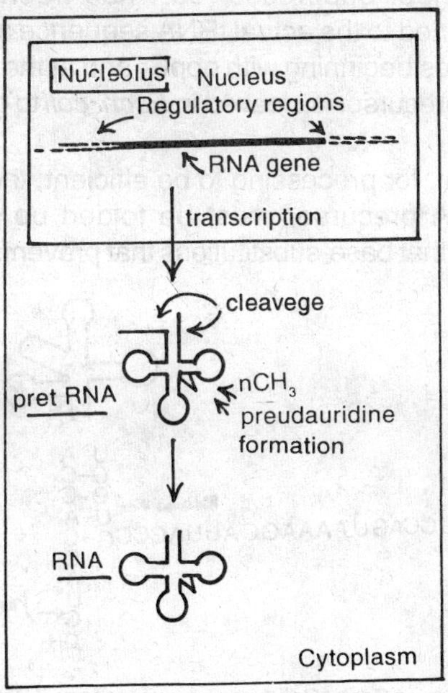

Fig. 14.3. A scheme the possible steps involved in the production of mammalian cell tRNA.

tRNA by the appropriate enzymes extends over a wide evolutionary range.

With the availability of an assay for the production of tRNA in frog oocytes, the sequence required for the transcription of several tRNA genes have been determined. The removal of either of two regions of a cloned tRNa gene greatly diminishes transcription. These two regions are located between positions +10 and +20 and between +50 and +60 of the mature tRNA molecule. These are correspond to the D and the TΨCG stem-loop structures, respectively, in the tRNA. The corresponding DNA sequences are highly conserved. Apparently these conserved sequences not only encode the structural components needed for the transfer of amino acids by tRNA molecules, but also constitute protein-binding sites in the DNA. Thus a dual role is now known for these highly conserved regions: one role in structuring functional tRNA and one role in promoting the transcription of the tRNA genes.

In *E.coli*, where there is also evidence for precursors, a sepcific study of *E.coli* tyrosine suppresor tRNA and its formation has been made by *Smith et. al.* A specific precursor to this special tRNA found in the number suppressor su_3^+ has been isolated and characterized. Joined to the actual tRNA sequence at its 5'-end are 41 extra nucleotides beginning with pppG, and at the 3'-end are two nucleotides. This precursor is cleaved in *Esch. coli* to give the mature tRNA sequence.

It appears that, for processing to be efficient, the tRNA part of the tyrosine tRNA precursor must be folded up in the correct conformation, and that base-substitutions that prevents this, or favour

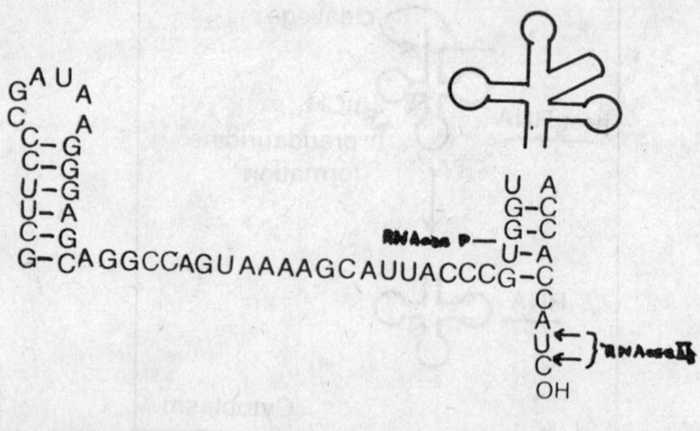

Fig. 14.4. The structure of Esch. Coli tyrosine suppressor tRNA, showing sits of cleavage (->).

an alternative structure, make processing inefficient. Nucleotide modifications apparently take place after the cleavages have occurred.

Biosynthesis of mRNA

Despite the great interest in the studies on the synthesis of rRNA and tRNA, it was considered of even greater importance to learn how mRNA in a cell determines which proteins the cell makes. The discoveries in the 1960s that tRNA and most rRNA were formed by the cleavage of precursor RNAs suggested that nuclear precursors to mRNA might also exist. Newly labeled nuclear RNA had been divided into pre-rRNA molecules (located in the nucleoplasm) and the hnRNA molecules (located in the nucleoplasm) whose lengths ranged from 2 to 30 kb. Also the hnRNA fraction had an average base composition like that of genomic DNA, suggesting that hnRNA was made up of many different molecules, as also might be expected of the total mRNA of a cell. Were these hnRNA molecules precursors to mRNAs? Early in the 1960s, cell free translation of eukaryotic mRNA had not been achieved, so direct tests of mRNA function could not be attempted. In their place, however, experiments designed to detect mRNA as a labeled RNA species were begun, so that the sizes and average base compositions of mRNA and of hnRNA could be compared.

In eukaryotes, heterogeneous nuclear RNA (HnRNA) also known as high molecular weight RNA (HMWRNA) or DNA like RNA (dRNA) is first synthesized in the nucleus. HnRNA molecules are much bigger in size than mRNA molecules which are utilized in cytoplasm. In the early 1960s, an enzyme termed poly A polymerase had been isolated; this enzyme did not copy a template, but simply added adenylate residues to the 3' end of RNA chains. The length of a poly A segment in the nucleus is quite consistent, it is between about 200 and 250 nucleotides. In the cells of lower plants and animals, the initial length of poly A may be shorter than in mammalian cells, but the size of the segment appears to be fairly uniform in each species. The specificity of poly A length is only one example of the nuclear RNA processing events that have been clearly demonstrated but that are not yet fully understood. After the poly A containing mRNA reaches the cytoplasm, the poly A segment is shortened so that its size is not longer uniform, but varies between 30-250 nucleotides.

Poly A at the 3' end of eukaryotic mRNA is called the tail and the other 5' terminus *cap.* The cap consists of a terminal nucleotide, 7-methylguanylate (m^7G) in a 5'→5' linkage with the initial nucleotide of mRNA chain. Thus there is no free 5' end of mRNA molecule; in fact it has 3' hydroxyl groups on the ribose sugar rug of the nucleotides at both ends. The cap probably aid in translation after a mRNA molecule is completely synthesized, but the structure is made in the nucleus by the series of enzymatic steps. It could be shown that removal of the cap leads to loss of translational activity due to loss of the formation of mRNA—ribosome complex. It suggests that cap can helps in recognition of ribosome.

15

PROTEIN SYNTHESIS

The proteins are highly organized and complex macromolecules. They have significant role in the structural and functional organisation of the cells. The structural proteins constitute various cellular components (such as plasma membrane, nuclear membrane, organelles etc.) as well as some extracellular parts such as cuticle, fibres etc. The functional proteins are enzymes and some hormones which control almost all the metabolic, regulatory, sensory and reproductive activities of the cells. They also control the expression of phenotypic traits.

Chemically, the proteins are the polymers of *amino acids* and at a primary level these can be considered as polypeptide chains. The polymer in a protein of amino acids linked with the help of peptide bonds (–Co –NH). The peptide bond consists of a convalent bond between carboxy group (COOH) of one amino acid and the amino group (NH_2) of another amino acid.

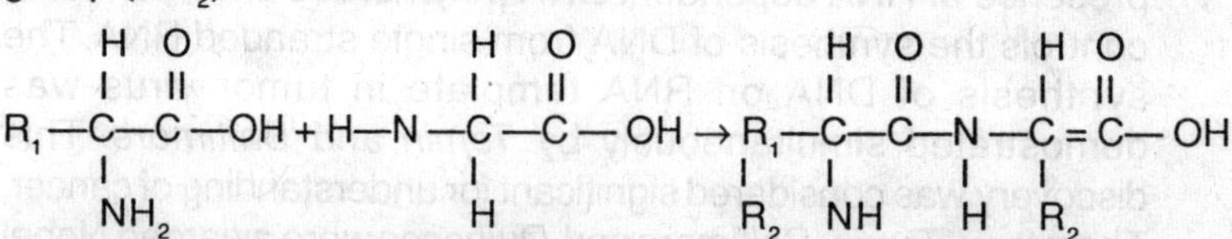

(I Aminoacid) (II Aminoacid) (Dipeptide)

Only 20 amino acids are biologically important and they participate in the protein synthesis forming polypeptide chains.

Deoxy ribonucleic acid (DNA) initiates the mechanism of protein synthesis since it carries the genetic message for protein synthesis. The information is transcribed or copied over messanger ribonucleic acid (mRNA) from one DNA strand. The mRNA translates the coded information by bringing amino acids in a particular sequence during protein synthesis through the agency of transfer ribonucleic acid (tRNA). It occurs over the ribosomes. The first step is called *transcription* which involves copying of genetic information from DNA into mRNA. The second step is called *translation* which involves the translation of information encoded on mRNA with newly synthesized protein molecules.

Flow of Genetic Informations

The protein synthesis is under direct control of DNA in most of the cases. Information of a polypeptide chain is stored in mRNA chain. The relationship between DNA, RNA and protein is called *central dogma.* From time to time dogma has been variously proposed.

1. *One way flow of information. Crick* in 1958 suggested that DNA transmits the genetic informations by transcribing mRNA. This mRNA translates the nucleotide sequence into the sequence of amino acid.

$$\text{DNA} \xrightarrow{\text{transcription}} \text{mRNA} \xrightarrow{\text{translation}} \text{RNA}$$

2. *Circular flow of information.* It was suggested by *Barry Commoner* in 1968. The DNA transcribes mRNA and mRNA translates information into protein. Further proteins directs the synthesis of mRNA and this mRNA dictates DNA synthesis.

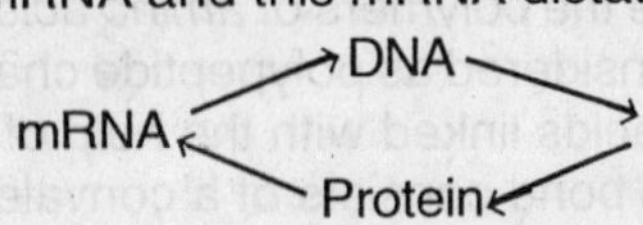

3. *Inverse flow of information. Temin* (1970) has described the presence of RNA dependent DNA polymerase enzyme, which controls the synthesis of DNA from single stranded RNA. The synthesis of DNA on RNA template in tumor virus was demostrated simultaneously by *Temin* and *Baltimore.* This discovery was considered significant for understanding of cancer. Therefore, *Temin, Baltimore* and *Dulbecco* were awarded Nobel Prize in medicine for 1975.

$$\text{DNA} \xrightarrow{\text{transcription}} \text{mRNA} \xrightarrow{\text{translation}} \text{Proteins}$$

Mechanism of Protein Synthesis in Prokaryotes

Transcription of mRNA from DNA

The copy of a complementary mRNA strand on DNA strand is called transcription. RNA polymerase is the main eyzyme involved in the transcription. All types of RNA (mRNA, t-RNA and rRNA) are synthesized with the help of this enzyme. In eukaryotic system three types of RNA polymerases are known to be present in the nucleus. In bacterial system only one type of RNA polymerase is known which is responsible for synthesis of all types of RNA. Besides RNA polymerase, transcription requires a template activated precursors and a divalent metal ion (Mg^{++} or Mn^{++}).

As earlier state that in bacterial system a single RNA polymerase is responsible for synthesis of all kinds of RNAs. It has been purified and its structure and function is now known. It consists of four polypeptide chains. There are two alpha (α) chains, one beta (β), one beta desh (β) and one sigma (σ) chains. The first-four form core enzyme and sigma factor units to form holoenzyme. The molecular weight of α-chain is 40,000 daltons, that of β chain 1,55,000 daltons, β' chain 1,65,000 daltons and σ chain is 95,000 daltons. The attachment of σ factor is not very firm sot hat core enzyme ($\alpha_2\beta\beta'$) can be easily isolated. Function of different polypetide chains is known β and β' help RNA polymerase in unwinding the DNA molecule for the process of transcription. The σ factor helps in recognition of start signals on DNA molecule and directs the enzyme in selecting the initiation sites. Once the mRNA synthesis is initiated, the σ factor dissociates and core enzyme brings about elongation of mRNA.

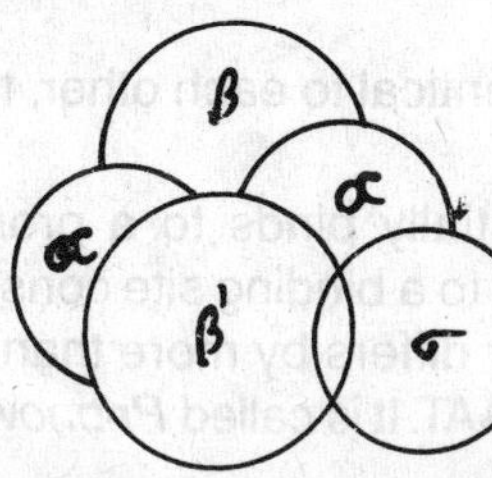

Fig. 15.1. A model of the structure of prokaryotic RNA polymerase showing association of five polypeptides (α_2 ββ σ)

For transcription entire chromosome or DNA molecule does not unwind but a local denaturation of the DNA helix takes place. Thus only those bases are exposed that constitute the gene to be transcribed. After unwinding only one of the two strands is copied. For synthesis of mRNA on DNA template, there are transcription units, each having a starting point and a termination signal. The start signals within DNA molecule are called *promctor sites.* These

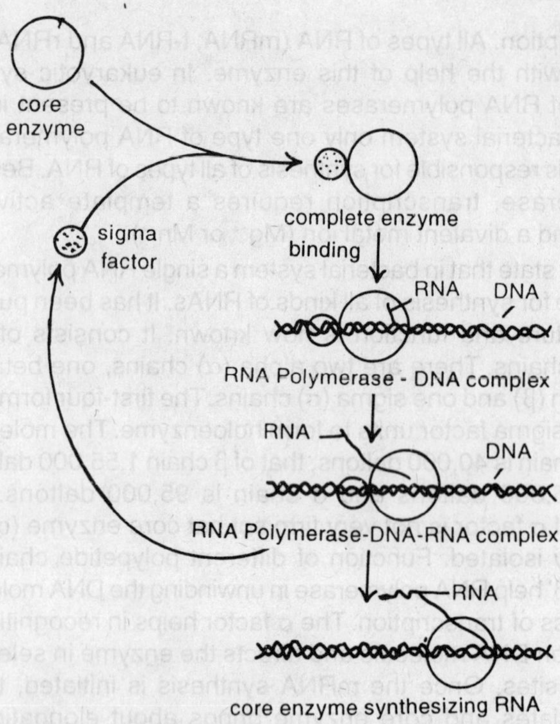

Fig. 15.2. Role of sigma factor and core enzyme of RNA polymerase during transcription.

promotor sites are not identical to each other, though all attract the same RNA polymerase.

RNA polymerase initially binds to a promotor site for RNA synthesis. It then diffuses to a binding site consisting of a sequence of six bases, which never differs by more than two bases from the following sequence 5'TATAAT. It is called *Pribnow box* and is available in almost all promotors.

The first base of RNA is always purine, either G or A. This means that the first base of the gene will be a pyrimidine either C or T in DNA. After synthesis the terminal triphosphate group may be cleaved by endonucleases. This results in RNA chain without terminal triphosphate groups starting with pyrimide nucleosides. The RNA chains can be formed *de novo* and do not require primers, as necessary for DNA replication.

Growth or elongation of RNA chain that proceeds in the 5'→3' direction, just as the growth of DNA chain does. Chain elongation takes place by addition of activated ribonucleoside triphosphates (ATP, GTP, CTP and UTP) to one strand of DNA template. From each activated ribonucleoside triphosphate, added to the growing mRNA chain, pyrophosphate (PP) is given out. During mRNA synthesis energy is released in the RNA synthesis is initiated, the σ factor dissociates and core enzyme brings about elongation of mRNA.

Termination of mRNA chain is brought about by certain *termination* or *stop signals* on DNA. Stop signals, like the promotor site, is a sequence of bases. Some of the stop signals may be recognized by the RNA polymerase, other appear to require the help of additional proteins. In *E. coli* such protein is known as *rho factor* (P). The P factor is discovered by *Roberts* (1962). It has a molecular weight of 2,00,000 daltons. It is made up of four polypeptide chains, each having a molecular weight of 50,000 daltons. It is however, not certain whether it also, or exclusively binds to DNA.

Translation in Prokaryotes

During translation the genetic information present in mRNA directs the order of specific amino acids to form protein molecule or polypeptide chain. This step is much more complex than transcription and involves the following steps:

1. Activation of amino acids

The 20 amino acids used in protein synthesis are activated by adenosine triphosphate (ATP). The activation is mediated by specific

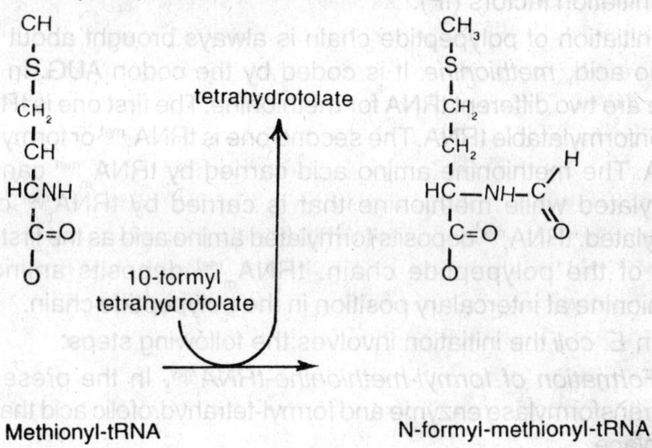

Fig. 15.3. The initiator N-formyl-methionyl-tRNA (1Met-tRNA$_1$) is generated by formylation of methionyl-tRNA$_1$ using formyl-tetrahydrofolate as cofactor.

activating enzyme known as *amino acyl RNA synthetases.* As a result of this reaction between amin acid and ATP, a complex is formed. This complex is known as amino acyl-AMP-enzyme or amino acyl AMP complex is formed. Pyrophosphates (PP) are released. It should be realized that amino acyl RNA synthetases are specific with respect to amino acids. Therefore, 20 such enzymes are required for 20 amino acids.

$$\underset{\text{Amino acid}}{AA_1} + ATP + \underset{\text{Enzyme}}{E_1} \longrightarrow \underset{\text{Amino acyl AMP enzyme complex}}{(AA_1 - AMP)\, E_1} + PP$$

2. Attachment of activated amino acid to tRNA

The complex formed in first step above, reacts with a particular tRNA and transfer the amino acid to tRNA with the release of AMP and enzyme amino acyl RNA synthetase.

$$(AA_1\ \text{-AMP})\ E_1 + tRNA_1 \longrightarrow AA_1\ tRNA_1 + AMP + E_1$$

For each amino acid, there is a specific tRNA. As there are 20 different amino acids so 20 different tRNAs are there. There are cases where more than one kind of tRNA may be available for the same amino acid, but no case is known where same tRNA can recognize two or more different amino acids.

3. Initiation of polypeptide chain

Initiation requires the ribosome subunits, mRNA, guanosine triphosphate (GTP) as energy source, amino acid t-RNA complex and initiation factors (IF).

Initiation of polypeptide chain is always brought about by the amino acid, *methionine.* It is coded by the codon AUG. In *E.coli,* there are two different tRNA for methionine. The first one is $tRNA_m^{met}$ or nonformylatable tRNA. The second one is $tRNA_1^{met}$ or formylatable tRNA. The methionine amino acid carried by $tRNA_m^{met}$ cannot be formylated while methionine that is carried by $tRNA_t^{met}$ can be formylated. $tRNA_t^{met}$ deposits formylated amino acid as the first amino acid of the polypeptide chain. $tRNA_m^{met}$ deposits amino acid methionine at intercalary position in the polypeptide chain.

In *E. coli* the initiation involves the following steps:

(i) *Formation of formyl-methionine-$tRNA_t^{met}$.* In the presence of transformylase enzyme and formyl-tetrahydrofolic acid that takes place.

$$\text{met-}tRNA_t^{met} + \text{formate} — \text{f-met-}tRNA_t^{met}$$

(ii) *Binding of 30 'S' subunit of ribosose with mRNA.* mRNA becomes attached to the 30 'S' subunit to form mRNA-30 'S' complex. This process requires the initian factor 3 (IF-3). Only one molecule of IF-3 binds to each 30 'S' subunit. (There are three 'F' in prokaryotes namely IF-1, IF-2 and IF-3. These factors are found in 30 'S' subunit of ribosomes. Their molecular weights are 9,200, 80,000 and 23,000 daltons respectively. These factors perform different functions). This 30 'S' subunit attaches at the 5' end of mRNA carrying AUG codon.

IF-3-30 'S' + mRNA ⟶ IF-3-30 'S' -mRNA

Besides binding 30 'S' subunit with mRNA, the IF-3 is also necessary for (a) affecting 30 'S' subunit conformation and (b) prevents reassociation of 30 'S' and 50 'S' subunits of ribosome.

(iii) *Binding of f-met-tRNA$_f^{met}$ with IF-3-30'S' -mRNA complex.* IF-2 and IF-1 are necessary for this step. Besides GTP is also necessary. IF-2 may either attached to f-met-tRNA$_t^{met}$ to form a binary complex or may directly bind to IF-3-30 'S' -mRNA complex. In the first case IF-2-f met-tRNA$_t^{met}$ may locate or recognize AUG codon on the IF-3-30 'S' mRNA complex and associate with it. In the second case IF-2 recognizes fmet-tRNA$_t^{met}$ and later associates to form the initiation complex.

IF-3-30 'S' mRNA + fmet-tRNA$_t^{met}$ + IF-2—

IF-3-30 'S' -mRNA-IF-2fmet-tRNAf^{met},

The mRNA has many AUG codons. Only one of these functions as an initiation codon. What factor determines that which AUG will function as the initiation codon? It is probably a specific sequence of nucleotides near AUG which is required for it to function as an initiation codon.

(iv) *Attachment of 50 'S' subunit of ribosome with above complex.* The complex formed in the above step associate with 50 'S' subunit of ribosome, to form a complete ribosome. This action depends on IF-2 activity. Energy is supplied by GTP. The IF-2 and IF-3 are now released.

IF3-30 'S' -mRNA - IF-2-fmet-tRNA$_t^{met}$ + GDP—

70 'S'-mRNA-fmet-tRNA$_t^{met}$ + GDP + P_i + IF-2 + IF-3

The bacterial ribosome has two binding sites for tRNA. The first one is *A-site* or *amino acyl site* or *acceptor site* and the

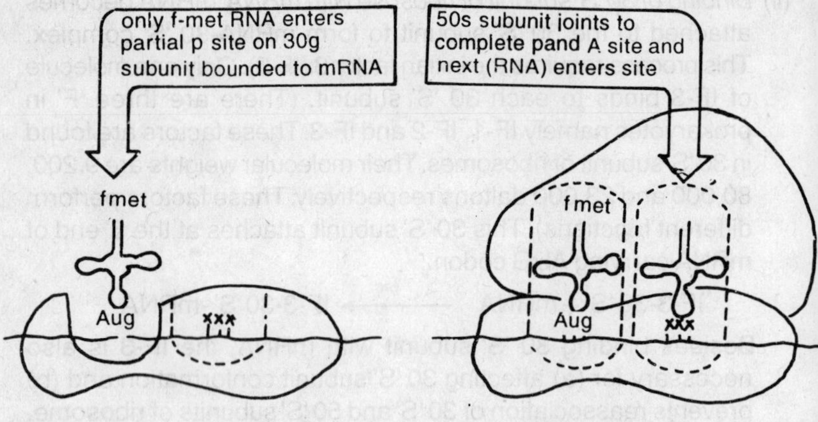

Fig. 15.4. Only tMet-tRNA can be used for initiation by 30S subunits; only aminoacyl-tRNAs can be used for elongation by 70S ribosomes.

other site is *P-site* or *peptidyl site* or *polymerisation site.* Thus each ribosome accomodates two codons or six nucleotides at a time. Both the sites are located in 50 'S' subunit of ribosome.

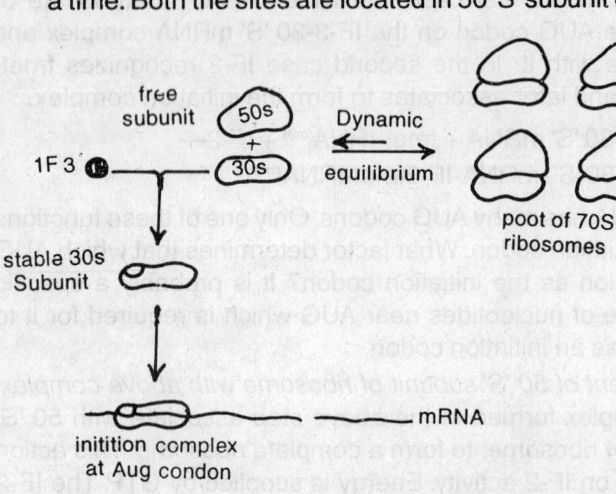

Fig. 15.5. IF3 is involved in upsetting the equilibrium with 70S ribosomes to generate stable 30S subunits with the ability to bind at Initiation sites on mRNA.

4. Elongation of polypeptide chain

After the formation of 70 'S' -mRNA-fmet-tRNA$_f^{met}$ complex, elongation of polypeptide chain takes place. Soluble protein *elongation* factors (EF) are required. Two EF are isolated called EF-T and EFG. Later EF-T was separated into two components, EF-Tu (unstable when heated) and EF-Ts (stable when heated). FE-Tu and FE-Ts are required for binding the amino acyl tRNA to the ribosomes. EF-G is involved in the translocation of mRNA.

The elongation process takes place in following steps:

(i) *Binding of AA_1tRNA, to the A-site of ribosome.* EF-Tu and EF-Ts are requried for binding the amino acyl tRNA (AA_1tRNA$_1$) to A-site of ribosome. EF-Tu forms a ternary complex with AA_1 -tRNA$_1$ and GTP.

$$\text{EF-Tu + GTP} \longrightarrow \underset{\text{(binary complex)}}{\text{EF-Tu-GTP}}$$

$$\text{EF-Tu-GTP + AA}_1\text{-tRNA}_1 \longrightarrow \underset{\text{(ternary complex)}}{\text{AA}_1\text{-tRNA}_1\text{-EF-Tu-GTP.}}$$

The P-site of ribosome being already occupied by f-met-tRNA$_f^{met}$ at the initial stage. Transfer of AA_1-tRNA$_1$ from ternary complex to A-site of ribosome now takes place.

$$\text{AA}_1\text{-tRNA}_1\text{-EF-Tu-GTP + Ribosome} \longrightarrow$$
$$\text{AA}_1\text{-tRNA}_1 \text{ Ribosome + EF -Tu -GDP + Pi}$$

EF-Ts displaces GDP and associates with Tu and then GTP can again associate with EF-Tu to start another cycle for binding of amino acyl tRNA.

$$\text{EF-Tu-GDP + EF-Ts} \longrightarrow \text{EFTu-EF-Ts + GDP}$$

$$\text{EF-Ts - EF-Tu + GTP} \longrightarrow \underset{\text{(Binary complex)}}{\text{EF-Tu - GTP}}$$

(ii) *Peptide bond formation.* Bond formation occurs by the transfer of the first amino acid i.e. f-met-tRNA$_f^{met}$ from its tRNA at the P-site to the amino acyl tRNA (AA_1–tRNA$_1$) just brought to the A-site. The peptide bond is formed between the carboxyl group of first amino acid (f-met) and the amino group of second one (AA_1). The reaction is catalysed by *peptidyl transferase*. This enzyme is the intergral part of the 50 'S' subunit of ribosome. After the formation of a peptide bond, the tRNA at P-site is deacylated and the tRNA at A-site now carries the polypeptide.

(iii) *Translocation.* The movement of ribosome relative to mRNA is called translocation. During translocation the ribosome moves along mRNA, a codon at a time, from 5' to 3' direction (5'→3'). The peptidyl tRNA present A-site now translocated to P-site. During this process, the deacylated tRNA is liberated out from

P-site. The translocation of peptidyl tRNA from A-site to P-site requires EF-G and GTP. Now the A-site of ribosome is occupied by another AA-tRNA-EF-Tu-GTP complex and the cycle continues till the ribosome comes across the termination codon i.e. UGA or UAA or UAG.

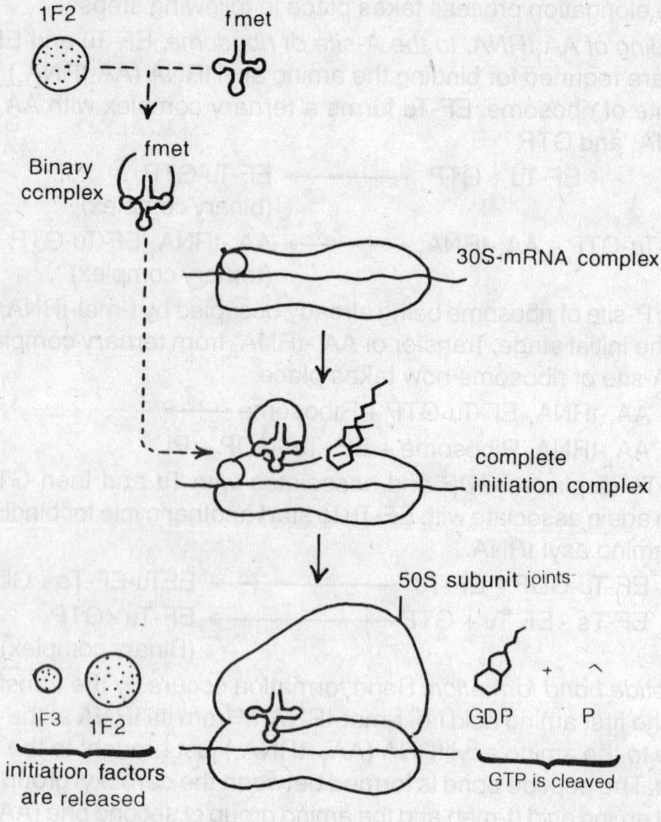

Fig. 15.6. Bacterial initiation factors are needed for 30S subunits to bind mRNA and for fMet-tRNA, to bind to the 30S-mRNA complex; they are released with cleavage of GTP when the 50S subunit arrives.

There exists a great ambiguity about relative movement of ribosome and mRNA. According to a view during translocation, the energy is provided by GTP which drive a pulsating ribosome contraction after the formation of a peptide bond between two amino acids. This contraction would move the ribosome along mRNA and change the status of two bound tRNAs is one

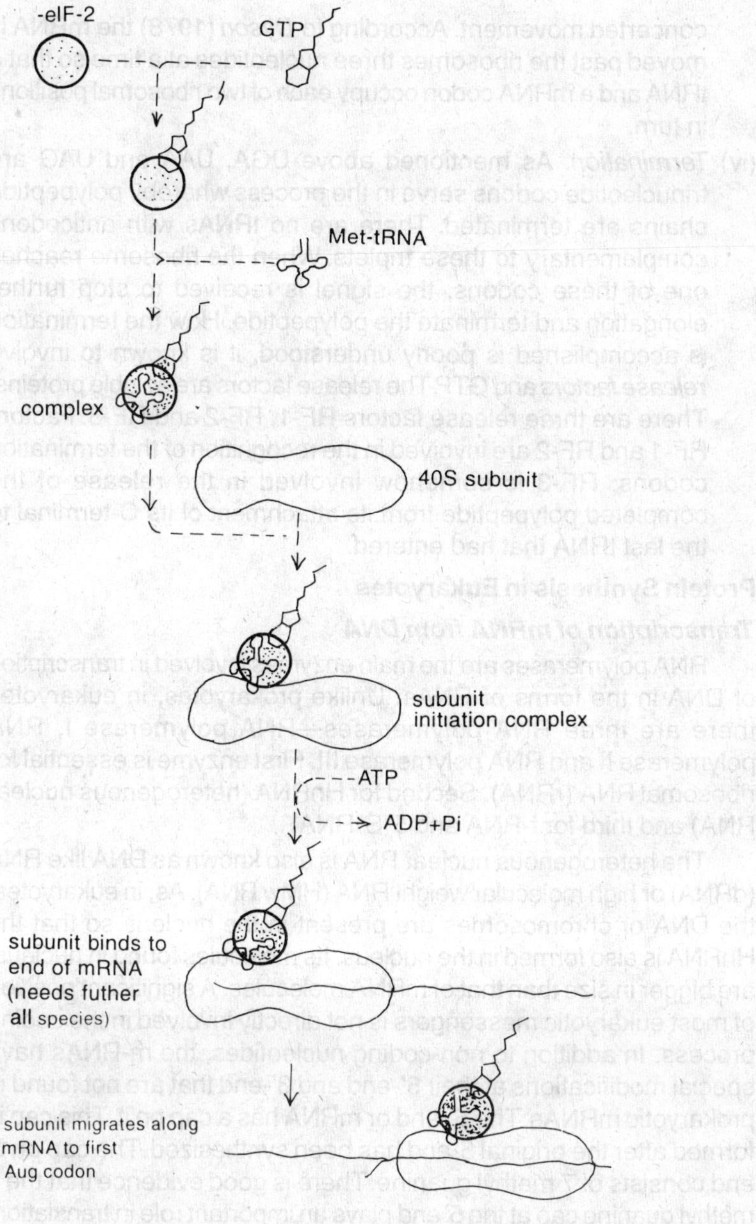

Fig. 15.7. In eukaryotic initiation, eIF$_2$ forms a ternary complex with tRNA, that binds to free 40S subunits; then the initiation plex binds to mRNA at its 5' end, from where it must migrates to the AUG initiation codon.

concerted movement. According to *Dyson* (1978) the mRNA is moved past the ribosomes three nucleotides at a time so that a tRNA and a mRNA codon occupy each of two ribosomal positions in turn.

(iv) *Termination.* As mentioned above UGA, UAA and UAG are trinucleotide codons serve in the process whereby polypeptide chains are terminated. There are no tRNAs with anticodons complementary to these triplets. When the ribosome reaches one of these codons, the signel is received to stop further elongation and terminate the polypeptide. How the termination is accomplished is poorly understood, it is known to involve *release factors* and GTP. The release factors are soluble proteins. There are three release factors RF-1, RF-2 and RF-3. Factors RF-1 and RF-2 are involved in the recognition of the termination codons. RF-3 is somehow involved in the release of the completed polypeptide from its attachment of its C-terminal to the last tRNA that had entered.

Protein Synthesis in Eukaryotes

Transcription of mRNA from DNA

RNA polymerases are the main enzymes involved in transcription of DNA in the forms of RNAs. Unlike prokaryotes, in eukaryotes there are three RNA polymerases—RNA polymerase I, RNA polymerase II and RNA polymerase III. First enzyme is essential for ribosomal RNA (rRNA). Second for HnRNA (heterogenous nuclear RNA) and third for t-RNA and 5 'S' RNA.

The heterogenous nuclear RNA is also known as DNA like RNA (dRNA) or high molecular weight RNA (HMw RNA). As, in eukaryotes, the DNA or chromosomes are present in the nucleus so that the HnRNA is also formed in the nucleus. Its molecules found in nucleus, are bigger in size than that of mRNA molecules. A significant position of most eukaryotic messengers is not directly involved in the coding process. In addition to non-coding nucleotides, the m-RNAs have special modifications at their 5'-end and 3'-end that are not found in prokaryotic mRNAs. The 5'-end or mRNA has a cap on it. This cap is formed after the original 5' end has been synthesized. The cap on 5' end consists of 7 methyl guanine. There is good evidence that the 7 methyl guanine cap at the 5' end plays an important role in translation. It could be shown that removal of cap bads to loss of translational activity due to loss of the formation of mRNA-ribosome complex. The base sequence of the 3' end of most mRNA is very unusual. It

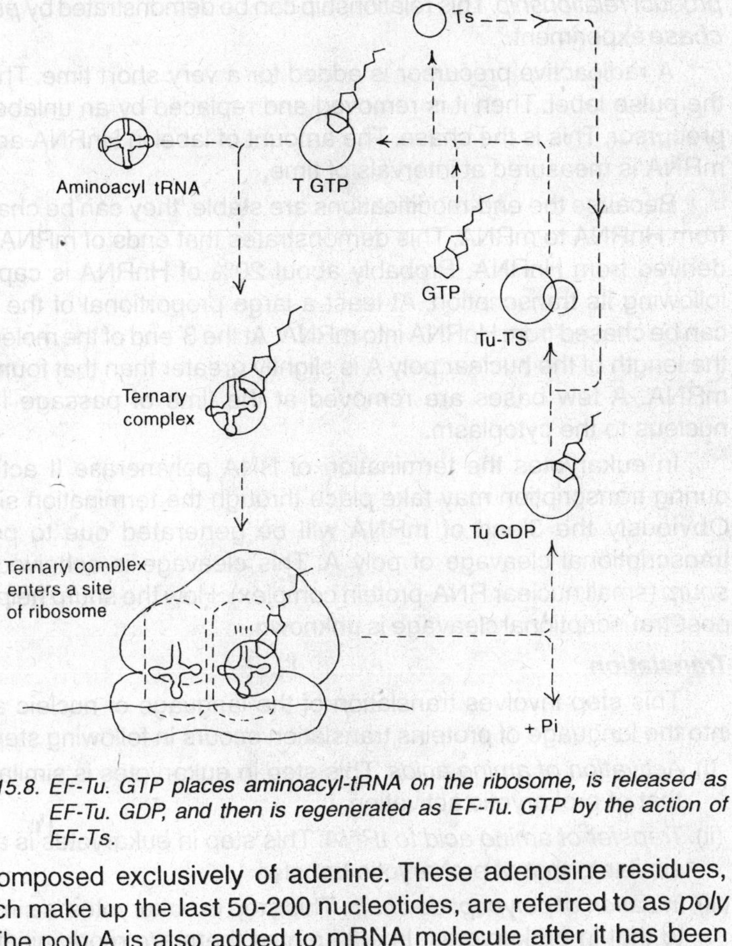

Fig. 15.8. EF-Tu. GTP places aminoacyl-tRNA on the ribosome, is released as EF-Tu. GDP, and then is regenerated as EF-Tu. GTP by the action of EF-Ts.

is composed exclusively of adenine. These adenosine residues, which make up the last 50-200 nucleotides, are referred to as *poly A.* The poly A is also added to mRNA molecule after it has been synthesized. The poly A at 3'-end is usually called *tail.* Conclusive evidence of function of tail has been very difficult to obtain. Some evidence point to a role in stabilizing the mRNA in cytoplasm, protecting it from premature degradation. Other evidence points out its role in processing the primary transcript within the nucleus.

mRNA is derived from HnRNA

The ends of some HnRNA molecules carry the same modifications that are seen in mRNA. The 5'-end gains a methylated cap and the 3'end is polyadenylated. The similarities of these end modifications suggest that HnRNA and mRNA may have a *precursor-*

product relationship. This relationship can be demonstrated by *pulse-chase* experiment.

A radioactive precursor is added for a very short time. This is the pulse label. Then it is removed and replaced by an unlabelled precursor. This is the chase. The amount of label in HnRNA and in mRNA is measured at intervals of time.

Because the end-modifications are stable, they can be chased from HnRNA to mRNA. This demonstrates that ends of mRNA are derived from HnRNA. Probably about 20% of HnRNA is capped following its transcription. At least a large proportional of the cap can be chased from HnRNA into mRNA. At the 3' end of the molecule the length of the nuclear poly A is slightly greater than that found in mRNA. A few bases are removed at the time of passage from nucleus to the cytoplasm.

In eukaryotes the termination of RNA polymerase II activity during transcription may take place through the termination sites. Obviously the 3'end of mRNA will be generated due to post-transcriptional cleavage of poly A. This cleavage is acheived by *snurp* (small nuclear RNA-protein complex). How the snurp helps in post-transcriptional cleavage is unknown.

Translation

This step involves translation of the language of nucleic acid into the language of proteins translation occurs in following steps:

(i) *Activation of amino acids.* This step in eukaryotes is similar to that of prokaryotic activation.

(ii) *Transfer of amino acid to tRNA.* This step in eukaryotes is also similar to that of prokaryotic transfer.

(iii) *Initiation of polypeptide chain.* This step in eukaryotes is similar to that in prokaryotes. In eukaryotes there are more initiation factors as compared to prokaryotes where three factors are present. They are named by putting prefix e to signify eukaryotic origin. These factors are eIF-1, eIF-2, eIF-3, eIF-4A, eIF-4B, eIF-4C, eIF-4D, eIF-5 and eIF-6.

In eukaryotes the methionine formylation does not occur. Following steps are involved in the initiation.

(a) *Formation of met-tRNA-elongation factor 2-GTP ternary complex.* This complex is formed in two steps—in the first GTP binds to eIF-2 forming binary complex

$$\text{GTP} + \text{EIF-2} \longrightarrow \text{GTP eIF-2}$$

In the second step above complex associates with met-tRNA

GTP-eIF-2 + met-tRNA ⟶ met-tRNA - eIF-2-GTP.

In mammals elf-2 has three subunits eIF-2 α-which binds to GTP eIF-2, γ-which binds to met-tRNA and eIF-2β which may be a recycling factor.

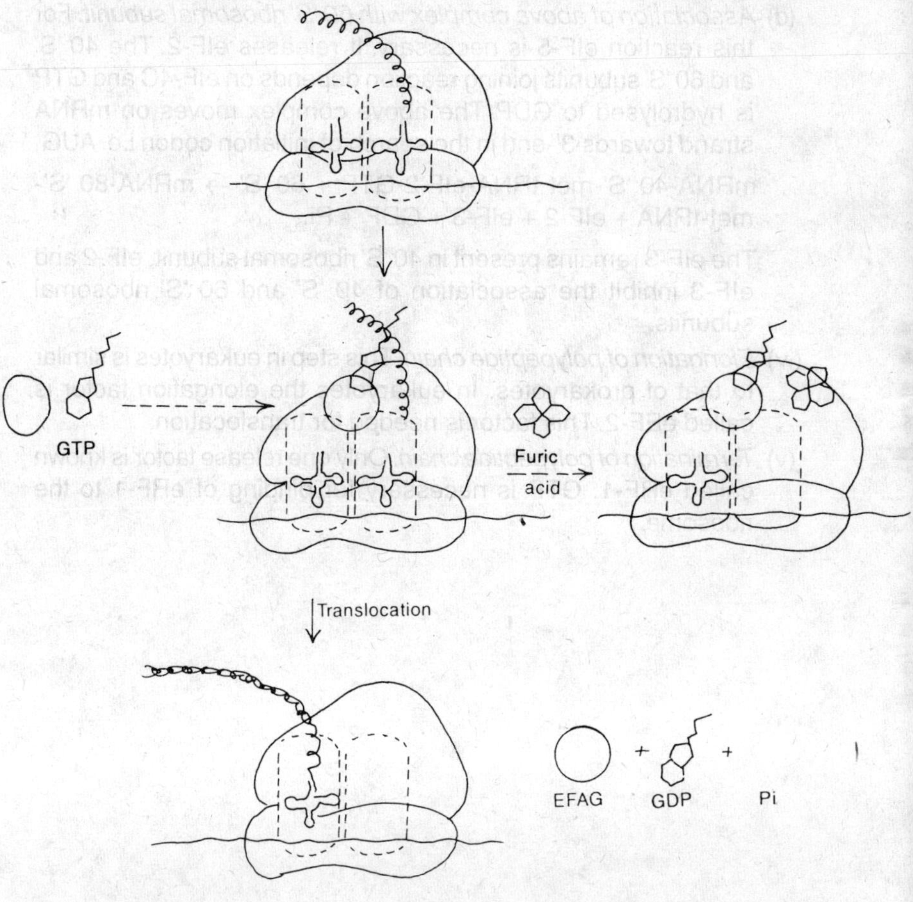

Fig. 15.9. After peptide bond formation, EF-G and GTP are needed for translocation, but then dissociate from the ribosome (unless fusidic acid is present to stabilize the post-translocation state).

(b) *Binding of above complex with 40 'S' ribosomal subunit.* The above complex associates with 40 'S' subunits of ribosome to form 40 'S' initiation complex.

met-tRNA-eIF-2-GTP + 40 'S' $\rightarrow$ 40 'S'-met-tRNA-eIF-2-GTP

(c) *Association of mRNA with 40 'S' initiation complex.* For this reaction eIF-3 is necessary. The binding of mRNA is assisted by other factors like eIF-1, eIF-4A, eIF-4B and ATP.

40 'S' -met-tRNA-eIF-2-GTP+mRNA→mRNA-40 'S'-met-tRNA-eIF-2-GTP.

(d) *Association of above complex with 60 'S' ribosomal subunit.* For this reaction eIF-5 is necessary. It releases eIF-2. The 40 'S' and 60 'S' subunits joining reaction depends on eIF-4C and GTP is hydrolysed to GDP. The above complex moves on mRNA strand towards 3'-end in the search of initiation codon i.e. AUG.

mRNA-40 'S'-met tRNA-eIF-2-GTP + 60 'S' → mRNA-80 'S'-met-tRNA + eIF-2 + eIF-3 + GDP + Pi.

The eIF-3 remains present in 40 'S' ribosomal subunit, eIF-2 and eIF-3 inhibit the association of 40 'S' and 60 'S' ribosomal subunits.

(iv) *Elongation of polypeptide chain.* This step in eukaryotes is similar to that of prokaryotes. In eukaryotes the elongation factor is called eEF-2. This factor is needed for translocation.

(v) *Termination of polypeptide chain.* Only one release factor is known called eRF-1. GTP is necessary for binding of eRF-1 to the ribosome.

16

BIOSYNTHESIS OF NUCLEOTIDES

A nucleotide consists of a nitrogenous base, a sugar, and one or more phosphate groups. The nitrogenous base is a *purine* or *pyrimidine* derivative. The two major purines are *adenine* and *guanine*, and the three major pyrimidines are *cytosine*, *uracil*, and *thymine*.

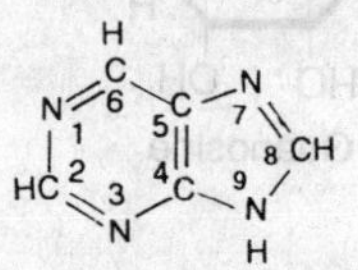

Purine

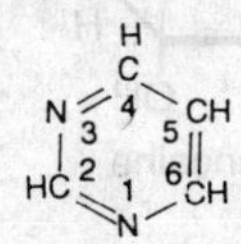

Pyrimidine

A *nucleoside* consists of a purine or pyrimidine base linked to a pentose. The pentose is *D-ribose* or *2-deoxy-D-ribose*. In a nucleoside,

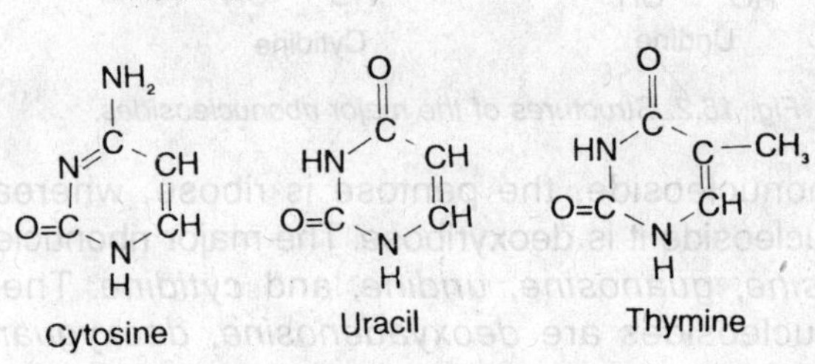

Fig. 16.1. Structures of the major purines and pyrimidines.

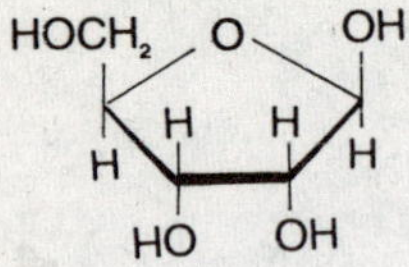

β - D - Ribose

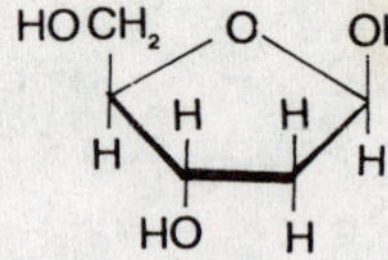

β - D - 2 - Deoxyribose

the glycosidic C-1 carbon atom of the pentose is bonded to N-1 of the pyrimidine or N-9 of the purine base. The configuration of this N-glycosidic linkage is β in all naturally occurring nucleosides.

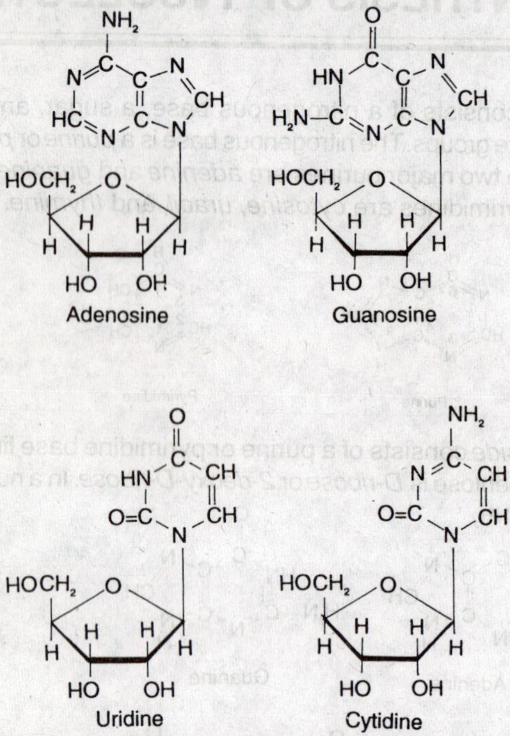

Fig. 16.2. Structures of the major ribonucleosides.

In a ribonucleoside, the pentose is ribose, whereas in a deoxyribonucleoside it is deoxyribose. The major ribonucleosides are *adenosine, guanosine, uridine,* and *cytidine.* The major deoxyribonucleosides are *deoxyadenosine, deoxyguanosine, deoxythymidine,* and *deoxycytidine.* Note that uracil is replaced by thymine, its methylated analog, in the deoxy series.

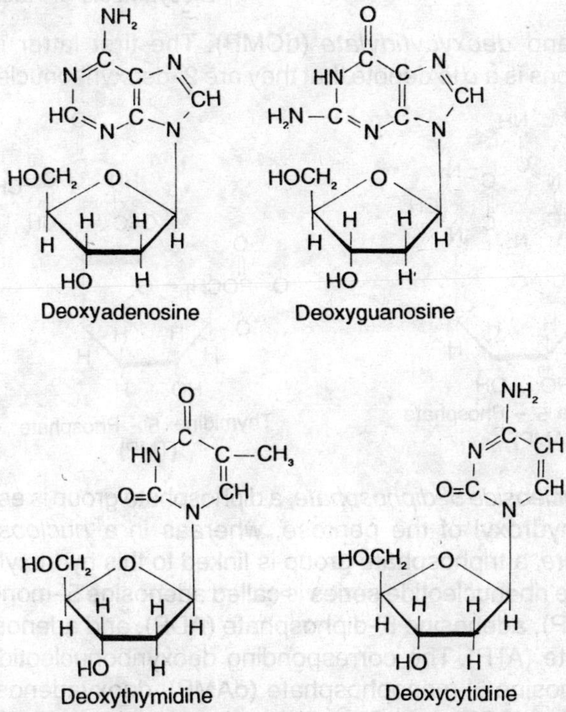

Fig. 16.3. Structures of the major deoxyribonucleosides.

A *nucleotide* is a phosphate ester of a nucleoside. At least one of the hydroxyl groups of the pentose moiety of a nucleotide is esterified. The most common site of esterification is the hydroxyl group attached to C-5 of the pentose. Such a compound is called a *nucleoside 5'-phosphate* or *5'-nucleotide.* A primed number designates an atom of the purine or pyrimidine ring. The type of pentose is denoted by the prefix in the terms *5'-ribonucleotide* and *5'-dexyribonucleotide.*

The nucleotide derived by esterification of the 5'-hydroxyl group of adenosine is *adenosine 5'-phosphate,* which is often called *adenylate.* The name adenylic acid is sometimes used, but adenylate is preferred because the phosphate group is ionized at physiological pH. The standard abbreviation for this compound is AMP (for adenosine monophosphate). The common names of the other major 5'-ribonucleotides are *guanylate* (GMP), *uridylate* (UMP), and *cytidylate* (CMP). The major 5'-deoxyribonucleotides are called *deoxyadenylate* (dAMP), *deoxyguanylate* (dGMP), *deoxythymidylate* (dTMP), and *deoxycytidylate* (dCMP). The first latter in their abbreviations is a *d* to denote that they are 2'-deoxyribonucleotides.

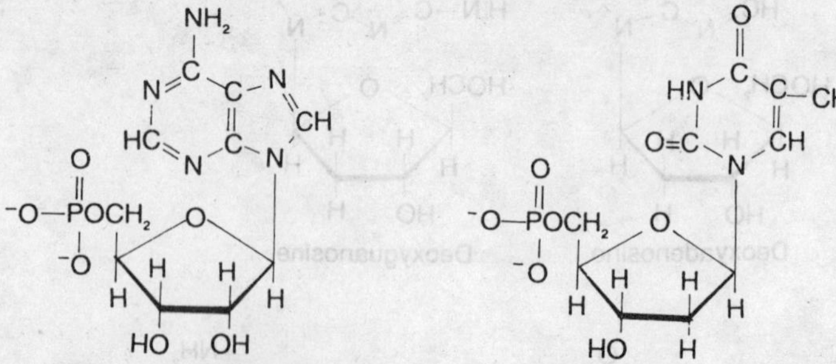

Adenosine 5' – Phosphate (AMP)   Thymidine 5' - Phosphate (TMP)

In a *nucleoside 5'-diphosphate,* a diphosphate group is esterified to the 5'-hydroxyl of the pentose, whereas in a *nucleoside 5'-triphosphate,* a triphosphate group is linked to this hydroxyl. Thus, the adenine ribonucleotide series is called adenosine 5'-monophosphate (AMP), adenosine 5'-diphosphate (ADP), and adenosine 5'-triphosphate (ATP). The corresponding deoxyribonucleotides are deoxyadenosine 5'-monophosphate (dAMP), deoxyadenosine 5'-diphosphate (dADP), and deoxyadenosine 5'-triphosphate (dATP). The nomenclature of bases, nucleosides, and nucleotides is summarized in table given below:

Table 16.1. Nomenclature of bases, nucleosides, and nucleotides

Base	*Ribonucleoside*	*Ribonucleotide (5-monophosphate)*
Adenine (A)	Adenosine	Adenylate (AMP)
Guanine (G)	Guanosine	Guanylate (GMP)
Uracil (U)	Uridine	Uridylate (UMP)
Cytosine (C)	Cytidine	cytidylate (CMP)
Base	*Deoxyribonucleoside*	*Deoxyribonucleotide (5'-monophosphate)*
Adenine (A)	Deoxyadenosine	Deoxyadenylate (dAMP)
Guanine (G)	Deoxyguanosine	Deoxyguanylate (dGMP)
Thymine (T)	Deoxythymidine	Deoxythymidylate (dTMP)
Cytosine (C)	Deoxycytidine	Deoxycytidylate (dCMP)

The Purine Ring is Synthesized from Amino Acids, Tetrahydrofolate Derivatives, and CO_2

The purine ring in purine nucleotides is assembled from a variety of precursors. *Glycine* provides C-4, C-5, and N-7. The N-1 atom comes from *aspartate.* The other two nitrogen atoms, N-3 and N-9, come from the amide group of the side chain of *glutamine.* Activated derivatives of *tetrahydrofolate* furnish C-2 and C-8, whereas CO_2 is the source of C-6.

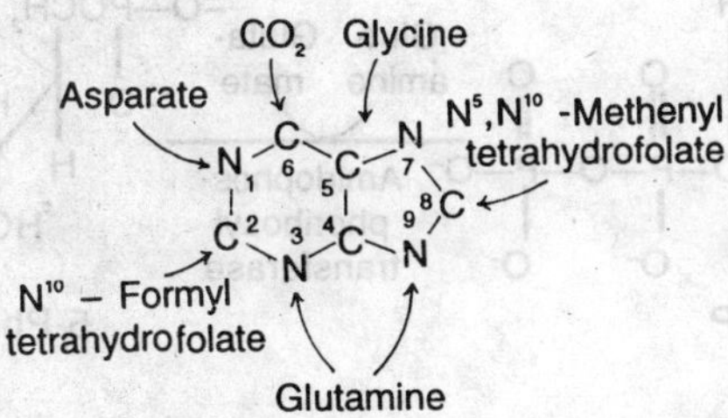

Fig. 16.4. Origins of the atoms in the purine ring.

Prpp is the Donor of the Ribose Phosphate Moiety of Nucleotides

The pathway of purine biosynthesis was elucidated in the 1950s by John Buchanan, G. Robert Greenberg, and others. The ribose phosphate portion of purine and pyrimidine nucleotides comes from *5-phosphoribosyl-1-pyrophosphate* (PRPP), a key intermediate in the biosynthesis of histidine and tryptophan. PRPP is synthesized from ATP and ribose 5-phosphate, which is primarily formed by the pentose phosphate pathway. The pyrophosphate group is transferred from ATP to C-1 of ribose 5-phosphate. PRPP has an a configuration.

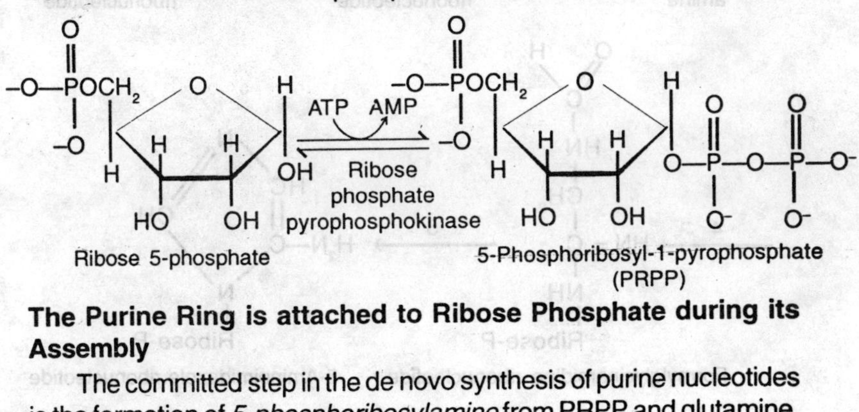

The Purine Ring is attached to Ribose Phosphate during its Assembly

The committed step in the de novo synthesis of purine nucleotides is the formation of *5-phosphoribosylamine* from PRPP and glutamine.

The amino group from the side chain of glutamine displaces the pyrophosphate group attached to C-1 of PRPP. The configuration at C-1 is inverted from α to β in this reaction. The resulting C-N glycosidic bond has the β configuration that is characteristic of naturally occurring nucleotides. This reaction is driven forward by the hydrolysis of pyrophosphate.

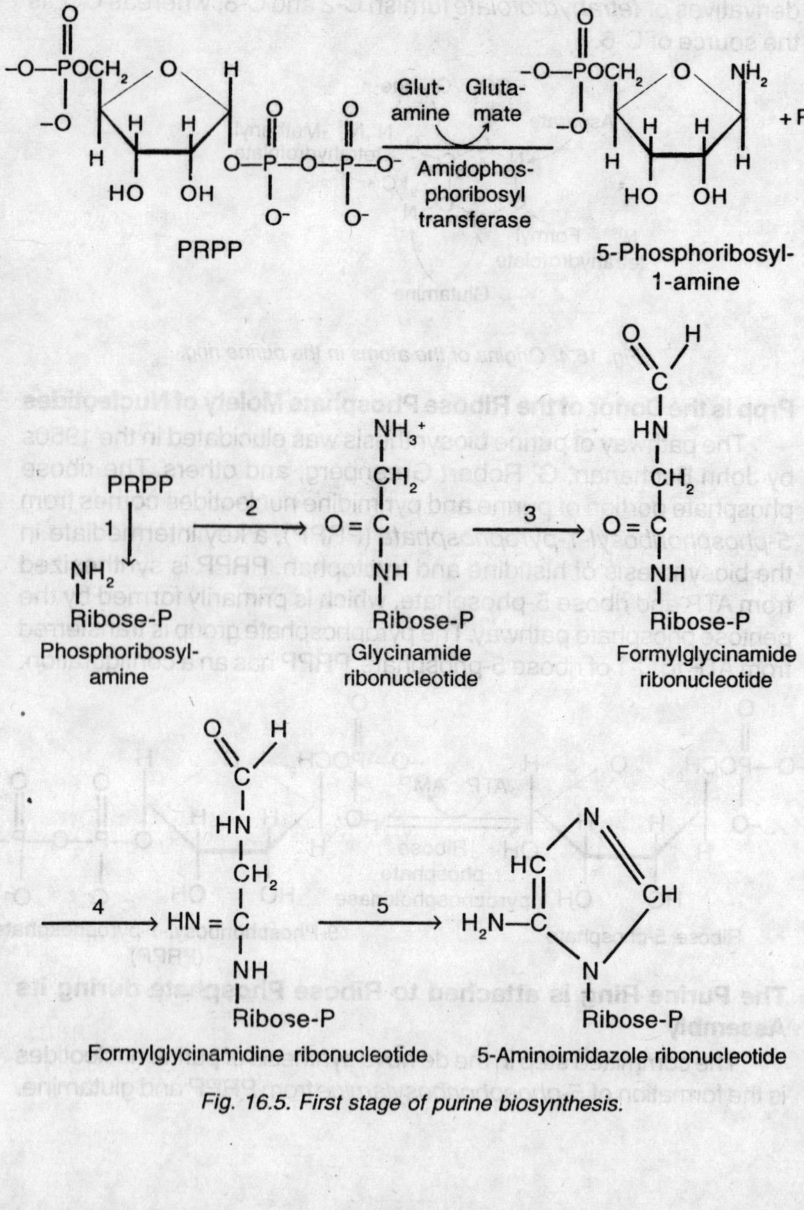

Fig. 16.5. First stage of purine biosynthesis.

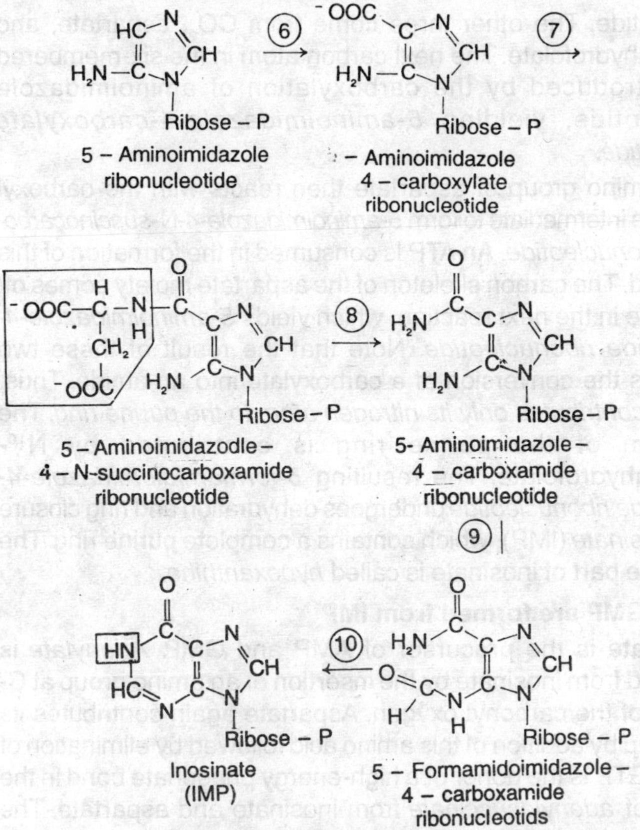

Fig. 16.6. Second stage of purine biosynthesis.

Glycine joins phosphoribosylamine to yield *glycinamide ribonucleotide.* An ATP is consumed in the formation of an amide bond between the carboxyl group of glycine and the amino group of phosphoribosylamine. The α-amino terminus of the glycine residue is then formylated by methenyltetrahydrofolate to give α-N-*formylglycinamide ribonucleotide.* The keto group in this compound is converted into an amidine group. The nitrogen atom is donated by the side chain of glutamine in a reaction that consumes an ATP. *Formylglycinamidine ribonucleotide then undergoes ring closure to form 5-aminoimidazole ribonucleotide.* This intermediate contains the complex five-membered ring of the purine skeleton.

The next phase in the synthesis of the purine skeleton, the formation of a six-membered ring, starts at this point. Three of the six atoms of this ring are already present in aminoimidazole ribonucleotide. The other three come from CO_2, aspartate, and formyltetrahydrofolate. The next carbon atom in the six-membered ring is introduced by the carboxylation of aminoimidazole ribonucleotide, yielding *5-aminoimidazole-4-carboxylate ribonucleotide.*

The amino group of aspartate then reacts with the carboxyl group of this intermediate to form *5-aminoimidazole-4-*N*-succinocarboxamide ribonucleotide.* An ATP is consumed in the formation of this amide bond. The carbon skeleton of the aspartate moiety comes off as fumarate in the next reaction, which yields *5-aminoimidazole-4-carboxamide ribonucleotide.* Note that the result of these two reactions is the conversion of a carboxylate into an amide. Thus, *aspartate contributes only its nitrogen atom to the purine ring.* The final atom of the purine ring is contributed by N^{10}-formyltetrahydrofolate. The resulting *5-formamidiomidazole-4-carboxamide ribonucleotide* undergoes dehydration and ring closure to form *inosinate* (IMP), which contains a complete purine ring. The purine-base part of inosinate is called *hypoxanthine.*

AMP and GMP are formed from IMP

Inosinate is the precursor of AMP and GMP. *Adenylate* is synthesized from inosinate by the insertion of an amino group at C-6 in place of the carbonyl oxygen. Aspartate again contributes its amino group by addition of this amino acid followed by elimination of fumarate. GTP is the donor of a high-energy phosphate bond in the synthesis of *adenylosuccinate* from inosinate and aspartate. The removal of fumarate from adenylosuccinate and from 5-aminoimidazole-4-N-succinocarboxamide ribonucleotide is catalyzed by the same enzyme.

Guanylate (GMP) is synthesized by the oxidation of inosinate, followed by the insertion of an amino group at C-2. NAD^+ is the hydrogen acceptor in the oxidation of inosinate to xanthylate (XMP). The amino group in the side chain of glutamine is then transferred to xanthylate. Two high-energy phosphate bonds are consumed in this reaction, because ATP is cleaved into AMP and PP_i, which is subsequently hydrolyzed.

In the conversion of inosinate into adenylate and into guanylate, a carbonyl oxygen atom is replaced by an amino group. A similar change occurs in the synthesis of formylglycinamide ribonucleotide from its amide precursor, in the formation of CTP from UTP, and in the conversion of citrulline into arginine in the urea cycle. *The common*

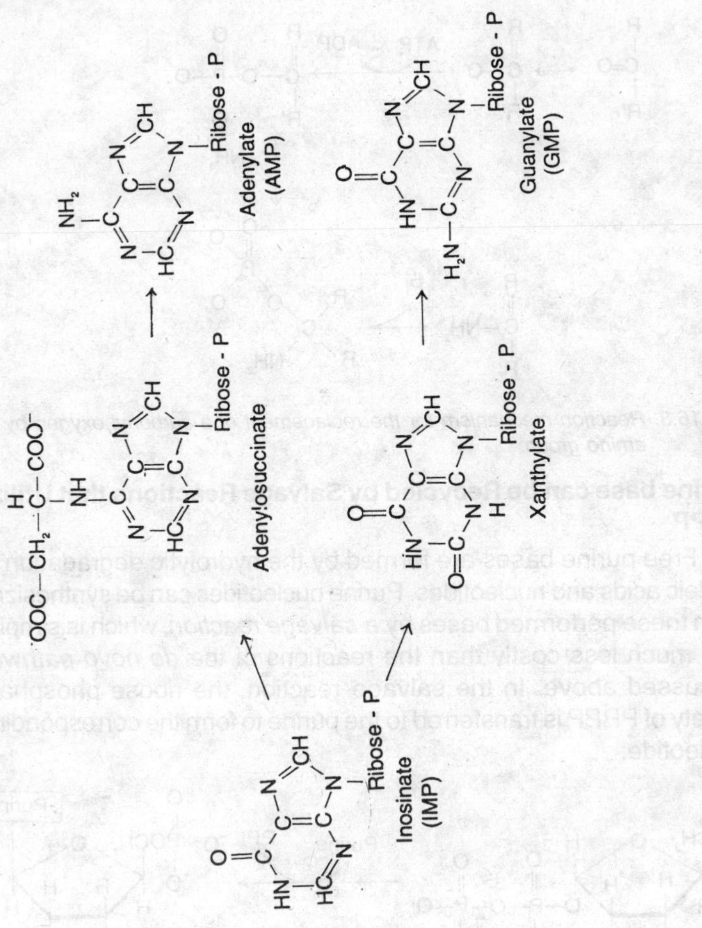

Fig. 16.7. AMP and GMP are synthesized from IMP.

mechanistic theme of these reactions is the conversion of the carbonyl oxygen into a derivative that can be readily displaced by an amino group. The tautomeric form of the carbonyl group reacts with ATP (or GTP) to form a phosphoryl ester, which is nucleophilically attached by an amine. Inorganic phosphate is then expelled from this tetrahedral adduct to complete the reaction. The attacking amine can be NH_3, the side-chain amide group of glutamine, or the α-amino group of aspartate. The leaving group in this class of reactions can be P_i, PP_i, or the AMP moiety. For example, PP_i is displaced by the amino group of glutamine in the synthesis of 5-phosphoribosyl-1-amine from PRPP.

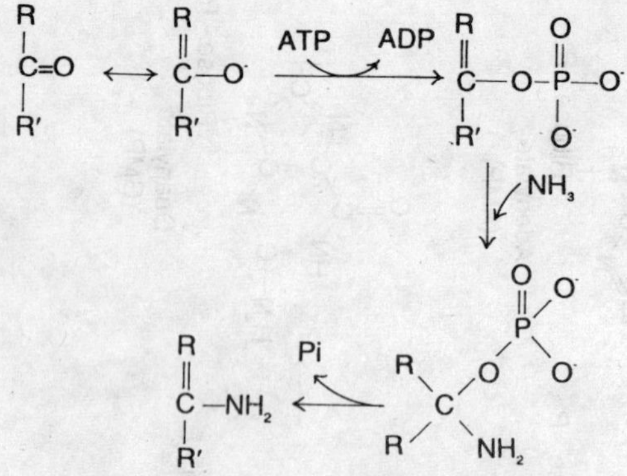

Fig. 16.8. Reaction mechanism for the replacement of a carbonyl oxygen by an amino group.

Purine base can be Recycled by Salvage Reactions that Utilize PRPP

Free purine bases are formed by the hydrolytic degradation of nucleic acids and nucleotides. Purine nucleotides can be synthesized from these performed bases by a *salvage reaction,* which is simpler and much less costly than the reactions of the *de novo pathway* discussed above. In the salvage reaction, the ribose phosphate moiety of PRPP is transferred to the purine to form the corresponding nucleotide:

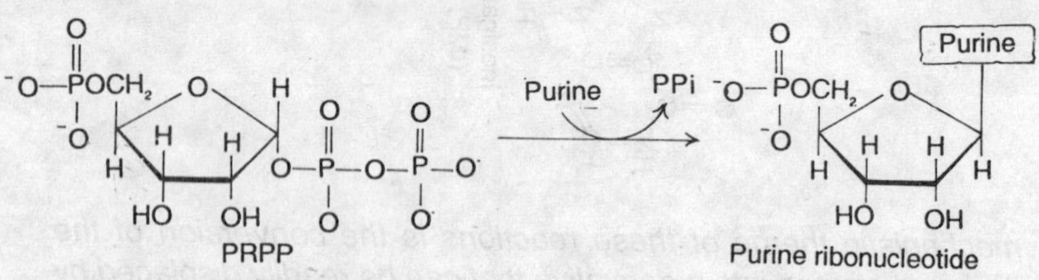

There are two salvage enzymes with different specificities. *Adenine phosphoribosyl transferase* catalyzes the formation of adenylate:

$$\text{Adenine} + \text{PRPP} \rightarrow \text{adenylate} + \text{PP}_i$$

whereas *hypoxanthine-guanine phosphoribosyl transferase* catalyzes the formation of inosinate and guanylate:

$$\text{Hypoxanthine} + \text{PRPP} \rightarrow \text{inosinate} + \text{PP}_i$$

$$\text{Guanine} + \text{PRPP} \rightarrow \text{guanylate} + \text{PP}_i$$

The versatile and efficient use of the purine ring is also evident in the biosynthesis of histidine. The six-membered portion of the purine ring of ATP contributes part of the imidazole ring of histidine. The rest of the purine skeleton is not discarded. Rather, it is conserved in 5-aminoimidazole-4-carboxamide ribonucleotide, an intermediate in the de novo pathway of purine biosynthesis.

AMP and GMP are Feedback Inhibitors of Purine Nucleotide Biosynthesis

The synthesis of purine nucleotide is controlled by feedback inhibition at several sites.

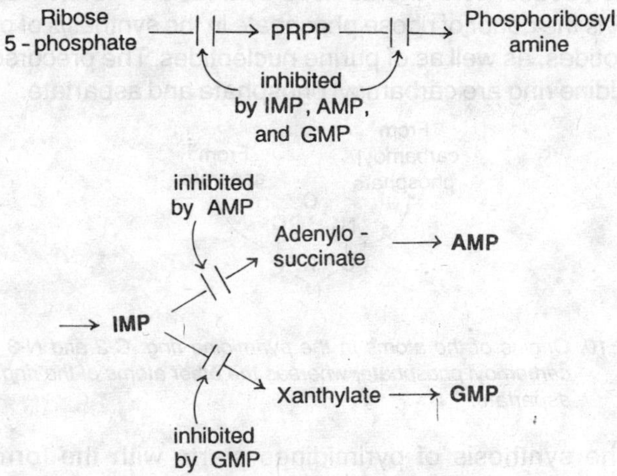

Fig. 16.9. Control of purine biosynthesis.

1. Feedback inhibition of *5-phosphoribosyl-1-pyrophosphate synthetase* by purine nucleotides regulates the level of PRPP. This synthetase is inhibited by AMP, GMP, and IMP.
2. The committed step in purine nucleotide biosynthesis is the conversion of PRPP into phosphoribosylamine by transfer of the side-chain amino group of glutamine. *Glutamine PRPP amidotransferase is feedback inhibited by many purine ribonucleotides.* It is noteworthy that AMP and GMP, the final products of the pathway, are synergistic in inhibiting this enzyme.
3. Inosinate is the branching point in the synthesis of AMP and GMP. *The reactions leading away from inosinate are site of feedback inhibition.* AMP inhibits the conversion of inosinate into adenylosuccinate, its immediate precursor. Similarly, GMP inhibits the conversion of inosinate into xanthylate, its immediate precursor.

4. GTP is a substrate in the synthesis of AMP, whereas ATP is a substrate in the synthesis of GMP. This *reciprocal substrate relation* tends to balance the synthesis of adenine and guanine ribonucleotides.

The Pyrimidine Ring is Synthesized from Carbamoyl Phosphate and Aspartate

The pyrimidine ring is assembled first and then linked to ribose phosphate to form a pyrimidine nucleotide, in contrast with the reaction sequence in the de novo synthesis of purine nucleotides. PRPP is the donor of ribose phosphate in the synthesis of pyrimidine nucleotides, as well as of purine nucleotides. The precursors of the pyrimidine ring are carbamoyl phosphate and aspartate.

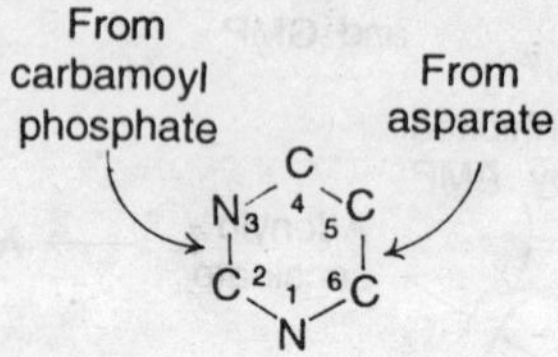

Fig. 16.10. Origins of the atoms in the pyrimidine ring. C-2 and N-3 come from carbamoyl phosphate, whereas the other atoms of the ring come from aspartate.

The synthesis of pyrimidines starts with the formation of *carbamoyl phosphate*, which is also an intermediate in the synthesis of urea. The synthesis of this activated carbamoyl donor is compartmentalized in eukaryotes. Carbamoly phosphate consumed in the synthesis of pyrimidines is formed in the cytosol, where as that used in the synthesis of urea is formed in mitochondria. There are two distinct carbamoyl phosphate synthetases. Another noteworthy difference is that glutamine rather than NH_4^+ is the nitrogen donor in the cytosol synthesis of carbamoyl phosphate.

$$\text{Glutamine} + 2\ \text{ATP} + HCO_3^- \rightarrow$$
$$\text{carbamoyl phosphate} + 2\ \text{ADP} + P_i + \text{glutamate}$$

The committed step in the biosynthesis of pyrimidine is the formation of N-carbamoylaspartate from aspartate and carbamoyl phosphate. This carbamoylation is catalyzed by *aspartate transcarbamoylase,* an especially interesting regulatory enzyme.

The pyrimidine ring is formed in the next reaction, in which carbamoyl aspartate cyclizes with loss of water to yield *dihydroorotate. Orotate* is then formed by dehydrogenation of dihydroorotate.

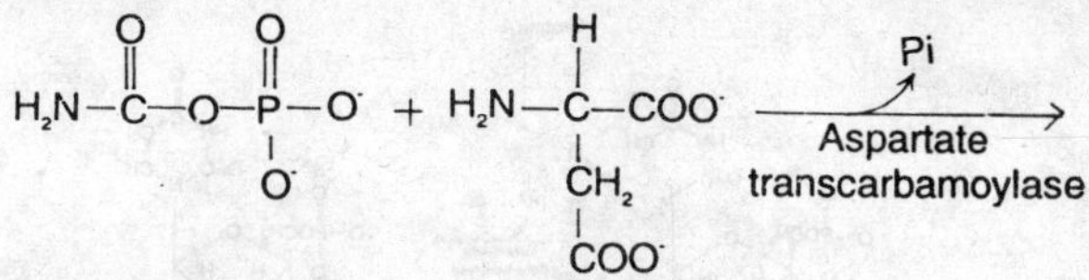

N- Carbamoylphosphate Aspartate

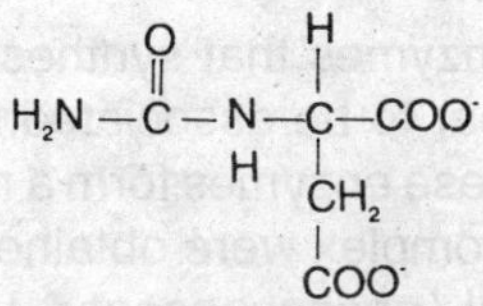

N – Carbampylaspartate

Orotate acquires a Ribose Phosphate Moiety from PRPP

The next step in the synthesis of pyrimidine nucleotides is the *acquisition of a ribose phosphate group.* Orotate (a free pyrimidine) reacts with PRPP to form *orotidylate* (a pyrimidine nucleotide). This reaction, which is catalyzed by orotidylate pyrophosphorylase, is driven forward by the hydrolysis of pyrophosphate. Orotidylate is then decarboxylated to yield *uridylate* (UMP), a major pyrimidine nucleotide.

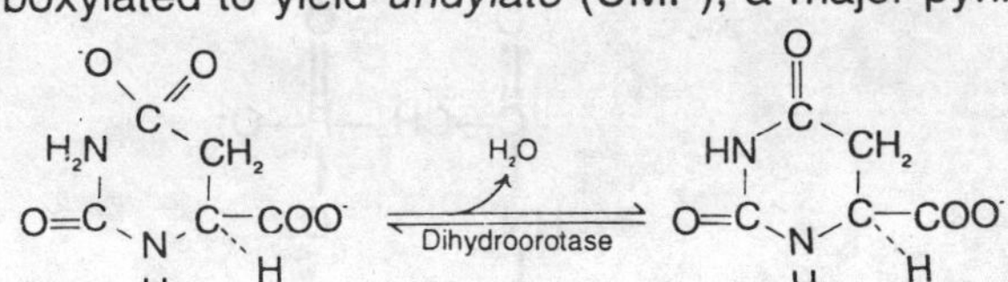

N – Carbamoylasparate Dihydroorotate

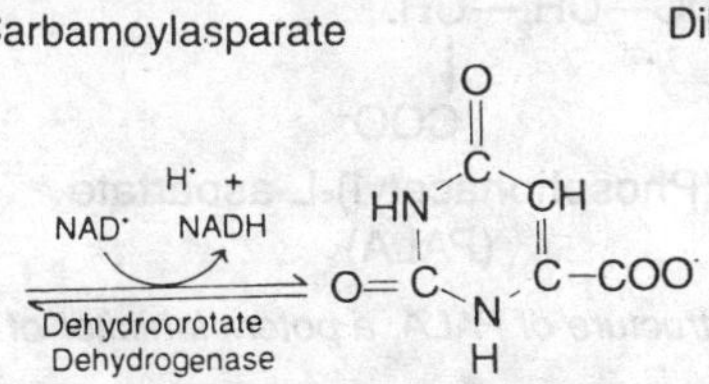

Orotate

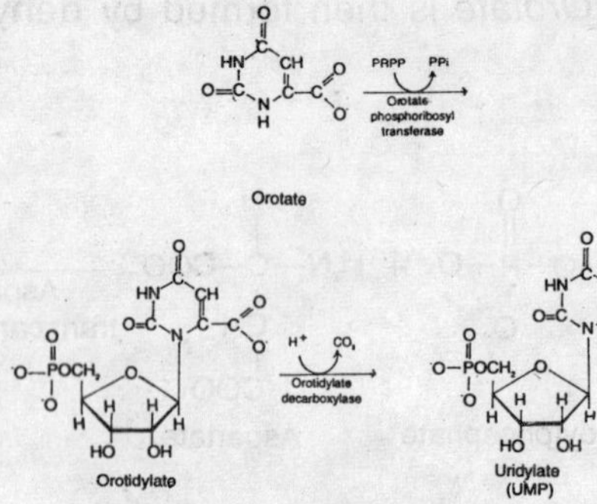

A single Polypeptide Chain contains the first three Enzymes of Pyrimidine Biosynthesis

In. *E.coli*, the six enzymes that synthesize UMP from simple precursors do not appear to be associated. In contrast, in higher organisms, several of these enzymes form a multienzyme complex. Large amounts of this complex were obtained by treating cultured mammalian cells with N-(*phosphonacetyl*)-L-*aspartate* (PALA), a potent inhibitor of aspartate transcarbamoylase (ATCase). PALA binds tightly to ATCase ($K_i = 10^{-8}M$) because it has some of the structural features of the transition state in catalysis. The surviving cells overcome the inhibitory effect of PALA by synthesizing 100-fold more ATCase than do normal cells. The levels of carbamoyl phosphate synthetase and of dihydroorotase were also elevated 100-fold, whereas there was little change in the amounts of the enzymes catalyzing the subsequent steps in pyrimidine biosynthesis. These observations led to the finding that *carbamoyl phosphate synthetase, aspartate transcarbamoylase, and dihydroorotase are covalently*

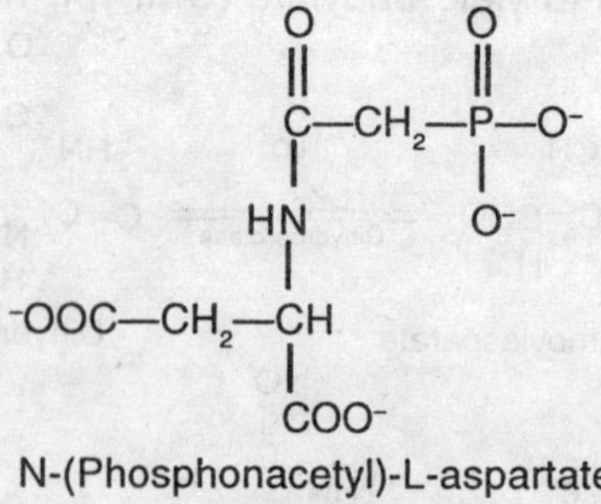

N-(Phosphonacetyl)-L-aspartate
(PALA)

Fig. 16.11. Structure of PALA, a potent inhibitor of ATCase.

joined on a single 200-kdal polypeptide chain. Orotate phosphoribosyltransferase and orotidylate decarboxylase, the enzymes catalyzing the last two steps in pyrimidine biosynthesis, from another complex. They, too may, be covalently joined. Recall that he fatty acid synthetase complex in yeast consists of two kinds of polypeptide chains, each containing several enzymes. *Covalent linkage of functionally related enzymes may be quite general in eukaryotes.* Such an arrangement would facilitate the assembly of a multienzyme complex. Another likely advantage of having several enzymes on a single polypeptide chain is that they would be synthesized in equimolar amounts.

Nucleoside Mono-, Di-, and Triphosphates are Interconvertible

The active forms of nucleotides in biosyntheses and energy conversions are the diphosphates and triphosphates. Nucleoside monophosphates are converted by specific *nucleoside monophosphate kinases* that utilize ATP as the phosphoryl donor. For example, UMP is phosphorylated by *UMP kinase.*

$$\text{UMP} + \text{ATP} = \text{UDP} + \text{ADP}$$

AMP, ADP, and ATP are interconverted by *adenylate kinase* (also called myokinase). The equilibrium constants of these reactions are close to 1.

$$\text{AMP} + \text{ATP} = \text{ADP} + \text{ADP}$$

Nucleoside diphosphates and triphosphates are interconverted by *nucleoside diphosphate kinase,* as enzyme that has broad specificity, in contrast with the monophosphate kinases. In the following equation, X and Y can be any of several ribonucleosides or deoxyribonucleosides.

$$\text{XDP} + \text{YTP} = \text{XTP} + \text{YDP}$$

For example

$$\text{UDP} + \text{ATP} = \text{UTP} + \text{ADP}$$

CTP is Formed by Amination of UTP

Cytidine triphosphate (CTP) is derived from uridine triphosphate (UTP), the other major pyrimidine ribonucleotide. The carbonyl oxygen at C-4 is replaced by an amino group. In mammals, the side chain of glutamine is the amino donor, whereas NH_4^+ is used in this reaction in *E.coli.* An ATP is consumed in both amination reactions.

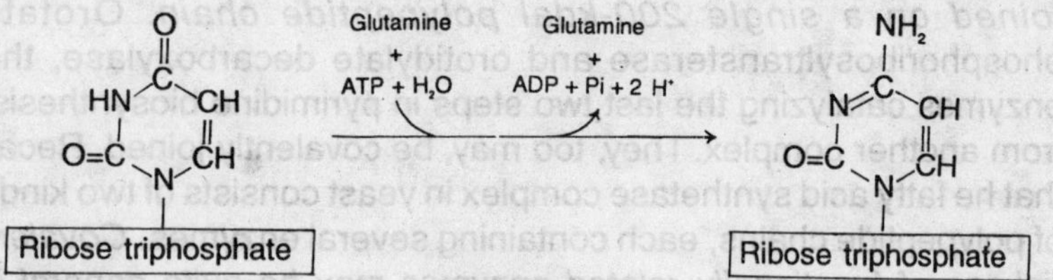

UTP CTP

Pyrimidine Nucleotide Biosynthesis is Regulated by Feedback Inhibition

The committed step in pyrimidine nucleotide biosynthesis in *E.coli* is the formation of N-carbamoylaspartate from aspartate and carbamoyl phosphate. *Aspartate transcarbamoylase, the enzyme that catalyzes this reaction, is feedback inhibited by CTP, the final product in the pathway.* A second control site is *carbamoyl phosphate synthetase*, which is feedback inhibited by UMP.

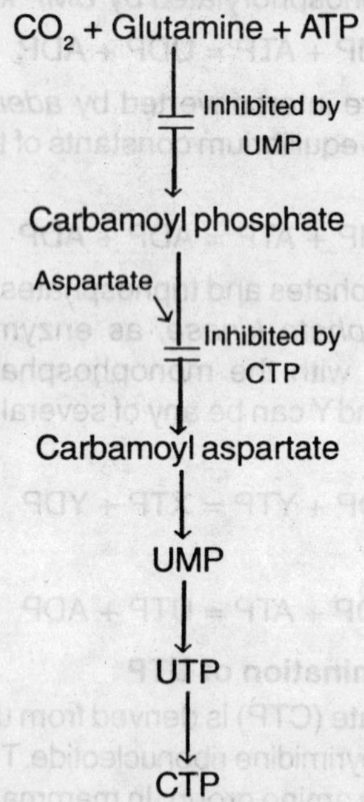

Fig. 16.12. Control of pyrimidine biosynthesis.

The allosteric properties of ATCase have been intensively investigated by John Gerhart and Howard Schachman. *The binding of carbamoyl phosphate and aspartate is cooperative,* as reflected in the sigmoidal dependence of the reaction velocity on substrate concentration. *CTP inhibits the enzyme by decreasing its affinity for substrates* without affecting its V_{max}. The extent of inhibition exerted by CTP, which may reach 90%, depends on the concentrations of the substrates. In contrast, *ATP is an activator of ATCase.* The affinity of the enzyme for its substrates is enhanced by ATP, whereas V_{max} is unaffected. Furthermore, the binding of ATP and CTP to the regulatory site of ATCase is competitive. High levels of ATP displace CTP from the enzyme so that it cannot exert its inhibitory effect.

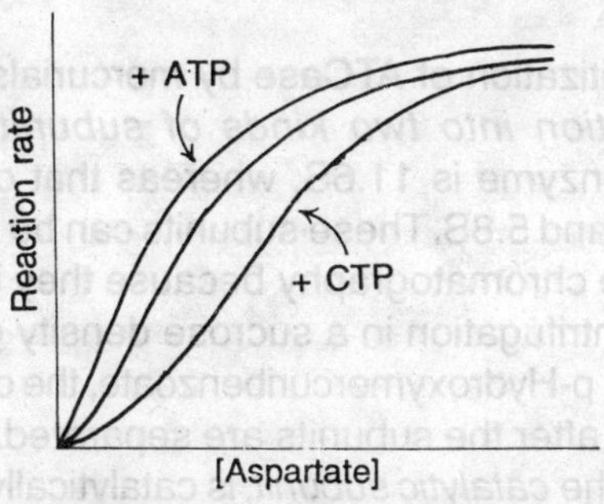

Fig. 16.13. Allosteric effects in aspartate transcarbamoylase. ATP is an activator, whereas CTP is an inhibitor.

The biological significance of the activation of ATCase by ATP is twofold. First, *it tends to equalize the rates of formation of purine and pyrimidine nucleotides. Comparable* quantities of these two types of nucleotides are needed for the synthesis of nucleic acids. Second, *activation by ATP signals its availability as a substrate* for some of the reactions of pyrimidine nucleotide biosynthesis, such as the synthesis of carbamoyl phosphate and the phosphorylations of UMP to UTP.

Aspartate Transcarbamoylase Consists of Separable Catalytic and Regulatory Subunits

The regulatory properties of ATCase vanish when the enzyme is treated with mercurials such as *p*-hydroxymercuribenzoate. ATP and CTP no longer have any effect on catalytic activity. Furthermore, the binding of substrates becomes noncooperative. However, the modified enzyme has full catalytic activity. This loss of regulatory properties with retention of enzymatic activity is called *desensitization.*

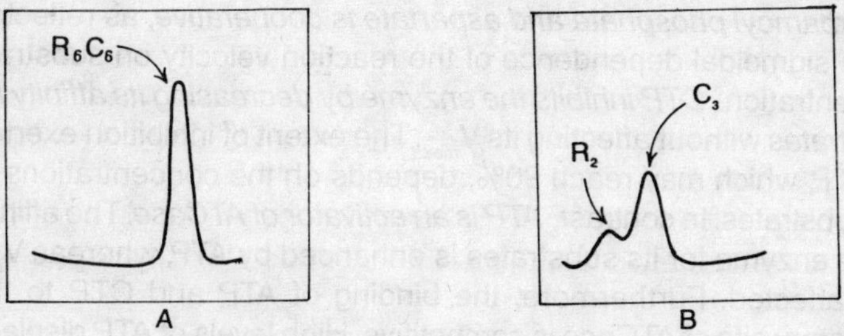

Fig. 16.14. Sedimentation velocity patterns of (A) native ATCase and (B) the enzyme dissociated by a mercurial into regulatory and catalytic subunits.

The desensitization of ATCase by mercurials is accompanied by its *dissociation into two kinds of subunits,* as shown by ultracentrifuge enzyme is 11.6S, whereas that of the dissociated subunits is 2.8S and 5.8S. These subunits can be readily separated by ion-exchange chromatography because they differ markedly in charge or by centrifugation in a sucrose density gradient because they differ in size. p-Hydroxymercuribenzoate, the dissociating agent, can be removed after the subunits are separated. The larger of the subunits, called the *catalytic subunit,* is catalytically active. However, the activity of the isolated catalytic subunit is not affected by ATP and CTP. The smaller of the subunits, called the *regulatory subunit,* is devoid of catalytic activity but contains specific binding sites for CTP and ATP. The catalytic subunit consists of three 34-kdal polypeptide chains, whereas the regulatory subunit is made up of two 17-kdal polypeptide chains.

The catalytic and regulatory subunits combine rapidly when they are mixed. The resulting complex has the same structure, R_6C_6, as that of the native enzyme.

$$3\,R_2 + 2\,C_3 \rightarrow R_6C_6$$

Furthermore, *the reconstituted enzyme has the same allosteric properties as those of the native enzyme.*

X-ray crystallographic studies of ATCase are in progress in William Lipscomb's laboratory. An electron-density map at 3.0-Å resolution shows that the two catalytic trimers (C_3) are above and below an equatorial belt of three regulatory dimers (R_2). A distinctive feature of the molecule is that it contains a large central cavity, which is accessible through several channels. It is interesting to note that the allosteric sites for CTP are far from the catalytic sites.

Deoxyribonucleotides and Synthesized by Reduction of Ribonucleoside Diphosphates

We turn now to the synthesis of deoxyribonucleotides. These precursors of DNA are formed by the reduction of ribonucleotides. The 2'-hydroxyl group on the ribose moiety is replaced by a hydrogen atom. In *E.coli* and in mammals, the substrates in this reaction are ribonucleoside diphosphates. The overall stoichiometry is

$$\text{Ribonucleoside diphosphate} + \text{NADPH} + H^+ \rightarrow \text{deoxyribonucleoside diphosphate} + \text{NADP}^+ \; H_2O$$

The actual reaction mechanism is more complex than implied by this equation. Peter Reichard has shown that in *E.coli* the electrons from NADPH are transferred to the substrate through a series of sulfhydryl groups. *Ribonucleotide reductase* (also called ribonucleoside diphosphate reductase) catalyzes the final stage, which has he stoichiometry

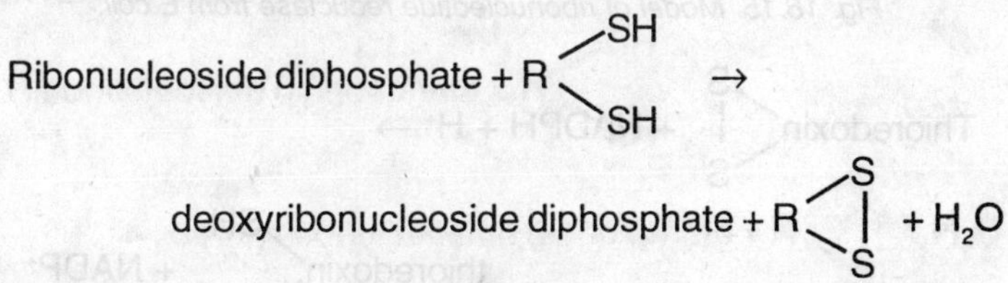

This enzyme consists of two subunits, B1 (a 160-kdal dimer) and B2 (a 78-kdal dimer). The B1 subunit contains the binding sites for ribonucleotide substrates and are allosteric effectors. In addition, B1 contains sulfhydryls that serve as the immediate electron donors in the reduction of the ribose unit. B2, an iron-sulfur protein, participates in catalysis by forming an unusual free radical on the aromatic ring of a tyrosine residue. The B1 and B2 subunits together form the active sites of the enzyme.

How are electrons transferred from NADPH to the sulfhydryl groups at the catalytic site of ribonucleotide reductase? One carrier of reducing power is *thioredoxin*, a 12-kdal protein with two cysteine residues in close proximity. These sulfhydryls are oxidized to a disulfide in the reaction catalyzed by ribonucleotide reductase. In turn, reduced thioredoxin is regenerated by the reaction of NADPH with oxidized thioredoxin. This reaction is catalyzed by *thioredoxin reductase,* a flavoprotein.

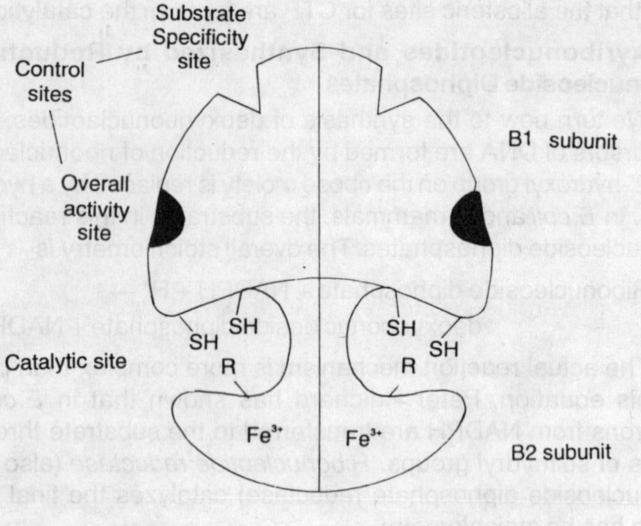

Fig. 16.15. Model of ribonucleotide reductase from E.coli.

$$\text{Thioredoxin}\begin{matrix} S \\ | \\ S \end{matrix} + \text{NADPH} + H^+ \rightarrow \text{thioredoxin}\begin{matrix} SH \\ \\ SH \end{matrix} + \text{NADP}^+$$

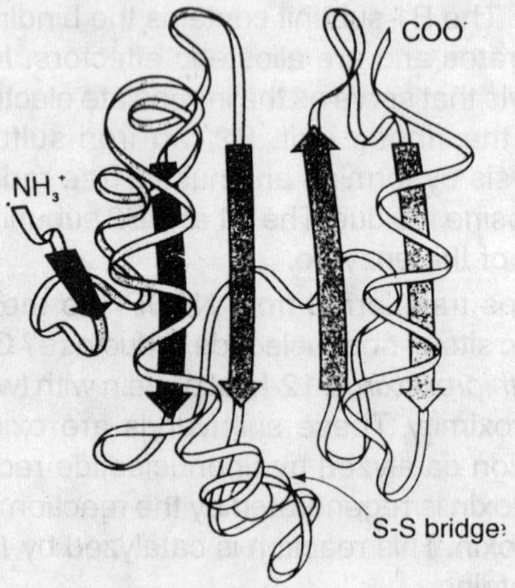

Fig. 16.16. Schematic diagram of the main-chain conformation of oxidized thioredoxin from E. coli.

It was thought for some time that thioredoxin is the only carrier of reducing power to ribonucleotide reductase. However, a mutant of *E.coli* totally devoid of thioredoxin was found to form deoxyribonucleotides. This surprising observation led to the isolation of a second carrier system. The electron donor in this mutant proved to be *glutathione,* a cysteine-containing tripeptide. As discussed previously, *glutathione reductase* catalyzes the reduction of oxidized glutathione (the disulfide form) by NADPH. In addition, *glutaredoxin*, a new protein, is needed to transfer the reducing power of glutathione to ribonucleotide reductase. The relative contributions of the thioredoxin and glutaredoxin systems in ribonucleotide reduction by normal cells are not yet known.

The reduction of ribonucleotide diphosphates is precisely controlled by allosteric interactions. The B1 subunit of ribonucleotide reductase contains two types of allosteric sites: one of them controls the *overall activity* of the enzyme, whereas the other regulates *substrate specificity.* The overall catalytic activity of ribonucleotide reductase is diminished by the binding of dATP, which signals an abundance of deoxyribonucleotides. This feedback inhibition is reversed by the binding of ATP. The binding of dATP to the substrate-specificity control sites enhances the reduction of UDP and CDP, the pyrimidine nucleotides. The reduction of GDP is promoted by the binding of dTTP, which also inhibits the further reduction of pyrimidine ribonucleotides. The subsequent increase in the level of dGTP leads to a stimulation of ADP reduction. It is evident that ribonucleotide reductase has a variety of conformational states, each with different catalytic properties. This complex pattern of regulation provides the appropriate supply of the four deoxyribonucleotide needed for the synthesis of DNA.

Deoxythymidylate is Formed by Methylation of Deoxyuridylate

Uracil is not a component of DNA. Rather, DNA contains *thymine,* the methylated analog of uracil. This finishing touch occurs at the level of the deoxyribonucleoside monophosphate: deoxyuridylate (dUMP) is methylated to deoxythymidylate (dTMP) by *thymidylate synthetase.* The methyl donor in this reaction is a tetrahydrofolate derivative rather than S-adenosylmethionine. Specifically, the methyl carbon comes from N^5,N^{10}-methylenetetrahydrofolate. Note that the methyl group inserted into deoxyuridylate is more reduced than the methylene group in this tetrahydrofolate derivative. What is the source of electrons for this reduction? The two electrons come from the

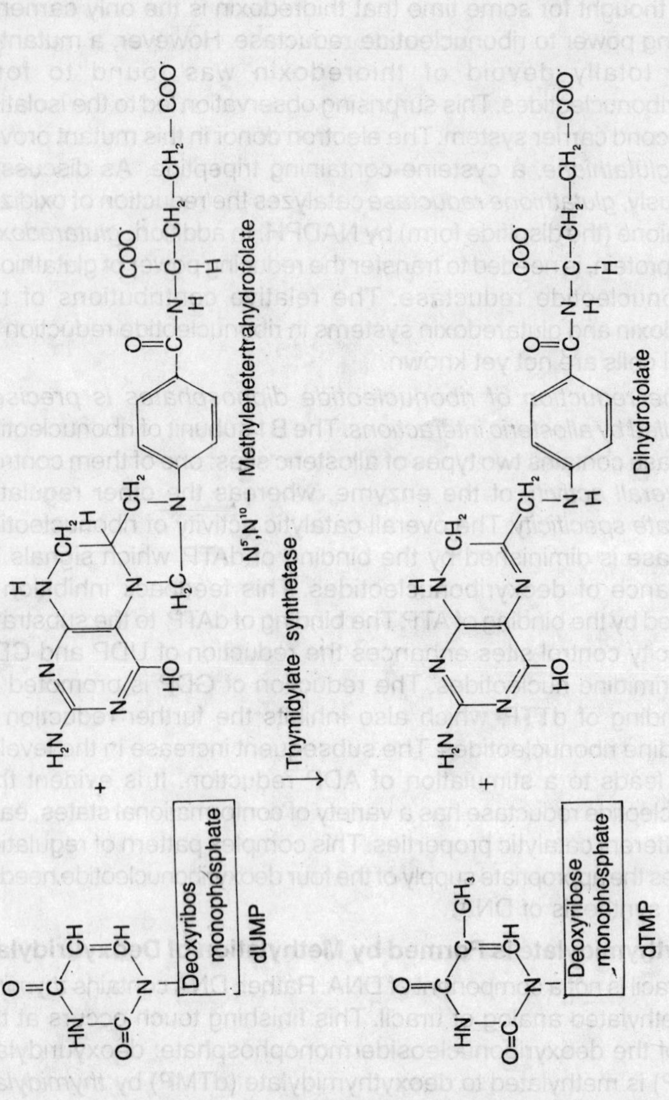

Fig. 16.17. Synthesis of dTMP from dUMP

tetrahydrofolate moiety itself in the form of a hydride ion (H^-), which is removed from the ring. This hydrogen becomes part of the methyl group of dTMP. In this reaction, tetrahydrofolate is oxidized to dihydrofolate. Thus N^5, N^{10}-methylenetetrahydrofolate serves both as an *electron donor* and as a *one-carbon* donor in the methylation reaction.

Recall that one-carbon transfers occur at the level of tetrahydrofolate rather than dihydrofolate. Hence, tetrahydrofolate must be regenerated. This is accomplished by *dihydrofolate reductase,* utilizing NADPH as the reductant.

Dihydrofolate + NADPH + $H^+ \rightarrow$ tetrahydrofolate + $NADP^+$

Several Anticancer Drugs Block the Synthesis of Deoxythymidylate

Rapidly dividing cell require an abundant supply of deoxythymidylate for the synthesis of DNA. The vulnerability of the cells to the inhibition of dTMP synthesis has been exploited in cancer chemotherapy. Thymidylate synthetase and dihydrofolate reductase are choice target enzymes. *Fluorouracil* (or fluorodeoxyuridine), a clinically useful anticancer drug, is converted in vivo into *fluorodeoxyuridylate* (F-dUMP). *This analog of dUMP irreversibly*

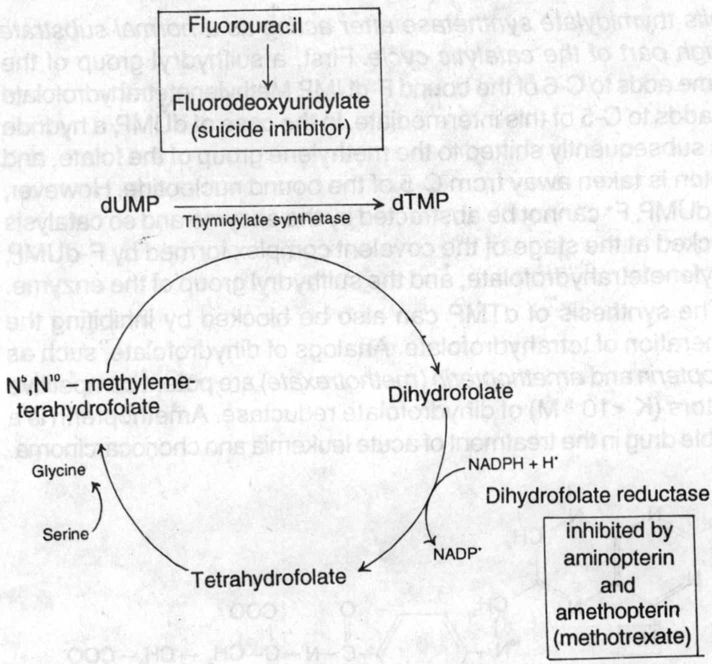

Fig. 16.18. Thymidylate synthetase and dihydrofolate reductase are target enzymes in cancer chemotherapy. Fluorodeoxyuridylate inhibits the methylation of dUMP. The folate analogs aminopterin and amethopterin (methotrexate) block the regeneration of tetrahydrofolate.

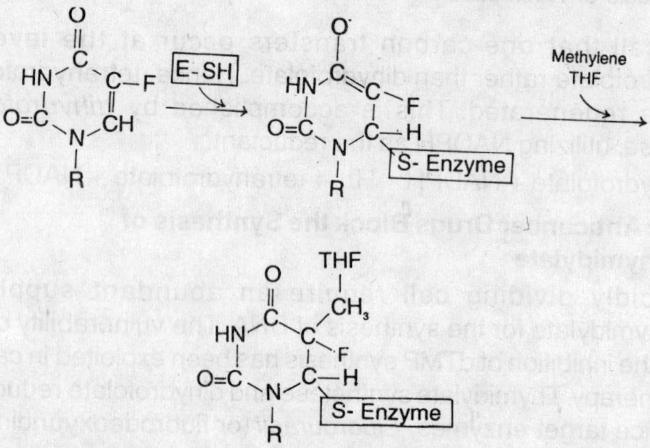

Fig. 16.19. Thymidylate synthetase is irreversibly inhibited by fluorodeoxyuridylate (F-dUMP). This analog forms a covalent complex with both a sulfhydryl residue of the enzyme and methylenetetrahydrofolate.

inhibits thymidylate synthetase after acting as a normal substrate through part of the catalytic cycle. First, a sulfhydryl group of the enzyme adds to C-6 of the bound F-dUMP. Methylenetetrahydrofolate then adds to C-5 of this intermediate. In the case of dUMP, a hydride ion is subsequently shifted to the methylene group of the folate, and a proton is taken away from C-5 of the bound nucleotide. However, for F-dUMP, F^+ cannot be abstracted by the enzyme and so catalysis is blocked at the stage of the covalent complex formed by F-dUMP, methylenetetrahydrofolate, and the sulfhydryl group of the enzyme.

The synthesis of dTMP can also be blocked by inhibiting the regeneration of tetrahydrofolate. Analogs of dihydrofolate, such as *aminopterin* and *amethopterin* (*methotrexate*) are potent competitive inhibitors ($K_i < 10^{-9}$ M) of dihydrofolate reductase. Amethopterin is a valuable drug in the treatment of acute leukemia and choriocarcinoma.

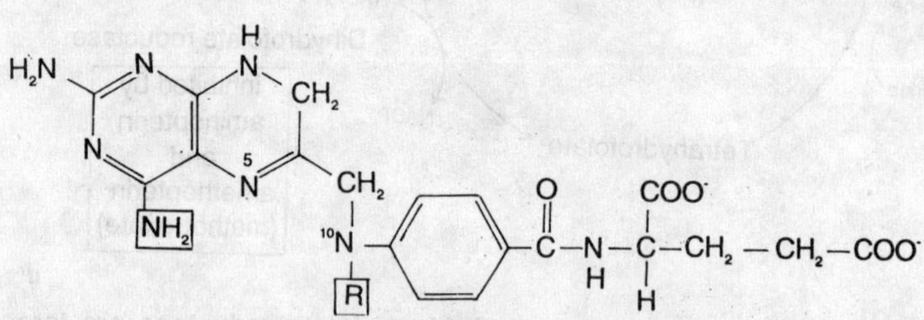

ATP is a Precursor of NAD$^+$, FAD, and Coenzyme A

The *biosynthesis of nicotinamide adenine dinucleotide* (NAD$^+$) starts with the formation of *nicotinate ribonucleotide* from nicotinate and PRPP. *Nicotinate* (also called *niacin*) is derived from tryptophan. Humans can synthesize the required amount of nicotinate if the supply of tryptophan in the diet is adequate. However, an exogenous supply of nicotinate is required if the dietary intake of tryptophan is low. *Pellagra* is a deficiency disease caused by a dietary insufficiency of tryptophan and nicotinate.

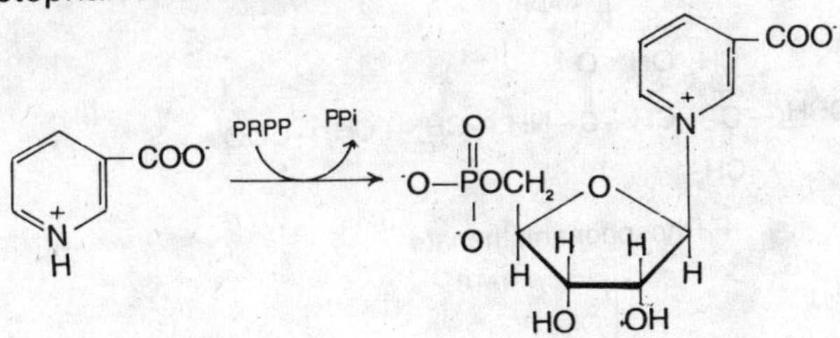

Nicotinate Nicotinate ribonucleotide

An AMP moiety is transferred from ATP to nicotinate ribonucleotide to form *desamido-NAD$^+$*. The final step is the transfer of the amide group of glutamine to the nicotinate carboxyl group to form NAD$^+$. NADP$^+$ is derived from NAD$^+$ by phosphorylation of the 2'-hydroxyl group of the adenine ribose moiety. This transfer of a phosphoryl group is catalyzed by NAD$^+$ kinase.

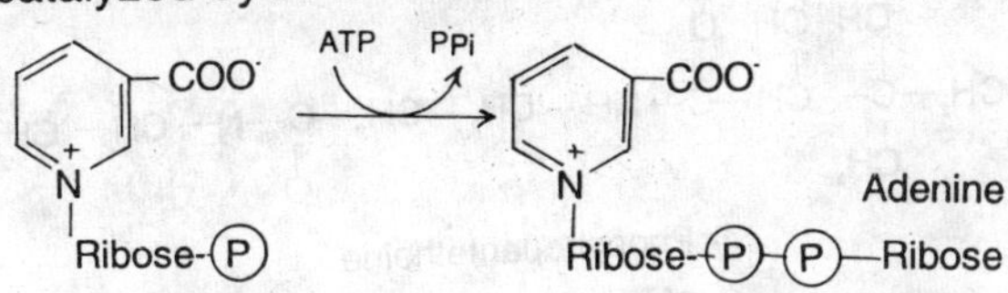

Nocotinate bonucleotide Desamido - NAD$^+$

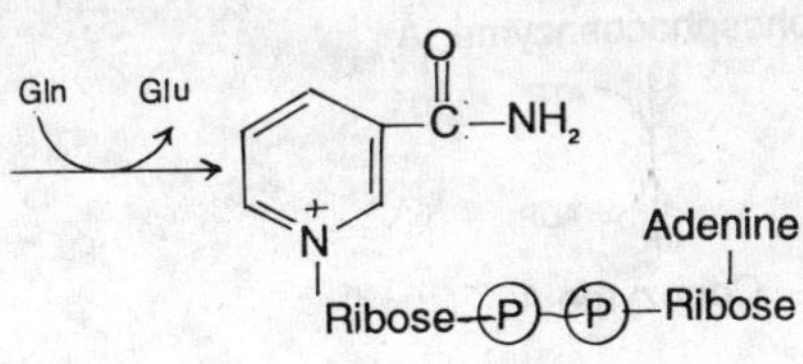

NAD$^+$

Fig. 16.20. Synthesis of NAD$^+$ from nicotinate ribonucleotide.

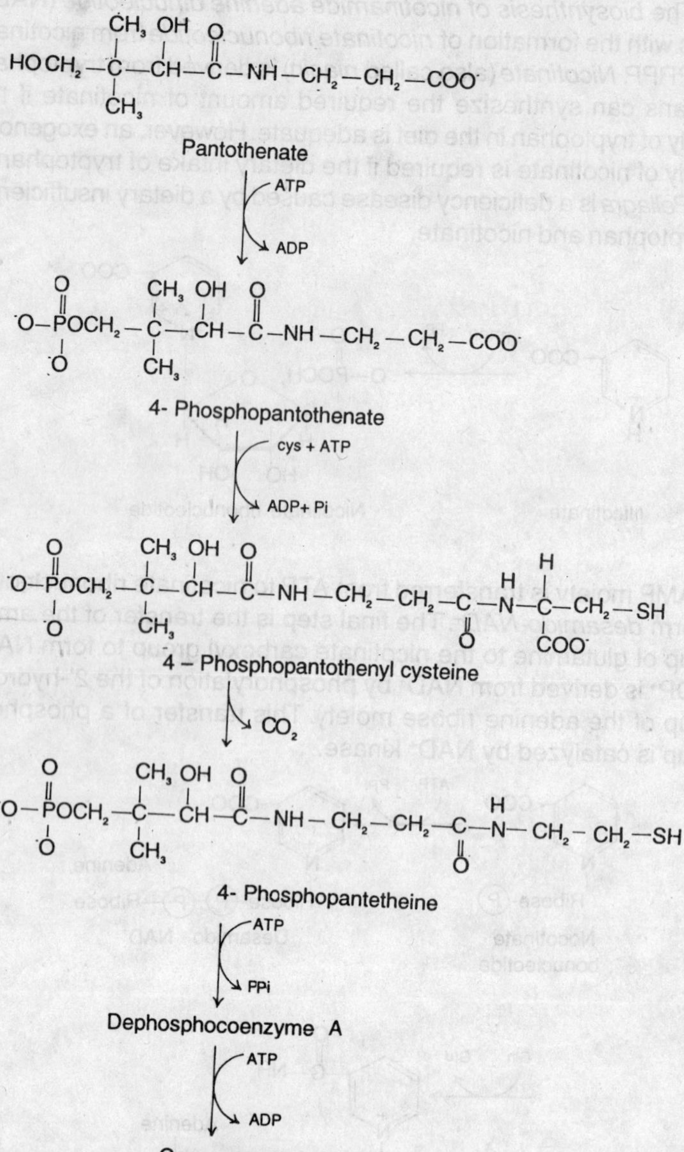

Fig. 16.21. Synthesis of coenzyme A from pantothenate.

Flavin adenine dinucleotide (FAD) is synthesized from riboflavin and two molecules of ATP. Riboflavin is phosphorylated by ATP to give *riboflavin 5-phosphate* (also called *flavin manonucleotide*). FAD is then formed by the transfer of an AMP moiety from a second molecule of ATP to riboflavin 5'-phosphate.

Riboflavin + ATP → riboflavin 5'-phosphate + ADP

Riboflavin 5'-phosphate + ATP =

flavin adenine dinucleotide + PP_i

The *synthesis of coenzyme A* (CoA) in animals starts with the phosphorylation of *pontothenate.* Pantothenate is required in the diet of animals, whereas it is synthesized by plants and microorganisms. A peptide bond is formed between the carboxyl group of 4'-phosphopantothenate and the amino group of cysteine. The carboxyl group of the cysteine moiety is lost, which results in *4'-phosphophantotheine.* The AMP moiety of ATP is then transferred to this intermediate to form *dephosphocoenzyme A.* Finally, phosphorylation of its 3'-hydroxyl group yields coenzyme A.

A common feature of the bioxyntheses of NAD^+, FAD, and CoA is the *transfer of the AMP moiety of ATP to the phosphate group of a phosphorylated intermediate.* The pyrophosphate formed in this reaction is hydrolyzed to orthophosphate. This is a recurring motif in biochemistry: *biosynthetic reactions are frequently driven by the hydrolysis of the released pyrophosphate.*

Purine are Degraded to Urate in Humans

The nucleotides of a cell undergo continuous turnover. Nucleotides are hydrolytically degraded to nucleosides by *nucleotidases.* Phosphorolytic cleavage of nucleosides to free bases and ribose 1-phosphate (or deoxyribose 1-phosphate) is catalyzed by *nucleoside phosphorylases.* Ribose 1-phosphate is isomerized by *phosphoribomutase* to ribose 5-phosphate, a substrate in the synthesis of PRPP. Some of the bases are reused to form nucleotides by salvage pathways.

The pathway for the degradation of AMP includes an additional step. AMP is deaminated to IMP by *adenylate deaminase.* The subsequent reactions leading to the free bases hypoxanthine follow the general pattern. *Xanthine oxidase,* a molybdenum and iron-containing flavoprotein, oxidizes hypoxanthine to *xanthine* and then a *urate.* Molecular oxygen, the oxidant in both reactions, is reduced to H_2O_2, which is decomposed to H_2O and O_2 by catalase. Xanthine is also an intermediate in the formation of urate from guanine. *In*

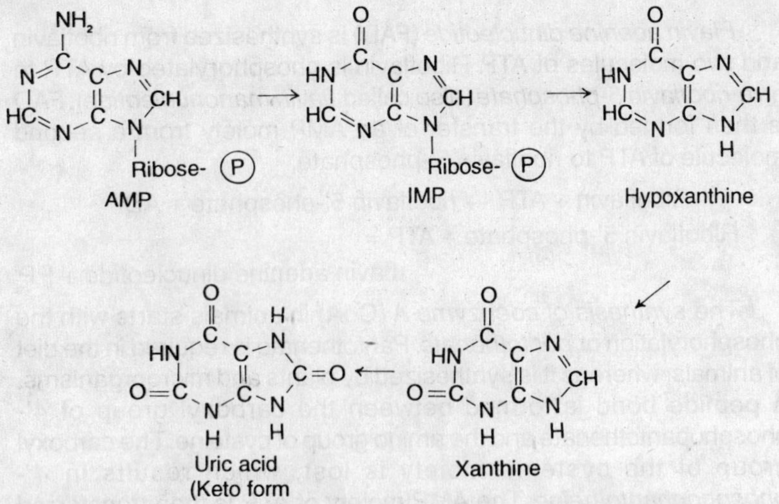

Fig. 16.22. Degradation of AMP to uric acid.

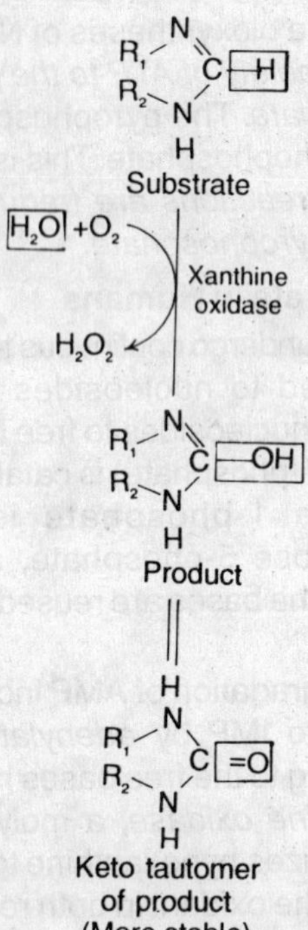

Fig. 16.23. Oxidation reaction catalyzed by xanthine oxidase.

humans, urate is the final product of purine degradation and is excreted in the urine.

Urate is further Degraded in some Organisms

The breakdown of purines proceeds further in some species. Mammals other than primates excrete *allantoin,* which is formed by oxidation of urate. Teleost fish excrete *allantoate,* which is formed by hydration of allantoin. The degradation proceeds a step further in amphibians and most fish. Allantoate is hydrolyzed to two molecules of *urea* and one of *glyoxylate.* Finally, some marine invertebrates hydrolyze urea to NH_4^+ and CO_2. It seems likely that the enzymes catalyzing these reactions were progressively lost in the evolution of primates.

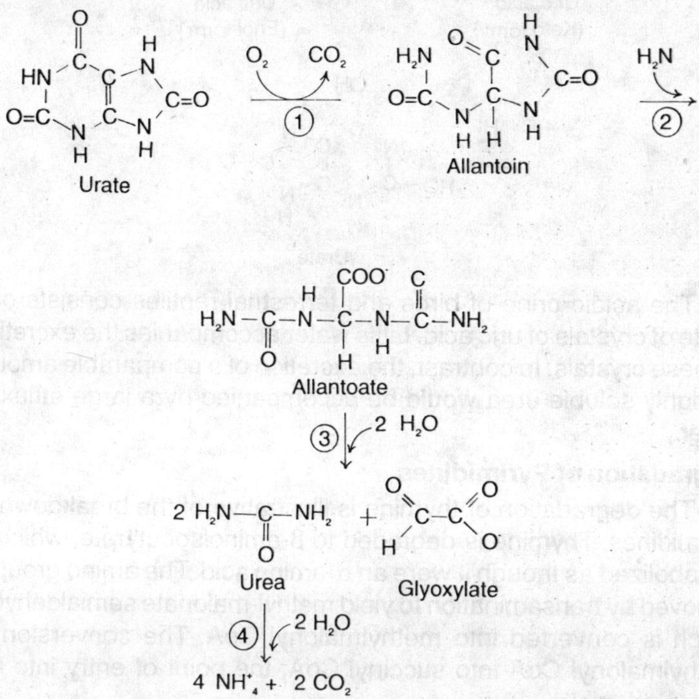

Fig. 16.24. Degradation of urate to NH_4^+ and CO_2. These reactions are catalyzed by (1) uricase, (2) allantoinase, (3) allantoicase, and (4) urease.

Birds and Terrestrial Reptiles Excrete Urate Instead of Urea of Conserve Water

In terrestrial reptiles and birds, urea is not the final product of amino nitrogen metabolism. These animals synthesize purines from their excess amino nitrogen and then degrade these purines to urate. *This circuitous path for the elimination of amino nitrogen serves a vital function: the conservation of water. Urate is the vehicle for the excretion of amino nitrogen because of its very low solubility at acid pH.* The pK_a of the most acidic group in uric acid is 5.4.

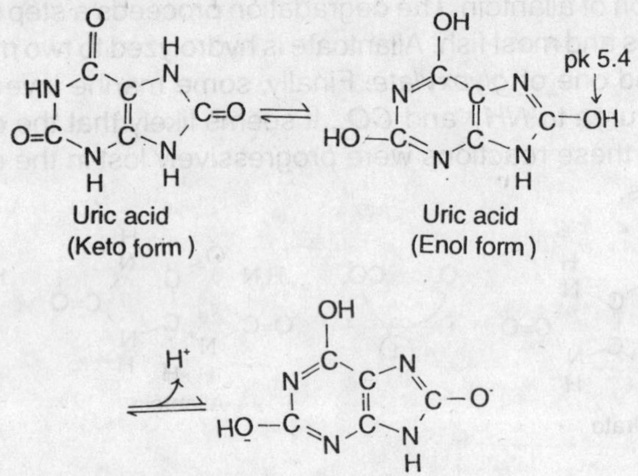

Uric acid (Keto form) ⇌ Uric acid (Enol form) ⇌ Urate

The acidic urine of birds and terrestrial reptiles consists of a paste of crystals of uric acid. Little water accompanies the excretion of these crystals. In contrast, the excretion of a comparable amount of highly soluble urea would be accompanied by a large efflux of water.

Degradation of Pyrimidines

The degradation of thymine is illustrative of the breakdown of pyrimidines. Thymine is degraded to β-aminoisobutyrate, which is metabolized as though it were an α-amino acid. The amino group is removed by transamination to yield methyl-malonate semialdehyde, which is converted into methylmalonyl CoA. The conversion of methylmalonyl CoA into succinyl CoA, the point of entry into the citric acid cycle.

Excessive Production of Urate is a cause of Gout

Gout is a disease that affects the joints and leads to arthritis. The major biochemical feature of gout is an *elevated level of urate in the serum.* Inflammation of the joints is triggered by the *precipitation of sodium urate crystals.* Kidney disease may also occur because of the deposition of urate crystals in that organ. Gout primarily affects

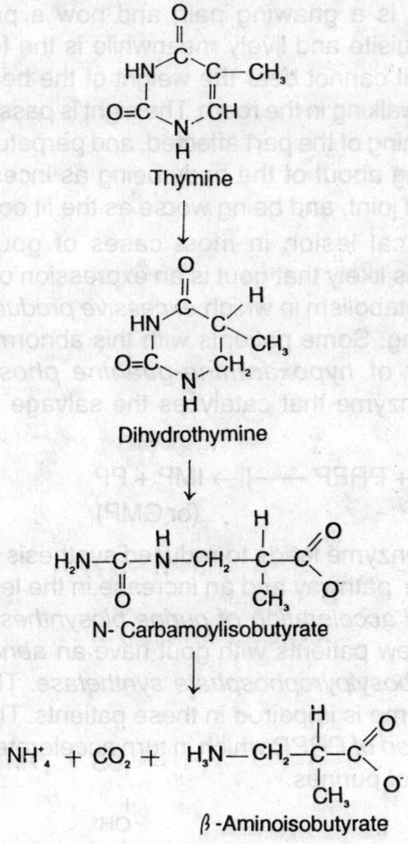

Fig. 16.25. Degradation of thymine.

adult males. A vivid description of an acute attack of gout was given by Thomas Sydenham, an outstanding seventeenth-century English physician, who himself was affected with this disease:

The vitamin goes to bed and sleeps in good health. About two o'clock in the morning he is awakened by a severe pain in the great toe; more rarely in the heel, ankle or instep. This pain is like that of a dislocation, and yet the parts feel as if cold water were poured over them. Then follow chills and shivers, and a little fever. The pain, which was at first moderate, becomes more intense. With its intensity the chills and shivers increase. After a time this comes to its height, accommodating itself to the bones and ligaments of the tarsus and metatarsus. Now it is a violent stretching and tearing of the ligaments—now it is a gnawing pain and now a pressure and tightening. So exquisite and lively meanwhile is the feeling of the part affected, that it cannot bear the weight of the bedclothes nor the jar of a person walking in the room. The night is passed in torture, sleeplessness, turning of the part affected, and perpetual change of posture; the tossing about of the body being as incessant as the pain of the tortured joint, and being worse as the fit comes on.

The biochemical lesion in most cases of gout has been elucidated. It seems likely that gout is an expression of a variety of inborn errors of metabolism in which *excessive production of urate* is a common finding. Some patients with this abnormality have a partial deficiency of *hypoxanthine-guanine phosphoribosyltransferase,* the enzyme that catalyzes the salvage synthesis of IMP and GMP.

$$\underset{\text{(or guanine)}}{\text{Hypoxanthine}} + \text{PRPP} \longrightarrow \underset{\text{(or GMP)}}{\text{IMP}} + \text{PP}_i$$

Deficiency of this enzyme leads to reduced synthesis of GMP and IMP by the salvage pathway and an increase in the level of PRPP. There is a *marked acceleration of purine biosynthesis by the de novo pathway.* A few patients with gout have an *abnormally high level of phosphoribosylpyrophosphate synthetase.* The allosteric control of this enzyme is impaired in these patients. This results in *excessive production of PRPP,* which in turn accelerates the rate of de novo synthesis of purines.

Allopurinol, an analog of hypoxanthine in which the positions of N7 and C8 are interchanged, is used to treat gout. The mechanism of action of allopurinol is very interesting: it acts *first as a substrate* and *then as an inhibitor* of xanthine oxidase. This enzyme hydroxylates allopurinol to *alloxantine,* which then remains tightly bound to the active site. The molybdenum atom of xanthine oxidase is kept in the +4 oxidation state by the binding of alloxanthine instead of returning to the +6 oxidation state as it does in a normal catalytic cycle. This mode of action of allopurinol is an example of *succide inhibition,* in which an enzyme converts a compound into a potent inhibitor that immediately inactivates the enzyme.

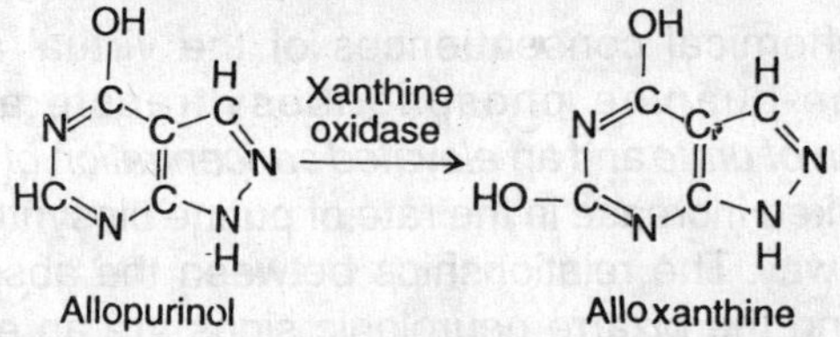

The synthesis of urate from hypoxanthine and xanthine decreases soon after the administration of allopurinol.

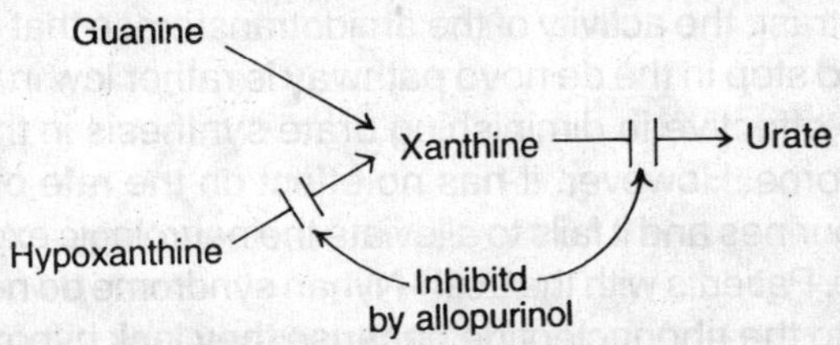

The serum concentrations of hypoxanthine and xanthine increase after administration of allopurinol, whereas that of urate drops. The formation of uric acid stones is virtually abolished by allopurinol, and there is some improvement in the arthritis. Also, *there is a decrease in the total rate of purine biosynthesis.* This inhibitory action of allopurinol depends on its reaction with PRPP to form the ribonucleotide. Consequently, the level of PRPP, the limiting substrate in the de novo synthesis of purines, is lowered. Furthermore, allopurinol ribonucleotide inhibits the conversion of PRPP into phosphoribosylamine by amidophosphoribosyltransferase.

Lesch-Nyhan Syndrome : Self-mutilation, mental Retardation, and Excessive Production of Urate

A nearly total absence of hypoxanthine-guanine phosphoribosyltransferase has devastating consequences. The most striking expression of this inborn error of metabolism, called the *Lesch-Nyhan syndrome,* is *compulsive self-destructive behaviour.* At age two or three, children with this disease begin to bite their fingers and lips. The tendency to self-mutilate is so extreme that it is necessary to protect these patients by such measures as wrapping their hands in gauze. Those afflicted also tend to be aggressive toward others. *Mental deficiency and spasticity* are other characteristics of the Lesh-Nyhan syndrome. Elevated levels of urate in the serum lead to the formation of stones early in life, followed by the symptoms of *gout* years later. The disease is inherited as a sex-linked recessive.

The biochemical consequences of the virtual absence of hypoxanthine-guanine phosphoribosyltransferase are an *overproduction of urate* and an *elevated concentration of PRPP.* Also, there is a marked increase in the rate of purine biosynthesis by the de novo pathway. The relationships between the absence of the transferase and the bizarre neurologic signs are an enigma. The brain may be very dependent on the salvage pathway for the synthesis of IMP and GMP. The normal level of hypoxanthine-guanine phosphoribosyltransferase is higher in the brain than in any other tissue. In contrast, the activity of the amidotransferase that catalyzes the committed step in the de novo pathway is rather low in the brain. Allopurinol is effective in diminishing urate synthesis in the Lesch-Nyhan syndrome. However, it has no effect on the rate of de novo synthesis of purines and it fails to alleviate the neurologic expressions of the disease. Patients with the Lesh-Nyhan syndrome do not convert allopurinol into the ribonucleotide because they lack hypoxanthine-guanine phosphoribosyltransferase. Hence, the administration of allopurinol does not lower the level of PRPP in these individuals and so de novo purine synthesis is not diminished.

The Lesch-Nyhan syndrome demonstrates that the salvage pathway for synthesis of IMP and GMP is not gratuitous. The salvage pathway evidently serves a critical role that is not yet fully understood. Furthermore, the interplay between the de novo and salvage pathways of purine synthesis remains to be elucidated. Moreover, the Lesch-Nyhan syndrome reveals that abnormal behaviour such as self-mutilation and extreme hostility can be caused by the absence of a single enzyme. This finding has important implications for the future development of psychiatry.